p-adic functional analysis

PURE AND APPLIED MATHEMATICS

A Program of Monographs, Textbooks, and Lecture Notes

LECTURE NOTES IN PURE AND APPLIED MATHEMATICS

1. *N. Jacobson,* Exceptional Lie Algebras
2. *L.-Å. Lindahl and F. Poulsen,* Thin Sets in Harmonic Analysis
3. *I. Satake,* Classification Theory of Semi-Simple Algebraic Groups
4. *F. Hirzebruch et al., Differentiable Manifolds and Quadratic Forms*
5. *I. Chavel,* Riemannian Symmetric Spaces of Rank One
6. *R. B. Burckel,* Characterization of C(X) Among Its Subalgebras
7. *B. R. McDonald et al.,* Ring Theory
8. *Y.-T. Siu,* Techniques of Extension on Analytic Objects
9. *S. R. Caradus et al.,* Calkin Algebras and Algebras of Operators on Banach Spaces
10. *E. O. Roxin et al.,* Differential Games and Control Theory
11. *M. Orzech and C. Small,* The Brauer Group of Commutative Rings
12. *S. Thomier,* Topology and Its Applications
13. *J. M. Lopez and K. A. Ross,* Sidon Sets
14. *W. W. Comfort and S. Negrepontis,* Continuous Pseudometrics
15. *K. McKennon and J. M. Robertson,* Locally Convex Spaces
16. *M. Carmeli and S. Malin,* Representations of the Rotation and Lorentz Groups
17. *G. B. Seligman,* Rational Methods in Lie Algebras
18. *D. G. de Figueiredo,* Functional Analysis
19. *L. Cesari et al.,* Nonlinear Functional Analysis and Differential Equations
20. *J. J. Schäffer,* Geometry of Spheres in Normed Spaces
21. *K. Yano and M. Kon,* Anti-Invariant Submanifolds
22. *W. V. Vasconcelos,* The Rings of Dimension Two
23. *R. E. Chandler,* Hausdorff Compactifications
24. *S. P. Franklin and B. V. S. Thomas,* Topology
25. *S. K. Jain,* Ring Theory
26. *B. R. McDonald and R. A. Morris,* Ring Theory II
27. *R. B. Mura and A. Rhemtulla,* Orderable Groups
28. *J. R. Graef,* Stability of Dynamical Systems
29. *H.-C. Wang,* Homogeneous Branch Algebras
30. *E. O. Roxin et al.,* Differential Games and Control Theory II
31. *R. D. Porter,* Introduction to Fibre Bundles
32. *M. Altman,* Contractors and Contractor Directions Theory and Applications
33. *J. S. Golan,* Decomposition and Dimension in Module Categories
34. *G. Fairweather,* Finite Element Galerkin Methods for Differential Equations
35. *J. D. Sally,* Numbers of Generators of Ideals in Local Rings
36. *S. S. Miller,* Complex Analysis
37. *R. Gordon,* Representation Theory of Algebras
38. *M. Goto and F. D. Grosshans,* Semisimple Lie Algebras
39. *A. I. Arruda et al.,* Mathematical Logic
40. *F. Van Oystaeyen,* Ring Theory
41. *F. Van Oystaeyen and A. Verschoren,* Reflectors and Localization
42. *M. Satyanarayana,* Positively Ordered Semigroups
43. *D. L Russell,* Mathematics of Finite-Dimensional Control Systems
44. *P.-T. Liu and E. Roxin,* Differential Games and Control Theory III
45. *A. Geramita and J. Seberry,* Orthogonal Designs
46. *J. Cigler, V. Losert, and P. Michor,* Banach Modules and Functors on Categories of Banach Spaces
47. *P.-T. Liu and J. G. Sutinen,* Control Theory in Mathematical Economics
48. *C. Byrnes,* Partial Differential Equations and Geometry
49. *G. Klambauer,* Problems and Propositions in Analysis
50. *J. Knopfmacher,* Analytic Arithmetic of Algebraic Function Fields
51. *F. Van Oystaeyen,* Ring Theory
52. *B. Kadem,* Binary Time Series
53. *J. Barros-Neto and R. A. Artino,* Hypoelliptic Boundary-Value Problems
54. *R. L. Sternberg et al.,* Nonlinear Partial Differential Equations in Engineering and Applied Science
55. *B. R. McDonald,* Ring Theory and Algebra III
56. *J. S. Golan,* Structure Sheaves Over a Noncommutative Ring
57. *T. V. Narayana et al.,* Combinatorics, Representation Theory and Statistical Methods in Groups
58. *T. A. Burton,* Modeling and Differential Equations in Biology
59. *K. H. Kim and F. W. Roush,* Introduction to Mathematical Consensus Theory
60. *J. Banas and K. Goebel,* Measures of Noncompactness in Banach Spaces

61. *O. A. Nielson,* Direct Integral Theory
62. *J. E. Smith et al.,* Ordered Groups
63. *J. Cronin,* Mathematics of Cell Electrophysiology
64. *J. W. Brewer,* Power Series Over Commutative Rings
65. *P. K. Kamthan and M. Gupta,* Sequence Spaces and Series
66. *T. G. McLaughlin,* Regressive Sets and the Theory of Isols
67. *T. L. Herdman et al.,* Integral and Functional Differential Equations
68. *R. Draper,* Commutative Algebra
69. *W. G. McKay and J. Patera,* Tables of Dimensions, Indices, and Branching Rules for Representations of Simple Lie Algebras
70. *R. L. Devaney and Z. H. Nitecki,* Classical Mechanics and Dynamical Systems
71. *J. Van Geel,* Places and Valuations in Noncommutative Ring Theory
72. *C. Faith,* Injective Modules and Injective Quotient Rings
73. *A. Fiacco,* Mathematical Programming with Data Perturbations I
74. *P. Schultz et al.,* Algebraic Structures and Applications
75. *L Bican et al.,* Rings, Modules, and Preradicals
76. *D. C. Kay and M. Breen,* Convexity and Related Combinatorial Geometry
77. *P. Fletcher and W. F. Lindgren,* Quasi-Uniform Spaces
78. *C.-C. Yang,* Factorization Theory of Meromorphic Functions
79. *O. Taussky,* Ternary Quadratic Forms and Norms
80. *S. P. Singh and J. H. Burry,* Nonlinear Analysis and Applications
81. *K. B. Hannsgen et al.,* Volterra and Functional Differential Equations
82. *N. L. Johnson et al.,* Finite Geometries
83. *G. I. Zapata,* Functional Analysis, Holomorphy, and Approximation Theory
84. *S. Greco and G. Valla,* Commutative Algebra
85. *A. V. Fiacco,* Mathematical Programming with Data Perturbations II
86. *J.-B. Hiriart-Urruty et al.,* Optimization
87. *A. Figa Talamanca and M. A. Picardello,* Harmonic Analysis on Free Groups
88. *M. Harada,* Factor Categories with Applications to Direct Decomposition of Modules
89. *V. I. Istrăţescu,* Strict Convexity and Complex Strict Convexity
90. *V. Lakshmikantham,* Trends in Theory and Practice of Nonlinear Differential Equations
91. *H. L. Manocha and J. B. Srivastava,* Algebra and Its Applications
92. *D. V. Chudnovsky and G. V. Chudnovsky,* Classical and Quantum Models and Arithmetic Problems
93. *J. W. Longley,* Least Squares Computations Using Orthogonalization Methods
94. *L. P. de Alcantara,* Mathematical Logic and Formal Systems
95. *C. E. Aull,* Rings of Continuous Functions
96. *R. Chuaqui,* Analysis, Geometry, and Probability
97. *L. Fuchs and L. Salce,* Modules Over Valuation Domains
98. *P. Fischer and W. R. Smith,* Chaos, Fractals, and Dynamics
99. *W. B. Powell and C. Tsinakis,* Ordered Algebraic Structures
100. *G. M. Rassias and T. M. Rassias,* Differential Geometry, Calculus of Variations, and Their Applications
101. *R.-E. Hoffmann and K. H. Hofmann,* Continuous Lattices and Their Applications
102. *J. H. Lightbourne III and S. M. Rankin III,* Physical Mathematics and Nonlinear Partial Differential Equations
103. *C. A. Baker and L. M. Batten,* Finite Geometrics
104. *J. W. Brewer et al.,* Linear Systems Over Commutative Rings
105. *C. McCrory and T. Shifrin,* Geometry and Topology
106. *D. W. Kueke et al.,* Mathematical Logic and Theoretical Computer Science
107. *B.-L. Lin and S. Simons,* Nonlinear and Convex Analysis
108. *S. J. Lee,* Operator Methods for Optimal Control Problems
109. *V. Lakshmikantham,* Nonlinear Analysis and Applications
110. *S. F. McCormick,* Multigrid Methods
111. *M. C. Tangora,* Computers in Algebra
112. *D. V. Chudnovsky and G. V. Chudnovsky,* Search Theory
113. *D. V. Chudnovsky and R. D. Jenks,* Computer Algebra
114. *M. C. Tangora,* Computers in Geometry and Topology
115. *P. Nelson et al.,* Transport Theory, Invariant Imbedding, and Integral Equations
116. *P. Clément et al.,* Semigroup Theory and Applications
117. *J. Vinuesa,* Orthogonal Polynomials and Their Applications
118. *C. M. Dafermos et al.,* Differential Equations
119. *E. O. Roxin,* Modern Optimal Control
120. *J. C. Díaz,* Mathematics for Large Scale Computing
121. *P. S. Milojević,* Nonlinear Functional Analysis
122. *C. Sadosky,* Analysis and Partial Differential Equations

123. *R. M. Shortt,* General Topology and Applications
124. *R. Wong,* Asymptotic and Computational Analysis
125. *D. V. Chudnovsky and R. D. Jenks,* Computers in Mathematics
126. *W. D. Wallis et al.,* Combinatorial Designs and Applications
127. *S. Elaydi,* Differential Equations
128. *G. Chen et al.,* Distributed Parameter Control Systems
129. *W. N. Everitt,* Inequalities
130. *H. G. Kaper and M. Garbey,* Asymptotic Analysis and the Numerical Solution of Partial Differential Equations
131. *O. Arino et al.,* Mathematical Population Dynamics
132. *S. Coen,* Geometry and Complex Variables
133. *J. A. Goldstein et al.,* Differential Equations with Applications in Biology, Physics, and Engineering
134. *S. J. Andima et al.,* General Topology and Applications
135. *P Clément et al.,* Semigroup Theory and Evolution Equations
136. *K. Jarosz,* Function Spaces
137. *J. M. Bayod et al., p*-adic Functional Analysis
138. *G. A. Anastassiou,* Approximation Theory
139. *R. S. Rees,* Graphs, Matrices, and Designs
140. *G. Abrams et al.,* Methods in Module Theory
141. *G. L. Mullen and P. J.-S. Shiue,* Finite Fields, Coding Theory, and Advances in Communications and Computing
142. *M. C. Joshi and A. V. Balakrishnan,* Mathematical Theory of Control
143. *G. Komatsu and Y. Sakane,* Complex Geometry
144. *I. J. Bakelman,* Geometric Analysis and Nonlinear Partial Differential Equations
145. *T. Mabuchi and S. Mukai,* Einstein Metrics and Yang–Mills Connections
146. *L. Fuchs and R. Göbel,* Abelian Groups
147. *A. D. Pollington and W. Moran,* Number Theory with an Emphasis on the Markoff Spectrum
148. *G. Dore et al.,* Differential Equations in Banach Spaces
149. *T. West,* Continuum Theory and Dynamical Systems
150. *K. D. Bierstedt et al.,* Functional Analysis
151. *K. G. Fischer et al.,* Computational Algebra
152. *K. D. Elworthy et al.,* Differential Equations, Dynamical Systems, and Control Science
153. *P.-J. Cahen, et al.,* Commutative Ring Theory
154. *S. C. Cooper and W. J. Thron,* Continued Fractions and Orthogonal Functions
155. *P. Clément and G. Lumer,* Evolution Equations, Control Theory, and Biomathematics
156. *M. Gyllenberg and L. Persson,* Analysis, Algebra, and Computers in Mathematical Research
157. *W. O. Bray et al.,* Fourier Analysis
158. *J. Bergen and S. Montgomery,* Advances in Hopf Algebras
159. *A. R. Magid,* Rings, Extensions, and Cohomology
160. *N. H. Pavel,* Optimal Control of Differential Equations
161. *M. Ikawa,* Spectral and Scattering Theory
162. *X. Liu and D. Siegel,* Comparison Methods and Stability Theory
163. *J.-P. Zolésio,* Boundary Control and Variation
164. *M. Křížek et al.,* Finite Element Methods
165. *G. Da Prato and L. Tubaro,* Control of Partial Differential Equations
166. *E. Ballico,* Projective Geometry with Applications
167. *M. Costabel et al.,* Boundary Value Problems and Integral Equations in Nonsmooth Domains
168. *G. Ferreyra, G. R. Goldstein, and F. Neubrander,* Evolution Equations
169. *S. Huggett,* Twistor Theory
170. *H. Cook et al.,* Continua
171. *D. F. Anderson and D. E. Dobbs,* Zero-Dimensional Commutative Rings
172. *K. Jarosz,* Function Spaces
173. *V. Ancona et al.,* Complex Analysis and Geometry
174. *E. Casas,* Control of Partial Differential Equations and Applications
175. *N. Kalton et al.,* Interaction Between Functional Analysis, Harmonic Analysis, and Probability
176. *Z. Deng et al.,* Differential Equations and Control Theory
177. *P. Marcellini et al.* Partial Differential Equations and Applications
178. *A. Kartsatos,* Theory and Applications of Nonlinear Operators of Accretive and Monotone Type
179. *M. Maruyama,* Moduli of Vector Bundles
180. *A. Ursini and P. Agliadò,* Logic and Algebra
181. *X. H. Cao et al.,* Rings, Groups, and Algebras
182. *D. Arnold and R. M. Rangaswamy,* Abelian Groups and Modules
183. *S. R. Chakravarthy and A. S. Alfa,* Matrix-Analytic Methods in Stochastic Models
184. *J. E. Andersen et al.,* Geometry and Physics
185. *P.-J. Cahen et al.,* Commutative Ring Theory

186. *J. A. Goldstein et al.*, Stochastic Processes and Functional Analysis
187. *A. Sorbi*, Complexity, Logic, and Recursion Theory
188. *G. Da Prato and J.-P. Zolésio*, Partial Differential Equation Methods in Control and Shape Analysis
189. *D. D. Anderson*, Factorization in Integral Domains
190. *N. L. Johnson*, Mostly Finite Geometries
191. *D. Hinton and P. W. Schaefer*, Spectral Theory and Computational Methods of Sturm–Liouville Problems
192. *W. H. Schikhof et al.*, *p*-adic Functional Analysis
193. *S. Sertöz*, Algebraic Geometry
194. *G. Caristi and E. Mitidieri*, Reaction Diffusion Systems
195. *A. V. Fiacco*, Mathematical Programming with Data Perturbations
196. *M. Křížek et al.*, Finite Element Methods: Superconvergence, Post-Processing, and A Posteriori Estimates
197. *S. Caenepeel and A. Verschoren*, Rings, Hopf Algebras, and Brauer Groups
198. *V. Drensky et al.*, Methods in Ring Theory
199. *W. B. Jones and A. Sri Ranga*, Orthogonal Functions, Moment Theory, and Continued Fractions
200. *P. E. Newstead*, Algebraic Geometry
201. *D. Dikranjan and L. Salce*, Abelian Groups, Module Theory, and Topology
202. *Z. Chen et al.*, Advances in Computational Mathematics
203. *X. Caicedo and C. H. Montenegro*, Models, Algebras, and Proofs
204. *C. Y. Yıldırım and S. A. Stepanov*, Number Theory and Its Applications
205. *D. E. Dobbs et al.*, Advances in Commutative Ring Theory
206. *F. Van Oystaeyen*, Commutative Algebra and Algebraic Geometry
207. *J. Kąkol et al.*, *p*-adic Functional Analysis

Additional Volumes in Preparation

p-adic functional analysis

edited by

J. Kąkol
Adam Mickiewicz University
Poznań, Poland

N. De Grande-De Kimpe
Free University of Brussels
Brussels, Belgium

C. Perez-Garcia
University of Cantabria
Santander, Spain

MARCEL DEKKER, INC. NEW YORK • BASEL

ISBN: 0-8247-8254-2

This book is printed on acid-free paper.

Headquarters
Marcel Dekker, Inc.
270 Madison Avenue, New York, NY 10016
tel: 212-696-9000; fax: 212-685-4540

Eastern Hemisphere Distribution
Marcel Dekker AG
Hutgasse 4, Postfach 812, CH-4001 Basel, Switzerland
tel: 41-61-261-8482; fax: 41-61-261-8896

World Wide Web
http://www.dekker.com

The publisher offers discounts on this book when ordered in bulk quantities. For more information, write to Special Sales/Professional Marketing at the headquarters address above.

Current printing (last digit)
10 9 8 7 6 5 4 3 2 1

PRINTED IN THE UNITED STATES OF AMERICA

Preface

The Fifth International Conference on p-adic Functional Analysis was held at the Adam Mickiewicz University of Poznań, Poland. Previous meetings of the same character took place in Laredo, Spain (1990), Santiago, Chile (1992), Clermount-Ferrand, France (1994), and Nijmegen, The Netherlands (1996).

This book contains original research articles presented at the conference in 30-minute talks by mathematicians from Europe, North and South America, Africa, and Japan. Various topics discussed by the authors in functional analysis over non-archimedean valued complete fields were inspired by recent designs for p-adic models in modern physics and in probability theory. This makes the book a reliable source for researchers in those areas.

Spaces of analytic functions and their properties are studied and discussed in several papers. This book also contains new non-archimedean results about Frechét (and locally convex) spaces with Schauder bases. A fundamental paper on Banach spaces over fields with an infinite rank valuation is presented.

In addition, spaces of continuous functions, isometrics, Banach Hopf algebras, summability methods, fractional differentation over local fields, and adelic formulas for gamma and beta functions in algebraic number theory are included.

We are grateful to the Faculty of Mathematics and Informatics at A. Mickiewicz University of Poznań, the Polish Academy of Sciences, and the President of Poznań, Dr. Sz. Kaczmarek, for financial support. We wish to thank Marcel Dekker, Inc., for publishing this volume and Mgr. B. Wilczyńska of the Faculty of Mathematics and Informatics in Poznań for secretarial help in producing the manuscript.

J. Kąkol
N. De Grande-De Kimpe
C. Perez-Garcia

Contents

Contributors

J. Aguayo Facultad de Ciencias Físicas y Matemáticas, Universidad de Concepción, Concepción, Chile

J. Araujo Facultad de Ciencias, Universidad de Cantabria, Santander, Spain

Edward Beckenstein St. John's University, Staten Island, New York

V. Benekas University of Ioannina, Ioannina, Greece

K. Boussaf Laboratoire de Mathématiques Pures, Université Blaise Pascal, Clermont-Ferrand, Aubiere, France

Abdelbaki Boutabaa Laboratoire de Mathématiques Pures, Université Blaise Pascal, Clermont-Ferrand, Aubiere, France

G. Christol Université Paris 6, Paris, France

N. De Grand-De Kimpe Vrije Universiteit Brussel, Brussels, Belgium

Stany de Smedt Vrije Universiteit Brussel, Brussels, Belgium

Bertin Diarra Laboratoire de Mathématiques Pures, Université Blaise Pascal, Clemont Ferrand, Aubière, France

Branko Dragovich Institute of Physics, Belgrade, Yugoslavia

Mikihiko Endo Rikkyo University, Tokyo, Japan

Alain Escassut Université Blaise Pascal, Clemont-Ferrand, Aubiere, France

Thomas Gilsdorf University of North Dakota, Grand Fork, North Dakota

M. Hemdaoui Faculté des Sciences, Université Mohammed I, Oujda, Morocco

J. Kąkol Faculty of Mathematics and Informatics, A. Mickiewicz University, Poznan, Poland

A. K. Katsaras University of Ioannina, Ioannina, Greece

A. Khrennikov University of Växjö, Växjö, Sweden

Anatoly N. Kochubei Institute of Mathematics, Ukrainian National Academy of Sciences, Kiev, Ukraine

A. Kubzdela Institute of Civil Engineering, University of Technology, Poznan, Poland

Nicolas Mainetti Laboratoire de Mathématiques Pures, Université Blaise Pascal, Clermont-Ferrand, Aubière, France

Z. Mebkhout Université Paris 7, Paris, France

Lawrence Narici St. John's University, Jamaica, New York

P. N. Natarajan Ramakrishna Mission Vivekananda College, Chennai, India

S. Navarro Facultad de Ciencias, Universidad de Santiago, Santiago, Chile

H. Ochsenius Facultad de Matemáticas, Universidad Católica de Chile, Santiago, Chile

C. Perez-Garcia Facultad de Ciencias, Universidad de Cantabria, Santander, Spain

W. H. Schikhof University of Nijmegen, Nijmegen, The Netherlands

L. van Hamme Faculty of Applied Sciences, Vrije Universiteit Brussel, Brussels, Belgium

Ann Verdoodt Faculty of Applied Sciences, Vrije Universiteit Brussel, Brussels, Belgium

Strict topologies and duals in spaces of functions

J AGUAYO Departamento de Matemática, Facultad de Ciencias Físicas y Matemáticas, Universidad de Concepción, Chile.[1]

N DE GRANDE-DE KIMPE Department of Mathematics, Vrije Universiteit Brussels, Pleinlaan 2, B-1050 Brussel, Belgium.

S NAVARRO Departamento de Matemática, Facultad de Ciencias, Universidad de Santiago, Chile.[2]

Abstract. Let X be a zero-dimensional Hausdorff topological space, $\mathbb{K}$ a complete non-archimedean valued field with a non-trivial valuation, and E a Hausdorff locally $\mathbb{K}$-convex space. Some locally convex topologies on the space $BC(X,E)$ of all bounded continuous functions defined from X to E, are studied. These topologies are known as strict topologies. The density of the space $RC(X,E)$, of all functions in $BC(X,E)$ with relatively compact image, for some strict topologies is shown. We also describe the duals of these spaces as spaces of certain E'– or $\mathbb{K}$-valued measures.

1 INTRODUCTION AND NOTATIONS

Throughout this paper $(X,\mathcal{T})$ is a zero-dimensional Hausdorff topological space, $(\mathbb{K},|.|)$ a non-archimedean, non-trivially valued, complete field and E a Hausdorff locally $\mathbb{K}$-convex space.

For a subset A of X we denote by $\mathcal{X}_A$ the $\mathbb{K}$-characteristic function on A, that is $\mathcal{X}_A(x) = 1 = 1_{\mathbb{K}}$ if $x \in A$ and $\mathcal{X}_A(x) = 0 = 0_{\mathbb{K}}$ if $x \notin A$. If f is a function from X

[1] This research is supported by Fondecyt N°1950546 and Proyecto No. 96.015.007-1, Dirección de Investigación, Universidad de Concepción.

[2] This research is supported by Fondecyt N°1950546 and Dicyt, USACH.

to E and p is a seminorm on E, we define $||f||_{A,p}$ and $||f||_p$ by

$$||f||_{A,p} = \sup\{p(f(x)) : x \in A\};$$
$$||f||_p = ||f||_{X,p}\,.$$

If $E = \mathbb{K}$, we define

$$||f||_A = \sup\{|f(x)| : x \in A\};$$
$$||f|| = ||f||_X\,.$$

We denote by $BC(X,E)$ (resp. $RC(X,E)$) the vector space of the continuous functions from X to E such that $f(X)$ is bounded (resp. relatively compact) in E. Let Γ be an upward directed family of continuous non-archimedean seminorms on E generating its topology. The uniform topology τ_u on $BC(X,E)$ is defined by the family of seminorms $\left\{||.||_p\right\}_{p\in\Gamma}$. The compact open topology τ_c on $BC(X,E)$ is generated by the family of seminorms $\left\{||.||_{K,p}\right\}_{(K,p)\in\mathcal{K}\times\Gamma}$, where $\mathcal{K}$ is the class of all compact subsets of X.

Let $\beta_o X$ denote the Banaschewski compactification of X. For all facts concerning $\beta_o X$ we refer to [4] and [5]. We recall that if $f : X \to E$ is continuous, then f has a unique continuous extension $\widehat{f} : \beta_o X \to \beta_o E$. Moreover, if $f \in RC(X,E)$, then $\widehat{f}(\beta_o X) \subset E$ ([5], p. 28).

Now, let Ω be the family of all compact subsets of $\beta_o X \setminus X$ and Ω_u the subfamily of Ω consisting of those elements $K \in \Omega$ for which there exists a clopen partition $\{U_\imath\}_{\imath\in I}$ of X such that

$$\overline{U_\imath}^{\beta_o X} \cap K = \emptyset \quad \text{for all} \quad i \in I \qquad \text{(see [1]; Lemma 1)}.$$

The families Ω and Ω_u give rise to specific strict topologies on $BC(X,E)$ denoted by β, β', β_u and β'_u. The last two topologies are new and are defined by the authors. The others topologies were already introduced in [3]. We prove that $RC(X,E)$ is β_u-dense in $BC(X,E)$. Finally, we study the dual of the space $RC(X,E)$ with the topology β'_u.

2 THE TOPOLOGIES β_u AND β'_u ON $BC(X,E)$

For $K \in \Omega$, we denote by Φ_K the collection

$$\Phi_K = \left\{\varphi \in RC(X,\mathbb{K}) : \widehat{\varphi}_{|K} \equiv 0\right\}$$

and, for $\varphi \in \Phi_K$ and $p \in \Gamma$, we define a seminorm $||.||_{\varphi,p}$ on $BC(X,E)$ by

$$||f||_{\varphi,p} = \sup\{p(\varphi(x)f(x)) : x \in X\}.$$

For $\varepsilon > 0$, we denote by $V_{\varphi,p}(\varepsilon)$ the set

$$V_{\varphi,p}(\varepsilon) = \left\{f \in BC(X,E) : ||f||_{\varphi,p} \leqslant \varepsilon\right\}.$$

DEFINITION 1

1. For each $p \in \Gamma$ we define $\beta_{u,p}$ as the locally convex inductive limit of the topologies $\beta_{K,p}$, with $K \in \Omega_u$, where $\beta_{K,p}$ is the locally $\mathbb{K}$-convex topology on $BC(X,E)$ generated by the family of seminorms $\left(\|.\|_{\varphi,p}\right)_{\varphi \in \Phi_K}$.
2. We define β'_u as the projective limit of the topologies $\beta_{u,p}$, with $p \in \Gamma$.
3. We define β_u as the locally convex inductive limit of the topologies β_K, with $K \in \Omega_u$, where β_K is the topology on $BC(X,E)$ generated by the family of seminorms $\left(\|.\|_{\varphi,p}\right)_{(\varphi,p)\in\Phi_K\times\Gamma}$.

Note that, if we replace Ω_u by Ω in the above definitions we obtain the strict topologies β' and β defined by Katsaras [3]. Since $\Omega_u \subset \Omega$ we have $\beta \leqslant \beta_u$ and $\beta' \leqslant \beta'_u$.

Also, it is not hard to see that if X is pseudocompact we have that $\beta'_u = \beta_u = \tau_u$ (use [1], Th. 10)

PROPOSITION 2 *The topologies β'_u and β_u are Hausdorff and $\tau_c \leqslant \beta'_u \leqslant \beta_u \leqslant \tau_u$.*

Proof. Let C be a compact subset of X, $p \in \Gamma$, $\varepsilon > 0$ and let

$$M_p(C,\varepsilon) = \{f \in BC(X,E) : p(f(x)) \leqslant \varepsilon; \quad \text{for all } \; x \in C\}$$

be the corresponding zero-neighborhood for τ_c. We prove that $M_p(C,\varepsilon)$ is a β_{up}-zero-neighborhood or that for each $K \in \Omega_u$, $M_p(C,\varepsilon)$ is a $\beta_{K,p}-$zero-neighborhood.

Let $K \in \Omega_u$. Then there exists a clopen partition $\{U_\alpha\}_{\alpha\in I}$ of X such that $K \cap \overline{U_\alpha}^{\beta_o X} = \emptyset$ for all $\alpha \in I$, and so $K \subset \beta_o X \setminus \bigcup_{\alpha\in I} \overline{U_\alpha}^{\beta_o X}$. On the other hand, by compactness of C, there exists a finite subset F of I such that $C \subset A = \bigcup_{\alpha\in F} U_\alpha$. Put $\varphi = \mathcal{X}_A$. Since $\widehat{\varphi} \equiv \mathcal{X}_{\overline{A}^{\beta_o X}}$, and $\overline{A}^{\beta_o X} \cap K = \emptyset$ we obtain that $\varphi \in \Phi_K$. It is easy to prove that $V_{\varphi,p}(\varepsilon) \subset M_p(C,\varepsilon)$. Hence, $\tau_c \leqslant \beta'_u$.

The inequality $\beta'_u \leqslant \beta_u$ follows from the definitions.

Finally, for the proof of $\beta_u \leqslant \tau_u$, it is sufficient to prove that $\beta_K \leqslant \tau_u$ for each $K \in \Omega_u$. This follows directly from the fact that all $p \in \Gamma$ and $\varphi \in \Phi_K$ we have $\|f\|_{\varphi,p} \leqslant \|\varphi\| \, \|f\|_p$, $f \in BC(X,E)$.

REMARK 3 For the later use we recall the following result of Katsaras ([3] Theorem 3.4).

"If $K \in \Omega$ and $p \in \Gamma$, then the topology $\beta_{K,p}$ has a zero-neighborhood base consisting of the sets $W_p(B_n,\lambda_n)$, where (B_n) is an increasing sequence of clopen subsets of X with $\overline{B_n}^{\beta_o X} \cap K = \emptyset$ for all n, $(\lambda_n)_{n\in\mathbb{N}}$ is a sequence in K with $0 < |\lambda_1| < |\lambda_2| < \ldots <$ and $\lim_n |\lambda_n| = \infty$, and $W_p(B_n,\lambda_n)$ is defined by

$$W_p(B_n,\lambda_n) = \bigcap_{n=1}^{\infty} \left\{f \in BC(X,E) : \|f\|_{B_n,p} \leqslant |\lambda_n|\right\}.$$

3 DENSITY

In this section we prove the β_u−density of $RC(X,E)$ in $BC(X,E)$. This result will imply the density of $RC(X,E)$ in $BC(X,E)$ for other strict topologies.

THEOREM 4 *$RC(X,E)$ is β_u-dense in $BC(X,E)$.*

Proof. Fix $f \in BC(X,E)$ and U, an absolutely convex β_u−zero-neighborhood in $BC(X,E)$. We shall construct $g \in RC(X,E)$ such that $f - g \in U$. Since $\beta_u \leqslant \tau_u$ there exist $p \in \Gamma$ and $\varepsilon > 0$ such that

$$V_{p,\varepsilon} = \left\{ f \in BC(X,E) : \|f\|_p \leqslant \varepsilon \right\} \subset U$$

and keep these p and $\varepsilon > 0$ fixed as well. We define an equivalence relation $\sim_p$ on X by:

$$x \sim_p y \quad \text{if} \quad p(f(x) - f(y)) \leqslant \varepsilon,$$

and we denote by $(A^p_\alpha)_{\alpha \in I_p}$ the corresponding clopen partition of X. For every $\alpha \in I_p$ choose $x^p_\alpha \in A^p_\alpha$ and define

$$f^*_p = \sum_{\alpha \in I_p} f(x^p_\alpha) \mathcal{X}_{A^p_\alpha}.$$

Then $f^*_p \in BC(X,E)$ and $f - f^*_p \in V_{p,\varepsilon}$.

Put $K_p = \beta_o X \setminus Y_p$, where $Y_p = \bigcup_{\alpha \in I_p} \overline{A^p_\alpha}^{\beta_o X}$. Then K_p is a compact subset of $\beta_o X$ and $K_p \in \Omega_u$. By definition U is then a zero-neighborhood in β_{K_p}. Hence, there exists q in Γ such that U a zero-neighborhood in $\beta_{K_p,q}$ and we can assume that $q \geqslant p$. We also fix this $q \in \Gamma$.

Then (see Remark 3) there exists an increasing sequence (B_n) of clopen subsets of X with $\overline{B_n}^{\beta_o X} \cap K_p = \emptyset$, for all n and a sequence $(\lambda_n)_{n \in \mathbb{N}}$ in $\mathbb{K}$ with $0 < |\lambda_1| < |\lambda_2| < \ldots.$ and $\lim_{n \to \infty} |\lambda_n| = \infty$ such that

$$W_q(B_n, \lambda_n) = \bigcap_{n=1}^{\infty} \left\{ f \in BC(X,E) : \|f\|_{B_n,q} \leqslant |\lambda_n| \right\} \subset U.$$

For each n, $\overline{B_n}^{\beta_o X}$ is a compact set in $\beta_o X$ with $\overline{B_n}^{\beta_o X} \cap K_p = \emptyset$. It follows that for each n there exists a finite set $F^n_p \subset I_p$ such that

$$B_n \subset \bigcup_{\alpha \in F^n_p} A^p_\alpha.$$

Then, for any $n \in \mathbb{N}$, we have that $f^*_p(B_n)$ is a finite subset of E.

Now we define the desired function $g : X \to E$. First fix $N \in \mathbb{N}$ such that $|\lambda_N| > \max\left\{1, \|f\|_q\right\}$. Then define g by

$$g = f^*_p \mathcal{X}_{B_{N-1}} + \lambda_N^{-1} f^*_p \mathcal{X}_{B_N \setminus B_{N-1}}.$$

Clearly, $g \in RC(X, E)$.

We claim that $f - g \in U$. We have

$$f - g = f - f_p^* + f_p^* - g$$

and $f - f_p^* \in V_{p,\varepsilon} \subset U$.

So it is left to prove that $f_p^* - g \in U$. For this it is sufficient to show that $f_p^* - g \in W_q(B_n, \lambda_n)$ or that for all n

$$\sup_{x \in B_n} q(f_p^*(x) - g(x)) \leqslant |\lambda_n| \,. \tag{$*$}$$

If $n < N$, $g_{|B_n} = f^*_{p|B_n}$ and so $(*)$ is satisfied.

If $n = N$ we only have to consider $x \in B_N \setminus B_{N-1}$. For such x we have

$$\begin{aligned} q(f_p^*(x) - g(x)) &= q(f_p^*(x)(1 - \lambda_N^{-1})) = q(f_p^*(x)) \\ &\leqslant \|f\|_q \leqslant |\lambda_N| \,. \end{aligned}$$

Finally, if $n > N$, again we only have to consider $x \in B_N \setminus B_{N-1}$ and the conclusion follows easily.

COROLLARY 5 *$RC(X, E)$ is dense in $BC(X, E)$ for the topologies β', β, β'_u.*

4 DUAL OF THE SPACES WITH STRICTTOPOLOGIES

In this section we will consider the strict topologies on $RC(X, E)$ which are defined in the same way as in section 2.

Let $S(X)$ be the ring of the clopen subsets of X. We understand by a finitely-additive $\mathbb{K}$-valued measure (or scalar measure) any set function m from $S(X)$ into $\mathbb{K}$ which satisfies:

M1 - For any disjoint finite family $\{A_i\}_{i=1}^n$ of elements of $S(X)$,

$$m\Big(\bigcup_{i=1}^n A_i\Big) = \sum_{i=1}^n m(A_i).$$

A trivial consequence of this is that $m(\emptyset) = 0$.

M2 - $m(S(X))$ is a bounded subset of $\mathbb{K}$.

The set of all such finitely-additive $\mathbb{K}$-valued measures m is denoted by $M(X, \mathbb{K})$.

We understand by a finitely-additive E'-valued measure any set function m from $S(X)$ into E' which satisfies M1 and

M2' - $m(S(X))$ is an equicontinuous subset of E'.

The set of all such finitely-additive E'-valued measures m is denoted by $M(X, E')$. If $m \in M(X, E')$, then for each $s \in E$, the set function ms defined from $S(X)$ into $\mathbb{K}$ by $ms(A) = m(a)(s)$, is an scalar measure.

For $m \in M(X, E')$, $p \in \Gamma$ and $A \in S(X)$ we define

$$m_p(A) = \sup\{|m(B)(s)| : p(s) \leqslant 1, B \in S(X), B \subset A\}\,. \tag{1}$$

The M2' condition permits to assure that if $m \in M(X,E')$, then $m_p(X) < \infty$, for some $p \in \Gamma$. Therefore, if for every $p \in \Gamma$, we denote by $M_p(X,E')$ the set $\{m \in M(X,E') : m_p(X) < \infty\}$, then

$$M(X,E') = \bigcup_{p\in\Gamma} M_p(X,E').$$

If $m \in M(X,\mathbb{K})$, then (1) can be written by

$$|m|(A) = \sup\{|m(B)| : B \in S(X), B \subset A\}$$

We recall the following definitions from [3], p.24.

1. $m \in M(X,\mathbb{K})$ is called a $\tau-$additive measure if for any decreasing net $\{G_\alpha\}_{\alpha\in I}$ in $S(X)$ with $\bigcap_{\alpha\in I} G_\alpha = \emptyset$, $\lim_{\alpha\in I} m(G_\alpha) = 0$. The set of the all $\tau-$additive measures is denoted by $M_\tau(X,\mathbb{K})$. An $m \in M(X,E')$ is called a $\tau-$additive measure if for any $s \in E$, $ms \in M_\tau(X,\mathbb{K})$. The set of these measures is denoted by $M_\tau(X,E')$.
2. An $m \in M(X,\mathbb{K})$ is called a $\sigma-$additive measure if for any decreasing sequence $\{G_n\}_{n\in\mathbb{N}}$ in $S(X)$ with $\bigcap_{n\in\mathbb{N}} G_n = \emptyset$, $\lim_{n\to\infty} m(G_n) = 0$. The set of the all $\sigma-$additive measures is denoted by $M_\sigma(X,\mathbb{K})$. An $m \in M(X,E')$ is called a $\sigma-$additive measure if for any $s \in E$, $ms \in M_\sigma(X,\mathbb{K})$. The set of these measures is denoted by $M_\sigma(X,E')$.

REMARK 6 We also can get from [3] characterizations for a measure to be τ or σ additive. Those characterizations are the following: $m \in M(X,\mathbb{K})$ is a $\tau\,(\sigma)-$additive measure if and only if for any decreasing net $\{G_\alpha\}_{\alpha\in I}$ (sequence $\{G_n\}_{n\in\mathbb{N}}$) in $S(X)$ with $\bigcap_{\alpha\in I} G_\alpha = \emptyset$ $(\bigcap_{n\in\mathbb{N}} G_n = \emptyset)$, $|m|(G_\alpha) \to 0$ $(|m|(G_n) \to 0)$.

In this paper, we introduce a new type of a measure as follows:

DEFINITION 7 Let $m \in M(X,\mathbb{K})$. We say that m is a $u-$additive measure if for any clopen partition $\{U_\alpha\}_{\alpha\in I}$ of X, $|m|(X \setminus \bigcup_{\alpha\in F} U_\alpha) \to 0$, where the limit has to be taken over the directed set of all finite subsets F of I. The set of all $u-$additive measures will be denoted by $M_u(X,\mathbb{K})$. An $m \in M(X,E')$ is called a u$-$additive measure if for any $s \in E$, $ms \in M_u(X,\mathbb{K})$. The set of these measures is denoted by $M_u(X,E')$.

PROPOSITION 8 $M_\tau(X,E') \subset M_u(X,E') \subset M_\sigma(X,E')$.

Proof. Let $m \in M_\tau(X,E')$ and $s \in E$. Let $(U_\alpha)_{\alpha\in I}$ be a clopen partition of X. For every finite subset F of I, we consider $V_F = X \setminus \bigcup_{\imath\in F} U_\imath$. Clearly, each V_F is clopen, $V_{F_1} \subset V_{F_2}$, for $F_2 \subset F_1$, and $\bigcap_F V_F = \emptyset$. Then $|ms|(V_F) \to 0$. Hence, $m \in M_u(X,E')$.

Suppose now that $m \in M_u(X,E')$ and take $s \in E$. If we pick a decreasing sequence $\{A_n\}_{n\in\mathbb{N}}$ in $S(X)$ with $\bigcap_{n=1}^{\infty} A_n = \emptyset$, then we define

$$U_1 = X \setminus A_1; \;\; U_n = A_{n-1} \setminus A_n, \; n \geqslant 2.$$

Clearly, $\{U_n\}_{n=1}^{\infty}$ is a clopen partition of X and so,

$$|ms|\,(X \setminus \bigcup_{i=1}^{n} U_i) \to 0.$$

Hence, $m \in M_\sigma(X, E')$.

For $A \in S(X)$, $A \neq \emptyset$, we denote by α the set $\{A_1, A_2, ..., A_n; x_1, x_2, ..., x_n\}$, where $A_i \in S(X)$ for $i = 1, .., n$, $\{A_i\}_{i=1}^{n}$ is a partition of A, and $x_i \in A_i$. Let Ψ_A be the collection of such α. On Ψ_A we introduce a partial order as follows:

$\alpha_1 \leqslant \alpha_2$ if the partition of A in α_2 is finer than the one in α_1.

For this order the set $\Psi_A$becomes a directed set. Now, for a function $f : X \to E$, $\alpha \in \Psi_A$, and $m \in M(X, E')$, we define

$$\varpi_\alpha(f, m) = \sum_{i=1}^{n} m(A_i)(f(x_i))$$

and if $\lim_{\alpha \in \Psi_A} \varpi_\alpha(f, m)$ exists, we will say that f is $m-$integrable over $A \in S(X)$. In such a case, we will write:

$$\lim_{\alpha \in \Psi_A} \varpi_\alpha(f, m) = \int_A f dm$$

and $\int f dm$ if $A = X$. If $A = \emptyset$ we put $\int_A f dm = 0$.

The following are known (see [2]):

1. The set of all $m-$integrable functions over $A \in S(X)$ is a vector space over $\mathbb{K}$.
2. There exists $\rho \geqslant 1$ such that if $f : X \to E$ is $m-$integrable over $A \in S(X)$ and $p \in \Gamma$, then
$$\left|\int_A f dm\right| \leqslant \rho \, \|f\|_{A,p} \, m_p(A).$$
3. If $f \in RC(X, E)$, then f is $m-$integrable over each $A \in S(X)$.
4. For $m \in M(X, E')$, the mapping $T_m : RC(X, E) \to \mathbb{K}$ defined by $T_m(f) = \int f dm$, is an element of $(RC(X, E), \tau_u)'$.
5. The mapping $M(X, E') \to (RC(X, E), \tau_u)'$, $m \to T_m$, is an algebraic isomorphism. In the sequel we identify every $m \in M(X, E')$ with its image under this map.

DEFINITION 9 Let $m \in M_p(X, E')$; we say that m is a u_p-additive measure if for any clopen partition $\{U_\alpha\}_{\alpha \in I}$ of X, $m_p(X \setminus \bigcup_{\alpha \in F} U_\alpha) \to 0$, where the limit is taken over the directed set of all finite subsets F of I . The set of all u_p-additive measures will be denoted by $M_{u,p}(X, E')$. The set $\bigcup_{p \in \Gamma} M_{u,p}(X, E')$ will be denoted by $M'_u(X, E')$.

THEOREM 10 *Let $m \in M_p(X, E')$. The following are equivalent:*
1. *m is $\beta_{u,p}-$continuous.*
2. *m is u_p-additive.*

Proof. 1) $\Rightarrow$ 2) Let $m : (RC(X,E), \beta_{u,p}) \to \mathbb{K}$ be continuous. We prove that m is u_p-additive. Let $\{U_\imath\}_{\imath \in I}$ be a clopen partition of X. Then, $K = \beta_o X \setminus \bigcup_{\imath \in I} \overline{U_\imath}^{\beta_o X}$ is a compact subset of $\beta_o X \setminus X$ and belongs to Ω_u. Since $\beta_{u,p}$ is the inductive limit of $\{\beta_{Q,p}\}_{Q \in \Omega_u}$, m is $\beta_{K,p}-$continuous. Hence there exists $\varphi_\imath \in \Phi_K$ with $i = 1, .., n$, such that

$$RC(X,E) \cap \bigcap_{\imath=1}^{n} V_{\varphi_\imath, p}(1) \subset V = \left\{ f \in RC(X,E) : \left| \int f dm \right| \leqslant 1 \right\}.$$

Now, given $\varepsilon > 0$, we pick $\lambda \in \mathbb{K}$ with $|\lambda|^{-1} \leqslant \varepsilon$, and consider the set

$$B = \left\{ x \in \beta_o X : |\widehat{\varphi}_\imath(x)| \leqslant |\lambda|^{-1}, \ i = 1, .., n \right\}.$$

Clearly, B is clopen in $\beta_o X$, $K \subset B$ and $\beta_o X \setminus B \subset \bigcup_{\imath \in I} \overline{U_\imath}^{\beta_o X}$. Then, there exists a finite subset F_o of I such that

$$\beta_o X \setminus B \subset \bigcup_{\imath \in F_o} \overline{U_\imath}^{\beta_o X}.$$

Let F be a finite subset of I containing F_o and $A \in S(X)$ contained in $X \setminus \bigcup_{\imath \in F} U_\imath$. We pick $s \in E$ with $p(s) \leqslant 1$, and take $f_s = \mathcal{X}_A \otimes s \in RC(X,E)$.

Then for $x \in X$ and $i = 1, ..., n$,

$$p(\varphi_\imath(x) f_s(x)) = \begin{cases} |\varphi_\imath(x)| \, p(s) & \text{if } x \in A \\ 0 & \text{if } x \notin A \end{cases}$$

Thus, $\|f_s\|_{\varphi_\imath, p} \leqslant |\lambda|^{-1}$. Indeed,

$$\begin{aligned} x \in A &\Rightarrow x \notin U_\imath; \ \forall i \in F \Rightarrow x \notin \overline{U_\imath}^{\beta_o X}; \ \forall i \in F \\ &\Rightarrow x \notin \bigcup_{\imath \in F} \overline{U_\imath}^{\beta_o X} \Rightarrow x \notin \beta_o X \setminus B \Rightarrow x \in B \\ &\Rightarrow |\varphi_\imath(x)| \leqslant |\lambda|^{-1}. \end{aligned}$$

Hence,

$$\|f_s\|_{\varphi_\imath, p} = \sup_{x \in X} p(\varphi_\imath(x) f_s(x)) = \sup_{x \in A} |\varphi_\imath(x)| \, p(s) \leqslant |\lambda|^{-1}.$$

Now as $\|f_s\|_{\varphi_\imath, p} \leqslant |\lambda|^{-1}$, we have $\|\lambda f_s\|_{\varphi_\imath, p} \leqslant 1$ for all $i = 1, ..., n$ and so $\lambda f_s \in V$. On the other hand, $\int_A f_s dm = m(A)(s)$. It follows that $|\lambda m(A)(s)| \leqslant 1$ or that

$|m(A)(s)| \leqslant |\lambda|^{-1} \leqslant \varepsilon$. Since $A \in S(X)$ with $A \subset X \setminus \bigcup_{i \in F} U_i$ and $s \in E$ with $p(s) \leqslant 1$ are arbitraries , we have that, for any finite subset F of I containing F_o,

$$m_p\left(X \setminus \bigcup_{i \in F} U_i\right) \leqslant \varepsilon.$$

This proves that m is u_p−additive.

2) ⇒ 1) We prove that $V = \left\{f \in RC(X,E) : \left|\int f dm\right| \leqslant 1\right\}$ is a $\beta_{u,p}$− zero-neighborhood.

Take $K \in \Omega_u$ and let $\{U_i\}_{i \in I}$ be a clopen partition of X such that

$$\overline{U_i}^{\beta_o X} \cap K = \emptyset; \; for \; all \; i \in I.$$

Let $\rho > 1$ be as in property 2 before Definition 9. Pick $d > 0$ and choose $\lambda, \gamma \in \mathbb{K} \setminus \{0\}$ such that $|\lambda| \geqslant \rho d$ and $\rho |\gamma| m_p(X) \leqslant 1$. Since m is u_p−additive, for $\varepsilon = |\lambda|^{-1}$, there exists a finite subset F_o of I such that, for all finite subset $F \subset I$, $F_o \subset F$,

$$m_p\left(X \setminus \bigcup_{i \in F} U_i\right) \leqslant \varepsilon.$$

Take $D = \bigcup_{i \in F_o} U_i$ and define the set

$$W = \left\{f \in RC(X,E) : \|f\|_p \leqslant d \quad \text{and } \|f\|_{D,p} \leqslant |\gamma|\right\}.$$

If $f \in W$,

$$\begin{aligned}
\left|\int fdm\right| = \left|\int_D fdm + \int_{X\setminus D} fdm\right| &\leqslant \max\left\{\left|\int_D fdm\right|, \left|\int_{X\setminus D} fdm\right|\right\} \\
&\leqslant \max\left\{\rho \|f\|_{D,p} |m|(D), \rho \|f\|_{X\setminus D,p} m_p(X \setminus D)\right\} \\
&\leqslant \max\left\{\rho \|f\|_{D,p} m_p(D), \rho \|f\|_p m_p(X \setminus D)\right\} \\
&\leqslant \max\left\{\rho |\gamma| m_p(D), \rho d m_p(X \setminus D)\right\} \\
&\leqslant \max\left\{1, |\lambda| m_p(X \setminus D)\right\} \leqslant 1.
\end{aligned}$$

Therefore $f \in V$.

By Theorem 4.1 of [3] we obtain that V is a $\beta_{K,p}$ zero-neighborhood and since $K \in \Omega_u$ was arbitrary, we conclude that V is a $\beta_{u,p}$−zero neighborhood.

COROLLARY 11 *Let m∈ $M(X,\mathbb{K})$. The following are equivalent:*

1. m is β_u−continuous.

2. m is u-additive.

Proof. It follows from Theorem 10 by taking $E = \mathbb{K}$ and $p = |.|$.

COROLLARY 12 *The dual of $(RC(X,\mathbb{K}), \beta_u)$ is $M_u(X,\mathbb{K})$.*

Proof. It follows directly from Corollary 11.

THEOREM 13 *The dual of* $(RC(X,E),\beta'_u)$ *is* $M'_u(X,E')$.

Proof. Take $m \in (RC(X,E),\beta'_u)'$. Then, there exists $p \in \Gamma$ such that $m_p(X) < \infty$ and $m \in (RC(X,E),\beta_{u,p})'$. By Theorem 10, 1) $\Rightarrow$ 2), $m \in M_{u,p}(X,E')$. Hence $m \in M'_u(X,E')$.
Conversely, take $m \in M'_u(X,E')$. Then, there exists $p \in \Gamma$ such that $m \in M_{u,p}(X,E')$. By Theorem 10, 2) $\Rightarrow$ 1), $m \in (RC(X,E),\beta_{u,p})'$. Hence $m \in (RC(X,E),\beta'_u)'$.

REFERENCES

[1] J Aguayo, N De Grande De Kimpe, S Navarro. Zero-dimensional pseudocompact and ultraparacompact spaces. In: WH Schikof, C Pérez-Garcia, J Kakol, ed. p-adic Functional Analysis. New York: Marcel Dekker, 1997, pp 11-18.

[2] AK Katsaras. Duals of non-archimedean Vector-Valued Functions Spaces. Bull of the Greek Math Soc, 22:25–43, 1981.

[3] AK Katsaras. Strict Topologies in non-archimedean Functions Spaces. J Math Sci 7:23–33, 1984.

[4] JB Prolla. Topics in Functional Analysis over Valued Division Rings. Amsterdam: North Holland, 1982.

[5] AC van Rooij. Non-Archimedean Functional Analysis. New York: Marcel Dekker, 1978.

Ultrametric weakly separating maps with closed range

J ARAUJO[1] Departamento de Matemáticas, Estadística y Computación,
Facultad de Ciencias, Universidad de Cantabria,
39071 Santander, Spain.

In this paper we give some results concerning automatic continuity of weakly separating maps between spaces of bounded continuous functions over a nonarchimedean field. Separating maps were introduced in [6] as a useful tool for the study of linear isometries in the nonarchimedean context, and were also used in [5] with a similar aim. But they have also played a fundamental role in the study of automatic continuity of maps between spaces of continuous functions, not only when dealing with nonarchimedean fields but also in the real or complex cases (see for instance [1, 7, 8, 2, 3] and [4] for some recent references).

From now on $\mathbf{K}$ will be a commutative complete nonarchimedean valued field endowed with a nontrivial valuation. X and Y will be $\mathbf{N}$-compact spaces, and we will denote by $\beta_0 X$ and $\beta_0 Y$ their Banaschewski compactifications. If U is a clopen (this is open and closed) subset of X, ξ_U will stand for the characteristic function on U, and if $a \in \mathbf{K}$, $\mathbf{a} := a\xi_X$. $C^*(X)$ and $C^*(Y)$ will be the spaces of $\mathbf{K}$-valued bounded continuous functions on X and Y, respectively, which are assumed to be endowed with the sup norm. For $f \in C^*(X)$ we will denote by $c(f)$ its cozero set $\{x \in X : f(x) \neq 0\}$, and by $f^{\beta_0 X}$ its continuous extension from $\beta_0 X$ into $\beta_0 \mathbf{K}$. Finally for $T : C^*(X) \to C^*(Y)$, we define $Y_0 := \bigcup_{f \in C^*(X)} c(Tf)$.

DEFINITION 1 A map $T : C^*(X) \to C^*(Y)$ is said to be weakly separating if whenever U and V are clopen disjoint subsets of X, and $f, g \in C^*(X)$ satisfy $c(f) \subset U$ and $c(g) \subset V$, then $c(Tf) \cap c(Tg) = \emptyset$.

Given $y_0 \in Y_0$, a point $x_0 \in \beta_0 X$ is said to be a support point of y_0 if for every neighborhood U of x_0 in $\beta_0 X$, there exists $f \in C^*(X)$ such that $c(f) \subset U$

[1] Research partially supported by the Spanish Direccion General de Investigacion Cientıfica y Tecnıca (DGICYT PB95-0582)

and $(Tf)(y_0) \neq 0$. The proof of the existence and uniqueness of support point for each $y_0 \in Y_0$ when we deal with a weakly separating linear map can be given as in [7, p. 260]. As a consequence of this fact we can define a (continuous) map $h : Y_0 \to \beta_0 X$ sending each point of Y_0 into its support point. Finally, if T is injective, h has dense range in $\beta_0 X_0$ so we can extend it to a continuous map from $\beta_0 Y_0$ onto $\beta_0 X_0$.

DEFINITION 2 Given a linear map $T : C^*(X) \to C^*(Y)$ and $M > 0$, a point $x \in \beta_0 X$ is said to satisfy the property T_M if there exists a neighborhood U of x in $\beta_0 X$ such that $\|Tf\| \leq M$ for every $f \in C^*(X)$ satisfying $\|f\| \leq 1$ and $c(f) \subset U$.

LEMMA 3 *Suppose that $T : C^*(X) \to C^*(Y)$ is an injective weakly separating linear map. Then there exist $M > 0$ and $x_1, x_2, \ldots, x_n$ in $\beta_0 X - X$ such that each $x \in \beta_0 X - \{x_1, x_2, \ldots, x_n\}$ satisfies the property T_M.*

Also, if $f \in C^*(X)$ satisfies $\|f\| \leq 1$, then $|(Tf)(y)| \leq M$ for every $y \in Y_0$ such that $h(y) \notin \{x_1, x_2, \ldots, x_n\}$.

Proof. Suppose on the contrary that there exist $\alpha \in \mathbb{K}$, $|\alpha| > 1$, and a sequence (z_n) of different points of $\beta_0 X$ such that for each $n \in \mathbb{N}$, z_n does not satisfy the property $T_{|\alpha|^{2n}}$. Then consider a sequence (U_n) of pairwise disjoint clopen subsets of $\beta_0 X$ and a sequence (f_n) in $C^*(X)$ such that for every $n \in \mathbb{N}$, $c(f_n) \subset U_n$, $\|f_n\| \leq 1$ and $\|Tf_n\| > |\alpha|^{2n}$. Then it is clear that $f_0 := \sum_{n=1}^{\infty} \alpha^{-n} f_n$ belongs to $C^*(X)$. Also, if for each $n \in \mathbb{N}$, $y_n \in Y$ satisfies $|(Tf_n)(y_n)| > |\alpha|^{2n}$, then

$$\begin{aligned}(Tf_0)(y_n) &= \alpha^{-n}(Tf_n)(y_n) + (T(\sum_{k \neq n} \alpha^{-k} f_k))(y_n) \\ &= \alpha^{-n}(Tf_n)(y_n),\end{aligned}$$

and consequently Tf_0 is not bounded, which is absurd. This implies that there exist $N > 0$ and $x_1, x_2, \ldots, x_n, \ldots, x_{n+r} \in \beta_0 X$ such that every $x \in \beta_0 X - \{x_1, x_2, \ldots, x_{n+r}\}$ satisfies the property T_N.

Then define $M := \max\{N, \|T\mathbf{1}\|\}$ and suppose that $x_i \in \beta_0 X - X$ for $i = 1, \ldots, n$ and $x_i \in X$ for $i = n+1, \ldots, n+r$.

On the other hand, if $\|f\| \leq 1$ and $h(y) \notin \{x_1, x_2, \ldots, x_n, \ldots, x_{n+r}\}$, then we take a clopen neighborhood U of $h(y)$ in $\beta_0 X$ such that $\|Tg\| \leq M$ whenever $\|g\| \leq 1$ and $c(g) \subset U$. Since T is weakly separating, we have that $|(Tf)(y)| = |(T(f\xi_U))(y)| \leq M$. Finally if $\|f\| \leq 1$ and $h(y) = x_j$ for some $j \in \{n+1, \ldots, n+r\}$, we have that $(f - f(x_j)\mathbf{1})(x_j) = 0$ which implies by [1, Lemma 2.5] that $|(Tf)(y)| = |f(x_j)(T\mathbf{1})(y)| \leq M$, as we wanted to prove.

LEMMA 4 *Suppose that $x_1, x_2, \ldots, x_n \in \beta_0 X - X$, and that $T : C^*(X) \to C^*(Y)$ is an injective weakly separating linear map with closed range. If $(Tf)(y) = 0$ whenever $h(y) \neq x_1, x_2, \ldots, x_n$, then $Tf = 0$.*

Proof. If $f \neq 0$, then we can take a clopen neighborhood U of $\beta_0 X$ such that $f\xi_U \neq 0$ and $x_i \notin U$ for $i = 1, 2, \ldots, n$. Then, since T is injective, by [1, Lemma 2.5] there exists $y \in Y$ such that $h(y) \notin \{x_1, x_2, \ldots, x_n\}$ and $(T(f\xi_{X \cap U}))(y) \neq 0$.

Also since T is weakly separating, $(Tf)(y) = (T(f\xi_{X\cap U}))(y) \neq 0$. The conclusion is easy.

THEOREM 5 *If $T : C^*(X) \to C^*(Y)$ is an injective weakly separating linear map with closed range, then it is continuous.*

Proof. Since the range of T is closed, it is enough to prove that T is a closed operator. Suppose that (f_n) is a sequence in $C^*(X)$ converging to zero and such that (Tf_n) converges to $g \in C^*(Y)$. By Lemma 3, we know that there exist $x_1, x_2, \ldots, x_n \in \beta_0 X - X$ such that $g(y) = 0$ for every $y \in Y_0 \cap h^{-1}(\beta_0 X - \{x_1, x_2, \ldots, x_n\})$. By Lemma 4, this implies that $g = 0$. We conclude that T is closed and the theorem is proved.

We see in the next results that for the case when X is compact we can also describe the general form of these maps.

LEMMA 6 *Assume that X is compact. Suppose that $T : C(X) \to C^*(Y)$ is an injective weakly separating linear map. Then $Y_0 = c(T\mathbf{1})$.*

Using Lemma 6, with a proof quite similar to that of [1, Theorem 3.2], the following result can be proven.

THEOREM 7 *Assume that X is compact. Suppose that $T : C(X) \to C^*(Y)$ is an injective weakly separating linear map with closed range. Then there exists a nonvanishing continuous map $a : Y_0 \to \mathbb{K}$ such that for every $y \in Y_0$ and $f \in C(X)$,*

$$(Tf)(y) = a(y)f(h(y)).$$

REMARK In Theorem 5, if we drop the condition of injectivity on T, it may not be continuous. To see this, take $X := \mathbf{N}$, $Y := \{0\}$. Consider a ultrafilter F_0 in the set of all parts of $\mathbf{N}$ containing $\mathbf{N} - \{n\}$ for all $n \in \mathbf{N}$. Let V be the linear space of all $f \in C^*(X)$ such that $X - c(f)$ belongs to F_0. Take $\alpha \in \mathbf{K}$, $0 < |\alpha| < 1$, and $f_0 := \sum_{n=0}^{\infty} \alpha^n \xi_{\{n\}}$. If $\mathcal{B}$ is a base of V as a linear space over $\mathbf{K}$, it is clear that we can extend $\mathcal{B} \cup \{f_0\}$ to a base $\mathcal{B}_0$ of $C^*(X)$ as a linear space over $\mathbf{K}$. Then define $T : \mathcal{B}_0 \to \mathbf{K}$ as $Tf_0 := \mathbf{1}$, and $Tf := 0$ otherwise. Since we can identify $\mathbf{K}$ with $C(Y)$, then we can extend by linearity to a map $T : C^*(X) \to C(Y)$. Now suppose that $f, g \in C^*(X)$ satisfy $c(f) \cap c(g) = \emptyset$. Then either $c(f) \in F_0$ or $X - c(f) \in F_0$. In the first case $X - c(g)$ contains $c(f)$ and then we deduce that $X - c(g) \in F_0$ and $Tg = 0$. In the other case $Tf = 0$. We conclude that T is weakly separating, but is is clear that it is not continuous.

REMARK When $\mathbf{K}$ is not locally compact and X is nor compact, the same hypothesis as in Theorem 7 do not lead in general to the fact that $h(y)$ belongs to X for every $y \in Y$. For instance, assume that $\mathbf{K}$ is spherically complete. Consider X the unit ball of $\mathbf{C}_p$ and $Y = X \cup \{x_0\}$ for some prime p and some $x_0 \in \mathbf{C}_p$, $|x_0| > 1$. Now take $y_0 \in \beta_0 X - X$. Now we take A as the (closed) linear subspace of $C^*(X)$ of those functions admitting a continuous extension (into K) to y_0, and define a continuous linear functional $F : A \to \mathbf{K}$ as $F(f) := f^{\beta_0 X}(h(y_0))$, $f \in C^*(X)$. According to [9, Theorem 4.15], we can extend F to a continuous linear functional $G : C^*(X) \to \mathbf{K}$. Then we define $T : C^*(X) \to C^*(Y)$ as $(Tf)(y) = f(y)$ if $y \in X$, and $(Tf)(x_0) = G(f)$. It is easy to see that T is linear and weakly separating, and its range is closed. Moreover $h(x_0) \notin X$. On the other hand, this implies that there exists a sequence (U_n) of clopen neighborhoods of $h(y_0)$ in $\beta_0 X$ such that U_{n+1} is strictly contained in U_n for every $n \in \mathbf{N}$, $U_1 = \beta_0 X$, and $X \cap \bigcap_{n=1}^{\infty} U_n = \emptyset$. Define $V_n := U_n - U_{n+1}$ for every $n \in \mathbf{N}$. Taking into account that $\mathbf{K}$ is not locally compact, there exists a sequence (α_n) in $\mathbf{K}$ such that $1 - 1/(n+1) \leq |\alpha_n| \leq 1$ for every $n \in \mathbf{N}$, and $|\alpha_n - \alpha_m| \geq 1 - 1/(n+1)$ for every $n, m \in \mathbf{N}$, $n > m$. We define $f_0 \in C^*(X)$ by the requirement that $f_0(x) := \alpha_n$ for $x \in V_n$. It is clear that f_0 does not admit a continuous extension to the point $h(y_0)$, which implies that we cannot extend in general the expression of Tf_0 as a weighted composition map to describe its value at every point of Y.

REFERENCES

[1] J Araujo. N-compactness and automatic continuity in ultrametric spaces of bounded continuous functions. To appear in Proc Amer Math Soc.

[2] J Araujo, E Beckenstein and L Narici. Biseparating maps and homeomorphic realcompactifications. J Math Anal App 192:258–265, 1995.

[3] J Araujo, E Beckenstein and L Narici. Separating maps and the nonarchimedean Hewitt theorem. Ann Math Blaise Pascal 2:19–27, 1995.

[4] J Araujo, E Beckenstein and L Narici. When is a separating map biseparating? Archiv Math 67:395–407, 1996.

[5] J Araujo and J Martinez-Maurica. The nonarchimedean Banach-Stone theorem. In: F Baldassari, S Bosch, B Dwork, ed. p-Adic Analysis. Berlin: Springer-Verlag, 1990, pp 64–79.

[6] E Beckenstein and L Narici. A nonarchimedean Stone-Banach theorem. Proc Amer Math Soc 100:242–246, 1987.

[7] E Beckenstein, L Narici and AR Todd. Automatic continuity of linear maps on spaces of continuous functions. Manuscripta Math 62:257–275, 1988.

[8] K Jarosz. Automatic continuity of separating linear isomorphisms. Canad Math Bull 33:139–144, 1990.

[9] ACM van Rooij. Nonarchimedean Functional Analysis. New York: Marcel Dekker, 1978.

Analytic spectrum of an algebra of strictly analytic p-adic functions

K BOUSSAF Laboratoire de Mathématiques Pures, Université Blaise Pascal, Clermont-Ferrand, Complexe Scientifique des Cézeaux, F 63177 Aubiere Cedex, France.

M HEMDAOUI Département de Mathématiques, Université Mohammed I, Faculté des Sciences, 60000 Oujda, Maroc.

1 PRELIMINARIES

Let K be an algebraically closed field, complete for an ultrametric absolute value. We recall some standard notations and definitions.

Given $a \in K$ and $r > 0$, $d(a,r)$ (resp. $d(a,r^-)$, resp. $C(a,r)$) denotes the *circumferenced disk* $\{x \in K | \ |x-a| \leqslant r\}$ (resp. the *non-circumferenced disk* $\{x \in K | \ |x-a| < r\}$, resp. the *circle* $\{x \in K | \ |x-a| = r\}$). We call *a class of* $d(a,r)$ any non-circumferenced disk $d(b,r^-)$ with $|a-b| \leq r$.

Let $r' > 0$ and $r'' > r'$. $\Gamma(a,r',r'')$ denotes the annulus $\{x \in K | \ r' < |x-a| < r''\}$. Given A, $B \subset K$, we denote by $\delta(A,B)$ the distance from A to B. Also $diam(A)$ denotes the diameter of A.

Let D be an infinite set in K, and let $a \in D$. If D is bounded of diameter r, we denote by $\widetilde{D}$ the disk $d(a,r)$, and if D is not bounded, we put $\widetilde{D} = K$. It is known that $\widetilde{D} \setminus \overline{D}$ admits a unique partition of the form $(d(a_i, r_i^-))_{i \in I}$, with $r_i = \delta(a_i, D)$ for each $i \in I$. The disks $d(a_i, r_i^-)_{i \in I}$ are named *the holes of* D [3].

$R(D)$ denotes the set of rational functions $h \in K(x)$ with no poles in D. This is a K-subalgebra of the algebra K^D of all functions from D into K. We endow $R(D)$ with the topology $\mathcal{U}_D$ of uniform convergence on D and $H(D)$ denotes the completion of $R(D)$ for this topology and its elements are named *the analytic elements on* D [3].

1.1 Monotonous filters

Let $a \in \widetilde{D}$ and $S \in \mathbb{R}_+^*$ be such that $\Gamma(a,r,S) \cap D \neq \emptyset$ whenever $r \in]0,S[$ (resp. $\Gamma(a,S,r) \cap D \neq \emptyset$ whenever $r > S$). We call *an increasing* (resp. *a decreasing*) *filter of center a and diameter S on D*, the filter $\mathcal{F}$ on D that admits for base the family of sets $\Gamma(a,r,S) \cap D$ (resp. $\Gamma(a,S,r) \cap D$). For every sequence $(r_n)_{n \in \mathbb{N}}$ such that $r_n < r_{n+1}$ (resp. $r_n > r_{n+1}$) and $\lim_{n \to \infty} r_n = S$, the sequence $\Gamma(a,r_n,S) \cap D$ (resp. $\Gamma(a,S,r_n) \cap D$) is a base of $\mathcal{F}$ and such a base is called *a canonical base* [3].

We call *a decreasing filter with no center and diameter $S > 0$ on D*, a filter $\mathcal{F}$ on D that admits for base a sequence $(D_n)_{n \in \mathbb{N}}$ of the form $D_n = d(a_n, r_n) \cap D$ with $D_{n+1} \subset D_n$, $r_{n+1} < r_n$, $\lim_{n \to \infty} r_n = S$, and $\bigcap_{n \in \mathbb{N}} d(a_n, r_n) = \emptyset$. The sequence $(D_n)_{n \in \mathbb{N}}$ is called *a canonical base of* $\mathfrak{F}$.

We call *a monotonous filter on D* a filter which is either an increasing filter or a decreasing filter (with or without a center).

Given an increasing (resp. a decreasing) filter $\mathcal{F}$ on D of center a and diameter r, we denote by $\mathcal{P}(\mathcal{F})$ the set $\{x \in D \mid |x-a| \geqslant r\}$ (resp. the set $\{x \in D \mid |x-a| \leqslant r\}$). Further, $\mathcal{P}(\mathcal{F})$ is called *the D-beach of* $\mathcal{F}$.

Given a monotonous filter $\mathcal{F}$ we denote by $diam(\mathcal{F})$ its diameter.

The field K is said to be *spherically complete* if every decreasing filter on K has a center in K. (The field $\mathbb{C}_p$, for example, is not spherically complete). However, every algebraically closed complete ultrametric field admits a spherically complete and algebraically closed extension [3].

Let $\mathcal{F}$ be an increasing (resp. a decreasing) filter of center a and diameter S on D. $\mathcal{F}$ is said *to be pierced* if for every $r \in]0,S[$ (resp. $r > S$), $\Gamma(a,r,S)$ (resp. $\Gamma(a,S,r)$) contains some hole of D. A decreasing filter with no center $\mathcal{F}$ and canonical base $(D_n)_{n \in \mathbb{N}}$ on D is said *to be pierced* if for every $m \in \mathbb{N}$, $\widetilde{D}_m \setminus \widetilde{D}_{m+1}$ contains some hole T_m of D.

1.2 Monotonous distances holes sequences

Let $a \in \widetilde{D}$. Let $(T_{m,i})_{\substack{1 \leq i \leq s(m) \\ m \in \mathbb{N}}}$ be a sequence of holes of D which satisfies $\delta(a, T_{m,i}) = d_m$ $(1 \leq i \leq s(m), m \in \mathbb{N})$, $d_m < d_{m+1}$ (resp. $d_m > d_{m+1}$), and $\lim_{m \to \infty} d_m = R > 0$. The sequence $(T_{m,i})_{\substack{1 \leq i \leq s(m) \\ m \in \mathbb{N}}}$ is called *an increasing (resp. decreasing) distances holes sequence that runs the increasing (resp. decreasing) filter $\mathcal{F}$ of center a and diameter R*. The filter $\mathcal{F}$ is called *the increasing (resp. decreasing) filter associated to the sequence* $(T_{m,i})_{\substack{1 \leq i \leq s(m) \\ m \in \mathbb{N}}}$. The D-beach of $\mathcal{F}$ is also named *the D-beach of* $(T_{m,i})_{\substack{1 \leq i \leq s(m) \\ m \in \mathbb{N}}}$.

Finally, an increasing (resp. decreasing) distances holes sequence is called *a monotonous distances holes sequence* and the sequence $(d_m)_{m \in \mathbb{N}}$ is called *the monotony of the monotonous distances holes sequence.*

Let $(T_{m,i})_{\substack{1 \leq i \leq s(m) \\ m \in \mathbb{N}}}$ be a monotonous distances holes sequence and for every (m, i)

$(i \in \{1, .., s(m)\}, m \in \mathbb{N})$, let $\rho_{m,i} = diam(T_{m,i})$. The number $\liminf_{m\to\infty}(\min_{1\le i\le s(m)}(\rho_{m,i}))$ (resp. $\limsup_{m\to\infty}(\max_{1\le i\le s(m)}(\rho_{m,i}))$ is called *inferior limit-piercing* (resp. *superior limit-piercing*) of the sequence $(T_{m,i})_{\substack{1\le i\le s(m)\\ m\in\mathbb{N}}}$.

If a monotonous holes sequence of diameter r has an inferior limit-piercing $\rho > 0$ and a superior limit-piercing $\rho' < r$, it is called *correctly pierced.*

A set D is said to be *correctly pierced* if every monotonous distances holes sequence of D with a non empty D-beach is correctly pierced.

A set D is said to be *well pierced* if $\delta(\overline{D}, K \setminus \overline{D}) > 0$, i.e. the set of diameters of the holes of D has a strictly positive lower bound.

1.3 Weighted sequences

We call a *weighted sequence* a sequence $(T_{m,i}, q_{m,i})_{\substack{1\le i\le s(m)\\ m\in\mathbb{N}}}$ with $(T_{m,i})_{\substack{1\le i\le s(m)\\ m\in\mathbb{N}}}$ a monotonous distances holes sequence and $(q_{m,i})_{\substack{1\le i\le s(m)\\ m\in\mathbb{N}}}$ a sequence of non-negative integers. The D-beach of $(T_{m,i})_{\substack{1\le i\le s(m)\\ m\in\mathbb{N}}}$ is also named *the D-beach of* $(T_{m,i}, q_{m,i})_{\substack{1\le i\le s(m)\\ m\in\mathbb{N}}}$.

A weighted sequence $(T_{m,i}, q_{m,i})_{\substack{1\le i\le s(m)\\ m\in\mathbb{N}}}$ is said to be *idempotent* if $q_{m,i} = 0$ or 1 for all (m, i) $(1 \le i \le s(m), m \in \mathbb{N})$ and $q_{m,i} = 1$ for infinitely many (m, i).

1.4 T-sequences

Let $\mathcal{S} = (T_{m,i}, q_{m,i})_{\substack{1\le i\le s(m)\\ m\in\mathbb{N}}}$ be an increasing (resp. decreasing) weighted sequence and for all (m, i) $(1 \le i \le s(m))$, let $T_{m,i} = d(a_{m,i}, \rho^-{}_{m,i})$ and let $q_m = \sum_{i=1}^{s(m)} q_{m,i}$.

The weighted sequence $\mathcal{S}$ is said to be a *T-sequence* if it satisfies:

$$\lim_{m\to\infty}\left(\sup_{1\le j\le s(m)}\left[\left(\frac{d_m}{\rho_{m,j}}\right)^{q_{m,j}}\prod_{\substack{i\ne j\\ 1\le i\le s(m)}}\left(\frac{d_m}{|a_{m,i}-a_{m,j}|}\right)^{q_{m,i}}\right]\prod_{n=1}^{m-1}\left(\frac{d_n}{d_m}\right)^{q_n}\right) = 0$$

(resp.

$$\lim_{m\to\infty}\left(\sup_{1\le j\le s(m)}\left[\left(\frac{d_m}{\rho_{m,j}}\right)^{q_{m,j}}\prod_{\substack{i\ne j\\ 1\le i\le s(m)}}\left(\frac{d_m}{|a_{m,i}-a_{m,j}|}\right)^{q_{m,i}}\right]\prod_{n=1}^{m-1}\left(\frac{d_m}{d_n}\right)^{q_n}\right) = 0).$$

REMARK A weighted sequence $(T_{m,i}, q_{m,i})_{\substack{1\leq i\leq s(m)\\ m\in\mathbb{N}}}$ is a T-sequence if and only if

$$\lim_{m\to\infty}\left(-\sup_{1\leq j\leq s(m)}\left[q_{m,j}(\log d_m - \log\rho_{m,j}) + \sum_{\substack{i\neq j\\ 1\leq i\leq s(m)}} q_{m,i}(\log d_m - \log|a_{m,i}-a_{m,j}|)\right]\right.$$
$$\left. + \sum_{n=1}^{m-1} q_n|\log d_m - \log d_n|\right) = +\infty.$$

DEFINITION A set D is said to be *analytic* if for every $f \in H(D)$, for every $a \in D$ and $r > 0$, the property $f(x) = 0$ whenever $x \in d(a,r)\cap D$ implies $f(x) = 0$ whenever $x \in D$.

2 STRICTLY ANALYTIC FUNCTIONS

2.1 Polar and T-polar sequences

Let D be infraconnected, let $a \in \widetilde{D}$, let $r > 0$ and let $\rho \in]0, r[$. We call *an increasing (resp. decreasing) polar sequence*, of center a, of diameter r and separation ρ, any sequence of the form $(b_{m,i})_{\substack{1\leq i\leq k(m)\\ m\in\mathbb{N}}}$ satisfying: $b_{m,i} \in \widetilde{D}\setminus D forall(m,i)$ $(m \in \mathbb{N}, i \in \{1,..,k(m)\})$, $|b_{m,i}-a| = |b_{m,j}-a| = d_m$ whenever $i,j \in \{1,..,k(m)\}$ and $|b_{m,i}-a| < |b_{m+1,j}-a|$ (resp. $|b_{m,i}-a| > |b_{m+1,j}-a|$) whenever $1\leq i\leq k(m)$ and $1\leq j\leq k(m+1)$, $\lim_{m\to\infty} d_m = r$ and $\inf_{(m,i)\neq(n,j)} |b_{m,i}-b_{n,j}| = \rho$. The sequence $(d_m)_{m\in\mathbb{N}}$ is called *the monotony of the polar sequence.* We call *the D-beach of* $(b_{m,i})_{\substack{1\leq i\leq k(m)\\ m\in\mathbb{N}}}$ the set $D\cap(K\setminus d(a,r^-))$ (resp. $D\cap d(a,r)$).

REMARK An element of a polar sequence is either an element of $\overline{D}\setminus D$ or an element of a hole of D.

A polar sequence $(b_{m,i})_{\substack{1\leq i\leq k(m)\\ m\in\mathbb{N}}}$ is called *T-polar sequence* if for some $\sigma \in]0,\rho]$ there exists a family $(q_{m,i})_{\substack{1\leq i\leq k(m)\\ m\in\mathbb{N}}}$ of nonnegative integers such that $(d(b_{m,i},\sigma^-), q_{m,i})_{\substack{1\leq i\leq k(m)\\ m\in\mathbb{N}}}$ is a T-sequence.

2.2 Definition of an analoid

D is said to be *an analoid* if D satisfies:

1) D is open.

2) Every monotonous distances holes sequence with a non empty D-beach has a superior limit-piercing strictly smaller than its diameter.

3) Every T-polar sequence admits an empty D-beach.

2.3 D-admissible sets

A hole $T = d(a, r^-)$ of D is said to be *circled* if $d(a, r) \cap D = \emptyset$.

D is said to be *peripherally circled* if $D \cap C(a, diam(D)) = \emptyset$, for some $a \in D$ (i.e. D is included in a class of $d(a, diam(D))$.

Let D be an analoid and let $U \subset D$. Then U is said to be D-*admissible* if U is empty or if it satisfies the following conditions:

a) U is closed and bounded.
b) U is analytic and correctly pierced.
c) U is well pierced.
d) Every circled hole of D included in $\widetilde{U}$ is strictly included in a hole of U.
e) If D is peripherally circled then $\widetilde{U} \subsetneq \widetilde{D}$.

REMARK Since D-admissible sets are well pierced, they are open too.

Henceforth, K is supposed to be topologically separable. It is well known that such a field is not spherically complete ([4]) *and that $\mathbb{C}_p$ satisfies such a condition.*

In [1] we proved that given an analoid D, the system of D-admissible sets and finite covering by D-admissible sets form a Grothendieck topology on D. Further, we showed that the presheaf which to a D-admissible set U associates the Banach algebra $H(U)$ of analytic elements on U is a sheaf. Hence, we defined strictly analytic functions on analoids.

DEFINITION AND NOTATION Let D be an analoid and let $f \in K^D$. Then we say that f is a *strictly analytic function on* D if for any D-admissible set U the restriction $f|U$ of f to U is an element of $H(U)$. We denote by $\mathcal{A}(D)$ the algebra of strictly analytic functions on D and it will be endowed with the projective topology which will be denoted by $\tau(D)$. By Proposition 20 of [1], we have $H(D) \subset \mathcal{A}(D)$.

3 CONTINUOUS MULTIPLICATIVE SEMI-NORMS ON $\mathcal{A}(D)$

In this section we characterize the set of continuous multiplicative semi-norms of an algebra $\mathcal{A}(D)$ of strictly analytic functions on an analoid D by a family of circular filters.

DEFINITIONS AND NOTATIONS

Let $a \in \widetilde{D}$, let $\rho = \delta(a, D)$ and let $S > 0$ be such that $\rho \leqslant S \leqslant diam(D)$. We call *circular filter of center a and diameter S on D*, the filter ${}_D\mathcal{F}$ which admits as a generating system the family of sets $\Gamma(\alpha, r', r'') \cap D$ with $\alpha \in d(a, S)$ and $r' < S < r''$, i.e. ${}_D\mathcal{F}$ is the filter which admits for base the family of sets of the form $D \cap \left(\bigcap_{i=1}^{q} \Gamma(\alpha_i, r'_i, r''_i)\right)$ with $\alpha_i \in d(a, S), r'_i < S < r''_i$ $(1 \leqslant i \leqslant q\ ,\ q \in \mathbb{N})$.

A decreasing filter with no center of canonical base $(D_n)_{n \in \mathbb{N}}$ is also called *circular filter on D with no center of canonical base* $(D_n)_{n \in \mathbb{N}}$.

The filter of neighbourhoods on D of a point a in D is called *circular filter of center a and diameter* 0 *on D*. It is also named *Cauchy circular filter* of center a on D and will be denoted by $\mathfrak{F}_a$.

Finally we call *circular filter on D* any filter of one of the three previous types. Given a circular filter ${}_D\mathfrak{F}$ on D, its diameter will be denoted by $diam({}_D\mathfrak{F})$. A circular filter on D will be said to be *large* if it has diameter different from 0. The set of the circular filters on D will be denoted by $\Theta(D)$. Each circular filter ${}_D\mathfrak{F}$ on D is induced by a unique circular filter on K secant with D that we denote by $\mathfrak{F}$ (see chapter 3 of [3]).

We denote by $Mult(\mathcal{A}(D), \tau(D))$ the set of the continuous multiplicative semi-norms on the K-algebra $\mathcal{A}(D)$ with respect to the topology $\tau(D)$. $Mult(\mathcal{A}(D), \tau(D))$ will be called *analytic spectrum of $\mathcal{A}(D)$* and it will be endowed with the topology of simple convergence.

Also, we denote by $Mult(H(D), \mathfrak{U}_D)$ the set of the continuous semi-norms ψ on the K-vector space $H(D)$ that satisfy $\psi(fg) = \psi(f)\psi(g)$ whenever $f, g \in H(D)$ are such that $fg \in H(D)$. Recall that given a set $D \subset K$, $H(D)$ is a K-algebra if and only if $\overline{D} \setminus D \subset \overset{\circ}{\overline{D}}$ and $\widetilde{D} \setminus D$ is bounded. Hence, we see that there exist analoid sets D such that $H(D)$ is not a K-algebra. So, to define $Mult(H(D), \mathfrak{U}_D)$ we don't require $H(D)$ to be a K-algebra.

Circular filters on D are used to characterize elements of $Mult(H(D), \mathfrak{U}_D)$ in the following way:

THEOREM 0 (see [3]) *Let ${}_D\mathcal{F}$ be a circular filter on D. For every $f \in H(D)$, $|f(x)|$ admits a limit along ${}_D\mathcal{F}$, and this limit, denoted by ${}_D\varphi_{\mathcal{F}}(f)$, defines an element of $Mult(H(D), \mathcal{U}_D)$. Further, the mapping Φ from $\Theta(D)$ into $Mult(H(D), \mathcal{U}_D)$ defined as $\Phi({}_D\mathcal{F}) = {}_D\varphi_{\mathcal{F}}$ is a bijection.*

DEFINITIONS Let D be infraconnected. We will call *uncircled filter on D*, every circular filter on D with center which is not less thin than a monotonous filter on D with an empty D-beach and every decreasing filter on D with no center.

We call *prepierced filter on D* every monotonous filter on D with center, less thin than a polar sequence (for the concept "less thin" see [3], p.18).

REMARK Let ${}_D\mathfrak{F}$ be a circular filter on D of center a and diameter r. Then ${}_D\mathfrak{F}$ is uncircled if and only if $d(a, r^-)$ is not a circled hole of D and if D is peripherally circled it satisfies $r < diam(D)$. We also see that ${}_D\mathfrak{F}$ is uncircled if and only if there exist $b \in d(a, r) \cap D$ and $c \in D$ such that $|b - c| \geq r$.

THEOREM 1 *Let D be an analoid and let ${}_D\mathcal{F}$ be an uncircled filter on D. Then, there exists a D-admissible set U such that ${}_D\mathcal{F}$ is secant with U.*

Proof. Let $r = diam({}_D\mathcal{F})$. First, suppose that ${}_D\mathcal{F}$ has a center. Then, since ${}_D\mathcal{F}$ is uncircled, according to the previous remark we see that ${}_D\mathcal{F}$ has a center $a \in D$ and that there exists $b \in D$ such that $|a - b| \geq r$. Let $\lambda \in]0, \min(r, \delta(a, K \setminus D))[$ be such that $\lambda \notin |K|$.

Since K is separable, let $(d(a_n, \rho_n^-))_{n\in\mathbb{N}}$ be the family of holes of D, included in $d(a, r^-)$, of diameter bigger or equal to λ, let $(d(b_n, \mu_n^-))_{n\in\mathbb{N}}$ be the family of holes of D included in $d(a, r^-)$, of diameter strictly smaller than λ, let $(c_n)_{n\in\mathbb{N}}$ be the family of centers of prepierced filters on D secant with $d(a, r^-)$, of diameter λ, and let $(d(d_n, \lambda^-))_{n\in\mathbb{N}}$ be the family of disks of diameter λ included in $d(a, r^-)$ and containing an element of $\overline{D} \setminus D$.

For all $n \in \mathbb{N}$, we put $v_n = |a - a_n|$. If $d(a_n, \rho_n^-)$ is circled, we have $v_n > \rho_n$. So we may choose $\varepsilon_n \in]0, \frac{1}{n+1}[$ satisfying $\rho_n(1 + \varepsilon_n) < v_n$. If $d(a_n, \rho_n^-)$ is not circled, then we put $\varepsilon_n = 0$.

We define $U_{a,r}$ as follows

$$U_{a,r} = [d(a, r^-)] \setminus [\bigcup_{n\in\mathbb{N}} (d(a_n, \rho_n(1 + \varepsilon_n)^-) \cup d(b_n, \lambda^-) \cup d(c_n, \lambda^-) \cup d(d_n, \lambda^-))].$$

In Lemma 18 of [1] we showed that $U_{a,r}$ is a D-admissible set of diameter r and containing a. Then, by Proposition 3.14 of [3], ${}_D\mathcal{F}$ is secant with $U_{a,r}$.

Now, suppose that ${}_D\mathcal{F}$ is without center. Then, obviously there exist a, $b \in D$ such that $|a - b| > r$ and a canonical base $(d(u_m, \nu_m) \cap D)_{m\in\mathbb{N}}$ of ${}_D\mathcal{F}$ such that for all $m \in \mathbb{N}$, $d(u_m, \nu_m) \subset d(a, |a - b|^-)$.

Let $\lambda \in]0, \min(\delta(a, K \setminus D), r)[$ be such that $\lambda \notin |K|$. Let $(d(a_n, \rho_n^-))_{n\in\mathbb{N}}$ be the family of holes of D included in $d(a, |a - b|^-)$, of diameter bigger or equal to λ, let $(d(b_n, \mu_n^-))_{n\in\mathbb{N}}$ be the family of holes of D included in $d(a, |a - b|^-)$, of diameter strictly smaller than λ, let $(c_n)_{n\in\mathbb{N}}$ be the family of centers of prepierced filters of diameter λ which are secant with $d(a, |a-b|^-)$, and let $(d(d_n, \lambda^-))_{n\in\mathbb{N}}$ be the family of disks of diameter λ included in $d(a, |a - b|^-)$ and containing elements of $\overline{D} \setminus D$.

Since ${}_D\mathcal{F}$ is without center, for all $n \in \mathbb{N}$ there exists $m_n \in \mathbb{N}$ such that $a_n \notin d(u_{m_n}, \nu_{m_n})$. By ultrametricity, for all $m \geq m_n$ we have $|a_n - u_m| = |a_n - u_{m_n}|$.

For all $n \in \mathbb{N}$, we put $v_n = |a_n - u_{m_n}|$. If $d(a_n, \rho_n^-)$ is circled, then clearly $\rho_n < v_n$, and then we may choose $\varepsilon_n \in]0, \frac{1}{n+1}[$ such that $\rho_n(1+\varepsilon_n) < v_n$; otherwise we put $\varepsilon_n = 0$.

We define $U_{a,|a-b|}$ as follows

$$U_{a,|a-b|} = [d(a, r^-)] \setminus [\bigcup_{n\in\mathbb{N}} (d(a_n, \rho_n(1 + \varepsilon_n)^-) \cup d(b_n, \lambda^-) \cup d(c_n, \lambda^-) \cup d(d_n, \lambda^-))].$$

In Lemma 18 of [1], we showed that $U_{a,|a-b|}$ is a D-admissible set of diameter $|a - b|$ and that its holes belong to the set $\{d(a_n, \rho_n(1 + \varepsilon_n)^-) | n \in \mathbb{N}\} \cup \{d(b_n, \lambda^-) | n \in \mathbb{N}\} \cup \{d(c_n, \lambda^-) | n \in \mathbb{N}\} \cup \{d(d_n, \lambda^-) | n \in \mathbb{N}\}$. We will show that ${}_D\mathcal{F}$ is secant with $U_{a,|a-b|}$. Indeed, let us suppose that there exists $m_0 \in \mathbb{N}$ such that $d(u_{m_0}, \nu_{m_0}) \cap U_{a,|a-b|} = \emptyset$. So $d(u_{m_0}, \nu_{m_0})$ is included in a hole T of $U_{a,|a-b|}$, and more precisely, as $r > \lambda$, T is of the form $d(a_n, \rho_n(1 + \varepsilon_n)^-)$. Then we see that for all $m > m_0$, $d(u_m, \nu_m)$ is strictly included in $d(a_n, \rho_n(1 + \varepsilon_n)^-)$ and consequently $\nu_m < \rho_n(1 + \varepsilon_n)$. This is impossible because for $m \geq \max(m_n, m_0)$ we have $|a_n - u_{m_n}| \geq \rho_n(1 + \varepsilon_n)$.

PROPOSITION 2 *Let D be an analoid and let U, V be two D-admissible sets. Then, there exists a D-admissible set W which contains U and V.*

Proof. Since U and V are D-admissible sets, there exist $a, b \in D$ such that U and V are included in $d(a, |a-b|)$. As U and V are well pierced, let $\lambda_0 > 0$ be such that the diameters of holes of U and V are strictly bigger than λ_0. Put $r = |a-b|$ and let $\mathcal{I} = (\Lambda_m)_{m \in I}$ be the family of classes of $d(a,r)$ which have a non empty intersection with $U \cup V$ (I is not an empty subset of $\mathbb{N}$ because K is separable). For each $m \in \mathbb{N}$, let $v_m \in \Lambda_m \cap (U \cup V)$.

Now let $\lambda \in]0, \min(r, \lambda_0)[$ be such that $\lambda \notin |K|$. Since $\lambda < \lambda_0$, it is obvious that $\lambda < \delta(v_n, K \setminus D)$ for all $n \in \mathbb{N}$.

Let m be fixed in I. Let $(d(a_n, \rho_n^-))_{n \in \mathbb{N}}$ be the family of holes of D included in $d(v_m, r^-)$, of diameter bigger or equal to λ, let $(d(b_n, \mu_n^-))_{n \in \mathbb{N}}$ be the family of holes of D included in $d(v_m, r^-)$, of diameter strictly smaller than λ, let $(c_n)_{n \in \mathbb{N}}$ be the family of centers of prepierced filters of diameter λ which are secant with $d(v_m, r^-)$, and let $(d(d_n, \lambda^-))_{n \in \mathbb{N}}$ be the family of disks of diameter λ included in $d(v_m, r^-)$ and containing elements of $\overline{D} \setminus D$.

For all $n \in \mathbb{N}$, we put $u_n = \delta(a_n, U \cup V)$. If $d(a_n, \rho_n^-)$ is circled, then since U and V are D-admissibles we have $\rho_n < u_n$ and we may choose $\varepsilon_n \in]0, \frac{1}{n+1}[$ such that $\rho_n(1+\varepsilon_n) < u_n$; otherwise we put $\varepsilon_n = 0$.

We define $U_{v_m,r}$ as follows

$$U_{v_m,r} = \left[d(v_m, r^-)\right] \setminus \left[\bigcup_{n \in \mathbb{N}} (d(a_n, \rho_n(1+\varepsilon_n)^-) \cup d(b_n, \lambda^-) \cup d(c_n, \lambda^-) \cup d(d_n, \lambda^-))\right].$$

By Lemma 18 of [1], for all $m \in I$, $U_{v_m,r}$ is a D-admissible set of diameter r, without increasing T-sequences, and diameters of holes of $U_{v_m,r}$ are lower bounded by λ. We see that by construction U and V are included in the set $W = \cup_{m \in I} U_{v_m,r}$. Now we cheek that W is D-admissible. Indeed, we clearly see that a hole of W is either a class of $d(a,r)$ or a hole of certain $U_{v_m,r}$. Then, on one hand, since diameters of holes of $U_{v_m,r}$ (for all $m \in I$) are lower bounded by λ, we see that W is well pierced. On the other hand, since for all $m \in I$, $U_{v_m,r}$ is D-admissible, we see that a circled hole of D included in $\widetilde{W}$ is strictly included in a hole of W. Now let $\mathcal{S}$ be a monotonous distances holes sequence of W. As, for all $n \neq m$ $\delta(U_{v_n,r}, U_{v_m,r}) = r$, we see that there exists $n_{\mathcal{S}} \in I$ such that $\mathcal{S}$ is a monotonous distances holes sequence of $U_{v_{n_\mathcal{S}},r}$. Then, since every $U_{v_m,r}$ doesn't admit an increasing T-sequence, W doesn't admit an increasing T-sequence too. Further, obviously if $\mathcal{S}$ is a decreasing distances sequence of holes of W with a non empty W-beach then $\mathcal{S}$ is also a decreasing distances sequence of holes of $U_{v_{n_\mathcal{S}},r}$ with a non empty $U_{v_{n_\mathcal{S}},r}$-beach. Thus, since for all $m \in I$, $U_{v_m,r}$ is analytic and correctly pierced, we see that W doesn't admit decreasing T-sequences with a non empty W-beach and that every decreasing distances sequence of holes of W with a non empty beach is correctly pierced. So, W is analytic and further as, by Lemma 18 of [1], for all $m \in I$, every increasing distances sequence of holes of $U_{v_m,r}$ is correctly pierced, then W is correctly pierced too. This proves that W is D-admissible.

NOTATIONS $\mathcal{A}(D)$ will be endowed with the projective topology denoted by $\tau(D)$ with respect to which all mapping Φ_U which to $f \in \mathcal{A}(D)$ associates its restriction to U are continuous, where U runs the family of D-admissible sets.

Let $f \in \mathcal{A}(D)$. For every D-admissible set U and for every $\varepsilon > 0$, we put $V(f, U, \varepsilon) = \{h \in \mathcal{A}(D) | \|f - h\|_U < \varepsilon\}$. Then we see that the family of finite intersections of these sets form a fundamental system of neighbourhoods of f. Recall that by Proposition 2, given two D-admissible sets U and V, there exists a D-admissible W which contains both U and V. Therefore, given $\varepsilon_1, \varepsilon_2 > 0$, we see that $V(f, W, \min(\varepsilon_1, \varepsilon_2)) \subset V(f, U, \varepsilon_1) \cap V(f, V, \varepsilon_2)$. Thus, the set of neighbourhoods of the form $V(f, U, \varepsilon)$ where U runs the family of D-admissible sets and $\varepsilon > 0$, defines a fundamental system of neighbourhoods of f.

Let ${}_D\mathfrak{F}$ be a filter on D which is secant with a subset A of D, i.e. for every $F \in {}_D\mathcal{F}$, $F \cap A$ is not empty. Then we denote by ${}_A\mathcal{F}$ the filter induced by ${}_D\mathfrak{F}$ on A.

THEOREM 3 *Let D be an analoid and let ${}_D\mathfrak{F}$ be an uncircled filter. Then, for every $f \in \mathcal{A}(D)$ and for every D-admissible set U such that ${}_D\mathfrak{F}$ is secant with U, $|f(x)|$ has a limit ${}_D\overline{\varphi}_{\mathfrak{F}}(f)$ along ${}_U\mathcal{F}$ which does not depend on U and ${}_D\overline{\varphi}_{\mathfrak{F}} \in Mult(\mathcal{A}(D), \tau(D))$.*

Proof. By definition of $\mathcal{A}(D)$, we see that $f|U \in H(U)$. Therefore, since ${}_D\mathcal{F}$ is secant with U, by Theorem 0, $|f(x)|$ has a limit along ${}_U\mathcal{F}$, denoted by ${}_U\varphi_{\mathcal{F}}(f)$. Let V be a second D-admissible set such that ${}_D\mathfrak{F}$ is secant with V. According to Proposition 2, there exists a D-admissible set W which contains both U and V. Since $f|W \in H(W)$ and $U \cup V \subset W$, it is clear that ${}_U\varphi_{\mathcal{F}}(f) = {}_W\varphi_{\mathcal{F}}(f) = {}_V\varphi_{\mathcal{F}}(f/V)$. Put ${}_D\overline{\varphi}_{\mathfrak{F}} = {}_U\varphi_{\mathfrak{F}}$, clearly ${}_D\overline{\varphi}_{\mathfrak{F}}$ is a multiplicative semi-norm on $\mathcal{A}\ (D)$ which satisfies ${}_D\overline{\varphi}_{\mathfrak{F}}\ (f) \leq \|f\|_U$ for all $f \in \mathcal{A}\ (D)$. Therefore we see that given $\varepsilon > 0$, for all $g \in V(f, U, \frac{\varepsilon}{2})$, we have $|{}_D\overline{\varphi}_{\mathfrak{F}}\ (f) - {}_D\overline{\varphi}_{\mathfrak{F}}\ (g)| < \varepsilon$. Thus ${}_D\overline{\varphi}_{\mathfrak{F}}$ is continuous on $\mathcal{A}\ (D)$ and consequently ${}_D\overline{\varphi}_{\mathfrak{F}} \in Mult(\mathcal{A}(D), \tau(D))$.

LEMMA 4 *Let $D \subset K$ and let E be a bounded closed infraconnected set satisfying $E \subset D$ and such that every hole of E contains a hole of D. Let $f \in H(E)$ and let $\varepsilon > 0$. Then, there exists $h \in R(D)$ such that $\|f - h\|_E < \varepsilon$.*

Proof. Let $(T_m)_{m \in \mathbb{N}}$ be the family of holes of E and for every $m \in \mathbb{N}$ let $\overline{f}_{T_m}$ be the Mittag-Leffler term of f associated to T_m [3], [4]. Let $f = \overline{f_0} + \sum_{m=1}^{\infty} \overline{f}_{T_m}$ be the Mittag-Leffler decomposition of f on E. We recall that $\overline{f_0} \in H(\widetilde{E})$ and that $\overline{f}_{T_m} \in H_0(K \setminus T_m)$, for all $m \geq 1$. Let $\varepsilon' \in]0, \varepsilon[$. Then there exists $t \in \mathbb{N}$ such that $\|f - (\overline{f_0} + \sum_{m=1}^{t} \overline{f}_{T_m})\|_E < \varepsilon'$. Also, since $\overline{f_0}$ belongs to $H(\widetilde{E})$, it may be expanded in a series of the form $\sum_{n=0}^{\infty} a_n(x-b)^n$ with $b \in E$ and so we may choose $s \in \mathbb{N}$ such that $\|\overline{f_0} - \sum_{n=0}^{s} a_n(x-b)^n\|_{\widetilde{E}} < \varepsilon'$. On other hand, since each hole T of E contains a hole $T' = d(a, \rho^-)$ of D, we may write the Mittag-Leffler term $\overline{f}_T$ of f on the form $\sum_{n=1}^{\infty} \frac{\lambda_n}{(x-a)^n}$ and we may choose $q \in \mathbb{N}$ such that $\|\overline{f}_T - \sum_{n=1}^{q} \frac{\lambda_n}{(x-a)^n}\|_{K \setminus T} < \varepsilon'$. Therefore, since $R_T = \sum_{n=1}^{q} \frac{\lambda_n}{(x-a)^n} \in R(D)$, we see that if we put $h = \sum_{n=0}^{s} a_n(x-b)^n + \sum_{m=1}^{t} R_{T_m}$, then $h \in R(D)$ and $\|f-h\|_E < \varepsilon$.

COROLLARY 5 *Let $D \subset K$ and let E be a bounded closed infraconnected set satisfying $E \subset D$. Let $f \in H(E)$ be such that for every hole T of E either T contains a hole of D or $\overline{f}_T = 0$. Then for all $\varepsilon > 0$, there exists $h \in R(D)$ such that $\|f-h\|_E < \varepsilon$.*

THEOREM 6 *Let D be an analoid. Then $R(D)$ is dense in $\mathcal{A}(D)$ with respect to the topology $\tau(D)$.*

Proof. We just have to show that given $f \in \mathcal{A}(D)$ and U a D-admissible set, then for every $\varepsilon > 0$, there exists $h \in R(D)$ such that $\|f-h\|_U < \varepsilon$. Indeed, let T be a hole of U. If T does not contain a hole of D, it is obvious that it is D-admissible. Therefore, by Proposition 2, there exists a D-admissible set W which contains both T and U. Consequently, $f \in H(W)$ and it follows that $\overline{f}_T = 0$. Thus by Corollary 5 we see that there exists $h \in R(D)$ such that $\|f-h\|_U < \varepsilon$.

LEMMA 7 *Every element ϕ of $Mult(\mathcal{A}(D), \tau(D))$ satisfies $\phi(f) \leq \|f\|_D$ whenever $f \in \mathcal{A}(D)$ and the restriction of ϕ to $H(D)$ is an element of $Mult(H(D), \mathcal{U}_D)$.*

Proof. Suppose that for some $f \in \mathcal{A}(D)$ we have $\phi(f) > \|f\|_D$. So, let $\lambda \in K$ be such that $\|f\|_D < |\lambda| < \phi(f)$. Then, since for $U \subset D$, U D-admissible, $\|f\|_U \leq \|f\|_D$, it is obvious that $\lim_{n\to\infty} (\frac{f}{\lambda})^n = 0$ with respect to the topology $\tau(D)$. Also, $\lim_{n\to\infty} \phi((\frac{f}{\lambda})^n) = +\infty$, which is in contradiction with the hypothesis "ϕ is continuous".

Hence, for $f \in H(D)$ we have $\phi(f) \leq \|f\|_D$ and consequently the restriction of ϕ to $H(D)$ belongs to $Mult(H(D), \mathcal{U}_D)$.

THEOREM 8 *Let D be an analoid. Then the mapping $\overline{\Phi}$ which to each ${}_D\mathcal{F}$ associates ${}_D\overline{\varphi}_{\mathcal{F}}$ is a bijection from the set of uncircled filters on D into $Mult(\mathcal{A}(D), \tau(D))$*

Proof. Let ${}_D\mathcal{F}$ be an uncircled filter on D. By Lemma 7 the restriction of ${}_D\overline{\varphi}_{\mathcal{F}}$ to $H(D)$ is an element of $Mult(H(D), \mathcal{U}_D)$ and consequently, by Theorem 0, $\overline{\Phi}$ is injective.

Now let $\psi \in Mult(\mathcal{A}(D), \tau(D))$. By Lemma 7, the restriction of ψ to $H(D)$ is an element of $Mult(H(D), \mathcal{U}_D)$. Hence by Theorem 0 there exists a circular filter ${}_D\mathcal{F}$ such that this restriction of ψ to $H(D)$ is equal to ${}_D\varphi_{\mathcal{F}}$. Let us suppose that ${}_D\mathcal{F}$ is not an uncircled filter on D of center a and diameter r such that $d(a, r^-)$ is a circled hole of D (resp. of diameter $diam(D)$ and D is peripherally circled). Then without loss of generality we may assume $a = 0$. Put $S = diam(D)$. Then let $(a_n)_{n\in\mathbb{N}}$ be a sequence of K satisfying

$$(\frac{|a_n|}{r^n})_{n\in\mathbb{N}} \text{ (resp. } (|a_n|S^n)_{n\in\mathbb{N}}) \text{ is unbounded} \tag{1}$$

$$\lim_{n\to\infty} \sqrt[n]{|a_n|} = r \text{ (resp. } \lim_{n\to\infty} \sqrt[n]{|a_n|} = \frac{1}{S}). \tag{2}$$

Then such a sequence satisfies $\lim_{n\to\infty}\frac{|a_n|}{s^n}=0$ for all $s>r$ (resp. $\lim_{n\to\infty}|a_n|s^n=0$ for all $s<S$). Hence the series $f=\sum_{n=0}^{\infty}\frac{a_n}{x^n}$ (resp. $f=\sum_{n=0}^{\infty}a_nx^n$) belongs to $\mathcal{A}(K\setminus d(0,r))$ (resp. $\mathcal{A}(d(0,S^-))$). Also, let $P_n=\sum_{i=0}^{n}\frac{a_i}{x^i}$ (resp. $P_n=\sum_{i=0}^{n}a_ix_i$). We see that if U is a D-admissible set then there exists $s>r$ (resp. $s<S$)) such that $U\subset K\setminus d(0,s)$ (resp. $U\subset d(0,s)$). So the sequence $(P_n)_{n\in\mathbb{N}}$ converges to f in $\mathcal{A}(D)$ endowed with the topology $\tau(D)$. Since ψ is continuous, it follows that the sequence $(\psi(P_n))_{n\in\mathbb{N}}$ converges to $\psi(f)$. But as $P_n\in H(D)$ and $\psi(P_n)={}_D\varphi_{\mathcal{F}}(P_n)=\max_{i\le n}(\frac{|a_i|}{r^i})$ (resp. $\psi(P_n)={}_D\varphi_{\mathcal{F}}(P_n)=\max_{i\le n}(|a_i|S^i)$), according to (1) we see that $\psi(f)$ is unbounded, which is absurd. Thus ${}_D\mathcal{F}$ is uncircled and it follows that the restriction of $\overline{\Phi}({}_D\mathcal{F})$ to $H(D)$ is equal to ${}_D\varphi_{\mathcal{F}}$. Finally, since $H(D)$ is dense in $\mathcal{A}(D)$, we have $\overline{\Phi}({}_D\mathcal{F})=\psi$.

REMARK $Mult(\mathcal{A}(D),\tau(D))$ endowed with the topology of simple convergence is not compact in general.

4 ARCWISE CONNECTNESS OF $Mult(\mathcal{A}(D),\tau(D))$

DEFINITIONS AND NOTATIONS Let $\mathfrak{F}$ be a circular filter on K of center a and diameter r. We denote by $\mathfrak{Q}(\mathfrak{F})$ the set of centers of $\mathfrak{F}$. The set $\mathfrak{Q}(\mathfrak{F})$ is called the *heart of* $\mathfrak{F}$. Here we have $\mathfrak{Q}(\mathfrak{F})=d(a,r)$. If $\mathfrak{F}$ is a circular filter without centers, we put $\mathfrak{Q}(\mathfrak{F})=\emptyset$.

Given two circular filters on K, $\mathcal{F}$ and $\mathcal{G}$, we say that $\mathcal{G}$ *surrounds* $\mathcal{F}$ if $\mathcal{F}$ is secant with $\mathcal{Q}(\mathcal{G})$ or if $\mathcal{F}=\mathcal{G}$. We write $\mathcal{F}\preceq\mathcal{G}$ when $\mathcal{G}$ surrounds $\mathcal{F}$. And we note $\mathcal{F}\prec\mathcal{G}$ when $\mathcal{G}$ *strictly surrounds* $\mathcal{F}$, i.e. if $\mathcal{F}\preceq\mathcal{G}$ and $diam(\mathcal{F})<diam(\mathcal{G})$; such a filter $\mathcal{G}$ clearly owns centers in K. It is easily seen that "$\preceq$" is a partial order relation on the set of circular filters on K.

In [2] we defined a partial order relation on $Mult(H(D),\mathfrak{U}_D)$, also denoted $\preceq$, as follows : given ${}_D\varphi_{\mathcal{F}}$, ${}_D\varphi_{\mathcal{G}}\in Mult(H(D),\mathfrak{U}_D)$, we write ${}_D\varphi_{\mathcal{F}}\preceq{}_D\varphi_{\mathcal{G}}$ when $\mathcal{F}\preceq\mathcal{G}$. We define in a natural way the segment $[{}_D\varphi_{\mathcal{F}},{}_D\varphi_{\mathcal{G}}]_{H(D)}$ of $Mult(H(D),\mathfrak{U}_D)$ as $[{}_D\varphi_{\mathcal{F}},{}_D\varphi_{\mathcal{G}}]_{H(D)}=\{{}_D\varphi_{\chi}\in Mult(H(D),\mathcal{U}_D)|\ {}_D\varphi_{\mathcal{F}}\preceq{}_D\varphi_{\chi}\preceq{}_D\varphi_{\mathcal{G}}\}$

In a similar way, here we define a partial order relation on $Mult(\mathcal{A}(D),\tau(D))$, also denoted $\preceq$, as follows : given ${}_D\overline{\varphi}_{\mathcal{F}}$, ${}_D\overline{\varphi}_{\mathcal{G}}\in Mult(\mathcal{A}(D),\tau(D))$, we write ${}_D\overline{\varphi}_{\mathcal{F}}\preceq{}_D\overline{\varphi}_{\mathcal{G}}$ when $\mathcal{F}\preceq\mathcal{G}$. We define the segment $[{}_D\overline{\varphi}_{\mathcal{F}},{}_D\overline{\varphi}_{\mathcal{G}}]_{\mathcal{A}(D)}$ of $Mult(\mathcal{A}(D),\tau(D))$ as $[{}_D\overline{\varphi}_{\mathcal{F}},{}_D\overline{\varphi}_{\mathcal{G}}]_{\mathcal{A}(D)}=\{{}_D\overline{\varphi}_{\chi}\in Mult(\mathcal{A}(D),\tau(D))|\ {}_D\overline{\varphi}_{\mathcal{F}}\preceq{}_D\overline{\varphi}_{\chi}\preceq{}_D\overline{\varphi}_{\mathcal{G}}\}$.

PROPOSITION 9 *Let D be an analoid. Let ${}_D\mathcal{F}$ and ${}_D\mathcal{G}$ be two uncircled filters on D such that $\mathcal{F}\preceq\mathcal{G}$ and let ${}_D\chi$ be a circular filter on D with $\mathcal{F}\preceq\chi\preceq\mathcal{G}$. Then ${}_D\chi$ is an uncircled filter on D.*

Proof. As the claims is obvious if $\chi=\mathfrak{F}$ or $\chi=\mathfrak{G}$, we may suppose that $\mathfrak{F}\prec\chi\prec\mathfrak{G}$. On one hand, since ${}_D\mathfrak{F}$ is uncircled, it owns a center in D. But as $\mathfrak{F}\prec\chi$, this center is also a center of ${}_D\chi$. On the other hand, since $\chi\prec\mathfrak{G}$,

$diam({}_D\chi) < diam({}_D\mathcal{G})$ and then $diam({}_D\chi) < diam(D)$. Consequently, ${}_D\chi$ is an uncircled filter on D.

REMARK In Lemma III.1 of [2] we showed that if ${}_D\mathcal{F}$ and ${}_D\mathcal{G}$ are two circular filters on D such that $\mathcal{F} \preceq \mathcal{G}$, then every circular filter ${}_D\chi$ such that $\mathcal{F} \preceq \chi \preceq \mathcal{G}$ is a circular filter on D.

PROPOSITION 10 *Let D be an analoid. Let ${}_D\mathcal{F}$ and ${}_D\mathcal{G}$ be two uncircled filters on D such that $\mathcal{F} \preceq G$. Then, there exists a D-admissible set W such that every circular filter χ on K with $\mathcal{F} \preceq \chi \preceq G$ is secant with W.*

Proof. By Theorem 1 there exist D-admissible sets U and V such that $\mathcal{F}$ and $\mathcal{G}$ are secant with U and V respectively. Further, by Proposition 2 there exists a D-admissible set W containing both U and V. Obviously, $\mathcal{F}$ and $\mathcal{G}$ are secant with W. Hence, by Lemma III.1 of [2] every circular filter χ on K satisfying $\mathcal{F} \preceq \chi \preceq G$ is secant with W.

According to Lemma 7, we denote by Ψ the mapping from $Mult(\mathcal{A}(D), \tau(D))$ into $Mult(H(D), \mathcal{U}_D)$ that to each ${}_D\overline{\varphi}_{\mathcal{F}} \in Mult(\mathcal{A}(D), \tau(D))$ associates its restriction ${}_D\varphi_{\mathcal{F}}$ to $H(D)$. Obviously Ψ is injective.

PROPOSITION 11 *Let D be an analoid. Let ${}_D\overline{\varphi}_{\mathcal{F}}, {}_D\overline{\varphi}_{\mathcal{G}} \in Mult(\mathcal{A}(D), \tau(D))$ be such that $\mathcal{F} \preceq \mathcal{G}$. Then Ψ is an homeomorphism from $[{}_D\overline{\varphi}_{\mathcal{F}}, {}_D\overline{\varphi}_{\mathcal{G}}]_{\mathcal{A}(D)}$ into $[{}_D\varphi_{\mathcal{F}}, {}_D\varphi_{\mathcal{G}}]_{H(D)}$ with respect to the topolologies of simple convergence.*

Proof. By Proposition 9, we see that Ψ is a bijection from $[{}_D\overline{\varphi}_{\mathcal{F}}, {}_D\overline{\varphi}_{\mathcal{G}}]_{\mathcal{A}(D)}$ into $[{}_D\varphi_{\mathcal{F}}, {}_D\varphi_{\mathcal{G}}]_{H(D)}$. Let ${}_D\overline{\varphi}_{\chi} \in [{}_D\overline{\varphi}_{\mathcal{F}}, {}_D\overline{\varphi}_{\mathcal{G}}]_{\mathcal{A}(D)}$, let $\varepsilon > 0$, let $h_i \in H(D)$ $(i = 1, .., n)$ and let $X = \{{}_D\varphi_{\mathcal{H}} \in [{}_D\varphi_{\mathcal{F}}, {}_D\varphi_{\mathcal{G}}]_{H(D)} \mid |{}_D\varphi_{\mathcal{H}}(h_i) - {}_D\varphi_{\chi}(h_i)| < \varepsilon \text{ for } i = 1, .., n\}$ be a neighbourhood of ${}_D\varphi_{\chi}$ with respect to the topology of simple convergence on $[{}_D\varphi_{\mathcal{F}}, {}_D\varphi_{\mathcal{G}}]_{H(D)}$. Put $Y = \{{}_D\overline{\varphi}_{\mathcal{H}} \in [{}_D\overline{\varphi}_{\mathcal{F}}, {}_D\overline{\varphi}_{\mathcal{G}}]_{\mathcal{A}(D)} \mid |{}_D\overline{\varphi}_{\mathcal{H}}(h_i) - {}_D\overline{\varphi}_{\chi}(h_i)| < \varepsilon \text{ for } i = 1, .., n\}$. Since for all ${}_D\varphi_{\mathcal{H}} \in [{}_D\varphi_{\mathcal{F}}, {}_D\varphi_{\mathcal{G}}]_{H(D)}$, ${}_D\mathcal{H}$ is an uncircled filter on D and since $h_i \in H(D)$, we see that ${}_D\varphi_{\mathcal{H}}(h_i) = {}_D\overline{\varphi}_{\mathcal{H}}(h_i)$ for all $i \in \{1, .., n\}$. Hence, we have $\Psi(Y) = X$ and consequently Ψ is continuous from $[{}_D\overline{\varphi}_{\mathcal{F}}, {}_D\overline{\varphi}_{\mathcal{G}}]_{\mathcal{A}(D)}$ into $[{}_D\varphi_{\mathcal{F}}, {}_D\varphi_{\mathcal{G}}]_{H(D)}$.

Now let ${}_D\varphi_{\chi} \in [{}_D\varphi_{\mathcal{F}}, {}_D\varphi_{\mathcal{G}}]_{H(D)}$, let $\varepsilon > 0$, let $f_i \in \mathcal{A}(D)$ $(i = 1, .., n)$ and let $V = \{{}_D\overline{\varphi}_{\mathcal{H}} \in [{}_D\overline{\varphi}_{\mathcal{F}}, {}_D\overline{\varphi}_{\mathcal{G}}]_{\mathcal{A}(D)} \mid |{}_D\overline{\varphi}_{\mathcal{H}}(f_i) - {}_D\overline{\varphi}_{\chi}(f_i)| < \varepsilon, \text{ for } i = 1, .., n\}$ be a neighbourhood of ${}_D\overline{\varphi}_{\chi}$ with respect to the topology of simple convergence on $[{}_D\overline{\varphi}_{\mathcal{F}}, {}_D\overline{\varphi}_{\mathcal{G}}]_{\mathcal{A}(D)}$.

By Proposition 10 there exists a D-admissible set U such that every circular filter χ on K with $\mathcal{F} \preceq \chi \preceq \mathcal{G}$ is secant with U. Then, since $R(D)$ is $\tau(D)$-dense in $\mathcal{A}(D)$ for all $i \in \{1, .., n\}$ there exists $h_i \in R(D)$ such that

$$\|f_i - h_i\|_U < \frac{\varepsilon}{4}. \tag{3}$$

Put $Y = \{{}_D\varphi_{\mathcal{H}} \in [{}_D\varphi_{\mathcal{F}}, {}_D\varphi_{\mathcal{G}}]_{H(D)} \mid |{}_D\varphi_{\mathcal{H}}(h_i) - {}_D\varphi_{\chi}(h_i)| < \frac{\varepsilon}{4}\}$. For every ${}_D\varphi_{\mathcal{H}} \in Y$ we see that $|{}_D\overline{\varphi}_{\mathcal{H}}(f_i) - {}_D\overline{\varphi}_{\chi}(f_i)| \leq |{}_D\overline{\varphi}_{\mathcal{H}}(f_i) - {}_D\overline{\varphi}_{\mathcal{H}}(h_i)| + |{}_D\overline{\varphi}_{\mathcal{H}}(h_i) - {}_D\overline{\varphi}_{\chi}(h_i)| + |{}_D\overline{\varphi}_{\chi}(h_i) - {}_D\overline{\varphi}_{\chi}(f_i)|$. Then, since $\mathcal{H}$ and χ are secant with

U and according to (3) we see that $|{}_D\overline{\varphi}_{\mathcal{H}}(f_i) - {}_D\overline{\varphi}_{\mathcal{H}}(h_i)| \leq \frac{\varepsilon}{4}$ and that $|{}_D\overline{\varphi}_{\chi}(h_i) - {}_D\overline{\varphi}_{\chi}(f_i)| \leq \frac{\varepsilon}{4}$. Further, since ${}_D\varphi_{\mathcal{H}} \in Y$, we have $|{}_D\varphi_{\mathcal{H}}(h_i) - {}_D\varphi_{\chi}(h_i)| < \frac{\varepsilon}{4}$. Then we see that $|{}_D\overline{\varphi}_{\mathcal{H}}(f_i) - {}_D\overline{\varphi}_{\chi}(f_i)| < \varepsilon$ and consequently $\Psi^{-1}(Y) \subset X$. This proves that Ψ^{-1} is continuous.

Hence, Ψ is an homeomorphism from $[{}_D\overline{\varphi}_{\mathcal{F}}, {}_D\overline{\varphi}_{\mathcal{G}}]_{\mathcal{A}(D)}$ into $[{}_D\varphi_{\mathcal{F}}, {}_D\varphi_{\mathcal{G}}]_{H(D)}$.

THEOREM 12 *Let D be an analoid. Then $Mult(\mathcal{A}(D), \tau(D))$ is arcwise connected with respect to the topology of simple convergence.*

Proof. Let ${}_D\overline{\varphi}_{\mathcal{F}}, {}_D\overline{\varphi}_{\mathcal{G}} \in Mult(\mathcal{A}(D), \tau(D))$.

First, suppose that ${}_D\mathcal{F} \preceq {}_D\mathcal{G}$. Then, by Corollary II.2 of [2], the segment $[{}_D\varphi_{\mathcal{F}}, {}_D\varphi_{\mathcal{G}}]_{H(D)}$ of $Mult(H(D), \mathcal{U}_D)$ is arcwise connected with respect to the topology of simple convergence. Hence by Proposition 11, $[{}_D\overline{\varphi}_{\mathcal{F}}, {}_D\overline{\varphi}_{\mathcal{G}}]_{\mathcal{A}(D)}$ is arcwise connected too.

Now suppose that ${}_D\mathcal{F}$ and ${}_D\mathcal{G}$ are non comparable. Then by Lemma I.4 of [2], there exist disks $d(a,\rho) \in \mathcal{F}$, $d(b,\sigma) \in \mathcal{G}$ such that $d(a,\rho) \cap d(b,\sigma) = \emptyset$ and $a, b \in D$. The circular filter $\mathcal{M}$ on K of center a and diameter $|a - b|$ is, by Proposition 3.14 of [3], secant with D. Hence, we clearly see that ${}_D\mathcal{M}$ is uncircled and that ${}_D\mathcal{F} \preceq {}_D\mathcal{M}$ and ${}_D\mathcal{G} \preceq {}_D\mathcal{M}$. Therefore by the previous case, both $[{}_D\overline{\varphi}_{\mathcal{F}}, {}_D\overline{\varphi}_{\mathcal{M}}]_{\mathcal{A}(D)}$ and $[{}_D\overline{\varphi}_{\mathcal{G}}, {}_D\overline{\varphi}_{\mathcal{M}}]_{\mathcal{A}(D)}$ are arcwise connected. Hence there exists a continuous path f from $[0,1]$ into $Mult(\mathcal{A}(D), \tau(D))$ such that $f(0) = {}_D\overline{\varphi}_{\mathcal{F}}$ and $f(1) = {}_D\overline{\varphi}_{\mathcal{G}}$.

REFERENCES

[1] K Boussaf. Strictly analytic functions on p-adic analytic open sets, to appear in Publ Mat Barc.

[2] K Boussaf, N Mainetti and M Hemdaoui. Arc-connectness of the set of continuous multiplicative semi-norms of a Krasner algebra $H(D)$, to appear.

[3] A Escassut. Analytic elements in p-adic analysis, Singapore: World Scientific, (1995).

[4] Ph Robba. Fonctions analytiques sur les corps valués ultramétriques complets. Prolongement analytique et algébres de Banach ultramétriques, Astérisque, n.10:109–220 (1973).

An improvement of the p-adic Nevanlinna theory and application to meromorphic functions

ABDELBAKI BOUTABAA Laboratoire de Mathématiques Pures Université Blaise Pascal (Clermont-Ferrand), Les Cézeaux, 63177 Aubiere Cedex, France.
e-mail: boutabaa@ucfma.univ-bpclermont.fr

ALAIN ESCASSUT
escassut@ucfma.univ-bpclermont.fr

Abstract. In this article we give an improvement of the ultrametric Nevanlinna Theory and apply these results to a problem of uniqueness for ultrametric meromorphic functions. We obtain a new construction of bi-urs' of n elements for such functions for every $n \geqslant 5$. In particular we find bi-urs' of the form $(S_n, \{\infty\})$ where S_n is the set of the zeros of the polynomial $R_n(x) = \frac{(n-1)(n-2)}{2}x^n - n(n-2)x^{n-1} + \frac{n(n-1)}{2}x^{n-2} - c$ with $n \geqslant 5$.

INTRODUCTION AND RESULTS

First, our paper is aimed at showing an improvement of Theorem 1.8 of [1]. Next we will apply this improvement to find new bi-urs for p-adic meromorphic functions. We have to recall many pieces of notation.

NOTATION W will denote an algebraically closed field of characteristic zero and we denote by $\widehat{W}$ the one dimensional projective space over W: $\widehat{W} = W \cup \{\infty\}$.
$\mathbb{K}$ will denote a complete ultrametric algebraically closed field of characteristic zero, and $\mathbb{K}^*$ will denote $\mathbb{K} \setminus \{0\}$.
We denote by $\mathcal{A}(\mathbb{K})$ the ring of entire functions in K and by $\mathcal{M}(\mathbb{K})$ the field of meromorphic functions in all $\mathbb{K}$.
Let $f \in \mathcal{M}(\mathbb{K})$. We put $|f|(r) = \lim_{|x| \to r, |x| \neq r} |f(x)|$, [8]. We will denote by $\mathcal{P}(f, r)$ the set of poles of f in the disk $d(0, r) = \{x \in \mathbb{K} | \ |x| \leqslant r\}$.

Then the mapping ψ from $K((x))$ to $\mathbb{Z} \cup \{\infty\}$ defined as $\psi(f) = q(f)$ if $f \in K((x)) \setminus \{0\}$, and $\psi(0) = \infty$ is known to be a discrete ultrametric valuation. Since $\mathcal{M}(\mathbb{K})$ is clearly included in $\mathbb{K}((X))$, the restriction of ψ to $\mathcal{M}(\mathbb{K})$ defines a discrete ultrametric valuation. Besides, for each $a \in \mathbb{K}$, we may write $f \in \mathcal{M}(\mathbb{K})$ in the form $g(u) = f(a+u)$, and consider the valuation ω_a defined as $\omega_a(f) = \psi(g)$. Let log denotes the real logarithm function of base p. We put $\log^+ x = \max(0, \log x)$.

For $f \in \mathcal{M}(\mathbb{K})$ such that $f(0) \neq 0$, $f(0) \neq \infty$ and $r > 0$, by properties of the valuation function [8], we have:

$$\log|f(0)| = \log|f|(r) - \sum_{|t| \leqslant r} \omega_t(f) \log \frac{r}{|t|}. \tag{1}$$

As usual, the p-adic Nevanlinna functions are defined by:
$m(r, f) = \log^+(|f|(r)), \quad N(r, f) = -\sum_{t \in \mathcal{P}(f,r)} \omega_t(f) \log \frac{r}{|t|},$
$\overline{N}(r, f) = \sum_{t \in \mathcal{P}(f,r)} \log \frac{r}{|t|}$ and $T(r, f) = m(r, f) + N(r, f)$.
The function $\big(r \mapsto T(r, f)\big)$, called a *characteristic function* of f, satisfies the following relations:

$$T(r, fg) \leqslant T(r, f) + T(r, g), \tag{2}$$
$$T(r, f + g) \leqslant T(r, f) + T(r, g), \tag{3}$$
$$T(r, \frac{1}{f}) = T(r, f) - \log|f(0)|, \tag{4}$$

for every $f, g \in \mathcal{M}(\mathbb{K})$ such that $f(0) \neq 0, \infty$; $g(0) \neq 0, \infty$ and $(f + g)(0) \neq 0$. Then Formula (1) gives:

PROPOSITION A *Let $f \in \mathcal{M}(\mathbb{K})$ and $a \in \mathbb{K}$ be such that $f(0) \neq 0$, $f(0) \neq a$ and $f(0) \neq \infty$. We have:*

$$T(r, \frac{1}{f - a}) = T(r, f) + O(1), \qquad r \longrightarrow +\infty.$$

We are now able to state Theorem 1.

THEOREM 1 *Let $f \in \mathcal{M}(\mathbb{K})$ be non constant and $\delta > 0$. Let $q \in \mathbb{N} \setminus \{0, 1\}$ and let $a_1, ...a_q \in \mathbb{K}$ be such that $|a_i - a_j| \geqslant \delta$ for $1 \leqslant i \neq j \leqslant q$. Let $A = \{a_1, ..., a_q\}$. Suppose that $f(0) \neq 0$, $f(0) \neq \infty$ and $f(0) \neq a_i$ for every $i = 1, ..., q$. Then we have:*

$$\sum_{i=1}^{q} T(r, \frac{1}{f - a_i}) \leqslant T(r, f) + \overline{N}(r, f) + \sum_{i=1}^{q} \overline{N}(r, \frac{1}{f - a_i}) - N_0^A\left(r, \frac{1}{f'}\right) + S(f) - \log r,$$

where $S(f) = q \log^+(\frac{1}{\delta}) - \log|f'(0)|$ and $N_0^A\left(r, \frac{1}{f'}\right)$ is the counting function of zeros of f', excluding those which are zeros of $f - a_j$ for any $j = 1, ...q$.

As a consequence of Theorem 1 and Proposition A, we obtain Corollary 1:

COROLLARY 1 *Under the assumptions of Theorem 1, we have:*

$$(q-1)T(r,f) \leqslant \overline{N}(r,f)+\sum_{i=1}^{q}\overline{N}(r,\frac{1}{f-a_i})-N_0^A\left(r,\frac{1}{f'}\right)-\log r+O(1), \quad r \longrightarrow +\infty.$$

REMARK 1 The main improvement obtained in Theorem 1 is connected with the term $-\log r$ involved in the upper bound, comp. Theorem 1.8 of [1]. A similar result was already given in [2], part III, Th. 2.6 dedicated to projective curves. Here we give a new proof of this, fact while insisting on the term $\overline{N}(r,f)$ instead of $N(r,f)$. In [7], there is an improvement that also shows a term $-\log r$, but involving $N(r,f)$ instead of $\overline{N}(r,f)$.

Another way to improve the p-adic Nevanlinna Second Main Theorem could be possible by giving the term $N_0^A\left(r,\frac{1}{f'}\right)$ (a lower bound), but this is a very delicate question. (For instance in $\mathbb{C}$, the fonction sin with $A = \{1, -1\}$ provides an example which shows that such a lower bound does not hold when $n = 2$. But in a p-adic field sin is not an entire function).

We will now recall the definition of bi-urs' for meromorphic or rational functions.

For a subset S of W (resp. $\mathbb{K}$) and $f \in W(X)$, (resp. $f \in \mathcal{M}(\mathbb{K})$) we denote by $E(f,S)$ the set in $W\times \mathbb{N}^*$ (resp. $\mathbb{K}\times \mathbb{N}^*$):

$$\bigcup_{a\in S}\{(z,q) \in W\times \mathbb{N}^* \mid \omega_z(f-a) = q\}, \quad (\text{resp. } \bigcup_{a\in S}\{(z,q) \in \mathbb{K}\times \mathbb{N}^* | \omega_z(f-a) = q\}.$$

Moreover, given a subset S of $\widehat{W}$ (resp. $\widehat{\mathbb{K}}$) containing $\{\infty\}$, by $E(f,S)$ we denote the subset of $W\times \mathbb{N}^*$ (resp. $\mathbb{K}\times \mathbb{N}^*$) by the formula $E(f,s) = E(f,S\cap W) \cup \{(z,q) \; |\omega_z(f) = -q\}$ (resp. $E(f,s) = E(f,S\cap \mathbb{K}) \cup \{(z,q) \; |\omega_z(f) = -q\}$).

DEFINITION Let $\mathcal{F}$ be a non empty subset of $W(X)$ (resp. $\mathcal{M}(\mathbb{K})$). A subset S of $\widehat{W}$ (resp. $\widehat{\mathbb{K}}$) is called a *unique range set* (a *URS* in brief) *for* $\mathcal{F}$ if for any two non constant functions $f, g \in \mathcal{F}$ such that $E(f,S) = E(g,S)$, one has $f = g$.

In the same way, a couple of sets S, T in $\widehat{W}$ (resp. $\widehat{\mathbb{K}}$) such that $S \cap T = \emptyset$ will be called *a bi-urs for* $\mathcal{F}$ if for any two non-constant functions $f, g \in \mathcal{F}$ such that $E(f,S) = E(g,S)$ and $E(f,T) = E(g,T)$ one has $f = g$.

REMARK 2 If a set S is a urs for $\mathcal{A}(\mathbb{K})$ (resp. $\mathcal{M}(\mathbb{K})$), then for every non-constant affine (resp. linear fractional) function h, $h(S)$ also is a urs for $\mathcal{A}(\mathbb{K})$ (resp. for $\mathcal{M}(\mathbb{K})$). In the same way, if a couple of sets (S,T) is a bi-urs for $\mathcal{A}(\mathbb{K})$ (resp. $\mathcal{M}(\mathbb{K})$), then for every non-constant affine (resp. linear fractional) function h, the couple $(h(S),h(T))$ also is a bi-urs for $\mathcal{A}(\mathbb{K})$ (resp. $\mathcal{M}(\mathbb{K})$).

It was shown, cf. [4] that for every $n \geqslant 5$, for every θ in $\mathbb{K}$, there exist bi-urs of the form $(\{a_1, ...a_n\}, \{\theta\})$. Then, we used a method that involved properties of analytic elements on infraconnected sets. Here we will obtain new bi-urs' of the form $(\{a_1, ...a_n\}, \{\theta\})$ for every $n \geqslant 5$, by a shorter method that is just based upon Theorem 1.

For every $n \geqslant 3$ and $c \in W \setminus \{0, \frac{1}{2}, 1\}$, (resp. $c \in \mathbb{K} \setminus \{0, \frac{1}{2}, 1\}$), we consider the polynomial R_n introduced in [10]:

$R_n(x) = \frac{(n-1)(n-2)}{2} x^n - n(n-2)x^{n-1} + \frac{n(n-1)}{2} x^{n-2} - c$.

We denote by $S_n(c)$ the set of zeros of $R_n(x)$. According to [10], $S_n(c)$ has n elements. Indeed it is easily checked that:

(α) $R'_n(x) = \frac{n(n-1)(n-2)}{2}(x-1)^2 x^{n-3}$;

(β) $R_n(x) + c - 1 = (x-1)^3 Q_{n-3}(x)$ with $Q_{n-3}(1) \neq 0$, $\deg Q_{n-3}(x) = n-3$;

(γ) $R_n(x) + c = x^{n-2} Q_2(x)$ with $Q_2(0) \neq 0$ and $\deg Q_2(x) = 2$.

Therefore, for every $d \in W \setminus \{c, c-1\}$, the polynomial $R_n(x) + d$ only has simple zeros. In particular, this is true for $R_n(x)$.

We can now state the following results.

THEOREM 2 *For every $n \geqslant 5$, $(S_n(c), \{\infty\})$ is a bi-urs for $\mathcal{M}(\mathbb{K})$.*

COROLLARY 2 *For every $n \geqslant 5$, and for every $\theta \in \mathbb{K}$, there exist bi-urs' for $\mathcal{M}(\mathbb{K})$ of the form $(S, \{\theta\})$, where S is a subset of $\mathbb{K}$ of cardinality n.*

Proof. By Remark 2, it is sufficient to consider a linear fractional transformation h such that $h(\infty) = \theta$ and take $S = S_n(c)$.

Corollaries 3 and 4 are easily shown by an embedding from a finite extension over $\mathbb{Q}$ into $\mathbb{C}_p$, as in [5] and [9].

COROLLARY 3 *For every $n \geqslant 5$, $(S_n(c), \{\infty\})$ is a bi-urs for $W(X)$.*

COROLLARY 4 *For every $n \geqslant 5$ and for every $\theta \in W$ there exist bi-urs' for $W(X)$ of the form $(S, \{\theta\})$, where S is a subset of W of cardinality n.*

REMARK 3 In [9], it was shown that there exist no bi-urs of the form $(\{a_1, a_2, a_3\}, \{\theta\})$. This was also mentioned in [13]. The argument used in the proof of Theorem 2 leads also to problems of [4]: The existence of bi-urs for p-adic meromorphic functions of the form $(\{a_1, a_2, a_3, a_4, a_5\}, \{\theta\})$. None of these two methods enable us to find bi-urs of the form $(\{a_1, a_2, a_3, a_4\}, \{\theta\})$. Thus, the question remains open whether there exist bi-urs of the form $(\{a_1, a_2, a_3, a_4\}, \{\theta\})$.

THE PROOFS

For the proof of Theorem 1, we need the following lemmas:

LEMMA B *Let $f \in \mathcal{M}(\mathbb{K})$ be non constant and such that $f(0) \neq 0$, $f(0) \neq \infty$*

and $f'(0) \neq 0$, $f'(0) \neq \infty$. *Then we have*

$$m(r, \frac{f'}{f}) = 0 \quad \text{and} \quad m(r, \frac{f}{f'}) \geqslant \log r, \quad \forall r > 0.$$

Proof. Indeed, by classical results concerning a meromorphic function f, it is well known that for every $r > 0$, we have $\log(\left|\frac{f'}{f}\right|(r)) \leqslant -\log r$, [4], [8]. Then, Lemma B is an obvious consequence.

Lemmas C, D, E are well-known and will be used in the sequel.

LEMMA C *Let* $f \in \mathcal{M}(\mathbb{K})$ *be non constant and such that* $f(0) \neq 0$, $f(0) \neq \infty$ *and* $f'(0) \neq 0$, $f'(0) \neq \infty$. *Then we have:*

$$N(r, \frac{f'}{f}) - N(r, \frac{f}{f'}) = N(r, \frac{1}{f}) - N(r, f) + N(r, f') - N(r, \frac{1}{f'}) \ \ \forall r > 0.$$

LEMMA D *Let* $f \in \mathcal{M}(\mathbb{K})$ *be non-constant and such that* $f(0) \neq 0$, $f(0) \neq \infty$ *and* $f^{(n)}(0) \neq 0$, $f^{(n)}(0) \neq \infty$. *Then we have:*

$$N(r, f^{(n)}) = N(r, f) + n\overline{N}(r, f) \quad \forall r > 0,$$

and

$$T(r, f^{(n)}) \leqslant T(r, f) + n\overline{N}(r, f) \ \ \forall r > 0.$$

LEMMA E *Let* A *be a finite subset of* $\mathbb{K}$ *and let* $f \in \mathcal{M}(\mathbb{K})$ *be non constant and such that* $f(0) \neq 0$, $f(0) \neq \infty$ *and* $f(0) \neq a$, *for every* $a \in A$. *Then we have:*

$$\sum_{a \in A} N(r, \frac{1}{f-a}) - \overline{N}(r, \frac{1}{f-a}) = N(r, \frac{1}{f'}) - N_0^A(r, \frac{1}{f'}), \quad \forall r > 0.$$

LEMMA F *Let* $f \in \mathcal{M}(\mathbb{K})$ *be non constant and* $q \in \mathbb{N}^*$. *Let* $a_1, ...a_q \in \mathbb{K}$ *be distinct and* $g = \sum_{j=1}^{q} \frac{1}{f-a_j}$. *Then* $m(r, g) \leqslant m(r, \frac{1}{f}) + m(r, \frac{f}{f'}) - \log r \ \ \forall r > 1$.

Proof. Of course, we have

$$m(r, g) \leqslant m(r, \frac{1}{f}) + m(r, fg). \tag{5}$$

Now we consider $m(r, fg)$, when $r > 1$. Then

$$\begin{aligned} m(r, fg) = m(r, (\frac{f}{f'})(f'g)) = \max\big(0, \log|\frac{f}{f'}f'g|(r)\big) = \\ \max\big(0, \log|\frac{f}{f'}|(r) + \log|f'g|(r)\big). \end{aligned} \tag{6}$$

But by Lemma B, we have

$$m(r, \frac{f}{f'}) = \log\left|\frac{f}{f'}\right|(r) \geqslant \log r > 0, \tag{7}$$

and similarly, for each $j = 1, ..., q$, we have $|\frac{f'}{f-a_j}|(r) \leqslant \frac{1}{r}$, hence

$$\log |f'g|(r) \leqslant -\log r. \tag{8}$$

Now, by (6), (7) and (8), we have

$$m(r, fg) \leqslant \max(0, \log|\frac{f}{f'}| - \log r) = \log|\frac{f}{f'}| - \log r = m(r, \frac{f}{f'}) - \log r. \tag{9}$$

Finally, we obtain our claim by (5) and (9).

Proof of Theorem 1. Let $g = \sum_{i=1}^{q} \frac{1}{f-a_i}$. First we show that

$$m(r, g) \geqslant \sum_{i=1}^{q} m(r, \frac{1}{f-a_i}) - q\log^+(\frac{1}{\delta}). \tag{11}$$

For this, we distinguish two cases:
1°) *Suppose that* $|f - a_j|(r) < \delta$ *for some* j. Then, for $i \neq j$, we have $|f - a_i|(r) = |a_i - a_j| \geqslant \delta$. Hence, in this case, we have $|g|(r) = |\frac{1}{f-a_j}|(r) > \frac{1}{\delta}$ and $|\frac{1}{f-a_i}|(r) \leqslant \frac{1}{\delta}$ $\forall i \neq j$. So we obtain $m(r, g) = m(r, \frac{1}{f-a_j}) \geqslant \log^+(\frac{1}{\delta})$ and $m(r, \frac{1}{f-a_i}) \leqslant \log^+(\frac{1}{\delta})$ $\forall i \neq j$. We easily deduce relation (1) in this case.
2°) *Suppose that* $|f - a_i|(r) \geqslant \delta$, $\forall i$. Then for every i we have $m(r, \frac{1}{f-a_i}) \leqslant \log^+(\frac{1}{\delta})$, and hence $m(r, g) \geqslant 0 \geqslant \sum_{i=1}^{q} m(r, \frac{1}{f-a_i}) - q\log^+(\frac{1}{\delta})$. Hence relation (11) is proved.
Now, by inequality (11) and Lemma F, we have

$$m(r, f) + \sum_{i=1}^{q} m(r, \frac{1}{f-a_i}) \leqslant q\log^+(\frac{1}{\delta}) + m(r, f) + m(r, \frac{1}{f}) + m(r, \frac{f}{f'}) - \log r.$$

Then, by Relation (4), we obtain

$$m(r, f) + \sum_{i=1}^{q} m(r, \frac{1}{f-a_i}) \leqslant q\log^+(\frac{1}{\delta}) + \{T(r, f) - N(r, f)\}$$
$$+\{T(r, f) - \log|f(0)| - N(r, \frac{1}{f})\} + \{T(r, \frac{f}{f'}) - N(r, \frac{f}{f'})\} - \log r.$$

So we have

$$m(r, f) + \sum_{i=1}^{q} m(r, \frac{1}{f-a_i}) \leqslant q\log^+(\frac{1}{\delta}) + 2T(r, f) - N(r, f) - N(r, \frac{1}{f}) - \log|f(0)|$$
$$+\{T(r, \frac{f'}{f}) - \log\frac{|f'(0)|}{|f(0)|} - N(r, \frac{f}{f'})\} - \log r,$$

hence

$$m(r,f)+\sum_{i=1}^{q}m(r,\frac{1}{f-a_i})\leqslant q\log^+(\frac{1}{\delta})+2T(r,f)-N(r,f)-N(r,\frac{1}{f})-\log|f'(0)|$$
$$+\{m(r,\frac{f'}{f})+N(r,\frac{f'}{f})-N(r,\frac{f}{f'})\}-\log r.$$

Then by Lemmas B and C, we have

$$m(r,f)+\sum_{i=1}^{q}m(r,\frac{1}{f-a_i})\leqslant q\log^+(\frac{1}{\delta})+2T(r,f)-N(r,f)-N(r,\frac{1}{f})-\log|f'(0)|$$
$$+\{N(r,\frac{1}{f})-N(r,f)+N(r,f')-N(r,\frac{1}{f'})\}-\log r.$$

So, putting $Q(r,f)=2N(r,f)+N(r,\frac{1}{f'})-N(r,f')$ we have

$$m(r,f)+\sum_{i=1}^{q}m(r,\frac{1}{f-a_i})\leqslant 2T(r,f)-Q(r,f)+S(f)-\log r. \tag{12}$$

Now we will show

$$N(r,f)-\overline{N}(r,f)+\sum_{i=1}^{i=q}\{N(r,\frac{1}{f-a_i})-\overline{N}(r,\frac{1}{f-a_i})\}\leqslant Q(r,f)-N_0^A(r,\frac{1}{f'}). \tag{13}$$

By Lemma E, we have

$$N(r,f)-\overline{N}(r,f)+\sum_{i=1}^{i=q}\{N(r,\frac{1}{f-a_i})-\overline{N}(r,\frac{1}{f-a_i})\}$$
$$=N(r,f)-\overline{N}(r,f)+N(r,\frac{1}{f'})-N_0^A(r,\frac{1}{f'}).$$

Then, by Lemma D, we deduce that

$$N(r,f)-\overline{N}(r,f)+\sum_{i=1}^{i=q}\{N(r,\frac{1}{f-a_i})-\overline{N}(r,\frac{1}{f-a_i})\}\leqslant Q(r,f)-N_0^A(r,\frac{1}{f'}).$$

Now by relations (12) and (13) we obtain

$$m(r,f)+\sum_{i=1}^{q}m(r,\frac{1}{f-a_i})+\sum_{i=1}^{i=q}\{N(r,\frac{1}{f-a_i})-\overline{N}(r,\frac{1}{f-a_i})\}+N(r,f)-\overline{N}(r,f)$$
$$\leqslant 2T(r,f)-N_0^A(r,\frac{1}{f'})-\log r+S(f).$$

Then applying Proposition A we complete the proof of Theorem 1.

In order to prove Theorem 2, we need the following lemma, given in [1] (Theorem II.1)

LEMMA G *Let* $P(x) \in \mathbb{K}[x]$ *be a polynomial of degree* n *and* $f \in \mathcal{M}(K)$. *Then*

$$T(r, P(f)) = nT(r,f) + O(1), \quad r \to \infty.$$

Proof of Theorem 2. Let $f, g \in \mathcal{M}(\mathbb{K}) \setminus \mathbb{K}$ be such that $E(f, S_n(c)) = E(g, S_n(c))$ and $E(f, \{\infty\}) = E(g, \{\infty\}$. Then

$$R_n(f) = \lambda R_n(g), \quad \text{with some } \lambda \in \mathbb{K}^*, \tag{14}$$

and

$$R_n(f) + \lambda(c-1) = \lambda\big[R_n(g) + c - 1\big]. \tag{15}$$

First we will show that $\lambda = 1$.
Suppose that $\lambda \neq 1$. Then we have $\lambda(c-1) \neq c-1$ and $\frac{1}{\lambda}(c-1) \neq c-1$. On the other hand, since either $\lambda(c-1) \neq c$ or $\frac{1}{\lambda}(c-1) \neq c$ (because $c \neq \frac{1}{2}$), then we can assume that $\lambda(c-1) \neq c$. This implies that the polynomial $R_n(x) + \lambda(c-1)$ has n distinct zeros $\alpha_1, ..., \alpha_n$.
Using relations (β) and (15) we deduce that

$$(f-\alpha_1)...(f-\alpha_n) = \lambda(g-1)^3 Q_{n-3}(g). \tag{16}$$

Finally using Theorem 1, Lemma G and Proposition A, we obtain

$$(n-2)T(r,f) \leqslant (n-2)T(r,f) - \log r + O(1), \quad r \to \infty.$$

This is a contradiction. Hence $\lambda = 1$, and then $R_n(f) = R_n(g)$.
Putting $h = \frac{f}{g}$, we have

$$\frac{(n-1)(n-2)}{2}(h^n - 1)g^2 - n(n-2)(h^{n-1}-1)g + \frac{n(n-1)}{2}(h^{n-2}-1) = 0. \tag{17}$$

We show that h is a constant. Indeed, suppose that h is not a constant. As in [10], we can write (17) in the form

$$\left(g - \frac{n}{n-1}\frac{h^{n-1}-1}{h^n-1}\right)^2 = -\frac{n(h-1)^4\phi(h)}{(n-1)^2(n-2)(h^n-1)^2}, \tag{18}$$

where $\phi(x)$ is a polynomial of degree $2n-6$ having only simple zeros and such that $\phi(1) \neq 0$.
Let us denote the zeros of $\phi(x)$ by $\beta_1, ..., \beta_{2n-6}$. Then applying Theorem 1 we have

$$(2n-7)T(r,h) \leqslant \overline{N}(r,h) + \sum_{i=1}^{2n-6} \overline{N}(r, \frac{1}{h-\beta_i}) - \log r + O(1), \qquad r \longrightarrow +\infty. \tag{19}$$

But from (18), we deduce that $\overline{N}(r, \frac{1}{h-\beta_i}) \leqslant \frac{1}{2}N(r, \frac{1}{h-\beta_i}) \leqslant \frac{1}{2}T(r, \frac{1}{h-\beta_i})$, $\forall i$.

Using Proposition A we obtain

$$(2n-7)T(r,h) \leqslant (n-2)T(r,h) - \log r + O(1), \qquad r \longrightarrow +\infty.$$

Hence we have

$$(n-5)T(r,h) \leqslant -\log r + O(1), \qquad r \longrightarrow +\infty. \tag{20}$$

This contradicts the fact that $n \geqslant 5$. Hence h is a constant.
Now using the fact that g is not constant, we deduce from (17) that $h^{n-1} - 1 = h^{n-2} - 1 = 0$. Then $h = 1$ and hence $f = g$. This completes the proof of Theorem 2.

REFERENCES

[1] A Boutabaa. Theorie de Nevanlinna p-adique. Manuscripta Mathematica 67:251–269, 1990.

[2] A Boutabaa. Sur la Théorie de Nevanlinna p-adique, Thèse de Doctorat, Université Paris 7, 1991.

[3] A Boutabaa, A Escassut, L Haddad. On uniqueness of p-adic entire functions. Indagationes Mathematicae 8:145–155, 1997.

[4] A Boutabaa, A Escassut. On uniqueness of p-adic meromorphic functions. Proceedings of the AMS, Vol 126, 9:2557–2568, 1998.

[5] A Boutabaa, A Escassut. Property $f^{-1}(S) = g^{-1}(S)$ for p-adic entire and meromorphic functions, (preprint).

[6] W Cherry, CC Yang. Uniqueness of non-archimedean entire functions sharing sets of values counting multiplicities, to appear in the Proceedings of the AMS.

[7] W Cherry, Zhuan Ye. Non-archimedean Nevanlinna Theory in several variables and the non-archimedean Nevanlinna inverse problem. Transaction of the AMS, Vol 349, 12:5043–5071, 1997.

[8] A Escassut. Analytic Elements in p-adic Analysis. World Scientific Publishing Co. Pte. Ltd. Singapore, 1995.

[9] A Escassut, L Haddad, R Vidal. Urs, ursim, and non-urs, to appear in the Journal of Number Theory.

[10] G Frank, M Reinders. A unique Range set for meromorphic functions with 11 eleven elements, to appear in Complex Variable Theory Appl.

[11] R Nevanlinna. Le théorème de Picard-Borel et la theorie des fonctions méromorphes. Gauthiers-Villars, Paris, 1929.

[12] HX Yi, CC Yang. Uniqueness theory of meromorphic functions, Pure and applied Math. Monographs n. 32, Science Press, 1995.

[13] CC Yang, XH Hua. Unique polynomials of entire and meromorphic functions. Matematicheskaia Fizika Analys Geometriye, v. 4, n.3:391–398, 1997.

An application of c-compactness

G CHRISTOL Université Paris 6, Arithmétique Case 247, 4 place Jussieu 75252, Paris Cedex05, France.

Z MEBKHOUT Université Paris 7, Mathématiques, 2 place Jussieu 75251, Paris Cedex05, France.

WH SCHIKHOF Department of Mathematics, University of Nijmegen, Toernooiveld 6525 ED Nijmegen, The Netherlands.

Abstract. We generalize to spherically complete fields a result proved in [1] for locally compact fields.

In this note, K is a field spherically complete for a non-archimedean dense valuation. We assume the characteristic of K to be 0 to avoid complications (that can be overcome). For general facts on normed spaces and locally convex inductive limits over K we refer to [4] and [2] respectively.

1 WHAT'S THE USE

1.1 Differential operators

Let $0 < r < R$, $I =]r, R[$ and let A be the ring of functions f with coefficients in K that are analytic in the annulus $|x| \in I$. The ring A is endowed with the locally convex topology T defined by the family of norms :

$$\left| \sum_{s \in \mathbb{Z}} \alpha_s \, x^s \right|_\rho = \max_{s \in \mathbb{Z}} |\alpha_s| \, \rho^s$$

for ρ in I. Actually, to define this topology it suffices to consider a countable family of norms (for instance $\rho \in I \cap \mathbb{Q}$), hence A is a complete metrizable space (Fréchet or F space).

Let $A\langle D\rangle$ be the left A-module of polynomials in D with coefficients in A endowed with the noncommutative ring structure defined by the rule

$$D\,a = a\,D + a'$$

for a in A and $a' = \frac{da}{dx}$.

For $\lambda \geqslant 1$ and ρ in I, one sets :

$$\left|\sum a_m\, D^m\right|_{\lambda,\rho} = \max_m(|m!\, a_m|_\rho\, \rho^{-\lambda m})^\dagger.$$

The family of these norms, for fixed λ and ρ varying in I, endows $A\langle D\rangle$ with a locally convex topology denoted by T_λ. As this topology can be defined by a countable family of norms, it is a metrizable one.

PROPOSITION 1 ([1]) *The space $(A\langle D\rangle, T_\lambda)$ is not an inductive limit of complete metrizable spaces. In particular it is not complete.*

Proof. Let $A\langle D\rangle_n$ be the subset of polynomials of degree strictly less than n. It is isomorphic to A^n hence endowed with a complete metrizable topology T^n, namely the product one. So, the inductive limit $A\langle D\rangle = \varinjlim A\langle D\rangle_n$ inherits a LF topology that we shall denote by T^∞.

The inclusion map $(A\langle D\rangle_n, T^\infty) \longrightarrow (A\langle D\rangle, T_\lambda)$ is obviously continuous. Hence the identity map $i : (A\langle D\rangle, T^\infty) \longrightarrow (A\langle D\rangle, T_\lambda)$ is continuous. Now, if T_λ were LF, the homomorphism theorem for LF spaces applied to i would assert that $T^\infty = T_\lambda$.

But $\varinjlim A\langle D\rangle_n$ is a strict inductive limit hence a bounded subset for T^∞ lies in $A\langle D\rangle_n$ for some n. On the other hand, the subset $\{x^{[\lambda m]}\, D^m/m!;\ m \in \mathbb{N}\}$ where $[x]$ denotes the integer part of x, is easily seen to be bounded for T_λ.

1.2 Differential modules

An A-differential module is a left $A\langle D\rangle$-module which is of finite rank when viewed as a (left) A-module.

Let M be an A-differential module of rank μ (over A). The $A\langle D\rangle$-module structure of M is determined by the action of D hence by the matrix G of D in some A-base of M. Actually, to give a base $(e_1, \ldots, e_\mu)$ of M amounts to give a "presentation" of M :

$$0 \longrightarrow A\langle D\rangle^\mu \xrightarrow{\;u\;} A\langle D\rangle^\mu \xrightarrow{\;p\;} M \longrightarrow 0$$

namely an exact sequence of left $A\langle D\rangle$-modules where u is the right multiplication by $D - G$ and where p is defined by $p(P_1, \ldots, P_\mu) = \sum P_i.e_i$.

By means of this presentation, the topology T_λ on $A\langle D\rangle$ provides a "quotient" topology $\overline{T}_\lambda$ on M. It is easily verified that it does not depend on the chosen presentation. The problem is to decide whether it is Hausdorff or not.

QUESTION [1]: A base of M provides an isomorphism between M and A^μ hence a "product" Hausdorff locally convex topology T^μ on M (actually independent on the base). Suppose $(M, \overline{T}_\lambda)$ to be Hausdorff. As the identity map $(M, T^\mu) \longrightarrow (M, \overline{T}_\lambda)$

is obviously continuous, and as (M, T^μ) is metrizable and complete, it is known ([2] 0.5.ii.1) that T^μ and $\overline{T}_\lambda$ have the same restriction on any absolutely convex closed compactoid subset of (M, T^μ). Then it seems likely that the "quotient" topology is Hausdorff if and only if it is equivalent to the "product" one. This is proved, by an entirely distinct way, for $\lambda > 1$. The case $\lambda = 1$ would be very interesting to understand because it should be connected with a "weight filtration" of M.

1.3 Results

Under some hypothesis on M, namely if M has a slope bigger than λ (see [1] for definitions) one constructs polynomials P_n in $A\langle D\rangle$ (actually in $K\langle D\rangle$) with the following properties :

1) The sequence $\{u_n\}_{n\in\mathbb{N}} = \{p(P_n, \ldots, P_n)\}_{n\in\mathbb{N}}$ is bounded and does not converge to 0 for the product topology T^μ of M.
2) The sequence $\{u_n\}_{n\in\mathbb{N}}$ converges to 0 for the quotient topology (i.e. the sequence $\{P_n\}_{n\in\mathbb{N}}$ converges to 0 for T_λ).

The aim of this note is to prove the following Proposition:

PROPOSITION 2 *Let E be a complete metrizable locally convex space over a spherically complete field K. Let $\{u_n\}_{n\in\mathbb{N}}$ be a sequence in E. Suppose there exist a compactoid set B in E, a constant $c > 0$ and a continuous seminorm p such that u_n lies in B and $p(u_n) \geqslant c$ for all n. Then, for $d < c$, there exist $\lambda_{i,n}$ in K $(0 \leqslant n \leqslant i \leqslant m_n)$ such that:*

1. $|\lambda_{i,n}| \leqslant 1, \ \ \sum_{i=n}^{m_n} \lambda_{i,n} = 1,$
2. *The sequence $\{\sum_{i=n}^{m_n} \lambda_{i,n}\, u_i\}_{n\in\mathbb{N}}$ converges to an element y in E for which $p(y) > d$.*

COROLLARY 3 *Suppose there exists a sequence $\{u_n\}_{n\in\mathbb{N}}$ satisfying conditions 1) and 2). Then M is not a Hausdorff space for the quotient topology.*

Proof. Let us apply the Proposition to the space $E = M$ endowed with the product topology and to the sequence $\{u_n\}$.

By condition 1), there exists a continuous seminorm p on E such that $p(u_n)$ does not converge to 0. Then, by restricting to a subsequence if necessary, one can suppose that $p(u_n) \geqslant c > 0$ for all n.

Choose a sequence of closed intervals $I_n = [r_n, R_n]$ with $r < r_{n+1} < r_n < R_n < R_{n+1} < R$ and such that $I =]r, R[= \bigcup I_n$. Let us denote by H_n the set of functions that are analytic with coefficients in K in the annulus $|x| \in I_n$. It is a Banach space for the maximum norm on the annulus. Set $M_n = H_n \otimes_A M$. It is well known that the inclusion $H_{n+1} \longrightarrow H_n$ is a compact map hence so is the inclusion $M_{n+1} \longrightarrow M_n$. Now, the topology T on A (resp. T^μ on M) is defined by the "semicompact" inverse limit $A = \varprojlim H_n$ (resp. $M = \varprojlim M_n$). Hence M is complete, and, each bounded set in M is compactoid. (If B is bounded in M it is bounded in M_{n+1} for all n hence it is compactoid in M_n for all n and then compactoid in M).

The condition 1) says precisely that the hypothesis of the Proposition are fulfilled.

Let us choose d such that $0 < d < c$ and let us use the notations of the Proposition. As the sequence $\{u_n\}$ converges to 0 for the topology $\overline{T}_\lambda$ and as $|\lambda_{i,n}| \leqslant 1$, the sequence $\sum_{i=n}^{m_n} \lambda_{i,n}\, u_i$ also converges to 0 for the locally convex topology $\overline{T}_\lambda$. Hence the limit y belongs to the closure of 0 for $\overline{T}_\lambda$. But $y \neq 0$ because $p(y) > d > 0$. Hence the topology $\overline{T}_\lambda$ is not Hausdorff.

2 PROOFS

Let E be a Hausdorff locally convex space over K.

DEFINITION 4 A *cluster* in E is a collection $C = (C_i)_{i\in I}$ of non-empty c-compact subsets of E such that each finite intersection contains a member of C.

By definition of c-compact sets, one has $\bigcap_{i\in I} C_i \neq \emptyset$.

LEMMA 5 *Let F be a Hausdorff locally convex space over K and let $u : E \to F$ be a linear continuous map. Then, for any cluster $(C_i)_{i\in I}$ in E, $\big(u(C_i)\big)_{i\in I}$ is a cluster in F and $u\left(\bigcap_{i\in I} C_i\right) = \bigcap_{i\in I} u(C_i)$.*

Proof. The inclusion $u\left(\bigcap_{i\in I} C_i\right) \subset \bigcap_{i\in I} u(C_i)$ is obvious.

Let $y \in \bigcap_{i\in I} u(C_i)$. Then $u^{-1}(\{y\}) \cap C_i \neq \emptyset$ for each i, so $\left(u^{-1}(\{y\}) \cap C_i\right)_{i\in I}$ is a cluster in E. Let x be in the intersection. Then x belongs to $\bigcap_{i\in I} C_i$ and $y = u(x)$ hence y belongs to $u\left(\bigcap_{i\in I} C_i\right)$.

LEMMA 6 *Let $(C_i)_{i\in I}$ be a cluster in E and let B be a closed and convex subset of E. Then $B + \bigcap_{i\in I} C_i = \bigcap_{i\in I}(B + C_i)$.*

Proof. The inclusion $B + \bigcap_{i\in I} C_i \subset \bigcap_{i\in I}(B + C_i)$ is obvious. To prove the opposite inclusion, let $y \in \bigcap_{i\in I}(B + C_i)$. Then $(y - B) \cap C_i \neq \emptyset$ and $((y - B) \cap C_i)_{i\in I}$ is a cluster. Let x be in the intersection. Then x belongs to $\bigcap_{i\in I} C_i$ and $y \in B + x$.

DEFINITION 7 Let X be a non-empty subset of some K-vector space F and let p be a seminorm on F. The *p-diameter of X* is defined by:

$$p\text{-diam}\, X = \sup\{p(x - y)\,;\, x, y \in X\}$$

(it can be infinite). Actually, for each a in X, one has

$$p\text{-diam}\, X = \sup\{p(x - a)\,;\, x \in X\}.$$

LEMMA 8 *Let $(C_i)_{i\in I}$ be a cluster in K. Then*

$$|\ \ |\text{-diam} \bigcap_{i\in I} C_i = \inf_{i\in I} \left(|\ \ |\text{-diam}\, C_i\right).$$

Proof. Each C_i is a disk, a singleton or K itself. Hence $(C_i)_{i\in I}$ is linearly ordered by inclusion. The formula follows.

LEMMA 9 *Let $(F, \|\ \ \|)$ be a finite dimensional normed space over K. Let $(C_\imath)_{\imath \in I}$ be a cluster in F. Then $\|\ \ \|$-diam $\bigcap_{\imath \in I} C_\imath = \inf_{\imath \in I} (\|\ \ \|\text{-diam}\, C_\imath)$.*

Proof. F has an orthogonal base $e_1, \ldots, e_\mu$. Let $f_1, f_2, \ldots, f_\mu$ be the continuous linear forms satisfying, for all x in E :

$$x = \sum_{k=1}^{\mu} f_k(x)\, e_k.$$

Let X be a non-emtpy subset of F and let a be in X. We find:

$$\begin{aligned}\|\ \ \|\text{-diam}\, X &= \sup_{x \in X} \|x - a\| = \sup_{x \in X} \max_{1 \leqslant k \leqslant \mu} |f_k(x-a)|\, \|e_k\| \\ &= \max_{1 \leqslant k \leqslant \mu} \sup_{x \in X} |f_k(x-a)|\, \|e_k\| = \max_{1 \leqslant k \leqslant \mu} (|\ \ |\text{-diam}\, f_k(X)\, \|e_k\|).\end{aligned}$$

Thus, using previous Lemmas 5 and 8, one obtains:

$$\begin{aligned}\|\ \ \|\text{-diam} \bigcap_{\imath \in I} C_\imath &= \max_{1 \leqslant k \leqslant \mu} (|\ \ |\text{-diam}\, f_k(\bigcap_{\imath \in I} C_\imath)\, \|e_k\|) \\ &= \max_{1 \leqslant k \leqslant \mu} \inf_{\imath \in I} (|\ \ |\text{-diam}\, f_k(C_\imath)\, \|e_k\|) \\ &= \inf_{\imath \in I} \max_{1 \leqslant k \leqslant \mu} (|\ \ |\text{-diam}\, f_k(C_\imath)\, \|e_k\|) \\ &= \inf_{\imath \in I} (\|\ \ \|\text{-diam}\, C_\imath).\end{aligned}$$

THEOREM 10 *Let $(C_\imath)_{\imath \in I}$ be a cluster in E. Let p be a continuous seminorm on E. Then p-diam $\bigcap_{\imath \in I} C_\imath = \inf_{\imath \in I} (p\text{-diam}\, C_\imath)$.*

Proof. The inequality p-diam $\bigcap_{\imath \in I} C_\imath \leq \inf_{\imath \in I} (p\text{-diam}\, C_\imath)$ is obvious. Then the theorem is proved when $\inf_{\imath \in I} (p\text{-diam}\, C_\imath) = 0$, so, to prove the opposite inequality, let us suppose $\inf_{\imath \in I} (p\text{-diam}\, C_\imath) > s > 0$; we prove that p-diam $\bigcap_{\imath \in I} C_\imath > s$. To this end we may assume that there exists an $i_0 \in I$ for which $C_\imath \subset C_{\imath_0}$ for all $i \in I$. The set $O := \{x \in E : p(x) \leq s\}$ is a convex zero neighbourhood. As K is spherically complete the c-compact set $C_{\imath_0}$ is a local compactoid i.e. there is a finite-dimensional space $F \subset E$ such that $C_{\imath_0} \subset F + O$. Let $F = F_1 \oplus (p^{-1}(0) \cap F)$ for some subspace F_1. Since $p^{-1}(0) \cap F \subset O$ we have $C_{\imath_0} \subset F_1 + O$. Hence, replacing F by F_1 if necessary we may assume that p is a norm on F. We have

$$C_\imath \subset F + O \qquad (i \in I), \tag{$*$}$$

and we now prove

$$p\text{-diam}\,(O + C_\imath) \cap F = p\text{-diam}\, C_\imath \qquad (i \in I). \tag{$**$}$$

In fact, let $x_1, x_2 \in C_\imath$ be such that $p(x_1 - x_2) > s$. From $(*)$ we obtain $x_1 = f_1 + t_1$, $x_2 = f_2 + t_2$ where $f_1, f_2 \in F$, $t_1, t_2 \in O$. Then $x_1 - t_1$ and $x_2 - t_2$ are in $(O + C_\imath) \cap F$ and their p-distance equals $p(x_1 - x_2)$ since $p(t_1 - t_2) \leq s$. It follows that p-diam $C_\imath \leq p$-diam $(O + C_\imath) \cap F$. The opposite inequality can be proved in a similar way. Now, by using (**), Lemma 9 and Lemma 6 we obtain $s < \inf_{\imath\in I}(p\text{-diam}\, C_\imath) = \inf_{\imath\in I}(p\text{-diam}\,(O + C_\imath) \bigcap F) = p\text{-diam} \bigcap_{\imath\in I}(O + C_\imath) \cap F = p\text{-diam}\,(O + \bigcap_{\imath\in I} C_\imath) \cap F \le p\text{-diam}\,(O + \bigcap_{\imath\in I} C_\imath)$. Thus there exist $x_1, x_2 \in \bigcap_{\imath\in I} C_\imath$ and $t_1, t_2 \in O$ such that $p(x_1 + t_1 - (x_2 + t_2)) > s$. But $p(t_1 - t_2) \le s$ so that $p(x_1 - x_2) > s$. It follows that $p\text{-diam} \bigcap_{\imath\in I} C_\imath > s$.

COROLLARY 11 *Let $(C_\imath)_{\imath\in I}$ be a cluster in E. Let p be a continuous seminorm on E. If for some $s > 0$ and for each i in I, $C_\imath \cap \{x \in E\,;\, p(x) > s\} \neq \emptyset$, then for $0 < t < s$, we have $\bigcap_{\imath\in I} C_\imath \cap \{x \in E\,;\, p(x) > t\} \neq \emptyset$.*

Proof. If there is an index i such that $C_\imath \subset \{x \in E\,;\, p(x) > t\}$ then $\emptyset \neq \bigcap_{\imath\in I} C_\imath \subset \{x \in E\,;\, p(x) > t\}$ and we are done. So we may assume $C_\imath \cap \{x \in E\,;\, p(x) \leqslant t\} \neq \emptyset$ for all $\imath$ hence $\bigcap_{\imath\in I} C_\imath \cap \{x \in E\,;\, p(x) \leqslant t\} \neq \emptyset$.

Each $C_\imath$ contains $a_\imath$ and $b_\imath$ for which $p(a_\imath) > s$ and $p(b_\imath) \leqslant t < s$ hence $p\text{-diam}\, C_\imath > s$. Thus, by Theorem 10

$$p\text{-diam} \bigcap_{\imath\in I} C_\imath = \inf_{\imath\in I}(p\text{-diam}\, C_\imath) \ge s.$$

Hence $\bigcap_{\imath\in I} C_\imath$ contains elements a and b with $p(a) \leqslant t$ and $p(b - a) > t$. Then $p(b) > t$.

PROOF OF PROPOSITION 2: Let C_n be the closure of the convex hull of $\{u_\imath\}_{\imath\geqslant n}$. Each C_n is convex, complete and compactoid hence it is c-compact ([3], Proposition 2.2). Thus $(C_n)_{n\in\mathbb{N}}$ is a cluster and by Corollary 11 one has $\bigcap_{n\in\mathbb{N}} C_n \cap \{x \in E\,;\, p(x) > d\} \neq \emptyset$. Choose y in this subset. Clearly $p(y) > d$. Moreover, if ρ is a metric on E inducing the topology, for each n there exist y_n in the convex hull of $\{u_\imath\}_{\imath\geqslant n}$ such that $\rho(y_n - y) < 1/n$. So the sequence y_n converges to y and, by definition of the convex hull, $y_n = \sum_{\imath=n}^{m_n} \lambda_{\imath,n}\, u_\imath$ with $|\lambda_{\imath,n}| \leqslant 1$ and $\sum_{\imath=n}^{m_n} \lambda_{\imath,n} = 1$.

REFERENCES

[1] G Christol, Z Mebkhout. Sur le théorème de l'indice des équations différentielles p-adiques III. preprint.

[2] N De Grande De Kimpe, J Kakol, C Perez-Garcia, WH Schikhof. p-Adic locally convex inductive limits. In: WH Schikhof, C Perez-Garcia, J Kakol, ed. p-Adic Functional Analysis. New York: Marcel Dekker, 1997, pp 159-222.

[3] WH Schikhof. Some properties of c-compact sets in p-adic spaces. Report 8632, Department of Mathematics, Catholic University, Nijmegen, The Netherlands:1-12, 1986.

[4] ACM van Rooij. Non-Archimedean Functional Analysis. New York: Marcel Dekker, 1978.

On the integrity of the dual algebra of some complete ultrametric Hopf algebras

BERTIN DIARRA Mathématiques Pures, Complexe Scientifique des Czeaux, 63177 Aubière Cedex, France.
e-mail: diarra@ucfma.univ-bpclermont.fr

Abstract. Let (H, m, c, η, σ) be a complete ultrametric Hopf algebra over the complete ultrametric valued field K. Then, the dual Banach space H' of H with the convolution $\mu * \nu = (\mu \otimes \nu) \circ c$ is a normed algebra. It is obvious that when H admits an integral ν such that $< \nu, e >= 1$, then ν being an idempotent is a zero divisor in H'. Aside from being without integral, on what other conditions a cocommutative Hopf algebra H is such that its dual H' is an integral domain? We have no general answer to this question. However, it is well known that if K is of residue characteristic $p \neq 0$, then the dual of the Hopf algebra $\mathcal{C}(\mathbb{Z}_p, K)$ is an algebra of power series, thus it is an integral domain Suppose that $\mathbb{Q}_p \subseteq K$ let V_q be the subgroup of the group of units of $\mathbb{Z}_p$ that is the closure of $\{q^n,\ n \geqslant 0\}$ where $|q| = 1$, q is not a root of unit. We show that the dual algebra $M(V_q, K)$ of the Hopf algebra structure on $\mathcal{C}(V_q, K)$ induced by the multiplication of V_q is an integral domain if and only if $q \equiv 1 \pmod{p}$ if $p \neq 2$, and for $p = 2$, $M(V_q, K)$ is always an integral domain. We also study complete divided powers coalgebras, which are a counterpart of purely algebraic theory.

INTRODUCTION

In Algebra, it is well known that the universal enveloping algebra of a (finite dimensional) Lie algebra over a field of characteristic zero is a Hopf algebra whose dual is isomorphic to a ring of formal series. More generally, equivalent conditions are provided for an irreducible cocommutative Hopf algebra to have its dual algebra to be an integral domain (cf. [1], Corollary 2.5.16); one of these conditions is that the dual algebra is isomorphic to a ring of formal series.

Now, let K be a complete ultrametric valued field. An ultrametric Banach space H over K is said to be a Banach coalgebra if there exist continuous linear maps $c: H \to H \widehat{\otimes} H$, called the coproduct of H, and σ: $H \to K$, called the counit of H, such that

(i) $(c \otimes 1_H) \circ c = (1_H \otimes c) \circ c$

(ii) $(1_H \otimes \sigma) \circ c = 1_H = (\sigma \otimes 1_H) \circ c$, and $||\sigma|| = 1$,

where 1_H is the identity map of H.

It follows that for $a \in H$, one has $||a|| \leqslant ||c(a)|| \leqslant ||c||\ ||a||$ and c is isometric if and only if $||c|| = 1$.

Furthermore, the Banach coalgebra H is said to be a Banach bialgebra, if it is an unitary Banach algebra with multiplication $m : H \widehat{\otimes} H \to H$ such that c and σ are algebra homomorphisms ; and the Banach bialgebra H is said to be a complete Hopf algebra if there exists a continuous linear map $\eta : H \to H$, called the antipode or inversion of H, such that

(iii) $m \circ (1_H \otimes \eta) \circ c = k \circ \sigma = m \circ (\eta \otimes 1_H) \circ c$,

where k is the canonical map of K into H.

For any Banach coalgebra H, the dual Banach space H' of H, with the convolution product $\mu * \nu = (\mu \otimes \nu) \circ c$, is an unitary normed algebra with unit σ and $||\mu * \nu|| \leqslant ||c||\ ||\mu||\ ||\nu||$.

As usual, the complete tensor product of two ultrametric Banach spaces E and F over K is the completion $E \widehat{\otimes} F$ of the algebraic tensor product $E \otimes F$ with respect to the tensor norm $||z|| = \inf_{\Sigma x_j \otimes y_j = z} \left(\max_j ||x_j||\ ||y_j|| \right)$. In the sequel all Banach spaces are ultrametric.

Let (H_1, c_1, σ_1) and (H_2, c_2, σ_2) be two Banach coalgebras. Let $\tau(a_1 \otimes a_2) = a_2 \otimes a_1$ be the twist operator of $H_1 \widehat{\otimes} H_2$ onto $H_2 \widehat{\otimes} H_1$. Setting $c = (1_{H_1} \otimes \tau \otimes 1_{H_2}) \circ (c_1 \otimes c_2)$ and $\sigma = \sigma_1 \otimes \sigma_2$, one defines on $H = H_1 \widehat{\otimes} H_2$ a structure of Banach coalgebra. Moreover, if H_1 and H_2 are complete bialgebras (resp. Hopf algebras) with multiplications respectively m_1 and m_2 (resp. and antipodes η_1 and η_2), one has on H a structure of complete bialgebra (resp. Hopf algebra) with multiplication $m = (m_1 \otimes m_2) \circ (1_{H_1} \otimes \tau \otimes 1_{H_2})$ (resp. and antipode $\eta = \eta_1 \otimes \eta_2$).

We have shown in [3] that the Banach dual of a divided powers coalgebra is isomorphic to a subring of the ring of power series. We give here complements on those coalgebras and show that the dual Banach algebra, studied in the same paper, of the Hopf algebra $\mathcal{C}(V_q, K)$ is an integral domain if and only if $q \equiv 1$ (mod. p) for $p \neq 2$ and is always integral for $p = 2$.

1 DIVIDED POWERS BANACH COALGEBRAS

1.1 Free ultrametric Banach spaces (cf. [3])

Let H be an ultrametric Banach space over the complete valued field K. A family $(e_j)_{j \in I} \subset H, e_j \neq 0$, is an *orthogonal base* of H if any $a \in H$ can be written in the unique form $a = \sum_{j \in I} \alpha_j e_j, \alpha_j \in K$, $\lim_j |\alpha_j|\ ||e_j|| = 0$ and $||a|| = \sup_{j \in I} |\alpha_j|\ ||e_j||$. One says that the Banach space H is *free* if it has an orthogonal base.

Let H be a free Banach space with base $(e_j)_{j \in I}$ and let E be an other Banach space. Then one has the following facts.

Fact 1 Any element $z \in H \widehat{\otimes} E$ (resp. $E \widehat{\otimes} H$) can be written in the unique form $z = \sum_{j \in I} e_j \otimes x_j$ (resp. $\sum_{j \in I} x_j \otimes e_j$), $x_j \in E$, $\lim_j ||e_j||\ ||x_j|| = 0$ and $||z|| =$

$\sup_{j \in I} \|e_j\| \, \|x_j\|$.

In other words, setting $\rho_j = \|e_j\|$, $j \in I$, and $c_0(I, (\rho_j)_{j\in I}, E) = \{(x_j)_{j\in I} \subset E \;/\; \lim_{j\in I} \rho_j \|x_j\| = 0\}$ with the norm $\|(x_j)\| = \sup_{j\in I} \rho_j \|x_j\|$, the Banach space $H\widehat{\otimes}E$ (resp. $E\widehat{\otimes}H$) is isometrically isomorphic to $c_0(I, (\rho_j)_{j\in I}, E)$.

If the space E is also a free Banach space with orthogonal base $(f_\ell)_{\ell\in L}$, then $H\widehat{\otimes}E$ is a free Banach space with orthogonal base $(e_j \otimes f_\ell)_{(j,\ell)\in I\times L}$.

Fact 2 Let $\mathcal{L}(H,E)$ be the Banach space of the continuous linear maps of H into E. Put $H' = \mathcal{L}(H,K)$ the Banach dual of H ; for $j \in I$, one has $e'_j \in H'$ defined by $< e'_j, e_\ell >= \delta_{j,\ell}$, $\forall \ell \in I$, and $\|e'_j\| = \dfrac{1}{\|e_j\|}$. Moreover, if $\mu \in H'$ and $x \in E$, one defines $\mu \otimes x \in \mathcal{L}(H,E)$ by setting for $a \in H$, $(\mu \otimes x)(a) \;= \;< \mu, a > x$.

Hence, one sees that any $u \in \mathcal{L}(H,E)$ can be written uniquely as a pointwise convergent sum $u = \sum_{j\in I} e'_j \otimes u(e_j)$ with $\|u\| = \sup_{j\in I} \dfrac{\|u(e_j)\|}{\|e_j\|}$. In other words, setting $\rho'_j = \|e'_j\| = \dfrac{1}{\|e_j\|}$ and $\ell^\infty(I, (\rho'_j)_{j\in I}, E) = \{(x_j)_{j\in I} \subset E \;/\; \sup_{j\in I} \|x_j\| \rho'_j < +\infty\}$ with the norm $\|(x_j)_j\| = \sup_{j\in I} \|x_j\| \rho'_j$, then the Banach spaces $\mathcal{L}(H,E)$ and $\ell^\infty(I, (\rho'_j)_{j\in I}, E)$ are isometrically isomorphic.

Fact 3 Furthermore if (H, c, σ) is an ultrametric coalgebra, one has for any $a \in H$, $c(a) = \sum_{j\in I} A_j(a) \otimes e_j$ with $\lim_j \|A_j(a)\| \, \|e_j\| = 0$, $\|c(a)\| = \sup_{j\in I} \|A_j(a)\| \, \|e_j\|$.

Moreover, $A_j = (1_H \otimes e'_j) \circ c \in \mathcal{L}(H,H)$; $\|c\| = \sup_{j\in I} \|e_j\| \, \|A_j\|$ and $1 \leqslant \|e_j\| \, \|A_j\|$ for $j \in I$.

Setting for $j, \ell \in I$, $A_\ell(e_j) = a_{j\ell} \in H$, one has $\|a_{j\ell}\| \, \|e_\ell\| \leqslant \|c\| \, \|e_j\|$ and $c(e_j) = \sum_{\ell\in I} a_{j\ell} \otimes e_\ell$.

Summarizing, one obtains :

THEOREM 1 [3]. *Let (H, c, σ) be an ultrametric Banach coalgebra which is a free Banach space with orthogonal base $(e_j)_{j\in I}$.*

Any element μ of H' can be written in the unique form of pointwise (= weak) convergent sum $\mu = \sum_{j\in I} \alpha_j e'_j$, where $(\alpha_j)_{j\in I} \subset K$ is such that $\sup_{j\in I} \dfrac{|\alpha_j|}{\|e_j\|} < +\infty$.*

Moreover $\alpha_j =< \mu, e_j >$ and $\|\mu\| = \sup_{j\in I} \dfrac{|\alpha_j|}{\|e_j\|}$.

The multiplication of the dual algebra H' is uniquely determined on the weak base $(e'_j)_{j\in I}$ of H', with for $j, \ell \in I$, $e'_j * e'_\ell = \sum_{i\in I} < e'_j, a_{i\ell} > e'_i$, where the family $(a_{i\ell})_{(i,\ell)\in I\times I} \subset H$ is defined by $c(e_i) = \sum_{\ell\in I} a_{i\ell} \otimes e_\ell$.*

Proof. See [3], Theorem 3 and Corollary 3.

1.2 Divided powers Banach coalgebras

Examples of Banach coalgebras which are free Banach spaces can be provided as follows.

Let $(\rho_n)_{n\geqslant 0} \subset \mathbb{R}_+^*$ be such that $\rho_0 = 1$, $\rho_n\rho_k \leqslant \rho_{n+k}$, $n,\ k \geqslant 0$. Let $H_0 = c_0(\mathbb{N}, (\rho_n)_{n\geqslant 0}, K)$ be the free Banach space with orthogonal base $(e_n)_{n\geqslant 0}$ such that $e_n = (\delta_{n,k})_{k\geqslant 0}$ and $\|e_n\| = \rho_n$. Let $\sigma = e'_0$ and c be the continuous linear map of H_0 into $H_0\widehat{\otimes}H_0$ such that $c(e_n) = \sum_{i+j=n} e_i \otimes e_j$; one has $\|\sigma\| = 1$ and $\|c\| = 1$. One verifies that (H_0, c, σ) is a cocommutative Banach coalgebra, called a divided powers coalgebra; *cocommutative* means that $\tau \circ c = c$.

Let $(e'_n)_{n\geqslant 0}$ be the weak* base of H'_0 defined by $< e'_n, e_\ell >= \delta_{n,\ell}$. It is readily seen that $e'_n * e'_\ell = e'_{n+\ell}, n, \ell \geqslant 0$; hence $e'_n = e_1'^n, n \geqslant 0$.

Applying Theorem 1, one has

Fact 4 The Banach algebra H'_0, the dual of H_0, is isometrically isomorphic to the algebra of power series

$$K < X, (\rho_n)_{n\geqslant 0} >= \left\{ S = \sum_{n\geqslant 0} \alpha_n X^n \in K[[X]] \ / \ \|S\| = \sup_{n\geqslant 0} \frac{|\alpha_n|}{\rho_n} < +\infty \right\}.$$

Therefore H'_0 is an integral domain (cf. [3], Corollary 5).

REMARK 1 More generally let $(\rho_{i,n})_{1\leqslant i\leqslant r, n\geqslant 0} \subset \mathbb{R}_+^*$ be such that for $1 \leqslant i \leqslant r$, $\rho_{i,0} = 1$ and $\rho_{i,n} \cdot \rho_{i,k} \leqslant \rho_{i,n+k}$. Let $H_r = c_0(\mathbb{N}^r, (\rho_{i,n})_{\substack{1\leqslant i\leqslant r \\ n\geqslant 0}}, K)$ be the free Banach space with orthogonal base $(e_\alpha)_{\alpha\in\mathbb{N}^r}$ such that $e_\alpha = (\delta_{\alpha,\beta})_{\beta\in\mathbb{N}^r}$ and $\|e_\alpha\| = \rho_\alpha$, where $\rho_\alpha = \prod_{i=1}^{r} \rho_{i,\alpha(i)}$. Setting $\sigma_r = e'_0$ and $c_r(e_\alpha) = \sum_{\beta+\delta=\alpha} e_\beta \otimes e_\delta$, one gets that (H_r, c_r, σ_r) is a Banach coalgebra whose dual algebra H'_r is isometrically isomorphic to the algebra $K < X_1, \ldots, X_r; (\rho_{1,n})_{n\geqslant 0}, \ldots, (\rho_{r,n})_{n\geqslant 0} >=$ $\left\{ S = \sum_{\alpha\in\mathbb{N}^r} a_\alpha X^\alpha \in K[[X_1, \ldots, X_r]] \ / \ \|S\| = \sup_{\alpha\in\mathbb{N}^r} \frac{|a_\alpha|}{\rho_\alpha} < +\infty \right\}$. Hence H'_r is an integral domain.

Let us consider the divided powers Banach coalgebras $(H_{0,i}, c_i, \sigma_i)$, $1 \leqslant i \leqslant r$, where $H_{0,i} = c_0(\mathbb{N}, (\rho_{i,n})_{n\geqslant 0}, K)$. One has on the tensor product $C_r = \widehat{\bigotimes_{1\leqslant i\leqslant r}} H_{0,i}$ a structure of Banach coalgebra with coproduct defined inductively by $\Delta_r = (1_{C_{r-1}} \otimes \tau \otimes 1_{H_{0,r}}) \circ (\Delta_{r-1} \otimes c_r)$ and counit $\varepsilon_r = \sigma_1 \otimes \ldots \otimes \sigma_r$. Setting for $\alpha = (\alpha(i))_{1\leqslant i\leqslant r} \in \mathbb{N}^r$, $f_\alpha = \bigotimes_{1\leqslant i\leqslant r} e_{i,\alpha(i)}$, where $(e_{i,n})_{n\geqslant 0}$ is the orthogonal base of $H_{0,i}$, then $(f_\alpha)_{\alpha\in\mathbb{N}^r}$ is an orthogonal base of C_r. Moreover $\Delta_r(f_\alpha) = \sum_{\beta+\gamma=\alpha} f_\beta \otimes f_\gamma$ and the coalgebras C_r and H_r are isometrically isomorphic.

Fact 5 Let $\rho \in \mathbb{R}_+^*$ and $\rho_n = \rho^n$, $n \geqslant 0$. Let $H_0(\rho) = H_0 = c_0(\mathbb{N}, (\rho^n)_{n\geqslant 0}, K)$ be the divided powers Banach coalgebra associated with $(\rho^n)_{n\geqslant 0}$. Let $(e_n)_{n\geqslant 0} \subset$

$H_0(\rho)$ be the canonical orthogonal base such that $\|e_n\| = \rho^n$, $n \geqslant 0$. One defines on $H_0(\rho)$, by setting for $n,k \geqslant 0$, $m(e_n \otimes e_k) = \binom{n+k}{k} e_{n+k}$, a structure of commutative Banach algebra with unity e_0. Together with the cocommutative coalgebra structure, $H_0(\rho)$ becomes a complete Hopf algebra whose antipode is defined by $\eta(e_n) = (-1)^n e_n$, $n \geqslant 0$.

If $\rho = |\lambda| \in |K^*|$, one sees that the sequence $\varphi_n = \dfrac{e_n}{\lambda^n}$, $n \geqslant 0$, is an orthonormal base of $H_0(\rho)$ and one can replace in the definition of c and m, $(e_n)_{n\geqslant 0}$ by $(\varphi_n)_{n\geqslant 0}$.

LEMMA 1 (i) *If K is of characteristic zero, then the Banach algebra $H_0(\rho)$ is an integral domain.*

(ii) *If K is of characteristic $p \neq 0$, the Banach algebra $H_0(\rho)$ is a local ring with the maximal ideal the augmentation ideal $ker\sigma$. Moreover, any $a \in ker\sigma$ is nilpotent such that $a^p = 0$ and $Hom.alg(H_0(\rho), K) = \{\sigma\}$.*

Proof. Since by definition $e_n \cdot e_k = \binom{n+k}{k} e_{n+k}, n,k \geqslant 0$, one has $e_n^2 = \binom{2n}{n} e_{2n}$ and by induction one obtains $e_n^k = \prod_{j=1}^{k} \binom{jn}{n} \cdot e_{kn} = \dfrac{(kn)!}{(n!)^k} \cdot e_{kn}$. In particular $e_1^n = n! e_n$.

(i) If K is of characteristic zero, one has $e_n = \dfrac{e_1^n}{n!}$. Therefore any $a = \sum_{n\geqslant 0} a_n e_n \in H_0(\rho)$ can be expanded in the Taylor form $a = \sum_{n\geqslant 0} \dfrac{a_n}{n!} e_1^n = \sum_{n\geqslant 0} b_n e_1^n$ with $\lim_{n\to+\infty} |b_n| \, \|e_1^n\| = 0$,

$\|a\| = \sup_{n\geqslant 0} |b_n| \, \|e_1^n\|$; i.e. $(e_1^n)_{n\geqslant 0}$ is an orthogonal base of $H_0(\rho)$. Hence $H_0(\rho)$ is isomorphic to a subalgebra of the algebra of formal power series $K[[X]]$ and therefore $H_0(\rho)$ is an integral domain.

(ii) Let v_p be the p-adic valuation. It is well known that for $n = \sum_{i=0}^{t} a_i p^i \in \mathbb{N}$, one has $v_p(n!) = \dfrac{n - S_p(n)}{p-1}$, where $S_p(n) = \sum_{i=0}^{t} a_i$. Hence for any $n \geqslant 1$, $v_p\left(\dfrac{(pn)!}{(n!)^p}\right) = S_p(n) \geqslant 1$.

It follows that if K is of characteristic $p \neq 0$, one has $e_n^p = \dfrac{(pn)!}{(n!)^p} \cdot e_{pn} = 0$ for $n \geqslant 1$. Hence any $b = \sum_{n\geqslant 1} b_n e_n \in ker\sigma$ is such that $b^p = \sum_{n\geqslant 1} b_n^p e_n^p = 0$.

Let $a \in H_0(\rho)$. Then $a = a_0 e_0 + b$, where $a_0 \in K$ and $b \in ker\sigma$. It is readily seen that $a \notin ker\sigma$ if and only if $a_0 \neq 0$; in this condition a is invertible with the inverse $a^{-1} = \sum_{j=0}^{p-1} (-1)^j \, a_0^{-j-1} b^j$. Therefore $H_0(\rho)$ is a local ring with the maximal ideal $ker\sigma$. From what, one deduces that $Hom.alg(H_0(\rho), K) = \{\sigma\}$ which was described incompletely in [3].

REMARK 2 Let $(\rho_i)_{1\leqslant i\leqslant r} \subset \mathbb{R}_+^*$ and $H_0(\rho_1, \ldots, \rho_r) = c_0(\mathbb{N}^r, (\rho_i^n)_{\substack{1\leqslant i\leqslant r \\ n\geqslant 0}}, K) \simeq$

$\widehat{\bigotimes_{1\leqslant i\leqslant r}} H_0(\rho_i)$ be the Banach coalgebra associated with $(\rho_i^n)_{\substack{1\leqslant i\leqslant r\\ n\geqslant 0}}$ (cf. Remark 1). If for $\alpha,\beta\in\mathbb{N}^r$, $e_\alpha\cdot e_\beta = m(e_\alpha\otimes e_\beta) = \binom{\alpha+\beta}{\beta} e_{\alpha+\beta}$ where $\binom{\alpha+\beta}{\beta} = \prod_{i=1}^{r}\binom{\alpha(i)+\beta(i)}{\beta(i)}$, the coalgebra $H_0(\rho_1,\ldots,\rho_r)$ becomes a complete Hopf algebra with antipode η_r defined by $\eta_r(e_\alpha) = (-1)^{|\alpha|}e_\alpha$ where $|\alpha| = \sum_{i=1}^{r}\alpha(i)$.

It is readily seen that for $k\geqslant 0$ and $\alpha\in\mathbb{N}^r$, $e_\alpha^k = \prod_{j=1}^{k}\prod_{i=1}^{r}\frac{(k\alpha(i))!}{(\alpha(i)!)^k}\cdot e_{k\alpha}$. Hence as in Lemma 1, if K is of characteristic $p\neq 0$, $H_0(\rho_1,\ldots\rho_r)$ is a local ring, etc.

Identifying e_α with $\bigotimes_{1\leqslant i\leqslant r} e_{i,\alpha(i)}$ and setting $e^\alpha = \bigotimes_{1\leqslant i\leqslant r} e_{i,1}^{\alpha(i)}$, one verifies that $e^\alpha = \alpha! e_\alpha$, where $\alpha! = \prod_{i=1}^{r}\alpha(i)!$. If K is of characteristic zero, one gets that $e^\alpha\cdot e^\beta =$
$\alpha!\beta! e_\alpha\cdot e_\beta = (\alpha+\beta)! e_{\alpha+\beta} = e^{\alpha+\beta}$. Moreover $(e^\alpha)_{\alpha\in\mathbb{N}^r}$ is an orthogonal base of $H_0(\rho_1,\cdots,\rho_r)$ and as in Lemma 1, $H_0(\rho_1,\ldots,\rho_r)$ is an integral domain isomorphic to a subalgebra of $K[[X_1,\ldots,X_r]]$.

The continuous dual of a Banach coalgebra is an algebra. In the opposite the continuous dual H' of an infinite dimensional unitary Banach algebra (H,m,k) is by no natural mean a coalgebra. However (cf. [2]), H' contains a bigest closed subspace which is a coalgebra. Namely, identifying $H'\widehat{\otimes}H'$ as a subspace of $(H\widehat{\otimes}H)'$ (cf. 2.1) and setting $H^{\odot} = {}^tm^{-1}(H'\widehat{\otimes}H')\subset H'$, one gets that $c_0 = {}^tm|_{H^{\odot}}$ is a coproduct on $H^{\odot}$ and together with $\sigma_0 = {}^tm$, $(H^{\odot},c_0,\sigma_0)$ is a Banach coalgebra.

Moreover, if (H,m,c,σ) is a complete bialgebra (resp. a Hopf algebra with antipode η), $H^{\odot}$ is an unitary subalgebra of H' and setting $m_0 = {}^tc|_{H^{\odot}\widehat{\otimes}H^{\odot}}$, one sees that $(H^{\odot},m_0,c_0,\sigma_0)$ is a complete bialgebra (resp. a Hopf algebra with antipode $\eta_0 = {}^t\eta|_{H^{\odot}}$). Furthermore, the bialgebras H and $H^{\odot}$ are in the duality, i.e. considering the continuous bilinear forms $\beta : H\times H^{\odot}\to K$ defined by $\beta(a,\mu) = <\mu,a>$ and $\beta_2 : (H\widehat{\otimes}H)\times(H^{\odot}\widehat{\otimes}H^{\odot})\to K$ such that $\beta_2(a\otimes b,\mu\otimes\nu) = <\mu,a><\nu,b>$, one has

$$\text{(i)}\qquad \beta(m(a\otimes b),\mu) = \beta_2(a\otimes b, c_0(\mu))$$
$$\text{(ii)}\qquad \beta(a,m_0(\mu\otimes\nu)) = \beta_2(c(a),\mu\otimes\nu)$$

Let H° be the closure of ${}^tm^{-1}(H'\otimes H')$ in H'; then $H^\circ\subset H^{\odot}$ and H° is a closed subcoalgebra (resp. subbialgebra, resp. sub-Hopf algebra) of $H^{\odot}$. Furthermore if H is a bialgebra, the bialgebras H and H° are in the duality.

More generally, if two complete bialgebras have a pairing in such a way that their morphisms of structure satisfy (i) and (ii), one says that they are in the duality.

Now, let $H_0(\rho) = c_0(\mathbb{N},(\rho^n)_{n\geqslant 0},K)$ be the above divided powers Hopf algebra with orthogonal base $(e_n)_{n\geqslant 0}$; the Hopf algebra structure operators are defined by $c(e_n) == \sum_{i+j=n} e_i\otimes e_j$, $\sigma(e_n) = \delta_{0,n}$, $e_n\cdot e_k = m(e_n\otimes e_k) = \binom{n+k}{k} e_{n+k}$ with unity e_0, and $\eta(e_n) = (-1)e_n$.

In the sequel we put $e'_n = \varepsilon_n$. Using Fact 2, one sees that $(\varepsilon_n)_{n\geqslant 0}$ (resp. $(\varepsilon_k \otimes \varepsilon_\ell)_{k\geqslant 0,\ell\geqslant 0}$) is an orthogonal family of $H_0(\rho)'$ (resp. $(H_0(\rho)\widehat{\otimes}H_0(\rho))'$) which is a weak* base. In particular, for $n \geqslant 0$, ${}^t\eta(\varepsilon_n) = \sum\limits_{j\geqslant 0} < {}^t\eta(\varepsilon_n), e_j > \varepsilon_j = \sum\limits_{j\geqslant 0} < \varepsilon_n, (-1)^j e_j > \varepsilon_j = (-1)^n \varepsilon_n$. On the other hand, one has ${}^t m(\varepsilon_n) = \sum\limits_{k,\ell} < {}^t m(\varepsilon_n);\ e_k \otimes e_\ell > \varepsilon_k \otimes \varepsilon_\ell = \sum\limits_{k,l} < \varepsilon_n, \binom{k+\ell}{\ell} e_{k+\ell} > \varepsilon_k \otimes \varepsilon_\ell = \sum\limits_{k+\ell=n} \binom{n}{\ell} \varepsilon_k \otimes \varepsilon_\ell$.

Hence ${}^t m(\varepsilon_n) = \sum\limits_{k+\ell=n} \binom{n}{\ell} \varepsilon_k \otimes \varepsilon_\ell \in H_0(\rho)' \otimes H_0(\rho)'$ for all $n \geqslant 0$, and $\varepsilon_n \in {}^t m^{-1}(H_0(\rho)' \otimes H_0(\rho)') \subset H_0(\rho)^\circ$.

Let us remind that $\|\varepsilon_n\| = \rho^{-n}$ and that $\varepsilon_n * \varepsilon_k = \varepsilon_{n+k}$; hence $\varepsilon_n = \varepsilon_1^n$, $n \geqslant 0$ and $\|\varepsilon_n\| = \|\varepsilon_1\|^n$.

Let $B_0(\rho) = \overline{span}\ \{\varepsilon_n,\ n \geqslant 0\}$ be the closed subspace of $H_0(\rho)'$ spanned by $(\varepsilon_n)_{n\geqslant 0}$. If $x \in B_0(\rho)$, one has $x = \sum\limits_{n\geqslant 0} x_n \varepsilon_1^n$ with $x_n \in K$, $\lim\limits_{n\to+\infty} |x_n|\rho^{-n} = 0$. This implies that $B_0(\rho)$ is a Banach subalgebra of $H_0(\rho)'$ isomorphic to the Tate algebra $K\{X, \rho^{-n}\} = \left\{ \sum\limits_{n\geqslant 0} x_n X^n \in K[[X]] \ / \ \lim\limits_{n\to+\infty} |x_n|\rho^{-n} = 0 \right\}$.

It is readily seen that $B_0(\rho)$ is a closed sub-Hopf algebra of $H(\rho)^\circ$, and it is the smallest closed sub-Hopf algebra containing $(\varepsilon_n)_{n\geqslant 0}$.

The duality between $H_0(\rho)$ and $H_0(\rho)^\odot$ induces a separating duality between the Hopf algebras and free Banach spaces $H_0(\rho)$ and $B_0(\rho)$: $\beta\left(\sum\limits_{n\geqslant 0} a_n e_n, \sum\limits_{n\geqslant 0} x_n \varepsilon_n\right) = \sum\limits_{n\geqslant 0} a_n x_n$.

Let $\varphi : H_0(\rho) \to B_0(\rho)'$ be defined by $\varphi(a)(x) = \beta(a, x)$ and $\psi : B_0(\rho) \to H_0(\rho)'$ by $\psi(x)(a) = \beta(a, x)$, for $a \in H_0(\rho)$, $x \in B_0(\rho)$. It is obvious that ψ coincides with the inclusion map of $B_0(\rho)$ into $H_0(\rho)'$. On the other hand, φ is an isometrical homomorphism of algebras and is a Hopf algebras morphism of $H_0(\rho)$ into $B_0(\rho)^\odot$. It is readily seen that $\varphi(e_n) = \varepsilon'_n$, $n \geqslant 0$.

Identifying ε'_n with e_n, $n \geqslant 0$, any element a of $B_0(\rho)'$ is a pointwise convergent sum $a = \sum\limits_{n\geqslant 0} a_n e_n$ with $a_n \in K$ and $\|a\| = \sup\limits_{n\geqslant 0} |a_n|\rho^n$.

The closure $B_0(\rho)^\circ$ of ${}^t m_0^{-1}(B_0(\rho)' \otimes B_0(\rho)')$ in $B_0(\rho)'$, as already said, is a sub-Hopf algebra of $B_0(\rho)^\odot$.

THEOREM 2 *The complete Hopf algebras $H_0(\rho)$ and $B_0(\rho)$ are in a separating duality.*

Moreover:

(i) *The Banach algebra $H_0(\rho)'$ is an integral domain and $B_0(\rho)$ is a closed sub-Hopf algebra of $H_0(\rho)^\circ$.*

(ii) *The divided powers Banach coalgebra $H_0(\rho)$ is a closed sub-Hopf algebra of $B_0(\rho)^\odot$ contained in $B_0(\rho)^\circ$.*

(iii) *If K is of characteristic zero, $B_0(\rho)'$ is an integral domain.*

(iv) *If K is of characteristic $p \neq 0$, $B_0(\rho)'$ and its subalgebra $H_0(\rho)$ are both local ring such that any element a in their maximal ideal satisfies $a^p = 0$.*

Proof.

(i) This follows from Theorem 1 and the definition of $B_0(\rho)$.

(ii) It suffices to prove that $e_n \in {}^t m_0^{-1}(B_0(\rho)' \otimes B_0(\rho)')$. But ${}^t m_0(e_n) = \sum_{i,j} < {}^t m_0(e_n), \varepsilon_i \otimes \varepsilon_j > e_i \otimes e_j = \sum_{i,j} < e_n, \varepsilon_{i+j} > e_i \otimes e_j = \sum_{i+j=n} e_i \otimes e_j$. Therefore any $a = \sum_{n \geqslant 0} a_n e_n \in H_0(\rho)$ belongs to $B_0(\rho)^\circ$.

(iii) The proof runs over as in the proof of Lemma 1.

(iv) Consider, for $n \geqslant 1$, the subset b_n of the elements $a = \sum_{k \geqslant 0} a_k e_k \in B_0(\rho)'$ such that $a_k = 0$ for $k < n$. It is readily seen that b_n is an ideal of $B_0(\rho)'$. Furthermore, $b_n \cdot b_m \subset b_{n+m}$, $b_{n+1} \subset b_n$ and $\bigcap_{n \geqslant 1} b_n = \{0\}$. One has the direct sum decomposition $B_0(\rho)' = \left(\bigoplus_{j=0}^{n} K.e_j \right) \oplus b_n$. In particular, for any $a \in b_1$, one has $a = \sum_{j=1}^{n} a_j e_j + b_n$ with $a_j \in K$, $b_n \in b_n$. However $e_k^p = 0$ for any $k \geqslant 1$, hence $a^p = \sum_{j=1}^{n} a_j^p e_j^p + b_n^p = b_n^p$ with $b_n^p \in b_{np}$. Therefore $a^p \in \bigcap_{n \geqslant 1} b_{np} \subset \bigcap_{n \geqslant 1} b_n = (0)$, that is $a^p = 0$. One concludes as in the proof of Lemma 1 that b_1 is the unique maximal ideal of $B_0(\rho)'$.

REMARK 3 (1) Theorem 2 can be stated motus mutandis in the "many variables" case described in Remark 2.

(2) Let us consider on the set of formal series $K[[X]]$, instead of the Cauchy product, the Hurwitz product: for $S = \sum_{n \geqslant 0} a_n X^n$, $T = \sum_{n \geqslant 0} b_n X^n$, $S \circ T = \sum_{n \geqslant 0} (\sum_{i+j=n} \binom{n}{i} a_i b_j) X^n$. Then $(K[[X]], +, \circ)$ is an algebra, furthermore $H_0(\rho)$ and $B_0(\rho)'$ can be identified as subalgebras of $(K[[X]], +, \circ)$. As above, if K is of characteristic $p \neq 0$, then $(K[[X]], +, \circ)$ is a local ring such that any element $S = \sum_{n \geqslant 1} a_n X^n$ of its maximal ideal is nilpotent with $S^{\circ p} = 0$.

Notice that if K is of characteristic zero, the algebra of formal power series $K[[X]]$ is isomorphic to the Hurwitz algebra $(K[[X]], +, \circ)$.

THEOREM 3 *Assume that the field K is of characteristic zero with non-trivial valuation. Then the map g defined by $g(e_n) = \frac{\varepsilon_n}{n!}$, $n \geqslant 0$, can be extended as a continuous linear map of $H_0(\rho)$ into $B_0(\rho)$ if and only if $\rho \geqslant \gamma$, where $\gamma = |p|^{-\frac{1}{2(p-1)}}$ or 1 when K contains $\mathbb{Q}_p$ or is of residue characteristic zero.*

In these conditions $g : H_0(\rho) \to B_0(\rho)$ is an injectie morphism of complete Hopf algebras.

Moreover:

(i) *If K contains $\mathbb{Q}_p$, g is never an isomorphism.*

(ii) *If K is of residue characteristic zero, g is an isomorphism if and only if $\rho = 1$.*

Proof. If the map $g(e_n) = \frac{\varepsilon_n}{n!}$ extends as a continuous linear map of $H_0(\rho)$ into $B_0(\rho)$, there exists $\alpha > 0$ such that $\|g(e_n)\| = \frac{\|\varepsilon_n\|}{|n!|} = \frac{\rho^{-n}}{|n!|} \leqslant \alpha\|e_n\| = \alpha\rho^n$. Therefore $\alpha^{-\frac{1}{2n}}|n!|^{-\frac{1}{2n}} \leqslant \rho$ for $n \geqslant 1$ and $\lim_{n\to+\infty} |n!|^{-\frac{1}{2n}} \leqslant \rho$. If K is of residue characteristic zero, one has $|n!| = 1$ and $\rho \geqslant 1$. If $\mathbb{Q}_p \subseteq K$, then $|n!| = |p|^{\frac{n-S_p(n)}{p-1}}$ and $1 < \lim_{n\to\infty} |n!|^{-\frac{1}{2n}} = |p|^{-\frac{1}{2(p-1)}} \leqslant \rho$.

Conversely, if K is of residue characteristic zero and $1 \leqslant \rho$, then $\rho^{-n} \leqslant \rho^n$ and $\|g(e_n)\| = \frac{\rho^{-n}}{|n!|} = \rho^{-n} \leqslant \rho^n = \|e_n\|$. If $\mathbb{Q}_p \subseteq K$ and $\rho \geqslant |p|^{-\frac{1}{2(p-1)}}$; since $|p|^{\frac{S_p(n)}{p-1}} < 1$ for $n \geqslant 1$, then $|n!|^{-1} = |p|^{-\frac{n}{p-1}} \cdot |p|^{\frac{S_p(n)}{p-1}} < |p|^{-\frac{n}{p-1}} \leqslant \rho^{2n}$; hence $\|g(e_n)\| = |n!|^{-1}\rho^{-n} < \rho^n = \|e_n\|$ for $n \geqslant 1$ and $\|g(e_0)\| = 1 = \|e_0\|$.

In the both cases we have $\|g(e_n)\| = \frac{\|\varepsilon_n\|}{|n!|} \leqslant \|e_n\|$ and if $a = \sum_{n\geqslant 0} a_n e_n \in H_0(\rho)$, one has $|a_n|\frac{\|\varepsilon_n\|}{|n!|} \leqslant |a_n|\ \|e_n\|$ with $\lim_{n\to+\infty} |a_n|\ \|e_n\| = 0$. Hence $\lim_{n\to+\infty} \|a_n\frac{\varepsilon_n}{n!}\| = 0$ and $g(a) = \sum_{n\geqslant 0} \frac{a_n}{n!}\varepsilon_n$ converges in $B_0(\rho)$ and $\|g(a)\| = \sup_{n\geqslant 0} \frac{|a_n|}{|n!|}\|\varepsilon_n\| \leqslant \|a\|$ and g is injective.

On the other hand, for any $n \geqslant 0$ and $k \geqslant 0$ one has $g \circ m(e_n \otimes e_k) = g\left(\binom{n+k}{k} e_{n+k}\right) = \frac{(n+k)!}{n!k!} \cdot \frac{\varepsilon_{n+k}}{(n+k)!} = \frac{\varepsilon_n}{n!} \cdot \frac{\varepsilon_k}{k!} = g(e_n) \cdot g(e_k) = (g \otimes g) \circ (m_0 \otimes m_0)(e_n \otimes e_k)$ and $(g \otimes g) \circ c(e_n) = \sum_{i+j=n} g(e_i) \otimes g(e_j) = \frac{1}{n!} \sum_{i+j=n} \binom{n}{i} \varepsilon_i \otimes \varepsilon_j = \frac{1}{n!} c_0(\varepsilon_n) = c_0 \circ g(e_n)$; also $g \circ \eta(e_n) = (-1)^n g(e_n) = (-1)^n \frac{\varepsilon_n}{n!} = \eta_0\left(\frac{\varepsilon_n}{n!}\right) = \eta_0 \circ g(e_n)$ and $\sigma_0 \circ g(e_n) = \sigma_0\left(\frac{\varepsilon_n}{n!}\right) = \frac{\delta_{0,n}}{n!} = \delta_{0,n} = \sigma(e_n)$. By linearity one gets $g \circ m = (g \otimes g) \circ m_0$; $(g \otimes g) \circ c = c_0 \circ g$; $g \circ \eta = \eta_0 \circ g$ and $\sigma_0 \circ g = \sigma$, that is g is a morphism of Hopf algebras.

(i) Assume that $\mathbb{Q}_p \subseteq K$ and $\rho \geqslant |p|^{\frac{-1}{2(p-1)}}$. Suppose that the Hopf morphism $g : H_0(\rho) \to B_0(\rho)$ defined by $g(e_n) = \frac{\varepsilon_n}{n!}$ is an isomorphism with the inverse g^{-1}. Then $g^{-1}(\varepsilon_n) = n!e_n$ and there exists $\beta > 0$ such that $\|g^{-1}(x)\| \leqslant \beta\|x\|$, $x \in B_0(\rho)$, by Banach's isomorphism theorem. Hence $\|g^{-1}(\varepsilon_n)\| = |n!|\rho^n \leqslant \beta\rho^{-n}$, i.e. $\rho^{2n} \leqslant \beta|n!|^{-1}$. It follows that $\rho \leqslant |p|^{-\frac{1}{2(p-1)}}$ and $\rho = |p|^{-\frac{1}{2(p-1)}}$. Therefore $|n!| = \rho^{-2n}|p|^{-\frac{S_p(n)}{p-1}}$ and $\|g^{-1}(\varepsilon_n)\| = |n!|\rho^n = \rho^{-n}|p|^{-\frac{S_p(n)}{p-1}} \leqslant \beta\rho^n$ which implies $|p|^{-\frac{S_p(n)}{n-1}} \leqslant \beta$. Since $|p|^{-1} > 1$, one obtains $\limsup_{n\to+\infty} |p|^{-\frac{S_p(n)}{p-1}} = +\infty \leqslant \beta$; a contradiction. Hence g is not bijective.

ii) Assume that K is of residue characteristic zero and $\rho \geqslant 1$. As above, if $g : H_0(\rho) \to B_0(\rho)$ is bijective, then $\rho = 1$ and reciprocally. Then, as easilly seen the linear map $g : H_0(1) \to B_0(1)$ defined by $g(e_n) = \frac{\varepsilon_n}{n!}$, $n \geqslant 0$, is isometric and bijective.

1.3 A sub-bialgebra of $\mathcal{C}(\mathbb{Z}_p, K)^\circ$.

Let K be a field of residue characteristic $p \neq 0$. Let $(B_n)_{n\geqslant 0}$ be the Mahler base of the Banach space $\mathcal{C}(\mathbb{Z}_p, K)$ of the continuous functions on the p-adic ring $\mathbb{Z}_p$ with values in K; $B_0(s) = 1, B_n(s) = \dfrac{s(s-1)\dots(s-n+1)}{n!}$, $n \geqslant 1$, $s \in \mathbb{Z}_p$, are the binomial polynomials. If K is of characteristic p, we write again $B_n(s)$ the residue class in $\mathbb{F}_p$.

Note that for the additive group of $\mathbb{Z}_p$, the Banach algebra $\mathcal{C}(\mathbb{Z}_p, K)$ (for the multiplication m given by the usual product of functions) has a natural Hopf algebra structure with a coproduct c defined by $\pi \circ c(f)(s,t) = f(s+t)$, counit $\sigma(f) = f(0)$ and an antipode η such that $\eta(f)(s) = f(-s)$.

Then $c(B_n) = \sum\limits_{i+j=n} B_i \otimes B_j$, $n \geqslant 0$, so $\mathcal{C}(\mathbb{Z}_p, K)$ is a divided powers Banach coalgebra (not a divided powers bialgebra for the usual multiplication m above).

Therefore, $B'_n * B'_k = B'_{n+k}$, $n,\ k \geqslant 0$, which implies $B'_n = B_1'^n$ and the algebra $M(\mathbb{Z}_p, K) = \mathcal{C}(\mathbb{Z}_p, K)'$ is isometrically isomorphic to $K < X >= \{S : \sum\limits_{n\geqslant 0} a_n X^n \in K[[X]] \ /\ \|S\| = \sup\limits_{n\geqslant 0} |a_n| < +\infty\}$; a well known theorem.

Put $B'_1 = \omega$. Hence $B'_n = \omega^n$, $n \geqslant 0$.

LEMMA 2 ${}^t m(\omega^n) = \sum\limits_{\max(k,\ell)\leqslant n} \binom{n}{n-\ell}\binom{\ell}{k+\ell-n} \omega^k \otimes \omega^\ell$, *with the convention* $\binom{\ell}{\nu} = 0$ *if* $\nu \leqslant -1$, *for* $n \geqslant 0$.

Proof. Let $\Delta = \tau_1 - id$ be the difference operator on $\mathcal{C}(\mathbb{Z}_p, K)$, where $\tau_1 f(s) = f(s+1)$ and more generally $\tau_j f(s) = f(s+j), \forall \in \mathbb{N}$. For any $\ell, j \geqslant 0$, then $\Delta^j B_\ell = B_{\ell-j}$ if $j \leqslant \ell$ and $\Delta^j B_\ell = 0$ if $j \geqslant \ell + 1$. Since $\tau_j \circ \Delta = \Delta \circ \tau_j$, one has for $f, g \in \mathcal{C}(\mathbb{Z}_p, K), \Delta^n(fg) = \sum\limits_{i+j=n} \binom{n}{i} \tau_j \circ \Delta^i(f)\Delta^j(g)$ (Leibniz formula). In particular $\Delta^n(B_k B_\ell) = 0$ if $n > k + \ell$ and $\Delta^n(B_k B_\ell) = \sum\limits_{\substack{i+j=n \\ i\leqslant k, j\leqslant \ell}} \binom{n}{j} \tau_j(B_{k-i})B_{\ell-j}$ if $n \leqslant k+\ell$. Therefore $\Delta^n(B_k B_\ell)(0) = 0$ if $n > k+\ell$ and $\Delta^n(B_k B_\ell)(0) = \sum\limits_{\substack{i+j=n \\ i\leqslant k, j\leqslant \ell}} \binom{n}{i} B_{k-i}(j) B_{\ell-j}(0) = \binom{n}{n-\ell} B_{k+\ell-n}(\ell)$ if $k+\ell \leqslant n$. Hence one obtains $B_k B_\ell = \sum\limits_{n\geqslant 0} \Delta^n(B_k B_\ell)(0) B_n = \sum\limits_{n=\max(k,\ell)}^{k+\ell} \binom{n}{n-\ell}\binom{\ell}{k+\ell-n} B_n$. It follows that for $n, k, \ell \geqslant 0$, one has $< {}^t m(\omega^n), B_k \otimes B_\ell >= \sum\limits_{j=\max(h,\ell)}^{k+\ell} \binom{j}{j-\ell}\binom{\ell}{k+\ell-j} \delta_{n,}$ $= \binom{n}{n-\ell}\binom{\ell}{k+\ell-n}$ if $\max(k,\ell) \leqslant n \leqslant k+\ell$ and $< {}^t m(B'_n),\ B_k \otimes B_\ell >= 0$ otherwise.

Therefore ${}^t m(\omega^n) = \sum_{k,\ell} < {}^t m(\omega^n),\ B_k \otimes B_\ell > \ \omega^k \otimes \omega^\ell =$

$\sum_{\max(k,\ell)\leqslant n} \binom{n}{n-\ell} \binom{\ell}{k+\ell-n} \omega^k \otimes \omega^\ell$ with the convention $\binom{\ell}{\nu} = 0$ if $\nu \leqslant -1$

Lemma 2 says that for $n \geqslant 0$, one has $\omega^n \in {}^t m^{-1}(M(\mathbb{Z}_p, K) \otimes M(\mathbb{Z}_p, K))$. Then ω^n belongs to the closed sub-algebra $\mathcal{C}(\mathbb{Z}_p, K)^\circ$ of $\mathcal{C}(\mathbb{Z}_p, K)^\odot$.

Let $\mathcal{B} = \overline{span}\ \{\omega^n, n \geqslant 0\}$ be the closed subspace of $M(\mathbb{Z}_p, K)$ spanned by $(\omega^n)_{n\geqslant 0}$. Since $(\omega^n)_{n\geqslant 0}$ is an orthonormal family in $M(\mathbb{Z}_p, K)$, then for any $\mu \in \mathcal{B}$ one has $\mu = \sum_{n\geqslant 0} a_n \omega^n$ with $a_n \in K$, $\lim_{n\to+\infty} a_n = 0$ and $\|\mu\| = \sup_{n\geqslant 0} |a_n|$. Moreover, $\mathcal{B}$ is a subalgebra of $\mathcal{C}(\mathbb{Z}_p, K)^\circ$ isometrically isomorphic to the Tate algebra $K\{X\}$. By Lemma 2, we get $c_0(\omega^n) = {}^t m(\omega^n) \in \mathcal{B} \otimes \mathcal{B}$, hence $\mathcal{B}$ is a closed sub-bialgebra of $\mathcal{C}(\mathbb{Z}_p, K)^\circ$. Furthermore, $\mathcal{B}$ is different from $\mathcal{C}(\mathbb{Z}_p, K)^\circ$. Indeed, for $s \in \mathbb{Z}_p$, the Dirac measure ε_s is such that $c_0(\varepsilon_s) = {}^t m(\varepsilon_s) = \varepsilon_s \otimes \varepsilon_s$; hence $\varepsilon_s \in \mathcal{C}(\mathbb{Z}_p, K)^\circ$. However $\varepsilon_s = \sum_{n\geqslant 0} B_n(s)\omega^n$ and $(B_n(s))_{n\geqslant 0}$ does not converge to zero if $s \in \mathbb{Z}_p \setminus \mathbb{N}$. In this case $\varepsilon_s \notin \mathcal{B}$; it suffices to show that if $k \geqslant 1$, then $\limsup_{n\to+\infty} |B_n(-k)| = 1$. Notice that if $n \in \mathbb{N}$, one has $\varepsilon_n = \sum_{j=0}^{n} \binom{n}{j} \omega^j \in \mathcal{B}$ and $\omega^n = (\varepsilon_1 - \sigma)^n = \sum_{j=0}^{n} (-1)^{n-j} \binom{n}{j} \varepsilon_1^j$ with $\varepsilon_1^j = \varepsilon_j$.

THEOREM 4 *The natural pairing of $\mathcal{C}(\mathbb{Z}_p, K)$ and $M(\mathbb{Z}_p, K)$ induces a continuous bilinear form of $\mathcal{C}(\mathbb{Z}_p, K) \times \mathcal{B}$ into K such that the complete bialgebras $\mathcal{C}(\mathbb{Z}_p, K)$ and $\mathcal{B}$ are in a separating duality.*

Moreover, $\mathcal{B}$ is not a Hopf algebra. The dual Banach algebra $\mathcal{B}'$ of the coalgebra $\mathcal{B}$ contains $\mathcal{C}(\mathbb{Z}_p, K)$ and is isometrically isomorphic to the usual Banach algebra $\ell^\infty(\mathbb{N}, K)$ of bounded sequences.

Proof. The pairing of $\mathcal{C}(\mathbb{Z}_p, K)$ and $M(\mathbb{Z}_p, K)$ is given by $\beta(\sum_{n\geqslant 0} a_n B_n, \sum_{n\geqslant 0} b_n \omega^n) = \sum_{n\geqslant 0} a_n b_n$, whose restriction to $\mathcal{C}(\mathbb{Z}_p, K) \times \mathcal{B}$ is readily seen to be a separating continuous bilinear form that induces a duality of complete bialgebras.

The antipode of $\mathcal{C}(\mathbb{Z}_p, K)^\odot$ is the restriction η_0 of ${}^t\eta$. However, if $s \in \mathbb{Z}_p$, $f \in \mathcal{C}(\mathbb{Z}_p, K)$, one has $< \eta_0(\varepsilon_s), f >=< \varepsilon_s, \eta(f) >= \eta(f)(s) = f(-s) = \varepsilon_{-s}(f)$ i.e. $\eta_0(\varepsilon_s)$
$= \varepsilon_{-s}$. If $\mathcal{B}$ is a Hopf algebra, its antipode must be the restriction of η_0 with $\eta_0(\mathcal{B}) \subset \mathcal{B}$. But if $n \geqslant 1$, one has $\varepsilon_n = \sum_{j=0}^{n} \binom{n}{j} \omega^j \in \mathcal{B}$ while $\eta_0(\varepsilon_n) = \varepsilon_{-n} = \sum_{j\geqslant 0} \binom{-n}{j} \omega^j = \sum_{j\geqslant 0} (-1)^j \binom{n+j-1}{j} \omega^j \notin \mathcal{B}$. Therefore $\mathcal{B}$ is not a Hopf algebra.

Since for $n \geqslant 0$, $\varepsilon_n = \sum_{j=0}^{n} \binom{n}{j} \omega^j$ and $\omega^n = \sum_{j=0}^{n} (-1)^{n-j} \binom{n}{j} \varepsilon_j$, is an orthonormal base of $\mathcal{B}$, then $(\varepsilon_n)_{n\geqslant 0}$ is also an orthonormal base of $\mathcal{B}$. Hence, any $g \in \mathcal{B}'$ is a unique weak* convergent sum $g = \sum_{n\geqslant 0} < g, \varepsilon_n > \varepsilon'_n$, where $< \varepsilon'_n, \varepsilon_k >=$

$\delta_{n,k}$, $\|\varepsilon'_n\| = 1$ and $\|g\| = \sup_{n\geqslant 0} |< g, \varepsilon_n >|$. Setting $g(n) =< g, \varepsilon_n >$ one can identify g with an element of $\ell^\infty(\mathbb{N}, K)$ and reciprocally. Furthermore, if $g, g_1 \in \mathcal{B}'$, one has $g \cdot g_1 = (g \otimes g_1) \circ c_0$ and $(g \cdot g_1)(n) =< g \otimes g_1, \varepsilon_n \otimes \varepsilon_n >= g(n)g_1(n)$, $n \geqslant 0$. Therefore the algebras $\mathcal{B}'$ and $\ell^\infty(\mathbb{N}, K)$ are isometrically isomorphic.

On the other hand the map $\varphi : C(\mathbb{Z}_p, K) \to \mathcal{B}'$ defined by $\varphi(f)(\mu) = < \mu, f >$, $f \in C(\mathbb{Z}_p, K), \mu \in \mathcal{B}$ is an isometrical Banach algebra morphism such that $\varphi(B_n) = B''_n$, where $B''_n \in \mathcal{B}'$ is such that $< B''_n, \omega^k >= \delta_{n,k}$. Identifying B''_n with B_n, we get also that $g \in \mathcal{B}'$ in the unique weak* convergent sum $g = \sum_{n\geqslant 0} < g, \omega^n > B_n$ which extends the Mahler expansion of the elements of $C(\mathbb{Z}_p, K)$.

N.B. 1 (i) *Applying Lemma 2 and the formulas giving change of bases between* ω^n *and* ε_n, $n \geqslant 0$, *one sees that for* $0 \leqslant k, \ell \leqslant n$, $\sum_{j=\max(k,\ell)}^{n} (-1)^{n-j} \binom{n}{j}\binom{j}{k}\binom{j}{\ell} = \binom{n}{n-\ell}\binom{\ell}{k+\ell-n}$.

Also, since c_0 *is an algebra morphism, then* $c_0(\omega^n) = c_0(\omega)^n = (1 \otimes \omega + \omega \otimes 1 + \omega \otimes \omega)^n = \sum_{j_1+j_2+j_3=n} \frac{n!}{j_1!j_2!j_3} \omega^{j_1+j_3} \otimes \omega^{j_2+j_3}$.

(ii) *In* $\mathcal{B}'$, *for* $n \geqslant 0$, *one has the following weak* convergent sums:*

$$B_n = \sum_{\ell\geqslant n} \binom{\ell}{n} \varepsilon'_\ell \qquad \text{and} \qquad \varepsilon'_n = \sum_{\ell\geqslant n} (-1)^{\ell-n} \binom{\ell}{n} B_\ell.$$

REMARK 4 *Put* $\sigma = 1$. *Then* $\eta_0(\omega^n) \in C(\mathbb{Z}_p, K)^\circ$ *with* $\eta_0(\omega^n) = (-1)^n\omega^n \sum_{j\geqslant 0} (-1)^j \binom{n+j-1}{j} \omega^j = (-1)^n \frac{\omega^n}{(1+\omega)^n} \notin \mathcal{B}$ *for* $n \geqslant 1$.

Proof. Since $C(\mathbb{Z}_p, K)^\circ$ is a closed sub-Hopf algebra of $C(\mathbb{Z}_p, K)^\odot$, then $C(\mathbb{Z}_p, K)^\circ$ is stable by η_0. Hence, since $\omega^n \in \mathcal{B} \subset C(\mathbb{Z}_p, K)^\circ$, then $\eta_0(\omega^n) \in C(\mathbb{Z}_p, K)^\circ$. However, η_0 is an algebra morphism ; therefore $\eta_0(\omega^n) = \eta_0(\omega)^n = (\eta_0(\varepsilon_1 - 1))^n = (\varepsilon_{-1} - 1)^n = (\sum_{j\geqslant 1} (-1)^j\omega^j)^n = (-\omega)^n(\sum_{j\geqslant 0} (-1)^j\omega^j)^n =$
$(-\omega)^n \sum_{j\geqslant 0} (\sum_{j_1+ \quad +j_n=j} (-1)^{j_1+ \quad +j_n})\omega^j = (-\omega)^n \sum_{j\geqslant 0} (-1)^j \binom{j+n-1}{j} \omega^j = (-1)^n \frac{\omega^n}{(1+\omega)^n} \notin \mathcal{B}$ for $n \geqslant 1$.

Notice that one can prove the above formula by observing that for $k \geqslant 0$, $\eta(B_k)(s) = B_k(-s) = (-1)^k B_k(s + k - 1) = (-1)^k \sum_{j=1}^{k} \binom{k_{-1}}{k_{-j}} B_j(s)$ and $\eta_0(\omega^n) = \sum_{k\geqslant 0} < \omega^n, \eta(B_k) > \omega^k$.

QUESTION Determine the smallest closed sub-Hopf algebra of $C(\mathbb{Z}_p, K)^\circ$ that contains $\mathcal{B}$.

2 THE DUAL ALGEBRA OF THE HOPF ALGEBRA $\mathcal{C}(V_q, K)$

2.1 Dual of a tensor product of Hopf algebras

LEMMA 3 *Let E, F, G be three ultrametric Banach spaces over a complete valued field K.*

(i) *Let E' be the Banach dual of E. Then $E'\widehat{\otimes}\mathcal{L}(F,G)$ is isometrically isomorphic to a subspace of $\mathcal{L}(E\widehat{\otimes}F, G)$. In particular $E'\widehat{\otimes}F'$ can be identified with a closed subspace of $(E\widehat{\otimes}F)'$.*

(ii) *Moreover if E is finite dimensional, then $E'\widehat{\otimes}\mathcal{L}(F,G) = \mathcal{L}(E\widehat{\otimes}F, G)$. In particular $E'\widehat{\otimes}F' = E' \otimes F' = (E\widehat{\otimes}F)'$.*

Proof. (i) It is well known that if E and H are two Banach spaces, then $E'\widehat{\otimes}H$ is isometrically isomorphic to the space $C(E, H)$ of completely continuous linear maps of E into H (cf. [6], Theorem 4.41). Therefore $E'\widehat{\otimes}\mathcal{L}(F,G)$ is isometrically isomorphic to $C(E, \mathcal{L}(F,G) \subseteq \mathcal{L}(E, \mathcal{L}(F,G))$. On the other hand, the spaces $\mathcal{L}(E, \mathcal{L}(F,G))$, and the space $\mathcal{B}(E \times F, G)$ of the continuous bilinear maps of $E \times F$ into G are isometrically isomorphic. Since $\mathcal{B}(E \times F, G)$ is isometrically isomorphic to $\mathcal{L}(E\widehat{\otimes}F, G)$ we complete this part.

(ii) If E is finite dimensional, for any Banach space F, one has $E\widehat{\otimes}F = E \otimes F$. Furthermore, if $n = \dim E$ and $(e_j)_{1\leqslant j\leqslant n}$ is a basis of E, then $z = \sum_{j=1}^{n} e_j \otimes y_j$ for $z \in E\widehat{\otimes}F$. The canonical isometry φ of $E'\widehat{\otimes}\mathcal{L}(F,G)$ into $\mathcal{L}(E\widehat{\otimes}F, G)$ is defined by $\varphi(x' \otimes v)(x \otimes y) = < x', x > v(y)$, where $x' \in E', x \in E, y \in F$ and $v \in \mathcal{L}(F,G)$. Now, let $u \in \mathcal{L}(E\widehat{\otimes}F, G)$. For $z = \sum_{j=1}^{n} e_j \otimes y_j \in E\widehat{\otimes}F$, one has $u(z) = \sum_{j=1}^{n} u(e_j \otimes y_j)$. Setting for $1 \leqslant j \leqslant n$ and $y \in F$, $v_j(y) = u(e_j \otimes y)$, one gets that $v_j : F \to G$ is linear and continuous. Put $w = \sum_{j=1}^{n} e'_j \otimes v_j \in E'\widehat{\otimes}\mathcal{L}(F,G)$. Then $\varphi(w)(z) = \sum_{j=1}^{n}\sum_{\ell=1}^{n} < e'_j, e_\ell > v_j(y_\ell) = \sum_{j=1}^{n} v_j(y_j) = \sum_{j=1}^{n} u(e_j \otimes y_j) = u(z)$. It follows that φ is surjective.

LEMMA 4 *Let H_1 and H_2 be two Banach coalgebras with coproduct c_1 (resp. c_2) and counit σ_1 (resp. σ_2).*

Let $H = H_1\widehat{\otimes}H_2$ be a (Banach coalgebra) complete tensor product of the coalgebras (H_1, c_1, σ_1) and (H_2, c_2, σ_2).

Then the complete tensor product $H'_1\widehat{\otimes}H'_2$ of the dual Banach algebras H'_1 and H'_2 is a closed unitary subalgebra of the dual Banach algebra $(H_1\widehat{\otimes}H_2)'$ of $H = H_1\widehat{\otimes}H_2$ with unit $\sigma_1 \otimes \sigma_2$.

Furthermore, if H_1 is finite dimensional, then $H'_1\widehat{\otimes}H'_2 = (H_1\widehat{\otimes}H_2)'$.

Proof. The inclusion and the equality of Banach spaces are provided by Lemma 3.

The coproduct of $H = H_1\widehat{\otimes}H_2$ is given by $c = (1_{H_1} \otimes \tau \otimes 1_{H_2}) \circ (c_1 \otimes c_2)$. Let $\mu_1, \nu_1 \in H'_1$ and $\mu_2, \nu_2 \in H'_2$. By definition the multiplication of tensor product of algebras is given by $(\mu_1 \otimes \mu_2) \cdot (\nu_1 \otimes \nu_2) = (\mu_1 * \nu_1) \otimes (\mu_2 * \nu_2)$. On the other hand the multiplication induced by that of $(H_1\widehat{\otimes}H_2)'$ is given by $(\mu_1 \otimes \mu_2) * (\nu_1 \otimes \nu_2) =$

$[(\mu_1 \otimes \mu_2) \otimes (\nu_1 \otimes \nu_2)] \circ c$.

For $a \in H_1$ and $b \in H_2$ we have $c_1(a) = \sum_{j \geqslant 1} a_j^1 \otimes a_j^2$ and $c_2(b) = \sum_{\ell \geqslant 1} b_\ell^1 \otimes b_\ell^2$. Hence $< (\mu_1 \otimes \mu_2) * (\nu_1 \otimes \nu_2), a \otimes b >= \sum_{j \geqslant 1} \sum_{\ell \geqslant 1} < \mu_1, a_j^1 > < \mu_2, b_\ell^1 > < \nu_1, a_j^2 > < \nu_2, b_\ell^2 > =$ $= \sum_{j \geqslant 1} < \mu_1, a_j^1 > < \nu_1, a_j^2 > \sum_{\ell \geqslant 1} < \mu_2, b_\ell^1 > < \nu_2, b_\ell^2 >= < (\mu_1 * \nu_1) \otimes (\mu_2 * \nu_2), a \otimes b >$. Therefore $(\mu_1 \otimes \mu_2) * (\nu_1 \otimes \nu_2) = (\mu * \nu_1) \otimes (\mu_2 * \nu_2) = (\mu_1 \otimes \mu_2) \cdot (\nu_1 \otimes \nu_2)$.

2.2 On the integrity of $\mathcal{C}(V_q, K)'$, $p \neq 2$

Let us remind that if G is a zero dimensional compact group and K a complete ultrametric valued field, then the algebra $\mathcal{C}(G, K)$ of the continuous functions f on G with values in K is a complete Hopf algebra with a coproduct c such that $\pi \circ c(f)(s,t) = f(st)$, $s, t, \in G$, and an antipode η defined by $\eta(f)(s) = f(s^{-1})$ and a counit defined by $\sigma(f) = f(e)$. Moreover, if G_1 is another compact group, the Hopf algebras $\mathcal{C}(G \times G_1, K)$ and $\mathcal{C}(G, K)\widehat{\otimes}\mathcal{C}(G_1, K)$ are isometrically isomorphic, cf. Theorem 8 of [2].

As a consequence, setting $\mathcal{C}(G, K)' = M(G, K)$, the tensor product of Banach algebras $M(G, K)\widehat{\otimes}M(G_1, K)$ can be identified with a closed subalgebra of $M(G \times G_1, K)$.

LEMMA 5 *Let $\varphi : G \to G_1$ be a continuous homomorphism of zero dimensional compact groups. Then the map $\widetilde{\varphi}$ from $\mathcal{C}(G_1, K)$ into $\mathcal{C}(G, K)$, defined by $\widetilde{\varphi}(g) = g \circ \varphi$, is a morphism of complete Hopf algebras such that $\|\widetilde{\varphi}\| = 1$.*

Moreover if φ is an isomorphism of groups, then $\widetilde{\varphi}$ is an isometrical isomorphism of Hopf algebras whose transpose ${}^t\widetilde{\varphi}$ is an isometrical isomorphism of the Banach algebra $M(G, K)$ onto $M(G_1, K)$.

Proof. It is a routine verification, cf. the proof of Theorem 9 of [2].

In the rest of this section we assume that $\mathbb{Q}_p$ is a valued subfield of K and $p \neq 2$.

Let $q \in \mathbb{Z}_p, |q| = 1$ and assume that *q is not a root of unit.* The closure V_q of $\{q^n, n \geqslant 0\}$ in $\mathbb{Z}_p$ is an infinite compact subset of the group of units U_p of $\mathbb{Z}_p$; in fact V_q is a subgroup of U_p.

Fact 6 *Let $\xi = \lim_{\nu \to +\infty} q^{p^\nu}$ be the Teichmüller representative of q and let h be the order of the root of unit $\xi, h \mid p-1$. Put $R_h = \{\xi^j, 0 \leqslant j \leqslant h-1\}$ and let ℓ be the p-adic valuation of $q^h - 1$. Then $V_q = R_h \times (1 + p^\ell \mathbb{Z}_p)$ is a clopen subgroup of U_p.*

If ξ_0 is a primitive $(p-1)^{th}$ root of unit in U_p, setting $q_0 = \xi_0(1+p)$, one has $U_p = V_{q_0}$ (cf. [3], Lemma 4 and Corollary 5).

Put $\xi^{-1} q = q_1$. One has $q_1 = 1 + p^\ell a$ with $|a| = 1$ and $V_{q_1} = 1 + p^\ell \mathbb{Z}_p$. The map $\exp_{q_1}$ of $\mathbb{Z}_p$ into $\mathbb{Z}_p$ defined by $\exp_{q_1}(\alpha) = q_1^\alpha = \sum_{n \geqslant 0} \binom{\alpha}{n} a^n p^{n\ell}$, is an isomorphism of compact groups of the additive group $\mathbb{Z}_p$ onto V_{q_1} whose inverse map, denoted $\log_{q_1}$, is given by $\log_{q_1}(t) = \dfrac{\log t}{\log_{q_1}}$, where log is the p-adic logarithm.

We get another description of the Hopf algebra $\mathcal{C}(V_q, K)$ than the one given in [3]. Also we provide another orthonormal base different from the Van Hamme base $(Q_n)_{n\geqslant 0}$, $Q_n(s) = \dfrac{(s-1)(s-q)\dots(s-q^{n-1})}{(q^n-1)(q^n-q)\dots(q^n-q^{n-1})}$, $s \in V_q$, $n \geqslant 0$ (cf. [3], [4] or [7]).

Put $D_n = \widetilde{\log}_{q_1}(B_n)$, where $(B_n)_{n\geq 0}$ is the Mahler base of $\mathcal{C}(\mathbb{Z}_p, K)$, *i.e.* $D_n(t) = B_n(\log_{q_1}(t))$, $t \in V_{q_1}, n \geqslant 0$. Since $\widetilde{\log}_{q_1}$ is a bijective linear isometry of $\mathcal{C}(\mathbb{Z}_p, K)$ onto $\mathcal{C}(V_{q_1}, K)$, one sees that $(D_n)_{n\geq 0}$ is an orthonormal base of $\mathcal{C}(V_{q_1}, K)$. Notice that for $n \geqslant 1$, the function D_n is an analytic function on V_{q_1} that is not a polynomial, while the elements of the Van Hamme base in $\mathcal{C}(V_{q_1}, K)$ are polynomials.

On the other hand, let $(e_j)_{0\leqslant j\leqslant h-1}$ be the canonical orthonormal base of $\mathcal{C}(R_h, K)$ with $e_j(\xi^\ell) = \delta_{j,\ell}$, $0 \leqslant j,\ \ell \leqslant h-1$. Then $(e_j \otimes D_n)_{\substack{0\leqslant j\leqslant h-1\\ n\geqslant 0}}$ is an orthonormal base of $\mathcal{C}(R_h, K)\widehat{\otimes}\mathcal{C}(V_1, K) \simeq \mathcal{C}(V_q, K)$. If $a \in \mathcal{C}(R_h, K)$ and $g \in \mathcal{C}(V_{q_1}, K)$, the function on $V_q = R_h \times V_{q_1}$ corresponding to $a \otimes g$ will be denoted by $a\overline{\otimes}g$, i.e. $(a\overline{\otimes}g)\ (\xi^j t) = a(\xi^j)\ g(t)$, $0 \leqslant j \leqslant h-1$, $t \in V_{q_1}$.

THEOREM 5 *The Hopf algebra $\mathcal{C}(V_q, K)$ is isometrically isomorphic to the complete tensor product $\mathcal{C}(R_h, K)\widehat{\otimes}\mathcal{C}(V_{q_1}, K)$ of complete Hopf algebras.*

(i) *The family $(e_j\overline{\otimes}D_n)_{0\leq j\leqslant h-1, n\geqslant 0}$ is an orthonormal base of the Banach space $\mathcal{C}(V_q, K)$. Let c_2 be the coproduct of $\mathcal{C}(V_q, K)$, then for $0 \leqslant j \leqslant h-1$, $n \geqslant 0$, one has* $c_2(e_j\overline{\otimes}D_n) = \sum\limits_{k+\ell\equiv j(mod.h)} \sum\limits_{r+u=n} (e_k\overline{\otimes}D_r) \otimes (e_\ell\overline{\otimes}D_u)$.

(ii) *Let ψ_j be the character of R_h defined by $\psi_j(\xi^\ell) = \xi^{j\ell}$. Put $C_j = \psi_j\overline{\otimes}\mathcal{C}(V_{q_1}, K)$. Then C_j is a closed subcoalgebra of $\mathcal{C}(V_q, K)$ and one has the direct sum of coalgebras* $\mathcal{C}(V_q, K) = \bigoplus\limits_{j=0}^{h-1} C_j$.

Proof. Note that many parts of Theorem 5 are easy consequences of various general results already quoted.

(i) Let c (resp. c_0 and c_1) be the coproduct of $\mathcal{C}(\mathbb{Z}_p, K)$ (resp. $\mathcal{C}(R_h, K)$ and $\mathcal{C}(V_{q_1}, K)$). Since $\widetilde{\log}_{q_1}$ is a coalgebra morphism, one has $c_1(D_n) = c_1 \circ \widetilde{\log}_{q_1}(B_n) = \left(\widetilde{\log}_{q_1} \otimes \widetilde{\log}_{q_1}\right) \circ c(B_n) = \left(\widetilde{\log}_{q_1} \otimes \widetilde{\log}_{q_1}\right)\left(\sum\limits_{r+u=n} B_r \otimes B_u\right) = \sum\limits_{r+u=n} D_r \otimes D_u$. On the other hand, one verifies easily that $c_0(e_j) = \sum\limits_{k+\ell\equiv j(mod\ h)} e_k \otimes e_\ell$, $0 \leqslant j \leqslant h-1$.

Hence, one has $(c_0 \otimes c_1)(e_j \otimes D_n) = \left(\sum\limits_{k+\ell\equiv j(mod.h)} e_k \otimes e_\ell\right) \otimes \left(\sum\limits_{r+u=n} D_r \otimes D_u\right)$ and $c_2(e_j\overline{\otimes}D_n) = \sum\limits_{k+\ell\equiv j(mod.h)} \sum\limits_{r+u=n} (e_k\overline{\otimes}D_r) \otimes (e_\ell\overline{\otimes}D_u)$.

(ii) The characters ψ_j, $0 \leqslant j \leqslant h-1$, of R_h take their values in U_p and they are the group-like elements of $\mathcal{C}(R_h, K)$, i.e. $c_0(\psi_j) = \psi_j \otimes \psi_j$, $0 \leqslant j \leqslant h-1$. Moreover, since $\psi_j = \sum\limits_{\ell=0}^{h-1} \xi^{j\ell} e_\ell$ and $e_\ell = \dfrac{1}{h}\sum\limits_{j=0}^{h-1} \xi^{-j\ell}\psi_j$, then $(\psi_j)_{0\leqslant j\leqslant h-1}$ is an orthonormal base of $\mathcal{C}(R_h, K)$. Therefore, $\mathcal{C}(R_h, K)\widehat{\otimes}\mathcal{C}(V_{q_1}, K) = \bigoplus\limits_{0\leqslant j\leqslant h-1} \psi_j\overline{\otimes}\mathcal{C}(V_{q_1}, K)$. Setting $C_j = \psi_j \overline{\otimes} C(V_{q1}, K)$, one gets that C_j is isometrically isomorphic to the Banach space $C(V_{q1}, K)$. Therefore C_j is a closed subspace of $C(V_q, K)$. Furthermore, $C(V_q, K) = \bigoplus_{0 \leqslant j \leqslant h-1} C_j$ is an orthogonal direct sum.

On the other hand, from $c_2(\psi_j \overline{\otimes} D_n) = \sum_{r+u=n} (\psi_j \overline{\otimes} D_r) \otimes (\psi_j \otimes D_u)$ we deduce that $c_2(C_j) \subset C_j \widehat{\otimes} C_j$, so C_j is a subcoalgebra of $C(V_q, K)$. Furthermore, $\psi_j \overline{\otimes} D_0$ is a group-like element of C_j.

N.B. 2 *One has* $V_q = \bigcup_{j=0}^{h-1} O_j$, *where* $O_j = \xi^j V_{q1}$ *is a clopen subset of* V_q. *Furthermore,* $e_j \overline{\otimes} 1 = \chi_j$ *is the characteristic function of* O_j *and* $e_j \overline{\otimes} g = \chi_j \cdot g$ *for* $g \in C(V_{q1}, K)$.

One deduces from Theorem 5 and Lemma 4 that $M(V_q, K) = M(R_h, K) \widehat{\otimes} M(V_{q1}, K)$. The algebra $M(R_h, K)$ is the group algebra of R_h . Let $(\varepsilon_j)_{0 \leqslant j \leqslant h-1}$ be the dual base of $(e_j)_{0 \leqslant j \leqslant h-1}$ in $M(R_h, K)$. In fact ε_j is the Dirac measure ε_{ξ^j}. Put $a'_j = \frac{1}{h} \sum_{\ell=0}^{h-1} \xi^{-j\ell} \varepsilon_\ell, 0 \leqslant j \leqslant h-1$. One verifies that $a'_i * a'_j = \delta_{i,j} a'_i$, $0 \leqslant i,\ j \leqslant h-1$; $\sum_{j=0}^{h-1} a'_j = \sigma_0$ is the counit of $C(R_h, K)$. Moreover one has $< a'_j, \psi_\ell >= \delta_{j,\ell}$, where ψ_ℓ is the character defined by $\psi_\ell(\xi^k) = \xi^{\ell k}$.

Let σ_1 be the counit of $C(V_{q1}, K)$. Put $\mu_j = a'_j \otimes \sigma_1 \in M(V_q, K)$, $0 \leqslant j \leqslant h-1$. Then $\mu_i * \mu_j = \delta_{i,j} \mu_i$, $0 \leqslant i,\ j \leqslant h-1$ and $\sum_{j=0}^{h-1} \mu_j = \sigma$ is the counit of $C(V_q, K)$ and $< \mu_j, \psi_\ell \overline{\otimes} 1 >= \delta_{j,\ell}$, $0 \leqslant j, \ell \leqslant h-1$.

Since for the orthonormal base $(D_n)_{n \geqslant 0}$ we have $c_1(D_n) = \sum_{r+u=n} D_r \otimes D_u$, the Banach coalgebra $C(V_{q1}, K)$ is a divided powers coalgebra isomorphic to $C(\mathbb{Z}_p, K)$.

Hence, the weak* base $(D'_n)_{n \geqslant 0}$ of $M(V_{q1}, K)$ is such that $D'_n * D'_r = D'_{n+r}$ and setting $D'_1 = \omega_1$, we note $D'_n = \omega_1^n$, $n \geqslant 0$. Any $\nu \in M(V_{q1}, K)$ is a weak* sum $\nu = \sum_{n \geqslant 0} < \nu, D_n > \omega_1^n$ with $\|\nu\| = \sup_{n \geqslant 0} |< \nu, D_n >|$. Furthermore, by Fact 4, one has that $M(V_{q1}, K)$ is an integral domain isomorphic to the algebra

$$K < X >= \left\{ S = \sum_{n \geqslant 0} \alpha_n X^n \in K[[X]] \ / \ \|S\| = \sup_{n \geqslant 0} |\alpha_n| < +\infty \right\}.$$

THEOREM 6 (i) *The family* $(\varepsilon_j \otimes \omega_1^n)_{0 \leqslant j \leqslant h-1, n \geqslant 0}$ *is a weak* base of the Banach space* $M(V_q, K)$, *i.e. any* $\mu \in M(V_q, K)$ *is a weak* sum* $\mu = \sum_{0 \leqslant j \leqslant h-1} \sum_{n \geqslant 0} < \mu, e_j \overline{\otimes} D_n > \varepsilon_j \otimes \omega_1^n$ *with* $\|\mu\| = \sup_{(j,n) \in [0,h-1] \times \mathbb{N}} |< \mu, e_j \overline{\otimes} D_n >|$.

(ii) *Let* $A = \sigma_0 \otimes M(V_{q1}, K)$. *Then* A *is a closed unitary subalgebra of* $M(V_q, K)$ *which is an integral domain isomorphic to* $K < X >$

(iii) *Let* $\mathfrak{a}_j = \mu_j * A$, $0 \leqslant j \leqslant h-1$. *Then* $\mathfrak{a}_j$ *is a closed ideal of* $M(V_q, K)$. *Furthermore,* $M(V_q, K) = \bigoplus_{0 \leqslant j \leqslant h-1} \mathfrak{a}_j$ *with* $\mathfrak{a}_j * \mathfrak{a}_\ell = (0)$ *for* $0 \leqslant j \neq \ell \leqslant h-1$.

Proof. It remains only to prove (iii). One verifies that $(a'_j)_{0\leqslant j\leqslant h-1}$ is an orthonormal base of $M(R_h,K)$. If $\mu\in M(V_q,K)$, then$\mu=\sum_{j=0}^{h-1} a'_j\otimes\gamma_j=\sum_{j=0}^{h-1}\mu_j*\nu_j$ with $\nu_j=\sigma_0\otimes\gamma_j\in A$ and $\|\mu\|=\max_{0\leqslant j\leqslant h-1}\|a'_j\|\,\|\gamma_j\|=\max_{0\leqslant j\leqslant h-1}\|\nu_j\|$. On the other hand, since $\sigma=\sum_{j=0}^{h-1}\mu_j$ and $\mu_j*\mu_\ell=\delta_{j,\ell}\mu_j$, one has $\mu=\sum_{j=0}^{h-1}\mu_j*\mu$ and $\mu_j*\mu=\sum_{\ell=0}^{h-1}\mu_j*\mu_\ell*\nu_\ell=\mu_j*\nu_j\in\mu_j*A=a_j$. Hence $\mu=\sum_{j=0}^{h-1}\mu_j*\nu_j\in\sum_{j=0}^{h-1}a_j$.

Furthermore, if $\mu_j*\nu_j\in a_j$ and $\mu_\ell*\nu_\ell\in a_\ell$, then $(\mu_j*\nu_j)*(\mu_\ell*\nu_\ell)=\delta_{j,\ell}\mu_j*(\nu_j*\nu_\ell)=0$ if $j\neq\ell$. It follows that $a_j*a_\ell=\{0\}$ and $a_j\cap a_\ell=\{0\}$ for $j\neq\ell$. Therefore $M(V_q,K)=\bigoplus_{j=0}^{h-1}a_j$ is an orthogonal sum.

COROLLARY 1 *$C_j^\perp=\bigoplus_{\ell\neq j}a_\ell$ and $a_j^\perp=\bigoplus_{\ell\neq j}C_\ell$, where the orthogonality is taken with respect to the natural pairing of $\mathcal{C}(V_q,K)$ and $M(V_q,K)$.*

Proof. This follows easily from the relations $<\mu_j,\psi_\ell\overline{\otimes}1>=\delta_{j,\ell},\ 0\leqslant j,\ell\leqslant h-1$.

COROLLARY 2 *The Banach algebra $M(V_q,K)$ is an integral domain if only if $q\equiv 1(mod.p)$.*

Proof. If $q\equiv 1\pmod p$, one has $\xi=1$ and $q=q_1$. Hence $M(V_q,K)=M(V_{q_1},K)$ is an integral domain.

If $q\not\equiv 1$ (mod.p), one has $\xi\neq 1$, i.e. $h\geqslant 2$. The idempotents $\mu_j, 0\leqslant j\leqslant h-1$, of $M(V_q,K)$ are zero divisors and $M(V_q,K)$ is not an integral domain.

N.B. 3

(i) Let $\Delta f(\alpha)=f(\alpha+1)-f(\alpha)$, $\alpha\in\mathbb{Z}_p$, be the difference operator of $\mathcal{C}(\mathbb{Z}_p,K)$. Setting $\Delta_{q_1}=\widetilde{\log_{q_1}}\circ\Delta\circ\widetilde{\exp_{q_1}}$, one gets that $\Delta_{q_1}=\tau_{q_1}-$id is the degree 1 difference operator of $\mathcal{C}(V_{q_1},K)$, where $\tau_{q_1}(g)(t)=g(q_1t)$. If $u\in\mathcal{L}(\mathcal{C}(V_{q_1},K))$ satisfies $u\circ\tau_{q_1}=\tau_{q_1}\circ u$, then u can be written in two different ways:

(1) $\quad u=\sum_{n\geqslant 0}u(D_n)(1)\Delta_{q_1}^n$

and

(2) $\quad u=\sum_{n\geqslant 0}u(Q_{1,n})(1)D_{q_1}^{(n)}$, where $(Q_{1,n})_{n\geqslant 0}$ is the Van Hamme base of $\mathcal{C}(V_{q_1},K)$ and $D_{q_1}^{(n)}=(\tau_{q_1}-id)\circ\ldots\circ(\tau_{q_1}-q_1^{n-1}id)$ (cf. [8] or [3]).

(ii) One deduces from the well known formulas (14) and (15) of [3] that $D_{q_1}^{(n)}=\sum_{j=0}^{n}\left(\sum_{\ell=j}^{n}(-1)^{n-\ell}\binom{\ell}{j}q_1^{\frac{(n-\ell)(n-\ell-1)}{2}}\begin{bmatrix}n\\ \ell\end{bmatrix}_{q_1}\right)\Delta_{q_1}^j$ and $\Delta_{q_1}^n=\sum_{j=0}^{n}\left(\sum_{\ell=0}^{n}\binom{n}{\ell}\begin{bmatrix}\ell\\ j\end{bmatrix}_{q_1}\right)D_{q_1}^{(j)}$.

(iii) As already noticed, the base elements D_n are analytic functions on V_{q_1}. For example $D_1=\log_{q_1}$ has the following expansion in the Van Hamme base $(Q_{1,n})_{n\geqslant 0}$:

$$\log_{q_1} = \sum_{n\geqslant 1}\prod_{j=1}^{n-1}(1-q_1^j)Q_{1,n} = \sum_{n\geqslant 1}(-1)^{n-1}(q_1-1)^{n-1}[n-1]_{q_1}!Q_{1,n}.$$

2.3 The case $p = 2$

Let $\pm 1 \neq q \in U_2$ and let U_2 be the group of units of $\mathbb{Z}_2$. Then, as in 2.2, the closure V_q of $(q^n)_{n\geqslant 0}$ is a compact subgroup of $U_2 = 1 + 2\mathbb{Z}_2$.

Fact 7

(i) $U_2 = (1+4\mathbb{Z}_2)\cup(-1+4\mathbb{Z}_2) = R_2 \times V_5,\ R_2 = \{-1, 1\}$

(ii) *If* $q = 1 + 2^\ell a,\quad \ell \geqslant 2,\ |a| = 1,$ *then* $V_q = 1 + 2^\ell\mathbb{Z}_2$.

(iii) *If* $q = -1 + 2^\ell a,\quad \ell \geqslant 2,\ |a| = 1,$ *then* $V_q = (1 + 2^{\ell+1}\mathbb{Z}_2)\cup(-1+2^\ell U_2)$ *is a proper clopen subgroup of* U_2 *(even if* $\ell = 2$*). Moreover* $V_q = V_{q_1}\cup(2^\ell - 1)\cdot V_{q_1}$, *where* $q_1 = 1 + 2^{\ell+1}b,\ |b| = 1$, *for example* $q_1 = q^2$.

If $q = 1 + 2^\ell a,\ \ell \geqslant 2,\ |a| = 1$, then $exp_q : \mathbb{Z}_2 \to \mathbb{Z}_2,\ exp_q(\alpha) = q^\alpha = \sum_{n\geqslant 0}\binom{\alpha}{n}a^n 2^{n\ell}$, is a continuous isomorphism of the additive compact group $\mathbb{Z}_2$ onto V_q with inverse $\log_q$ defined by $\log_q(t) = \dfrac{\log t}{\log q}$, where log is the 2-adic logarithm.

2.3.1 If $\mathbb{Q}_2$ is a valued subfield of K and $q = 1 + 2^\ell a,\ \ell \geqslant 2,\ |a| = 1$, then $\widetilde{\log_q}$ is the isometrical isomorphism of Hopf algebra between $C(\mathbb{Z}_2, K)$ and $C(V_q, K)$. *Therefore, the Banach algebra* $M(V_q, K)$ *is an integral domain (since is isomorphic to* $M(\mathbb{Z}_2, K) \simeq K < X >$*).*

2.3.2 If $\mathbb{Q}_2 \subseteq K$, then (since $U_2 = R_2 \times V_5$), Theorem 5 remains true for $C(U_2, K)$ and Theorem 6 for $M(U_2, K)$.

2.3.3 It remains to study the case Fact 7, (iii), i.e. $q = -1 + 2^\ell a,\ \ell \geqslant 2,\ |a| = 1$ and $V_q = V_{q_1}\cup(2^\ell - 1)V_{q_1}$ where $q_1 = 1 + 2^{\ell+1}b,\ |b| = 1$. One has $[V_q : V_{q_1}] = 2$ and V_q is not isomorphic to the direct product $V_{q_1} \times V_q/_{V_{q_1}}$, otherwise one must have $-1 \in V_q$ which is impossible.

Let us consider the clopen subgroup $W_\ell = (1 + 2^\ell\mathbb{Z}_2)\cup(-1 + 2^\ell\mathbb{Z}_2) = R_2 \times (1 + 2^\ell\mathbb{Z}_2)$ of U_2.

Since $V_{q_1} = 1 + 2^{\ell+1}\mathbb{Z}_2$ and $V_q = (1+2^{\ell+1}\mathbb{Z}_2)\cup(-1+2^\ell U_2)$, then V_q is a subgroup of W_ℓ. On the other hand, it is easily seen that $1 + 2^\ell\mathbb{Z}_2 = V_{q_1}\cup(2^\ell + 1)V_{q_1}$ and $-1+2^\ell\mathbb{Z}_2 = -V_{q_1}\cup(2^\ell - 1)V_{q_1}$. Hence $W_\ell = V_{q_1}\cup(2^\ell+1)V_{q_1}\cup -V_{q_1}\cup(2^\ell - 1)V_{q_1} = V_q\cup(2^\ell + 1)V_{q_1}\cup -V_{q_1}$.

Since $\gamma = 1 - 2^{2\ell} \in V_{q_1}$ and $(2^\ell+1)(2^\ell-1) = -\gamma$, then $-V_{q_1} = (2^\ell+1)(2^\ell-1)V_{q_1}$. Therefore $(2^\ell + 1)V_{q_1}\cup -V_{q_1} = (2^\ell + 1)V_q$ and $W_\ell = V_q \cup (2^\ell + 1)V_q$. Also, since $2^\ell - 1 \in V_q$ and $2^\ell + 1 = \delta$ with $-\delta = (2^\ell - 1)^{-1}\gamma \in V_q$, then $(2^\ell + 1)V_q = -V_q$. It follows that $W_\ell = V_q \cup -V_q = R_2 \times V_q$.

Notice that the quotient group $W_\ell/_{V_{q_1}}$ is isomorphic to $R_2 \times R_2$.

Since $W_\ell = R_2 \times (1 + 2^\ell U_2)$ and $1 + 2^\ell U_2 = V_{q_0}$ for any $q_0 = 1 + 2^\ell d,\ |d| = 1$, then(comp. 2.2) $M(W_\ell, K) = M(R_2, K)\widehat{\otimes}M(V_{q_0}, K)$. For $\mathbb{Q}_2 \subseteq K$, $M(V_{q_0}, K)$ is an integral domain isomorphic to $K < X >$ (by 2.3.1). As in 2.2, for $a_0' =$

$\frac{1}{2}(\varepsilon_1 + \varepsilon_{-1})$ and $a'_1 = \frac{1}{2}(\varepsilon_1 - \varepsilon_{-1})$, we have $a'_i * a'_j = \delta_{i,j} a'_i$, $0 \leqslant i,\ j \leqslant 1$. Moreover, $M(W_\ell, K) = \bigoplus_{0 \leqslant j \leqslant 1} A_j$, where $A_j = \omega_j * \Omega$, $\Omega = \varepsilon_1 \otimes M(V_{q_0}, K)$ is an integral subalgebra of $M(W_\ell, K)$ and $\omega_j = a'_j * \varepsilon_1$, $0 \leqslant j \leqslant 1$, $\omega_i * \omega_j = \delta_{i,j}\omega_i$, $0 \leqslant i, j \leqslant 1$. Furthermore, $\omega_0 = \frac{1}{2}(\varepsilon_1 + \varepsilon_{-1})$ and $\omega_1 = \frac{1}{2}(\varepsilon_1 - \varepsilon_{-1})$.

For $0 \leqslant j \leqslant 1$, the ideal A_j of $M(W_\ell, K)$ is an unitary integral algebra with unit ω_j, isomorphic to Ω; moreover $A_j = \omega_j * M(W_\ell, K)$ and $A_0 \cap A_1 = (0)$.

We shall need the following two additional lemmas.

LEMMA 6 *Let $\nu \in M(W_\ell, K)$. Then ν is a zero divisor if and only if $\nu \in A_0$ or $\nu \in A_1$.*

Proof. Indeed, since $\nu \in M(W_\ell, K)$ is the unique sum $\nu = \nu_0 + \nu_1$ with $\nu_j \in A_j$, $0 \leqslant j \leqslant 1$, then $\nu * \rho = \nu_0 * \rho_0 + \nu_1 * \rho_1$ for $\rho = \rho_0 + \rho_1$. Therefore, if $\nu \neq 0$ is a zero divisor with $\nu * \rho = 0$, then $\nu_0 * \rho_0 = 0$ and $\nu_1 * \rho_1 = 0$. Hence, if $\nu_0 \neq 0$, then $\rho_0 = 0$; therefore $\rho = \rho_1 \neq 0$ and $\nu_1 = 0$, so is $\nu = \nu_0 \in A_0$. In the same way, if $\nu_1 \neq 0$, then $\nu = \nu_1 \in A_1$.

Consider the exact sequence of groups

$$\{1\} \to V_q \xrightarrow{i} W_\ell \xrightarrow{h} W_{\ell/V_q} = R_2 \longrightarrow \{1\}.$$

Consequently one gets the sequence of Hopf algebras

$$\mathcal{C}(R_2, K) \xrightarrow{\widetilde{h}} \mathcal{C}(W_\ell, K) \xrightarrow{\widetilde{i}} \mathcal{C}(V_q, K)$$

with $\widetilde{h}$ injective and $\widetilde{i}$ surjective by Urgsohn's lemma (or because V_q is a clopen subgroup of W_ℓ).

LEMMA 7

$${}^t\widetilde{i}(M(V_q, K)) \cap A_j = \{0\} \quad \text{for} \quad 0 \leqslant j \leqslant 1.$$

Proof. (i) Assume that ${}^t\widetilde{i}(M(V_q, K)) \cap A_0 \neq \{0\}$. Therefore, there exist $\mu \in M(V_q, K)$ and $\nu \in M(W_\ell, K)$ such that ${}^t\widetilde{i}(\mu) = \omega_0 * \nu$, with $\mu \neq 0$ and $\nu \neq 0$. It is readily seen that for any $s \in W_\ell$, $f \in \mathcal{C}(W_\ell, K)$ and $\nu \in M(W_\ell, K)$ one has $< \varepsilon_s * \nu, f > = < \nu, \tau_s f >$. Hence $< \omega_0 * \nu, f > = \frac{1}{2} < \nu, f > + \frac{1}{2} < \nu, \tau_{-1} f >$. Since $\mu \neq 0$, there exists $g \in \mathcal{C}(V_q, K)$ such that $< \mu, g > = 1$. For any $f \in \mathcal{C}(W_\ell, K)$ such that $f \circ i = g$ we have $1 = < \mu, g > = < {}^t\widetilde{i}(\mu), f > = < \omega_0 * \nu, f > = \frac{1}{2} < \nu, f > + \frac{1}{2} < \nu, \tau_{-1} f >$. Let $f : W_\ell \to K$ be the function defined by $f(s) = g(s)$ if $s \in V_q$ and $f(s) = -g(-s)$ if $s \in -V_q$. It is obvious that f is continuous and $f \circ i = g$. Moreover, $f(-s) = -f(s)$, $s \in V_q$. Therefore f is an odd function, so we have $\tau_{-1} f = -f$. Hence $1 = \frac{1}{2} < \nu, f > + \frac{1}{2} < \nu, \tau_{-1} f > = 0$, a contradtiction.

(ii) The same argument can be used for the case when ${}^t\widetilde{i}(M(V_q, K)) \cap A_1 \neq \{0\}$. Let $0 \neq \mu \in M(V_q, K)$ and $\nu \in M(W_\ell, K)$ be such that ${}^t\widetilde{i}(\mu) = \omega_1 * \nu$. There exists $g \in \mathcal{C}(V_q, K)$ such that $< \mu, g > = 1$. For any $f \in \mathcal{C}(W_\ell, K)$ such that $f \circ i = g$, we have $1 = < \mu, g > = \frac{1}{2} < \mu, f > - \frac{1}{2} < \nu, \tau_{-1} f >$. The continuous function $f : W_\ell \to K$, defined by $f(s) = g(s)$ if $s \in V_q$ and $f(s) = g(-s)$ if $s \in -V_q$, is such that $f \circ i = g$ and satisfies $f(-s) = f(s)$. Hence $\tau_{-1} f = f$. Hence $1 = < \mu, g > = \frac{1}{2} < \nu, f > - \frac{1}{2} < \nu, \tau_{-1} f > = 0$, a contradiction.

THEOREM 7 *Let K be a complete valued field, extension of $\mathbb{Q}_2$. Let $q = -1 + 2^\ell a, \ell \geqslant 2, |a| = 1$. Then the Banach algebra $M(V_q, K)$ is an integral domain.*

Proof. Suppose that there exist two non-zero elements μ and ν of $M(V_q, K)$ such that $\mu * \nu = 0$. Hence $0 = {}^t\widetilde{i}(\mu * \nu) = {}^t\widetilde{i}(\mu) * {}^t\widetilde{i}(\nu)$. By Lemma 6 we have ${}^t\widetilde{i}(\mu) \in A_0$ and ${}^t\widetilde{i}(\nu) \in A_1$ or ${}^t\widetilde{i}(\mu) \in A_1$ and ${}^t\widetilde{i}(\nu) \in A_0$. Since ${}^t\widetilde{i}$ is injective, ${}^t\widetilde{i}(M(V_q, K)) \cap A_0 \neq (0)$ and ${}^t\widetilde{i}(M(V_q, K)) \cap A_1 \neq (0)$, which contradicts Lemma 7.

COROLLARY *Let $\mathbb{Q}_2 \subseteq K$. For any $q \in U_2, q \neq \pm 1$, the Banach algebra $M(V_q, K)$ is an integral domain.*

Proof. This is an immediate consequence of 2.3.1 and Theorem 7.

REMARK 5 One can prove an analogue of Corollary 1.6 of [5], i.e., if a complete cocommutative Hopf algebra H is a Banach space of countable type such that the norm of the Banach algebra H' is multiplicative, then any comodule endomorphism $u \neq 0$ of H is surjective. Many Hopf algebras studied in this paper satisfy the above condition. In particular the q-differences operators $u \neq 0$ on the Hopf algebra $C(V_q, K)$ are surjective whenever $M(V_q, K)$ is an integral domain.

REFERENCES

[1] E Abe. Hopf Algebras, Cambridge University Press, 1980.

[2] B Diarra. Algèbres de Hopf et fonctions presque périodiques ultramétriques, Rivista di Matematica pura ed applicata 17:113–132, 1966.

[3] B Diarra. Complete ultrametric Hopf algebras which are free Banach spaces. In: WH Schikhof, C Perez-Garcia, J. Kąkol, ed. p-Adic Functional Analysis. New York: Marcel Dekker, 1997, pp 61–80.

[4] L van Hamme. Jackson's interpolation formula in p-adic analysis. Proceedings of the Conference on p-adic analysis, Report 7806, Nijmegen, June 1978, 119–125.

[5] M van der Put. Difference equations over p-adic fields. Math. Ann. 198:189–203, 1972.

[6] ACM van Rooij. Non Archimedean functional analysis, New York: Marcel Dekker, 1978.

[7] A Verdoodt. Normal bases for non-archimedean spaces of continuous functions, Publicacions Matemàtiques, 37:403–427, 1993.

[8] A Verdoodt. The use of operators for the construction of normal bases for the space of continuous functions on V_q. Bull. Belg. Math. Soc. 1:685–699, 1994.

On p-adic power series

BRANKO DRAGOVICH Institute of Physics, P.O.Box 57, 11001 Belgrade, Yugoslavia.

Abstract. We obtained the region of convergence and the summation formula for some modified generalized hypergeometric series (1.2). We also investigated rationality of the sums of the power series (1.3). As a result the series (1.4) cannot be the same rational number in all $\mathbb{Z}_p$.

1 INTRODUCTION

We are interested in investigation of various properties of some p-adic power series of the form

$$\sum_{n=0}^{\infty} A_n x^n \, , \tag{1.1}$$

where coefficients $A_n \in \mathbb{Q}$ and variable $x \in \mathbb{Q}_p$. Such series are often encountered in p-adic analysis [1] as well as in its applications in mathematical and theoretical physics (for a review, see, e.g. Refs. 2-5). Due to rationality of A_n, the series (1.1) can be simultaneously considered in all $\mathbb{Q}_p$ and in $\mathbb{R}$. It is of particular interest to find all rational points for some classes of the series (1.1). Some previous author's investigations on p-adic series of the form (1.1) were presented at the Fourth International Conference on p-Adic Analysis ([6] and references therein).

In this contribution we mainly consider some general properties of the series

$$\sum_{n=0}^{\infty} a_n R_{k,l}(n) x^n \, , \tag{1.2}$$

where a_n are coefficients of the generalized hypergeometric series and $R_{k,l}(n) = P_k(n)/Q_l(n)$ are rational functions in $n \in \mathbb{Z}_0 = \{0, 1, 2, \cdots\}$. We also examine in some detail the series

$$\sum_{n=0}^{\infty} n!P_k(n)x^n \ , \tag{1.3}$$

where $P_k(n)$ is a polynomial of degree k. In particular, we show that

$$\sum_{n=0}^{\infty} n! \tag{1.4}$$

cannot be the same rational number in $\mathbb{Z}_p$ for every p.

Note that in virtue of the non-archimedean properties of the p-adic valuation, the usual necessary condition is also sufficient for the series (1.1) to be convergent, *i.e.* (1.1) is p-adically convergent for some x iff

$$| A_n x^n |_p \to 0 \ , \qquad n \to \infty \ . \tag{1.5}$$

It is worth mentioning that Schikhof's book [1] contains an excellent introductory course to analysis of p-adic series and, if necessary, can be used to better understand some of our considerations.

2 GENERALIZED HYPERGEOMETRIC SERIES

Let $P_k(n)$ be a polynomial

$$P_k(n) = C_k n^k + C_{k-1} n^{k-1} + \cdots + C_0 \ , \quad 0 \neq C_k, C_{k-1}, \cdots, C_0 \in \mathbb{Q} \ , \tag{2.1}$$

in $n \in \mathbb{Z}_0$ of degree k. Let also $Q_l(n)$ be another polynomial of degree l,

$$Q_l(n) = D_l n^l + D_{l-1} n^{l-1} + \cdots + D_0 \ , \quad 0 \neq D_l, D_{l-1}, \cdots, D_0 \in \mathbb{Q} \ , \tag{2.2}$$

with restriction $Q_l(n) \neq 0$ for every $n \in \mathbb{Z}_0$.

We will call the R-modified generalized hypergeometric series

$$\begin{aligned} &{}_rF_s(\alpha_1, \alpha_2, \cdots, \alpha_r; \beta_1, \beta_2, \cdots, \beta_s; R_{k,l}; x) \\ &= \sum_{n=0}^{\infty} \frac{(\alpha_1)_n (\alpha_2)_n \cdots (\alpha_r)_n}{(\beta_1)_n (\beta_2)_n \cdots (\beta_s)_n} R_{k,l}(n) \frac{x^n}{n!} \ , \end{aligned} \tag{2.3}$$

where $R_{k,l}(n)$ is a rational function

$$R_{k,l}(n) = \frac{P_k(n)}{Q_l(n)} \tag{2.4}$$

with polynomials $P_k(n)$ and $Q_l(n)$ defined by (2.1) and (2.2), respectively, and $(u)_0 = 1$, $(u)_n = u(u+1)\cdots(u+n-1)$ for $n \geq 1$. When $R_{k,l}(n) \equiv 1$ one gets the standard definition of the generalized hypergeometric series.

PROPOSITION 1 *The R-modified hypergeometric series defined by (2.3), where* $\alpha_1, \alpha_2, \cdots, \alpha_r \in \mathbb{Z}_+ = \{1, 2, 3, \cdots\}$ *and* $\beta_1, \beta_2 \cdots \beta_s \in \mathbb{Z}_+$*, is p-adically convergent in the region*

$$| x |_p < p^{\frac{r-s-1}{p-1}} . \tag{2.5}$$

Proof. Note that

$$(u)_n = \frac{(u+n-1)!}{(u-1)!} \tag{2.6}$$

if $u \in \mathbb{Z}_+$. Then p-adic value of the general term in (2.3) can be written as

$$\begin{aligned} &\left| \frac{(\beta_1 - 1)! \cdots (\beta_s - 1)!}{(\alpha_1 - 1)! \cdots (\alpha_r - 1)} \right|_p \left| \frac{(\alpha_1 + n - 1)! \cdots (\alpha_r + n - 1)!}{(\beta_1 + n - 1)! \cdots (\beta_s + n - 1)!} \right|_p \\ &\times \left| \frac{P_k(n)}{Q_l(n)} \right|_p \left| \frac{x^n}{n!} \right|_p . \end{aligned} \tag{2.7}$$

Recall that

$$| m! |_p = p^{-\frac{m - \sigma_m}{p-1}} , \quad m \in \mathbb{Z}_+ , \tag{2.8}$$

where σ_m is the sum of digits in the expansion of m over the base p. Since the first factor does not depend on n and $|P_k(n)|_p / |Q_l(n)|_p$ is bounded it suffices to analyse

$$\left| \frac{(\alpha_1 + n - 1)! \cdots (\alpha_r + n - 1)!}{(\beta_1 + n - 1)! \cdots (\beta_s + n - 1)!} \right|_p \left| \frac{x^n}{n!} \right|_p . \tag{2.9}$$

For large enough n (2.9) behaves like

$$\left(p^{-\frac{r-s-1}{p-1}} | x |_p \right)^n , \tag{2.10}$$

which tends to zero as $n \to \infty$ if

$$p^{-\frac{r-s-1}{p-1}} | x |_p < 1 , \tag{2.11}$$

what just gives (2.5).

Note that (2.5) does not depend on the values of the parameters $\alpha_1, \alpha_2, \cdots, \alpha_r$ and $\beta_1, \beta_2, \cdots, \beta_s$ but only on their multiplicity r and s. For the Gauss series

$${}_2F_1(\alpha, \beta; \gamma; x) = \sum_{n=0}^{\infty} \frac{(\alpha)_n (\beta)_n}{(\gamma)_n n!} x^n \tag{2.12}$$

one obtains $| x |_p < 1$, like $| x |_\infty < 1$ in the real case.

Let us now turn to finding the corresponding summation formula.

PROPOSITION 2 *Let* ${}_rF_s(\alpha_1, \alpha_2, \cdots, \alpha_r; \beta_1, \beta_2, \cdots, \beta_s; R_{k,l}; x)$ *be an R-modified generalized hypergeometric series defined by (2.3) with the region of convergence given by (2.5). Then the following summation formula*

$$\begin{aligned} &\sum_{n=0}^{\infty} \frac{(\alpha_1)_n (\alpha_2)_n \cdots (\alpha_r)_n}{(\beta_1)_n (\beta_2)_n \cdots (\beta_s)_n n!} \left[\frac{(\alpha_1 + n)(\alpha_2 + n) \cdots (\alpha_r + n)}{(\beta_1 + n)(\beta_2 + n) \cdots (\beta_s + n)(n+1)} \right. \\ &\times \left. \frac{A_\mu(n+1)}{B_\nu(n+1)} x - \frac{A_\mu(n)}{B_\nu(n)} \right] x^n = -\frac{A_\mu(0)}{B_\nu(0)} \end{aligned} \tag{2.13}$$

is valid, where $A_\mu(n)$ and $B_\nu(n)$ are polynomials in $n \in \mathbb{Z}_0$ of the form (2.1) and (2.2), respectively.

Proof. The left hand side of (2.13) can be rewritten in the form

$$\sum_{n=1}^{\infty} \frac{(\alpha_1)_n(\alpha_2)_n \cdots (\alpha_r)_n}{(\beta_1)_n(\beta_2)_n \cdots (\beta_s)_n} \frac{A_\mu(n)}{n! B_\nu(n)} x^n$$
$$- \sum_{n=0}^{\infty} \frac{(\alpha_1)_n(\alpha_2)_n \cdots (\alpha_r)_n}{(\beta_1)_n(\beta_2)_n \cdots (\beta_s)_n} \frac{A_\mu(n)}{n! B_\nu(n)} x^n$$

which, by mutual cancellation of all terms except term for $n = 0$, gives just $-\frac{A_\mu(0)}{B_\nu(0)}$.

Although based on a simple derivation, (2.13) leads to the rather non-trivial results. Notice that always when

$$R_{k,l}(n) = \frac{(\alpha_1 + n)(\alpha_2 + n) \cdots (\alpha_r + n)}{(\beta_1 + n)(\beta_2 + n) \cdots (\beta_s + n)} \frac{t}{n+1} \frac{A_\mu(n+1)}{B_\nu(n+1)} - \frac{A_\mu(n)}{B_\nu(n)}, \tag{2.14}$$

where $A_\mu(n)$ and $B_\nu(n)$ are arbitrary polynomials defined like (2.1) and (2.2), respectively, if $x = t$ we have the resulting rational sum of (2.3), which does not depend on x and is equal to $-A_\mu(0)/B_\nu(0)$. Of course, the parameter t and the argument x belong to the region of convergence (2.5).

A generalized hypergeometric series is defined by its parameters, $\alpha_1, \alpha_2, \cdots, \alpha_r$ and $\beta_1, \beta_2, \cdots, \beta_s$. For a given generalized hypergeometric series there are many possibilities to choose rational functions $R_{k,l}(n)$ (2.14) with the corresponding rational sums $-A_\mu(0)/B_\nu(0)$. Let us notice some characteristic cases with $B_\nu(n) \equiv 1$. $A_\mu(n)$ may contain any partial or complete product of factors in the denominator: $\beta_1 + n - 1, \beta_2 + n - 1, \cdots, \beta_s + n - 1, n$. In the case when $A_\mu(n)$ includes n as a factor then $A_\mu(0) = 0$ and the sum of the corresponding series (2.13) will be also equal to zero. An extreme case is

$$A_\mu(n) = (\beta_1 + n - 1)(\beta_2 + n - 1) \cdots (\beta_s + n - 1) n B_\nu(n) \tag{2.15}$$

that gives in (2.14) the polynomial

$$P_k(n) = (\alpha_1 + n)(\alpha_2 + n) \cdots (\alpha_r + n) t A_\mu(n+1) - A_\mu(n) \tag{2.16}$$

instead of a rational function $R_{k,l}(n)$. Thus we have

$$\sum_{n=0}^{\infty} \frac{(\alpha_1)_n(\alpha_2)_n \cdots (\alpha_r)_n}{(\beta_1)_n(\beta_2)_n \cdots (\beta_s)_n} P_k(n) \frac{t^n}{n!} = 0 \tag{2.17}$$

if $P_k(n)$ has the form (2.16).

3 THE SERIES $\sum_{n=0}^{\infty} n!P_k(n)x^n$

This series can be regarded as a simple example of the R-modified generalized hypergeometric series, *i.e.*

$$ {}_2F_0(1,1;P_k;x) = \sum_{n=0}^{\infty} n!P_k(n)x^n \ . \tag{3.1}$$

Because of its relative simplicity the series (3.1) is suitable for examination of various p-adic properties. The power series (3.1) is divergent in the real case. From (2.5) it follows that its p-adic region of convergence is $\mid x \mid_p < p^{1/(p-1)}$ and it yields in $\mathbb{Q}_p$:

$$x \in \mathbb{Z}_p = \{x \in \mathbb{Q}_p : \mid x \mid_p \leq 1\} \ . \tag{3.2}$$

As a consequence of (3.2) we may take for x any integer and the series (3.1) will be p-adically convergent for every prime p.

The corresponding summation formula is

$$\sum_{n=0}^{\infty} n![(n+1)A_{k-1}(n+1)x - A_{k-1}(n)]x^n = -A_{k-1}(0) \ . \tag{3.3}$$

For $x = 1$ it can be rewritten in the more suitable form

$$\sum_{n=0}^{\infty} n!(n^k + u_k) = v_k \ , \tag{3.4}$$

where $(n+1)A_{k-1}(n+1) - A_{k-1}(n) = n^k + u_k$, $v_k = -A_{k-1}(0)$. One can easily see that $u_k = A_{k-1}(1) - A_{k-1}(0)$.

It is very useful to have expressions for finite (partial) sums of (3.4).

PROPOSITION 3 *If*

$$S_n^{(k)} = \sum_{i=0}^{n-1} i!i^k \ , \quad k \in \mathbb{Z}_0 \ , \tag{3.5}$$

then

$$S_n^{(k+1)} = -\delta_{0k} - kS_n^{(k)} - \sum_{l=0}^{k-1} \binom{k+1}{l} S_n^{(l)} + n!n^k \tag{3.6}$$

is a recurrent relation, where δ_{0k} *is the Kronecker symbol* ($\delta_{0k} = 1$ *if* $k = 0$ *and* $\delta_{0k} = 0$ *if* $k \neq 0$).

Proof.

$$\begin{aligned} S_n^{(k)} &= \delta_{0k} + \sum_{i=0}^{n-2} (i+1)!(i+1)^k = \delta_{0k} + \sum_{i=0}^{n-1} i!(i+1)^{k+1} - n!n^k \\ &= \delta_{0k} + \sum_{l=0}^{k+1} \binom{k+1}{l} S_n^{(l)} - n!n^k \ . \end{aligned}$$

Table 1 contains the first eleven values of u_k and v_k.

k	1	2	3	4	5	6	7	8	9	10	11
u_k	0	1	−1	−2	9	−9	−50	267	−413	−2180	17731
v_k	−1	1	1	−5	5	21	−105	141	777	−5513	13209

Applying successively the recurrent relation (3.6) we obtain summation formula of the form

$$\sum_{i=0}^{n-1} i!(i^k + u_k) = v_k + n!A_{k-1}(n) \ , \tag{3.7}$$

where $A_{k-1}(n)$ is a polynomial of degree $k-1$ in n with integer coefficients. As an illustration, here are the first four examples:

$$\begin{aligned} &(a) \ \sum_{i=0}^{n-1} i!i = -1 + n! \ , \\ &(b) \ \sum_{i=0}^{n-1} i!(i^2+1) = 1 + n!(n-1) \ , \\ &(c) \ \sum_{i=0}^{n-1} i!(i^3-1) = 1 + n!(n^2-2n-1) \ , \\ &(c) \ \sum_{i=0}^{n-1} i!(i^4-2) = -5 + n!(n^3-3n^2+5) \ . \end{aligned} \tag{3.8}$$

In a similar way to the Proposition 3 one can obtain recurrent relations for u_k and v_k:

$$\begin{aligned} u_{k+1} &= -ku_k - \sum_{l=1}^{k-1} \binom{k+1}{l} u_l + 1 \ , \quad u_1 = 0, \ k \geq 1 \ , \\ v_{k+1} &= -kv_k - \sum_{l=1}^{k-1} \binom{k+1}{l} v_l - \delta_{0k} \ , \quad k \geq 0 \ . \end{aligned} \tag{3.9}$$

It is worth noting that $i^k + u_k$ in (3.7) is a simplified form of $P_k(i)$ which gives rational sum of (3.1) if $x = 1$. Such $P_k(i) = i^k + u_k$ are suitable to obtain a general expression for the series (3.1) with rational sum at $x = 1$. In fact, the generalized form of (3.7) is

$$\sum_{i=0}^{n-1} i!P_k(i) = V_k + n!B_{k-1}(n) \ , \quad k \geq 1 \ , \tag{3.10}$$

where $P_k(i) = \sum_{r=0}^{k} C_r i^r$ with $C_0 = \sum_{r=1}^{k} C_r u_r$, $V_k = \sum_{r=1}^{k} C_r v_r$, $B_{k-1}(n) = \sum_{r=1}^{k} C_r A_{r-1}(n)$ and $C_1, C_2, \cdots, C_k \in \mathbb{Q}$.

The above consideration performed for $x = 1$ can be extended to other positive integers x with some other values of u_k and v_k.

Let us turn now to the sum of the power series (3.1) and investigate some of its rationality problem at $x \in \mathbb{Z}_+$. It is useful to start with the simplest case, *i.e.* $P_k(n) \equiv 1$.

THEOREM 1 *Let x be a given positive integer. If the p-adic sum of the power series*

$$\sum_{n=0}^{\infty} n!x^n \tag{3.11}$$

is a rational number then it cannot be the same in $\mathbb{Z}_p$ for every p.

Proof. Suppose there is $x = t \in \mathbb{Z}_+$ such that (3.11) rational is:

$$\sum_{n=0}^{\infty} n!t^n = \frac{a(t)}{b(t)} \qquad a(t) \in \mathbb{Z}\ ,\ b(t) \in \mathbb{Z}_+ \tag{3.12}$$

and the same for every p. Let $S_n(t)$ be

$$S_n(t) = \sum_{i=0}^{n-1} i!t^i\ . \tag{3.13}$$

Since $i!t^i < (n-1)!t^{n-1}$ when $0 \leq i \leq n-2$ one has $S_n(t) = 0! + 1!t + 2!t^2 + \cdots + (n-2)!t^{n-2} + (n-1)!t^{n-1} < (n-1)!(n-1)t^{n-1} + (n-1)!t^{n-1} = n!t^{n-1} \leq n!t^n$. Thus we have inequality

$$0 < S_n < n!t^n\ , \quad n > 2,\ t \geq 1\ . \tag{3.14}$$

For a fixed $b(t) \in \mathbb{Z}_+$ one can write $b(t)S_n(t) = b(t)[0! + 1!t + 2!t^2 + \cdots + (n-2)!t^{n-2} + (n-1)!t^{n-1}] < b(t)[(n-1)!t^{n-1} + (n-1)!t^{n-1}] = 2b(t)(n-1)!t^{n-1} < n!t^n$ if $2b(t) < nt$, *i.e.*

$$0 < b(t)S_n(t) < n!t^n\ , \quad 2b(t) < nt\ . \tag{3.15}$$

According to (3.12) one has $a(t) = b(t)S_n(t) + n!t^n b(t)[1 + (n+1)t + \cdots]$. Due to our assumption, $b(t)[1 + (n+1) + ...]$ must be the same rational integer in all $\mathbb{Z}_p$ and we get the congruence

$$a(t) \equiv b(t)S_n(t) (\text{mod } n!t^n)\ , \quad n \geq 0,\ t \geq 1\ . \tag{3.16}$$

The value of $a(t)$ belongs to the one of the following three possibilities: (i) $a(t) > 0$, (ii) $a(t) < 0$ and (iii) $a(t) = 0$. Consider each of these possibilities. According to (3.15) and (3.16) for large enough n we have:

$$\begin{aligned} (i) \quad & 0 < a(t) < n!t^n\ , \\ & 0 < b(t)S_n(t) < n!t^n\ , \\ & a(t) \equiv b(t)S_n(t) (\text{mod } n!t^n)\ ; \end{aligned}$$

$$\begin{aligned} (ii) \quad & -n!t^n < a(t) < 0\ , \\ & 0 < b(t)S_n(t) < n!t^n\ , \\ & a(t) \equiv b(t)S_n(t) (\text{mod } n!t^n)\ ; \end{aligned}$$

$$(iii)\quad \begin{aligned} & a(t) = 0 \ , \\ & 0 < b(t)S_n(t) < n!t^n \ , \\ & a(t) \equiv b(t)S_n(t) (\text{mod } n!t^n) \ . \end{aligned} \tag{3.17}$$

Analysing the conditions in (3.17) we find the following candidates for solution:

$$\begin{aligned} (i) \quad & a(t) = b(t)S_n(t) \ , \\ (ii) \quad & a(t) = b(t)S_n(t) - n!t^n \end{aligned} \tag{3.18}$$

and (iii) without solution. Since $a(t)$ must be a fixed integer we conclude that the solutions (3.18), which depend on n, are impossible.

As a particular case of the Theorem 1 we have that the sum of the series (1.4) cannot be the same rational number in all $\mathbb{Z}_p$. Note an earlier assertion (see [1], p.17) that $\sum_{n=0}^{\infty} n!$ cannot be rational in $\mathbb{Z}_n$ for every n.

THEOREM 2 *For fixed k and x the sum of the power series*

$$\sum_{n=0}^{\infty} n! n^k x^n \ , \qquad k \in \mathbb{Z}_0 \ , \ x \in \mathbb{Z}_+ \setminus \{1\} \ , \tag{3.19}$$

cannot be the same rational number in $\mathbb{Z}_p$ for every p.

Proof. When $k = 0$ it follows from Theorem 1. Deviding (3.3) by x, for $k \geq 1$ one has

$$\sum_{n=0}^{\infty} n![n^k + u_k(x)]x^n = v_k(x) \ , \quad x \in \mathbb{Z}_p \setminus \{0\} \ , \tag{3.20}$$

as a generalization of (3.4). Analysing the system of linear equations for coefficients of the polynomial $A_{k-1}(n)$, which follows from

$$(n+1)A_{k-1}(n+1) - \frac{A_{k-1}(n)}{x} = n^k + u_k(x) \ , \tag{3.21}$$

we conclude that $u_k(x)$ has the form

$$u_k(x) = \frac{-1 + xF_{k-1}(x)}{x^k} \ , \tag{3.22}$$

where $F_{k-1}(x)$ is a polynomial in x of degree $k-1$ with integer coefficients (for $k = 1, \cdots, 4$ see the Table 2). The series (3.19) might be the same rational number in all $\mathbb{Z}_p$ for some $x \in \mathbb{Z}_+$ iff

$$-1 + xF_{k-1}(x) = 0 \ . \tag{3.23}$$

However $F_{k-1}(x)$ is a polynomial with integer coefficients and eq. (3.23) has no solutions in $x \in \mathbb{Z}_+ \setminus \{1\}$.

Among the series of the form

$$\sum_{n=0}^{\infty} n! n^k \ , \quad k \in \mathbb{Z}_+ \ , \tag{3.24}$$

Table 2 Expressions for $u_k(x)$ and $v_k(x)$ $(k = 1, \cdots, 4)$ illustrate some of our conclusions.

k	1	2	3	4
$u_k(x)$	$\frac{-1+x}{x}$	$\frac{-1+3x-x^2}{x^2}$	$\frac{-1+6x-7x^2+x^3}{x^3}$	$\frac{-1+10x-25x^2+15x^3-x^4}{x^4}$
$v_k(x)$	$-\frac{1}{x}$	$\frac{-1+2x}{x^2}$	$\frac{-1+5x-3x^2}{x^3}$	$\frac{-1+9x-17x^2+4x^3}{x^4}$

it is easy to see (Table 1 and (3.8)) that

$$\sum_{n=0}^{\infty} n!n = -1 \tag{3.25}$$

in $\mathbb{Z}_p$ for every p. According to the Table 1 the sum of the series

$$\sum_{n=0}^{\infty} n!n^k \ , \quad k = 2, 3, \cdots, 11 \ , \tag{3.26}$$

cannot be the same rational number (for a fixed k) in all $\mathbb{Z}_p$.

PROPOSITION 4 *The sum of the series*

$$\sum_{n=0}^{\infty} n!n^{q+1} \ , \quad q = \textit{any of prime numbers} \ , \tag{3.27}$$

cannot be the same rational number (for a fixed q) in $\mathbb{Z}_p$ for every prime p.

Proof. According to the recurrent relations (3.9) one has for any prime number q that $u_{q+1} \equiv 1(\text{mod } q)$ and $v_{q+1} \equiv 1(\text{mod } q)$. Thus, $u_{q+1} \neq 0$ and v_{q+1} is a rational integer.

It is unlikely that $\sum_{n=0}^{\infty} n!n^k$ is a rational number if $k \neq 1$. Thus there is a sense to introduce the following

CONJECTURE The sum of the series

$$\sum_{n=0}^{\infty} n!n^k \ , \quad k \in \mathbb{Z}_0$$

is a rational number in all $\mathbb{Z}_p$ iff $k = 1$. Or, in the more general form, p-adic sum of the power series

$$\sum_{n=0}^{\infty} n!n^k x^n \ , \quad k \in \mathbb{Z}_0 \ , \ x \in \mathbb{Z}_+$$

is a rational number iff $k = x = 1$.

4 CONCLUDING REMARKS

It is worth noting that the p-adic power series

$$\sum_{j=0}^{\infty}(n+1)_j x^j \ , \quad x \in \mathbb{Z}_+$$

cannot be a rational integer in any $\mathbb{Z}_p$ as well as the same rational number in all $\mathbb{Z}_p$. This follows from identity

$$\sum_{n=0}^{\infty} n!x^n = S_n(x) + n!x^n \sum_{j=0}^{\infty}(n+1)_j x^j$$

and the proof of the Theorem 1.

It is clear that the p-adic hypergeometric series (2.12) satisfies the corresponding hypergeometric differential equation, i.e.

$$x(1-x)w'' + [\gamma - (\alpha+\beta+1)x]w' - \alpha\beta w = 0 \ ,$$

where $w = {}_2F_1(\alpha, \beta; \gamma; x)$. Let us also notice that the p-adic series

$$F_\nu(x) = \sum_{n=0}^{\infty} n!x^{n+\nu} \ , \quad \nu \in \mathbb{Z}_+ \ ,$$

is a solution of the following differential equation

$$\left(\frac{d^\nu}{dx^\nu} - \frac{1}{x^{2\nu}}\right)F_\nu(x) = f_\nu(x) \ , \quad x \in \mathbb{Z}_p \setminus \{0\} \ ,$$

where

$$f_\nu(x) = -\sum_{l=0}^{\nu-1} \frac{l!}{x^{\nu-l}} .$$

The series

$$F(x) = \sum_{n=0}^{\infty} n!x^n$$

may be regarded as an analytic solution of the differential equation

$$x^2F''(x) + (3x-1)F'(x) + F(x) = 0.$$

Many of the above results, obtained for $x \in \mathbb{Z}_+$, may be extended to $x \in \mathbb{Z} \setminus \{0\}$ and it will be done elsewhere.

ACKNOWLEDGMENTS The author wishes to thank the organizers of the Fifth International Conference on p-Adic Analysis for invitation and hospitality, Prof. L.

Van Hamme for discussions, and especially Prof. W. H. Schikhof for discussions and some informal communications.

REFERENCES

[1] WH Schikhof. Ultrametric Calculus - An Introduction to p-Adic Analysis. Cambridge: Cambridge University Press, 1984.

[2] L Brekke, PGO Freund. p-Adic Numbers in Physics. Phys Rep 233:1-66, 1993.

[3] VS Vladimirov, IV Volovich, EI Zelenov. p-Adic Analysis and Mathematical Physics. Singapore: World Scientific, 1994.

[4] A Khrennikov. p-Adic Valued Distributions in Mathematical Physics. Dordrecht: Kluwer Academic Publishers, 1994.

[5] A Khrennikov. Non-Archimedean Analysis: Quantum Paradoxes, Dynamical Systems and Biological Models. Dordrecht: Kluwer Academic Publishers, 1997.

[6] B Dragovich. On Some p-Adic Series with Factorials. In: WH Schikhof, C Perez-Garcia, J Kakol ed. p-Adic Functional Analysis. New York: Marcel Dekker, 1997, pp 95–105.

Hartogs-Stawski's theorem in discrete valued fields

MIKIHIKO ENDO* Dep. of Math., Rikkyo Univ., Tokyo 171-0021, Japan.

1 INTRODUCTION

Non-archimedean Hartogs theorem is one of the interesting topics in non-archimedean analysis. It was first proved by Stawski [8] in 1965.

Let K be a complete but not locally compact subfield of the p-adic complex numbers field C_p. It is well known that "*K is not locally compact*" iff "*$[K : Q_p]$ is infinite*". The latter is equivalent to "*the value group $|K^\times|$ is a dense subgroup of $R_+^\times$*" or "*the residue class field $k = O/M$ is infinite*". Stawski proved that, if $f(x_1, \ldots, x_n)$ is a function analytic for each variable defined on a product of open disks (poly disk), then f is an analytic function in the whole variables on the same domain. Stawski's proof is very interesting. But his proof has some defects in case the value group $|K^\times|$ is discrete. In fact the Theorem does not hold in that case [2].

In 1983, Stawski extended his previous results to a wider class of analytic functions, namely, to analytic elements whose domain of definition is an open annulus $U_{(R,S)} = \{x \in K \mid R < |x| < S\}$ (cf. Stawski [10]). In that paper, he found a new method for proving the analyticity of the coefficients of certain convergent sequences. But again I do not agree with his results. As an example shows (cf. the example in [2] or in Section 5 of this paper), the domain of convergence must be, in general, narrower than that of Stawski's theorem in case the value group $|K^\times|$ is discrete.

In the followings, we shall reproduce the proof in [9] again and show how far the Theorem holds. The main idea of the proof is the same as that of Stawski. But, we extend some results to a general complete subfield K of C_p. In fact, we extend the outer linear measure theory to general complete subfields K of C_p. Moreover, in our proof, we do not use the notion of regular sets and regular coverings. We mainly consider the case when the value group $|K^\times|$ is discrete. But it also gives an

* The research was partly suppoted by Grant-in-Aid for Sientific Research (No 40062616), Ministry of Education, Science and Culture, Japan.

alternative proof of the Theorem in case the value group $|K^\times|$ is dense (Theorem 6).

We notice that results in [10] also hold in case the value group $|K^\times|$ is dense, since the proof in [10] is essentially the same to that in [9]. When the value group $|K^\times|$ is discrete, we need to change the domain of analyticity in his Theorems.

Let $O = \{x \in K \big| |x| \leqslant 1\}$ be the ring of p-adic integers in K and $M = \{x \in K \big| |x| < 1\}$ be the maximal ideal of O. We assume that the value group $|K^\times|$ is discrete. Hence, the ring O is a principal ideal domain. We may write $M = \pi O$ for some $\pi \in O$. We call such an element π *a prime element* of K. Clearly, the value $|\pi|$ (< 1) is a generating element of the multiplicative group $|K^\times|$.

Let $f(x) = f(x_1, \ldots, x_n)$ be a function defined on a product $U_1 \times \cdots \times U_n$ of disks U_i. The function f is called *an analytic function* in the variables $x_1, \cdots, x_n$ on the domain $U_1 \times \cdots \times U_n$, if

(1) $f(x_1, \ldots, x_n)$ is expanded as a power series

$$f(x_1, \ldots, x_n) = \sum a_{i_1, \cdots, i_n} x_1^{i_1} \cdots x_n^{i_n},$$

and

(2) for any $(x_1, \cdots, x_n) \in U_1 \times \cdots \times U_n$,

$$a_{i_1, \cdots, i_n} x_1^{i_1} \cdots x_n^{i_n} \to 0 \quad \text{as} \quad i_1 + \cdots + i_n \to \infty.$$

The function f is called *a function analytic for each variable* or *a function holomorphic separately for each variable*, if, for any k $(1 \leqslant k \leqslant n)$ and for any $c_i \in U_i$ $(i \neq k)$, the function $f(c_1, \cdots, c_{k-1}, x_k, c_{k+1}, \cdots, c_n)$ is an analytic function in the variable $x_k \in U_k$.

Let D be a disk of the form $\{x \in K \big| |x - a| \leqslant r\}$ (resp. $\{x \in K \big| |x - a| < r\}$). We will denote the similar disk in C_p by $D^* = \{x^* \in C_p \big| |x^* - a| \leqslant r\}$ (resp. $\{x^* \in C_p \big| |x^* - a| < r\}$). In the same way, let D be an anulus in K of the form $\{x \in K \big| r' \leqslant |x - a| \leqslant r\}$ (resp. $\{x \in K \big| r' < |x - a| < r\}$). We will denote the annulus in C_p by $D^* = \{x^* \in C_p \big| r' \leqslant |x^* - a| \leqslant r\}$ (resp. $\{x^* \in C_p \big| r' < |x^* - a| < r\}$). If a function $f(x)$ is defined on a polydisk (resp. polyannulus) $D_1 \times D_2 \times \cdots \times D_n$, we denote by $f^* = f(x^*)$ the function defined on the polydisk (resp. polyannulus) $D_1^* \times D_2^* \times \cdots \times D_n^*$. If $f(x, y)$ is defined on the domain $C_1 \times \cdots \times C_m \times D_1 \times \cdots D_n$ $(x \in C_1 \times \cdots \times C_m$ and $y \in D_1 \times \cdots D_n)$, we also use the notation like $f(x, y^*)$ to denote the function defined on the domain $C_1 \times \cdots \times C_m \times D_1^* \times \cdots D_n^*$.

We also use the conventional notation $y^k = y_1^{k_1} \cdots y_n^{k_n}$ and $\|k\| = k_1 + \cdots + k_n$ for the variables $y = (y_1, \ldots, y_n)$ and indeces $k = (k_1, \ldots, k_n)$.

THEOREM 1 *Let K be a complete but not locally compact subfield of the p-adic complex number field C_p. We assume that the value group $|K^\times|$ is discrete. Let π be a prime element of K and $q = |\pi|^{-1} > 1$. Let $f(x_1, \ldots, x_n)$ be a function defined on the domain $|x_1| \leqslant R_1, \ldots, |x_n| \leqslant R_n$ (we assume that all R_i belong to the value group $|K^\times|$). If $f(x_1, \ldots, x_n)$ is analytic for each variable on the domain, then,* *for arbitrary indices j and k, the function f^* is an analytic function in the whole variables on the domain*

$$|x_j^*| \leqslant q^{-1}R_j, |x_k^*| < R_k, \quad \textit{and} \quad |x_i^*| < q^{-1}R_i(\textit{ for all other } i \neq j, k).$$

We may take the indeces j and k arbitrary and have the symmetric results. This gives us a useful result for the convergence radius of $f(x_1^*, \ldots, x_n^*)$ (cf. Theorem 7). We proceed the proof of Theorem 1 in Section 4.

We express our heartfelt thanks to Professor Alain Escassut for his various comments and kind advice. In particular, thanks to his comments, the logical structure of section 2 became clearer and neater.

We also express our cordial thanks to Professor Andrei Khrennikov for the fruitful discussion at Rikkyo University. He first informed me about Stawski's paper [9].

2 OUTER MEASURE THEORY

Let X be a non-archimedean metric space with the metric ρ. Let $Q = \{\rho(a,x) \big| x \in X\}$ be the set of distances. We assume that the set Q does not depend on the choice of the special element a. We call a subset C of X a *closed (resp. open) disk of radius R* if C is a set of the type $\{x \in X \big| \rho(a,x) \leqslant \text{(resp. } <) R\}$ for some $R \in Q$ and we put $r(C) = R$. Sometimes we call the element a a *center* of C. In non-archimedean geometry, " *a is a center of a disk* " simply means that " *a belongs to the disk* ". In this section we mainly investigate families $V = \{C_i \big| i \in I\}$ of disks which satisfy the following conditions :

(1) $r(C_i) > 0$ for all $i \in I$ (i.e. each C_i is not a single point),
(2) if $i \neq j$, then $C_i \cap C_j = \emptyset$ (mutually disjoint),
(3) $I \subset N$ (i.e. the index set I is a countable set).

Later we also assume the condition

(4) C_i are closed disks.

But, for a while, we only assume the conditions (1),(2) and (3).

For such a family V, we define *the radius* of V by $r(V) = \sum_{i \in I} r(C_i)$. In the following we mainly investigate the case when $r(V)$ is finite. We use the notation $u(V)$ to denote the union of the disks in V : $u(V) = \bigcup_{C \in V} C$. Thus, for a subset E of X, V is a covering of E iff $E \subset u(V)$.

We define *the (outer linear) measure* mes(E) of E (Stawski [8]) by

$$\text{mes}(E) = \inf_{E \subset u(V)} r(V).$$

If the set of distances Q is a dense subset of the positive real numbers $R_+^\times$ and $C = \{x \in X \big| \rho(a,x) < R\}$ is an open disk of radius $R = r(C)$, we put $\tilde{C} = \{x \in X \big| \rho(a,x) \leqslant R\}$ and call it the *envelope* of the disk C. This is the smallest "*closed*" disk containing C. Topologically, X is a totally disconnected space. Hence every open disk is a closed set. Therefore, in general, the *envelope* $\tilde{C}$ is bigger than the topological closure of the disk C.

If a family of disks $V = \{C_\imath \big| i \in I\}$ contains open disks, we exchange open disks $C_\imath$ by their envelope $\tilde{C}_\imath$ and get a new family $\tilde{V}$ of closed disks. It is clear that $r(C_\imath) = r(\tilde{C}_\imath)$. It may happen that a closed disk $\tilde{C}_\imath$ contains several open disks and so, in general, we have that

$$r(\tilde{V}) \leqslant r(V).$$

Given two families $V = \{C_\imath \big| i \in I\}$ and $W = \{D_\jmath \big| j \in J\}$, we define a new relation

$$V \prec W,$$

if, for any $C_\imath$ in V, there exists a disk $D_\jmath$ in W such that $C_\imath \subset D_\jmath$. Clearly the relation $\prec$ is different from the usual inclusion relation $\subset$.

If $V = \{C_\imath \big| i \in I\}$ and $W = \{D_\jmath \big| j \in J\}$ are two families of disks satisfying the above conditions (1),(2) and (3), we define *a strict union* $V \vee W$ as the smallest family *containing* V and W (i.e. $V \prec V \vee W$ and $W \prec V \vee W$) satisfying the conditions (1), (2) and (3). The existance of $V \vee W$ is clear. In fact, we can define the family $V \vee W$ as a subset of $V \cup W$ in the following way :

$$\begin{aligned} &\text{A disk } C_\imath \text{ belongs to } V \vee W \text{iff} && \text{(a) } C_\imath \cap D_\jmath = \emptyset \text{ for all } j \in J, \\ & && \text{or (b) } C_\imath \supset D_\jmath \text{ for some } j \in J. \\ &\text{As the same way,} && \text{(c) } C_\imath \cap D_\jmath = \emptyset \text{ for all } i \in I, \\ &\text{a disk } D_\jmath \text{ belongs to } V \vee W \text{iff} && \text{(d) } C_\imath \subsetneq D_\jmath \text{ for some } i \in I. \end{aligned}$$

It seems that the conditions (b) and (d) are not symmetric. But it is only outward appearance. If $C_\imath = D_\jmath$ for some $i \in I$ and $j \in J$, the procedure asks us to take $C_\imath$ as an element of $V \vee W$ and to remove $D_\jmath$ from $V \vee W$. Thus, if we change the symbol $\supset$ by $\supsetneq$ in (b) and the symbol $\subsetneq$ by $\subset$ in (d), we have the same family $V \vee W$.

We also define *a strict intersection* $V \wedge W$ by

$$V \wedge W = \{C_\imath \cap D_\jmath \big| C_\imath \cap D_\jmath \neq \emptyset\}.$$

It is clear that, if V and W are covering of a set E, then the strict intersection $V \wedge W$ is also a covering of E.

Let

$$V_1 \prec V_2 \prec \cdots \prec V_n \prec \cdots$$

be an increasing sequence of families of disks with respect to the relation $\prec$: $V_n = \{C_{n,\imath} \big| i \in I_n\}$. For simplicity, we assume that the radii $r(V_n)$ are bounded. For each $C_{n,\imath} \in V_n$, there exists a disk $C_{n+1,\jmath} \in V_{n+1}$ such that $C_{n,\imath} \subset C_{n+1,\jmath}$. Let $W = \bigcup_{n=1}^{\infty} V_n = \{C_{n,\imath} \big| i \in \bigcup_{n=1}^{\infty} I_n\}$ the family of all the disks $C_{n,\imath}$ in V_n $(n = 1, 2, \ldots)$. We define an equivalence relation $\sim$ in W in the following way:

$$C_{n,\imath} \sim C_{m,\jmath} \underset{def}{\Longleftrightarrow} \exists N \geqslant m, n;\ \exists k \in I_N;\ C_{n,\imath},\ C_{m,\jmath} \subset C_{N,k}.$$

i.e. there exists a large disk $C_{N,k}$ containing both disks $C_{n,k}$ and $C_{m,\jmath}$.

The transitivity of the relation $\sim$ is a consequence of the strong triangle inequality. Put $V = W/\sim = \{D_i \mid i \in I\}$, where each D_i is the union of all the disks in an equivalence class. We shall show that

(1) $r(D_i) > 0$,
(2) *the index set I is countable,*
(3) *$i \neq j$ implies $D_i \cap D_j = \emptyset$,*
(4) *D_i is an open disk or a closed disk.*

(1) and (2) are trivial. In fact, the index set I is a partition of the countable set $\cup_{n=1}^{\infty} I_n$ into equivalence classes. If we show (4), then (3) is also clear from the definition. So we only show (4). Let D be such a class D_i. Let $R = \sup_{x,y \in D} \rho(x,y)$.

[1] If there exist elements $x, y \in D$ for which $R = \rho(x,y)$, then $x \in C_{n,i}$ and $y \in C_{m,j}$ for some $C_{n,i}$, $C_{m,j}$ and $C_{n,i} \sim C_{m,j}$. There exists a disk $C_{N,k}$ which contains both $C_{n,i}$ and $C_{m,j}$. Then, from $r(C_{N,k}) \geqslant \rho(x,y)$, we see that $r(C_{N,k}) = R$. Thus in this case $D = C_{N,k}$ is a closed disk.

[2] On the other hand, if there does not exist such x, y in D, we take an arbitrary small positive number $\epsilon > 0$. There exist elements x, y in D such that $\rho(x,y) \geqslant R - \epsilon$. In the same way as above, we see that there exists a disk $C_{N,k}$ which contains both x, y. Thus $r(C_{N,k}) \geqslant R - \epsilon$. This shows that D is an open disk of radius R.

In both cases, it is clear that for any $\epsilon > 0$ there exists a sufficiently large $C_{N,k} \subset D$ for which $r(C_{N,k}) > r(D) - \epsilon$.

We call the system $V = W/\sim$ *the strict union of the increasing sequence* $V_1 \prec V_2 \prec \cdots \prec V_n \prec \cdots$ and denote it as $V = \vee_{n=1}^{\infty} V_n$. We now show that, if $r(V_n) \leqslant R$ for all n, then $r(V) \leqslant R$. On the contrary, if $r(V) = \sum_{i \in I} r(D_i) > R$, there exists a finite subset I_0 of I such that $\sum_{i \in I_0} r(D_i) > R$. Put $\delta = \sum_{i \in I_0} r(D_i) - R$ and $\epsilon = \frac{\delta}{M}$, where $M = \#(I_0)$ (the cardinal of the finite set I_0). For each $D_i (i \in I_0)$, take a disk $C_{n(i),i} \in I_{n(i)}$ such that $r(C_{n(i),i}) > r(D_i) - \epsilon$. Taking N so large that $N > n(i)$ for all $i \in I_0$, we see that $\sum_{i \in I_N} r(C_{N,i}) > \sum_{i \in I_0} r(D_i) - M \cdot \epsilon > R$. This contradicts the assumption $r(V_N) \leqslant R$. Putting these facts together, we get

LEMMA 1 *Let*

$$V_1 \prec V_2 \prec \cdots \prec V_n \prec \cdots$$

be an increasing sequence of families of closed disks. Then the strict union $V = \vee_{n=1}^{\infty} V_n$ is a system of open or closed disks. If $r(V_n) \leqslant R$ for all n, then $r(V) \leqslant R$.

LEMMA 2 *Let $V = \{C_i \mid i \in I\}$ and $W = \{D_j \mid j \in J\}$ be families of closed disks. Then $r(V \vee W) + r(V \wedge W) = r(V) + r(W)$.*

Proof. Let $C_i \in V$ and $D_j \in W$. Only three cases may happen:

[1] $C_i \cap D_j = \emptyset$,
[2] $C_i \supset D_j$,
[3] $C_i \subset D_j$.

This shows that, in any case, $r(C_i \cup D_j) + r(C_i \cap D_j) = r(C_i) + r(D_j)$. From this, Lemma 2 is easily deduced.

COROLLARY 1 $\mathrm{mes}(V \vee W) \leqslant r(V \vee W) = r(V) + r(W) - r(V \wedge W)$.

THEOREM 2 *Let $E_1 \subset E_2 \subset \cdots \subset E_n \subset \cdots$ be an increasing sequence of subsets of X. Then*

$$\mathrm{mes}(\bigcup_{n=1}^{\infty} E_n) = \lim_{n\to\infty} \mathrm{mes}(E_n).$$

Proof. Since $E_n \subset \bigcup_{i=1}^{\infty} E_i$, it is clear that

$$\mathrm{mes}(\bigcup_{i=1}^{\infty} E_i) \geqslant \lim_{n\to\infty} \mathrm{mes}(E_n).$$

Conversely, we shall show that, if $R = \lim_{n\to\infty} \mathrm{mes}(E_n)$, then $\mathrm{mes}(\bigcup_{i=1}^{\infty} E_i) \leqslant R$. We put $R_n = \mathrm{mes}(E_n)$. Then it is clear that

$$R_1 \leqslant R_2 \leqslant \cdots \leqslant R_n \leqslant \cdots.$$

For any positive $\epsilon > 0$, there exist coverings $V_n = \{C_{n,i} \big| i \in I_n\}$ of E_n such that

$$R_n = \mathrm{mes}(E_n) \leqslant r(V_n) \leqslant R_n + \frac{\epsilon}{2^n}.$$

By the previous Lemma, we see that

$$\begin{aligned} r(V_1 \vee V_2) &= r(V_1) + r(V_2) - r(V_1 \wedge V_2) \\ &= R_1 + \frac{\epsilon}{2} + R_2 + \frac{\epsilon}{2^2} - r(V_1 \wedge V_2). \end{aligned}$$

Since $E_1 \subset V_1 \wedge V_2$, $R_1 = \mathrm{mes}(E_1) \leqslant r(V_1 \wedge V_2)$. We have $r(V_1 \vee V_2) \leqslant R_2 + (1 - \frac{1}{2^2})\epsilon$.

Assume that there holds the inductive hypothesis

$$r(\vee_{i=1}^{n} V_i) \leqslant R_n + (1 - \frac{1}{2^n})\epsilon.$$

Then, by induction, we see that

$$\begin{aligned} r(\vee_{i=1}^{n+1} V_i) &= r(\vee_{i=1}^{n} V_i) + r(V_{n+1}) - r(V_{n+1} \wedge \vee_{i=1}^{n} V_i) \\ &\leqslant R_n + (1 - \frac{1}{2^n})\epsilon + R_{n+1} + \frac{\epsilon}{2^{n+1}} - R_n = R_{n+1} + (1 - \frac{1}{2^{n+1}})\epsilon. \end{aligned}$$

Put $W_n = \vee_{i=1}^{n} V_i$. Then $W_1 \prec W_2 \prec \cdots$. Put $V = \vee_{n=1}^{\infty} W_n$. Then, $V = \{D_i \big| i \in I\}$ is a covering of the set $\bigcup_{i=1}^{\infty} E_i$. Since $r(\vee_{i=1}^{n} V_i) \leqslant R + \epsilon$, $r(V) \leqslant R + \epsilon$ (Lemma 1), we see that

$$\mathrm{mes}(\bigcup_{n=1}^{\infty} E_n) \leqslant r(V) \leqslant R + \epsilon.$$

Finally, since $\epsilon > 0$ is arbitrary, this ends the proof.

REMARK In the proof of Theorem 2, we did not use the completeness of the space X. But, if the space X is not complete, it may happen that the unit disk (and then every disk in X) has measure 0. So we show here that in our case $X = K$ (K is a complete but not locally compact subfield of C_p), the measure of a disk is equal to its radius.

Let X be a complete metric space with metric ρ. We assume that X satisfies one of the following properties :

(1) The set of distances $Q = \{\rho(a,x) \mid x \in X\}$ is dense in the set of the positive real numbers $R_+^\times$ and Q does not depend on the choice of the element a.

(2) The set Q is discrete (that is, Q has no limit points except zero) and does not depend on the choice of the element a. If R_1 and R are numbers in Q such that $R_1 < R$, then every closed disk C of radius R contains infinitely many closed disks of radius R_1.

We first show that even in case (1) every closed disk C of radius R contains infinitely many closed disks of radius R_1.

LEMMA 3 *Every closed disk C of radius R contains at least two closed disks of radius R_1.*

Proof. Let $C = \{x \mid \rho(a,x) \leqslant R\}$ and $C_1 = \{x \mid \rho(a,x) \leqslant R_1\}$. Since R belongs to the set Q, C contains a point b such that $\rho(a,b) = R$. Then the disk $C_2 = \{x \mid \rho(b,x) \leqslant R_1\}$ is contained in C and disjoint with C_1 : $C_1 \cap C_2 = \emptyset$.

COROLLARY 2 *If the set Q of distances is dense in $R_+^\times$, then every closed disk C of radius R contains infinitely many closed disks of radius R_1.*

Proof. For any natural number n, take $R_2, R_3, \ldots, R_n$ in Q such that $R_1 < R_2 < \cdots < R_n < R$. By Lemma 3, each disk of radius R_i contains at least two disks of radius R_{i-1}. This shows that the disk C contains at least 2^n disks of radius R_1. Since n is arbitrary, Corollary 2 is proved.

LEMMA 4 *Let X be a complete metric space satisfying the above conditions (1) or (2). A closed disk C of radius R is not covered by a finite number of closed disks $\{C_i \mid i = 1, 2, \cdots, n\}$ of smaller radius : $r(C_i) < R$.*

Proof. Assume that C is covered by $\{C_i \mid i = 1, 2, \cdots, m\}$. Put $R_1 = \max_i r(C_i)$. Clearly $R_1 < R$. Corollary 2 says that there exist in C an infinite number of closed disks of radius R_1. And so C is not covered by the finite system $\{C_i\}$. A contradiction!

LEMMA 5 *Let C be a closed disk of radius R. Let $\{C_i \mid i = 1, 2, \cdots\}$ be a countable family of closed disks such that $r(C_i) < R$ for all i and $\sum_{i=1}^{\infty} r(C_i) < \infty$. Then C is not covered by the disks $\{C_i \mid i = 1, 2, \cdots\}$.*

Proof. Let $R_0 = \sum_{i=1}^{\infty} r(C_i)$. Choose $R_0 > R_1 > \cdots > R_n > \cdots$ in Q such that $\lim_{n\to\infty} R_n = 0$. Without loss of generality we may assume that $R_0 > R$.

Let n_1 be a natural number such that $\sum_{\imath=1}^{n_1} r(C_\imath) > R_0 - R_1$. Then, $\bigcup_{\imath=1}^{n_1} C_\imath$ does not cover C (by Lemma 4). And there exists a closed disk D_1 contained in C such that $r(D_1) = R_1$ and $D_1 \cap (\bigcup_{\imath=1}^{n_1} C_\imath) = \emptyset$. Next, take $n_2 > n_1$ such that $\sum_{\imath=1}^{n_2} r(C_\imath) > R_0 - R_2$. Since $\sum_{\imath=n_1+1}^{n_2} r(C_\imath) < R_1$, the union $\bigcup_{\imath=n_1+1}^{n_2} C_\imath$ does not cover the disk D_1. Again, by Lemma 4, we find a closed disk D_2 of radius R_2 such that $D_2 \subset D_1$ and $D_2 \cap (\bigcup_{\imath=n_1+1}^{n_2} C_\imath) = \emptyset$. Since $D_2 \subset D_1$, we have $D_2 \cap (\bigcap_{\imath=1}^{n_2} C_\imath) = \emptyset$. In this way, we get a sequence of closed disks

$$D_1 \supset D_2 \supset \cdots \supset D_n \supset \cdots$$

such that $D_k \cap (\bigcup_{\imath=1}^{n_k} C_\imath) = \emptyset$. Since the space X is complete and since $\lim_{k\to\infty} r(D_k) = 0$, the intersection $\bigcap_{k=1}^{\infty} D_k$ is not empty and does not belong to any of $C_\imath$.

This Lemma guarantees that, if the field K is complete but not locally compact, the measure of a closed disk is equal to the radius of the disk.

It is easy to see that, if the field K is locally compact, Stawski's outer measure is the invariant Haar measure of K. Thus, we have seen that, if K is a complete subfield of C_p, the linear outer measure of a closed disk is always the same as its radius.

We give one more elementary remark which is useful in the proof of main Theorem. Let C^* be a (closed or open) disk in C_p. Put $C = C^* \cap K$. Then, clearly, $r(C) \leqslant r(C^*)$. Of course, $r(C)$ is the radius of the disk C in the space K and $r(C^*)$ is the radius of the disk C^* in the space C_p. More generally, for $V^* = \{C_\imath^* \big| i \in I\}$ is a family of disks in C_p, we put

$$V = V^* \wedge \{K\} = \{C_\imath \equiv C_\imath^* \cap K \big| C_\imath^* \cap K \neq \emptyset\}.$$

Then it also holds that

$$r(V) = \sum_\imath r(C_\imath) \leqslant \sum_\imath r(C_\imath^*) = r(V^*).$$

Now, if V^* is a covering of a set E^* in the space C_p, then $V = V^* \wedge \{K\}$ is a covering of the subset $E = E^* \cap K$ in the space K. And so it holds that

$$\text{mes } E = \inf_V r(V) \leqslant \inf_{V^*} r(V^*) = \text{mes } E^*.$$

At the end of this section, we give an example such that the space X is not complete and the measure of the unit disk is zero.

EXAMPLE Let $X = Q$ be the rational number field with p-adic topology. So, X is not complete. Let $O = \{a \in Q \big| |a|_p \leqslant 1\}$ be the unit disk, i.e. the ring of the p-adic integers in Q. O is a countable set. Let $a_1, a_2, \cdots, a_n, \cdots$ be a numbering of the elements of O. Let r be an arbitrary positive number and C_n be the disk of the center a_n and radius $\frac{r}{2^n}$. From the covering $\{C_n \big| n \in N\}$ we select a proper covering $V = \{C_\imath \big| i \in I\}$ in the following way:

The first disk C_1 belongs to V. If $(\bigcup_{i=1}^{n-1} C_i) \cap C_n = \emptyset$, then C_n belongs to V. Else, $C_n \subset C_i$ for some $i < n$, so we remove it from V. In this way we get a proper covering V of O. Since

$$r(V) = \sum_{i \in I} r(C_i) \leqslant \sum_{n=1}^{\infty} r(C_n) = r,$$

and since r is arbitrary, the outer measure of O must be zero.

3 PREPARATION THEOREMS

THEOREM 3 (Weierstrass Preparation Theorem) *Let O be a complete local ring with the maximal ideal M. Let*

$$f(x) = \sum_{i=0}^{\infty} a_i x^i \in O[[x]]. \tag{1}$$

Assume that $a_0, \ldots, a_{n-1} \in M$ and $a_n \in O^\times = O \setminus M$ (units group). Then we can expand $f(x)$ as a product of a polynomial of degree n and of an unit $u(x)$ in $O[[x]]$ such that

$$f(x) = (x^n + b_{n-1}x^{n-1} + \cdots + b_0)u(x), \quad b_i \in M.$$

For the proof, one see, for example, Lang [5], Koblitz [4] (p. 97 Theorem 14) or Lazard [6] (p. 229, Proposition 2).

For our later use, we give here some elementary remarks.

[1] Let R be the convergence radius of $f(x)$ and r be a real number such that $r < R$. There exists a finite number of coefficients (in general, one coefficient) $a_n, \ldots, a_N$ such that

$$\|f\|_r = \max_k |a_k| r^k = |a_n| r^n = \cdots = |a_N| r^N$$

and $|a_k| r^k < \|f\|_r$ for $k < n$ or $k > N$. We put $n = n(r) = n(f, r)$ and $N = N(r) = N(f, R)$. The functions $n(r)$ and $N(r)$ are increasing step functions. $n(r)$ is a left continuous function of r and $N(r)$ is a right continuous function of r.

[2] If $r \in |K^\times|$, then $\|f\|_r = \max_k |a_k| r^k = \sup_{|x| \leqslant r} |f(x)| = \sup_{|x|=r} |f(x)|$ (cf. Schikhof [11], pp.121-123). If $f(x)$ has no zeros on the annulus $\Gamma_r = \{x \in C_p \big| |x| = r\}$, then $\|f\|_r = \max_n |a_n||x|^n = |f(x)|$ for any $x \in \Gamma_r$ (cf. Escassut [3] (Theorem 23.3)).

[3] Taking logarithm of $\|f\|_r = |a_N| r^N$, we get

$$\ln \|f\|_r = \ln |a_{N(r)}| + N(r) \ln r.$$

Let $r_1 < r_2 < \cdots < r_m < \cdots$ be the non continuous points of $N(r)$ (and of $n(r)$). Then, for $r_i < r < r_{i+1}$, $N(r) = n(r) =$ constant. Thus, for any s_1, s_2 such that $r_i < s_1 < s_2 < r_{i+1}$, we have

$$\begin{aligned} \ln \|f\|_{s_2} - \ln \|f\|_{s_1} &= N(s_2) \ln s_2 - N(s_1) \ln s_1 \\ &= N(s_1) \int_{s_1}^{s_2} \frac{1}{r} dr = \int_{s_1}^{s_2} \frac{N(r)}{r} dr. \end{aligned}$$

Especially, for any small ϵ, δ, it hold

$$\ln \|f\|_{r_\imath+1-\epsilon} - \ln \|f\|_{r_\imath+\delta} = \int_{r_\imath+\delta}^{r_\imath+1-\epsilon} \frac{N(r)}{r} dr.$$

If the value group $|K^\times|$ is dence, $\|f\|_r$ is a continuous function in the variable r. So, summing up term by term, we get

$$\ln \|f\|_{R_2} - \ln \|f\|_{R_1} = \int_{R_1}^{R_2} \frac{N(r)}{r} dr,$$

for any $R_1 < R_2$. When $f(0) \neq 0$, we get from this

$$\int_0^R \frac{N(r)}{r} dr = \ln \|f\|_R - \ln \|f\|_0 = \ln \|f\|_R - \ln |f(0)|. \tag{2}$$

When the value group $|K^\times|$ is discrete, $\|f\|_r$ is not continuous and the left hand side does not cancel. In this case we only have the inequality :

$$\int_0^R \frac{N(r)}{r} dr \leqslant \ln \|f\|_R - \ln \|f\|_0 = \ln \|f\|_R - \ln |f(0)|. \tag{3}$$

These results are also given in term of valuation functions in Escassut [3] (cf. Theorem 23.13).

THEOREM 4 (non-archimedean version of Cartan's theorem) *Let c be an arbitrary element in K and let $D_R = \{x \in K \mid |x - c| < R\}$ be an open disk of radius R. Given arbitrary $r < R$ and points $a_1, \ldots, a_n$ in D_R (we may allow repetition of the same $a_\imath$), there exists a system of open disks in D_R such that*

(1) the sum of the radii of disks is equal to r,

(2) if x does not belong to any of the disks, then

$$|x - a_1| \cdot |x - a_2| \cdots |x - a_n| > \left(\frac{r}{e}\right)^n.$$

This Theorem is a non-archimedean version of Theorem 10 in [7]. The proof is easier than in the archimedean case (cf. [9]). It is also easy to see that this Theorem holds in any ultrametric space X. Of couse, we must change $|x-a|$ by $\rho(x,a)$, where ρ is the metric in the space X.

THEOREM 5 (Hartogs series) *Let $f(x,y) = f(x, y_1, \cdots, y_m) = \sum_k f_k(x) y^k$ be a function analytic for each variable on the domain $|x| \leqslant R$ and $|y| \leqslant S$ ($|y_1| \leqslant S_1, \cdots, |y_m| \leqslant S_m$). If $f(x,y)$ is bounded on this domain, then the coefficients $f_k(x)$ are all analytic functions in the single variables x on the domain $|x| \leqslant R$. The functions $f_k(x^*)$ are also analytic on the domain $|x^*| \leqslant R$.*

Proof.

[1] We prove the Theorem in case y is a single variable ($m = 1$). General cases are easily deduced from this. Moreover, without loss of generality, we may assume that $R = 1$ and $S = 1$. Else, we change the variables x by Rx and y_i by Sy and get the desired condition. Let $|f(x,y)| < M$. For any fixed $x(|x| \leqslant R)$, the function $f(x,y) = \sum_{k=1}^{\infty} f_k(x)y^k$ is an analytic function in the variable y. Hence by the maximum principle, we have $|f_k(x)| < M$. Now putting $y = 0$, we see that $f_0(x)$ is analytic function in the variable x. Assume that $f_0(x), \cdots, f_{k-1}(x)$ are known to be analytic. Then, $f_0(x), f_1(x)y, \cdots, f_{k-1}(x)y^{k-1}$ are all analytic functions in the variables x, y. Now we get

$$\left| \frac{1}{y^k} \left[f(x,y) - \{f_0(x) + f_1(x)y + \cdots + f_{k-1}(x)y^{k-1}\} \right] - f_k(x) \right|$$
$$= |y| \, |f_{k+1}(x) + f_{k+2}(x)y + \cdots| < |y|M.$$

Thus, as a uniform limit of analytic functions, the function $f_k(x)$ itself is an analytic function.

[2] The general case. If $y = (y_1, \ldots, y_m)$, then we expand $f(x,y)$ as

$$f(x,y) = \sum_{l=1}^{\infty} \varphi_l(x, y_1, \ldots, y_{m-1}) y_m^l.$$

Then, above discussion shows that $\varphi_l(x, y_1, \ldots, y_{n-1})$ are bounded by M and are analytic for each variable. Then, by induction on the number m of variables, we prove that the coefficients $f_k(x)$ are all analytic functions.

[3] Let R be in the value group $|K^\times|$. Let $f(x) = \sum_k a_k x^k$ be an analytic function on the closed disk

$$D_R = \{x \in K \mid |x| \leqslant R\}.$$

Then

$$\lim_{k\to\infty} a_k x^k = 0 \quad \text{for all} \quad x \in D_R.$$

Since the values R is in $|K^\times|$, this condition is equivalent to the condition

$$\lim_{k\to\infty} |a_k| \cdot R^k = 0.$$

Thus, the function $f(x^*)$ is also analytic on the disk

$$D_R^* = \{x^* \in C_p \mid |x^*| \leqslant R\}.$$

Hence, the last assertion of theorem 5 follows.

In [10], Stawski found a more elegant method to prove the analyticity of the coefficients $f_k(x)$. It is very useful when we deal with analytic elements whose domain of definition is an annulus $U_{<R,S>}$ ($<R, S>$ means an open interval (R, S) or a closed interval $[R, S]$, as the case may be).

4 PROOF OF THEOREM

We shall proceed the proof of Theorem 1 by induction on the number n of variables.

In case $n = 1$, the Theorem is trivial.

Assume that the Theorem holds in n variables case.

Let $f(x, y)$ be a function of $n+1$ variables $x, y_1, \cdots, y_n$ which is analytic for each variable on the domain

$$|x| \leqslant R, |y| \leqslant S \quad (\text{ i.e. } \quad |y_1| < S_1, \ldots, |y_n| < S_n).$$

We fix the variable x in this domain. Then, the function $f(x, y)$ is a function analytic for each variable on the domain

$$|y_1| \leqslant S_1, \cdots, |y_n| \leqslant S_n.$$

By inductive hypothesis, the function $f(x, y^*)$ is an analytic function on the domain

$$|y_j^*| < S_j, |y_i^*| < \frac{1}{q} S_i \qquad (\text{ for other } \quad i \neq j).$$

Without loss of generality, we may assume that $j = n$.

Take a positive constant $\eta(< 1)$ and put

$$T_1 = \frac{\eta}{q} S_1, \cdots, T_{n-1} = \frac{\eta}{q} S_{n-1}, T_n = \eta S_n.$$

Let

$$U^* = \{y^* \in C_p^n \big| |y_1^*| \leqslant T_1, \ldots, |y_n^*| \leqslant T_n\}$$

and $U = U^* \cap K^n$. Then the function $f(x, y^*)$ is analytic in the variable y^* on the domain U^* and so we may expand the function as

$$f(x, y^*) = \sum_k f_k(x) y^{*k} \qquad \text{where} \quad y^* = y_1^{*k_1} \cdots y_n^{*k_n}.$$

Changing variables y^* by certain constant multiples (for example, y_1^* by $T_1 y_1^*$, etc.), we may assume that

$$T_1 = \cdots = T_{n-1} = T_n = 1.$$

In the same way, we may assume that $R_1 \equiv \frac{1}{q} R = 1$. We shall now show that $f(x^*, y^*)$ is analytic on the domain

$$|x^*| \leqslant 1, \ |y^*| < 1.$$

Since $\eta(< 1)$ is arbitrary, this will prove Theorem 1.

First, we shall show that there exist a point y_0 and a positive number ρ such that $f(x, y)$ is bounded on the domain $x \in O$ and $|y - y_0| \leqslant \rho$.

Assume that it does not hold.

For any natural number N, put $D_N = \{y \in U \big| |f(x,y)| \leqslant N (\forall x \in O)\}$. If D_1 covers U, then $f(x,y)$ is bounded : $|f(x,y)| \leqslant 1$. Thus, by assumption, D_1 does not cover whole U and $U \setminus D_1$ is a non-empty open set. Hence there exists an open disk $\sigma_1 \subset U \setminus D_1$. In the same way, D_2 does not cover σ_1 and therefore there exists an open disk $\sigma_2 \subset \sigma_1$ such that $\sigma_2 \subset U \setminus D_2$, and so on. We may assume that $\lim_{i\to\infty} r(\sigma_i) = 0$. Then the sequence $\sigma_1 \supset \sigma_2 \supset \cdots$ is a system of non-empty closed sets (an open disk is topologically a closed set). By the completeness of K, the intersection $\cap\, \sigma_i$ is not empty. Then for the element $y_0 \in \cap\, \sigma_i$, the values $|f(x,y_0)|$ $(x \in O)$ are not bounded. Since $f(x,y_0)$ is an analytic function in the variable x, this contradict the maximum principle (cf. [11] p.121 Theorem 42.2).

Changing y by $y - y_0$ and f by some constant multiple, we may assume that

$$|y| \leqslant \rho \quad \text{implies} \quad |f(x,y)| \leqslant 1 \quad (\text{for all} \quad x \in O).$$

By the maximum principle, this means that $|f_k(x)y^k| \leqslant 1$, i.e. $|f_k(x)| \leqslant \rho^{-\|k\|}$. Then, by Theorem 5, the coefficients $f_k(x)$ are analytic functions on O. And the functions $f_k(x^*)$ are analytic functions on O_p.

Further, we may assume $|f_k(0)| \geqslant e^{-\|k\|}$ for all k.

If not, let k' be these indices k for which the inequalities $f_k(0) \geqslant e^{-\|k\|}$ do not hold. We add $\sum_{k'} e^{-\|k'\|} y^{k'}$ to $f(x,y)$ and have a function satisfying the desired condition $|f_k(0)| \geqslant e^{-\|k\|}$ for all k.

Now, for simplicity, we put $N(f_k, r) = N_k(r)$. Then, by (2), we have

$$\ln \|f_k\|_R - \ln \|f_k\|_0 = \ln \|f_k\|_R - \ln |f_k(0)|$$

$$= \int_0^R \frac{N_k(r)}{r} dr \geqslant \int_{R_1}^R \frac{N_k(r)}{r} dr \geqslant \frac{R - R_1}{R} N_k(R_1).$$

Note that

$$\ln \|f_k\|_R \leqslant -\|k\| \ln \rho \quad \text{and} \quad \ln |f_k(0)| \geqslant -\|k\|.$$

Then

$$N_k(R_1) \leqslant \frac{R}{R - R_1}(-\ln \rho + 1)\|k\| \equiv A\|k\|, \tag{4}$$

for some constant A.

We now show that for any positive α there exists a large constant M such that, if $\|k\| \geqslant M$, then

$$|f_k(x)| \leqslant e^{\alpha\|k\|} \tag{5}$$

for all x satisfying $|x| \leqslant R_1 = 1$. Since K is not locally compact, this means that

$$\|f_k\|_1 \leqslant e^{\alpha\|k\|} \quad \text{i.e.} \quad |f_k(x^*)| \leqslant e^{\alpha\|k\|} \quad \text{for} \quad |x^*| \leqslant 1.$$

If the above inequality is proved, then the function $f(x^*, y^*)$ is analytic on the domain $|x^*| \leqslant R_1 = 1$ and $|y^*| < 1 = (1, \cdots, 1)$. In fact, take δ such that $\delta > \alpha > 0$. Then, for $|y^*| \leqslant e^{-\delta}$, it holds

$$|f_k(x^*) {y^*}^k| \leqslant e^{\alpha\|k\|} e^{-\delta\|k\|} = e^{(\alpha-\delta)\|k\|}.$$

And then, as a uniform limit of analytic functions, the function

$$f(x^*, y^*) = \sum_k f_k(x^*) {y^*}^k$$

is analytic (uniform closedness. cf. [11] p.123). We can take α and δ arbitrary small. This shows that $f(x^*, y^*)$ is analytic in the variables x^* and y^* on the domain $|x^*| \leqslant 1, |y^*| < 1$.

We now prove the inequality (5) by a reduction to an absurdity. If (5) does not holds, then there exists a positive α and an infinite subsequence $k_{i_1}, k_{i_2}, \ldots$ such that

$$||k_{i_1}|| < ||k_{i_2}|| < \cdots$$

satisfying

$$||f_k||_1 > e^{\alpha ||k||}. \tag{6}$$

Removing several k_i and renumbering the remaining $\{k_i\}$, we may assume that all k_l satisfy the inequality (6).

In the field C_p, the unit disk $O_p = \{x^* \in C_p \big| |x^*| \leqslant 1\}$ contains an infinite countable family of distinct open disks of radius 1, say $\{C_i^* \big| i \in I\}$. Put

$$J = \{i \in I \big| C_i = C_i^* \cap K \neq \emptyset\},$$
$$K = \{i \in I \big| C_i = C_i^* \cap K = \emptyset\}.$$

Since there are no elements $x \in K$ such that $q < |x| < 1$, J is the set of closed disks C_i of radius $\frac{1}{q}$ contained in O. We have pointed out in Section 2, the unit disk O contains infinitely many closed disks of radius $\frac{1}{q}$. Thus the set J is also countably infinite. Now, we see by (4) that $N_k(R_1) = N_k(1) \leqslant A||k||$. Thus, for any small positive ϵ, there exists a disk $C^* = C_{i_0}^*$ $(i_0 \in J)$ and infinitely many indices k_l such that the number of zeros of $f_{k_l}(x)$ which are in the disk C^* is not greater than $\epsilon ||k_l||$ (for all such k_l). We remove those indices k_j which do not satisfy this property. So we may assume that the number of zeros of each $f_{k_l}(x)$ is smaller than $\epsilon ||k_l||$. Now fix $r < \frac{1}{q}$. Since $\frac{r}{e} < 1$, we can take $\epsilon > 0$ so small that $\left(\frac{r}{e}\right)^\epsilon > e^{-\alpha}$. We may take C^* not equal to the open disk $\{x^* \in C_p \big| |x^*| < 1\}$.

Fix an index $k = k_l$. Let $a_{i,1}, \ldots, a_{i,s_i}$ be the roots of $f_k(x)$ in C_i^* $(i \in I)$. Put

$$P_i(x) = (x - a_{i,1}) \cdots (x - a_{i,s_i}).$$

Especially, we put

$$P(x) = P_{i_0}(x) \equiv (x - a_1) \cdots (x - a_s).$$

By the above discusion $s \equiv s_{i_0} < \epsilon ||k||$.Then

$$f_k(x) = P(x) P_{i_1}(x) \cdots P_{i_m}(x) g(x) = P(x) G_k(x),$$

where $g(x)$ has no roots in O_p. It is trivial that $P_{i_j}(x) \in O_p[x]$ and so $||P_{i_j}||_1 = 1 = |P_{i_j}(x)|$ for any $x \in C^*$. As the same way, $||P||_1 = 1$. Hence $||f||_1 = ||P||_1||G_k||_1 = ||G_k||_1 = ||g||_1$. Since $g(x)$ has no roots in O_p, $||g||_1 = |g(x)|$ for any $x \in C^*$, we have

$$|G_k(x)| = |P_{i_1}(x)\cdots P_{i_m}(x)g(x)| = ||G_k||_1 = ||f_k||_1 \quad (\forall x \in C^*)$$

(cf. Escassut [3], Lemma 4.2). Thus

$$|G_k(x)| > e^{\alpha||k||} \quad \text{for all} \quad x \in C^*.$$

Now put $C = C^* \cap K$. As we noted above, $r(C) = \frac{1}{q}$. Now, let

$$E_k = \{x \in C \big| |f_k(x)| \leqslant 1\}.$$

If $x \in E_k$, then

$$|P(x)| = \frac{|f_k(x)|}{|G_k(x)|} < e^{-\alpha||k||}. \tag{7}$$

By Cartan's theorem (Therorem 4), there exists a system V_k^* of closed disks in C^* such that $r(V_k^*) = r < \frac{1}{q}$ and

$$\prod_{j=1}^{s} |x - a_j| > (\frac{r}{e})^s \quad (\forall x \notin u(V_k^*)).$$

Put $V_k = V_k^* \wedge \{C\}$. If x does not belong to the union $u(V_k)$ of the disks in V_k, then

$$|P(x)| > (\frac{r}{e})^s.$$

Hence,

$$|P(x)| > (\frac{r}{e})^{\epsilon||k||} > e^{-\alpha||k||} \quad (\text{ for all } \; x \in C \setminus u(V_k)). \tag{8}$$

On the other hand by (7),

$$|P(x)| < e^{-\alpha||k||} \quad \text{if} \;\; x \in E_k.$$

Thus $E_k \subset V_k$ so that

$$\text{mes}(E_k) \leqslant r(E_k) \leqslant r(V_k) \leqslant r(V^*) = r < \frac{1}{q}.$$

If $s = 0$, that is, if $f_k(x)$ has no zeros on C^*, then

$$|f_k(x)| = ||f_k||_1 > e^{\alpha||k||} \quad (\text{ for all } \;\; x \in C^*).$$

Hence $E_k = \emptyset$. In this case, it trivially holds that $r(E_k) < r$ and mes $(E_{k_i}) \leqslant r < \frac{1}{q}$.
Thus, in either case,

$$\text{mes}\ (\bigcap_{i=m}^{\infty} E_{k_i}) \leqslant r.$$

Then, by Theorem 2, we have

$$\text{mes}\ (\bigcup_{m=1}^{\infty}\bigcap_{i=m}^{\infty} E_{k_i}) \leqslant r < \frac{1}{q}.$$

On the other hand, for any fixed element $x \in C$, $f(x, y^*)$ is an analytic function in the variables y^* on the domain $|y^*| \leqslant T = (1, \ldots, 1)$ and then

$$f(x, y^*) = \sum_k f_k(x) {y^*}^k$$

is a convergent power series for any $|y^*| \leqslant 1$. So, $f_k(x) \to 0$ as $||k|| \to \infty$. That is, any element $x \in C$ belongs to all E_k for sufficiently large $||k||$. Or, for any element $x \in C$, there exists a number m such that

$$x \in \bigcap_{i=m}^{\infty} E_{k_i}.$$

This means that

$$C \subset \bigcup_{m=1}^{\infty}\bigcap_{i=m}^{\infty} E_{k_i}.$$

But this contradict the fact that mes $(C) = r(C) = \frac{1}{q}$.
Thus the Theorem is proved.

We now sketch the proof when the value group $|K^\times|$ is dense.
Let $f(x, y)$ be a function defined on the domain

$$|x| < R^{(0)}, |y_1| < S_1^{(0)}, \cdots, |y_n| < S_n^{(0)},$$

which is analytic for each variable on this domain. Then we can show the analyticity of the function $f(x^*, y^*)$ on the same domain in the following way.
We take arbitrary $R, S_1, \cdots, S_n$ from the value group $|K^\times|$ satisfying

$$R < R^{(0)}, S_1 < S_1^{(0)}, \cdots, S_n < S_n^{(0)}.$$

Then $f(x, y)$ is analytic for each variable on the domain

$$|x| \leqslant R, |y_1| \leqslant S_1, \cdots, |y_n| \leqslant S_n.$$

We take arbitrary R_1 in the value group $|K^\times|$ such that $R_1 < R$. We also take an element q in $|K^\times|$ with $q > 1$. Then, by induction on the number of variables, we may proceed the proof in the same way. And we see that the function $f(x^*, y^*)$ is analytic in the variables x^*, y^* on the domain

$$|x^*| < R_1, |y_1^*| < \frac{1}{q} S_1, \cdots, |y_{n-1}^*| < \frac{1}{q} S_{n-1}, |y_n^*| < S_n.$$

Since $R_1, S_1, \cdots, S_n$ and $q > 1$ are arbitrary, the function $f(x^*, y^*)$ is in fact analytic on the whole domain

$$|x^*| < R^{(0)}, |y_1^*| < S_1^{(0)}, \cdots, |y_n^*| < S_n^{(0)}.$$

Thus we get

THEOREM 6 *Let K be a complete subfield of C_p, whose value group $|K^\times|$ is dense in $R_+^\times$. Let $f(x)$ be a function defined on the domain*

$$|x_1| < R_1, \cdots, |x_n| < R_n,$$

which is analytic for each variable. Then $f(x^*)$ is an analytic function in the whole variables on the domain

$$|x_1^*| < R_1, \cdots, |x_n^*| < R_n.$$

5 DOMAIN OF CONVERGENCE

As in the previous section, K denotes a complete and non-locally compact subfield of C_p with a discrete value group $|K^\times| = < q >$ ($q > 1$).

We have shown that, if $f(x_1, \cdots, x_n)$ is a function analytic for each variable on the domain

$$|x_1| \leqslant R_1, \cdots, |x_n| \leqslant R_n,$$

then the function $f(x_1^*, \cdots, x_n^*)$ is an analytic function in the whole variables on each of the following domains :

$$\begin{array}{lllll} |x_1^*| < R_1 , & |x_2^*| < \frac{1}{q}R_2, & |x_3^*| < \frac{1}{q}R_3, & \cdots & |x_n^*| < \frac{1}{q}R_n; \\ |x_1^*| < \frac{1}{q}R_1, & |x_2^*| < R_2 , & |x_3^*| < \frac{1}{q}R_3, & \cdots & |x_n^*| < \frac{1}{q}R_n; \\ & \cdots & & & \\ |x_1^*| < \frac{1}{q}R_1, & |x_2^*| < \frac{1}{q}R_2, & \cdots & |x_{n-1}^*| < \frac{1}{q}R_{n-1}, & |x_n^*| < R_n . \end{array} \qquad (*)$$

Thus, in these domains, we can expand the function $f(x^*)$ as the convergent power series :

$$f(x^*) = \sum_{k_1, \cdots, k_n} a_{k_1, \cdots, k_n} x_1^{*k_1} \cdots x_n^{*k_n}.$$

We want to extend a little more the domain of convergence of the function $f(x^*) = f(x_1^*, \cdots, x_n^*)$.

For any set of n positive numbers $r_1, r_2, \cdots, r_n$, put

$$\overline{r_i} = \max(\frac{1}{q}R_i, r_i) \qquad \text{(for all } i = 1, \cdots, n).$$

We shall now show that

THEOREM 7 *If $\overline{r}_1 \cdot \overline{r}_2 \cdots \overline{r}_n < \frac{1}{q^{n-1}} R_1 R_2 \cdots R_n$, then $f(x^*)$ is analytic in the whole variables x^* on the domain*

$$|x_1^*| \leqslant \overline{r}_1, |x_2^*| \leqslant \overline{r}_2, \ldots, |x_n^*| \leqslant \overline{r}_n. \qquad (9)$$

Proof. First, we take a positive number ρ $(0 < \rho < 1)$ satisfying

$$\bar{r}_1 \cdot \bar{r}_2 \cdots \bar{r}_n < \frac{\rho^n}{q^{n-1}} R_1 R_2 \cdots R_n < \frac{1}{q^{n-1}} R_1 R_2 \cdots R_n. \tag{10}$$

We shall show that $a_{k_1,\cdot,k_n} x_1^{*k_1} \cdots x_n^{*k_n}$ converges to zero on this domain (9) as $\|k\| \to \infty$. Without loss of generality, we may assume that $k_1 \leqslant k_2 \leqslant \cdots \leqslant k_n$. By the condition (*), the series

$$f(x_1^*, \cdots, x_n^*) = \sum_{k_1, \ ,k_n} a_{k_1, \ ,k_n} x_1^{*k_1} \cdots x_n^{*k_n}$$

converges whenever $|x_1^*| \leqslant \frac{\rho}{q} R_1, \ldots, |x|_{n-1}^* \leqslant \frac{\rho}{q} R_{n-1}, |x_n^*| \leqslant \rho R_n$. Thus

$$|a_{k_1,\cdot,k_n}| \frac{\rho^{k_1+ \cdot +k_{n-1}+k_n}}{q^{k_1+k_2+ \cdot +k_{n-1}}} R_1^{k_1} R_2^{k_2} \cdots R_n^{k_n} \to 0 \quad \text{as} \quad k_1 + \cdots + k_n \to \infty. \tag{11}$$

If $x^* = (x_1^*, \ldots, x_n^*)$ satisfies the condition (9) in Theorem, then

$$\begin{aligned} |a_{k_1, \ ,k_n} x_1^{*k_1} \cdots x_n^{*k_n}| &= |a_{k_1, \ ,k_n}| \cdot \bar{r}_1^{k_1} \cdots \bar{r}_n^{k_n} \\ &\leqslant |a_{k_1, \ ,k_n}| \cdot \left(\frac{q}{R_1}\bar{r}_1\right)^{k_1} \cdots \left(\frac{q}{R_n}\bar{r}_n\right)^{k_n} \frac{R_1^{k_1} \cdots R_n^{k_n}}{q^{k_1+ \ +k_{n-1}+k_n}}. \end{aligned}$$

Since $\frac{q}{R_\iota}\bar{r}_\iota \geqslant 1$, we may replace each term $\left(\frac{q}{R_\iota}\bar{r}_\iota\right)^{k_\iota}$ by $\left(\frac{q}{R_\iota}\bar{r}_\iota\right)^{k_n}$ in the above inequality. And then, we have

$$\begin{aligned} |a_{k_1, \ ,k_n} x_1^{*k_1} \cdots x_n^{*k_n}| &\leqslant |a_{k_1, \ ,k_n}| \frac{q^{nk_n} (\bar{r}_1 \cdots \bar{r}_n)^{k_n}}{(R_1 \cdots R_n)^{k_n}} \frac{R_1^{k_1} \cdots R_n^{k_n}}{q^{k_1+ \cdot\cdot k_{n-1}+k_n}} \\ &\leqslant |a_{k_1,\cdot\cdot,k_n}| \rho^{nk_n} q^{k_n} \cdot \frac{R_1^{k_1} \cdots R_n^{k_n}}{q^{k_1+\cdot \ +k_{n-1}+k_n}} \\ &\leqslant |a_{k_1,\cdot\cdot,k_n}| \rho^{k_1+ \ +k_n} \frac{R_1^{k_1} \cdots R_n^{k_n}}{q^{k_1+\cdots+k_{n-1}}}. \end{aligned}$$

The last term converges to zero by (11).

We have extended the domains $(*)$ a little more

COROLLARY 3 *The function f^* converges on each of the domains*

$$\begin{array}{llll} |x_1^*| < R_1, & |x_2^*| \leqslant \frac{1}{q} R_2, & \cdots, & |x_n^*| \leqslant \frac{1}{q} R_n; \\ |x_1^*| \leqslant \frac{1}{q} R_1, & |x_2^*| < R_2, & \cdots, & |x_n^*| \leqslant \frac{1}{q} R_n; \\ & \cdots & & \\ |x_1^*| \leqslant \frac{1}{q} R_1, & |x_2^*| \leqslant \frac{1}{q} R_2, & \cdots, & |x_n^*| < R_n. \end{array} \tag{**}$$

The following example was first given by Alain Escassut and me [2]. This shows that we cannot extend the domain of analyticity of f^* outside of the domain

$$|x_1^*| = \bar{r}_1, |x_2^*| = \bar{r}_2, \dots, |x_n^*| = \bar{r}_n,$$

with

$$\bar{r}_1 \cdot \bar{r}_2 \cdots \bar{r}_n = \frac{1}{q^{n-1}} R_1 R_2 \cdots R_n.$$

EXAMPLE Let $\{\alpha_i \big| i = 0, 1, 2, \cdots\}$ be a complete representative system of O modulo $M = \pi O$. We take $\alpha_0 = 0$. For any natural number m, put

$$b_m(x) = x^m \prod_{i=1}^{m} (x - \alpha_i)^m.$$

Let $\{a_m \big| m = 1, 2, \cdots\}$ be a series of elements in K such that $\lim_{m\to\infty} a_m = 0$. It is easy to see that the function

$$f(x_1, \dots, x_n) = \sum_m a_m \pi^{-(n-1)m} b_m(x_1) b_m(x_2) \cdots b_m(x_n) \tag{12}$$

defined on the product O^n is analytic for each variable (cf. [2]).

In fact, for any x in O, there exists a uniquely determined element α_i such that $x \equiv \alpha_i \pmod{M}$. Hence $\frac{1}{\pi^m} b_m(x)$ belongs to O for $m \geqslant i$. Thus, for any fixed $x_2, \cdots, x_n \in O$, the function

$$f(x_1, x_2, \dots, x_n) = \sum_{m=1}^{\infty} a_m \left(\frac{1}{\pi^m} b_m(x_2)\right) \cdots \left(\frac{1}{\pi^m} b_m(x_n)\right) b_m(x_1)$$

is an analytic function in the variable $x_1 \in O$.

It is easy to see that the dominant term of the coefficients of $x_1^m x_2^m \cdots x_n^m$ in (12) is $\frac{a_m}{\pi^{(n-1)m}}$. Hence the function is analytic in the variables $x_1, \dots, x_n$ on the domain

$$|x_1^*| \leqslant 1, |x_2^*| \leqslant |\pi| = \frac{1}{q}, \cdots, |x_n^*| \leqslant \frac{1}{q}.$$

Now given any $h > 1$, there exists a suitable sequence $\{a_m\}$ such that the function $f(x)$ is not analytic on the domain

$$|x_1^*| \leqslant 1, |x_2^*| \leqslant \frac{h}{q}, |x_3^*| \leqslant \frac{1}{q}, \cdots, |x_n^*| \leqslant \frac{1}{q} \quad (h > 1).$$

In fact, we can take $\{a_m\}$ satisfying

$$\lim_{m\to\infty} a_m = 0, \quad \text{but} \quad |a_m| \geqslant h^{-\frac{m}{2}}.$$

Then, for $|x_1| = 1$, $\frac{h^{\frac{1}{2}}}{q} \leqslant |x_2^*| \leqslant \frac{h}{q}, |x_3^*| = \frac{1}{q}, \cdots, |x_n^*| = \frac{1}{q}$, we have

$$|a_m \pi^{(n-1)m} x_1^{*m} x_2^{*m} \cdots x_m^{*\,m}| \geqslant h^{-\frac{m}{2}} q^{(n-1)m} h^{\frac{m}{2}} (\frac{1}{q})^{(n-1)m} \geqslant 1.$$

Changing variable x_i by $q^{-1}x_i$, we see that, if $f(x_1, \ldots, x_n)$ is analytic for each variables on the domain

$$|x_1| \leqslant q, |x_2| \leqslant q, \cdots, |x_n| \leqslant q,$$

But, in general, we cannot extend the domain of analyticity of f^* outside of the domain

$$|x_1^*| \leqslant q, |x_2^*| \leqslant 1, \cdots, |x_n^*| \leqslant 1.$$

We have shown (Theorem 7 or Corollary 4) that the function f^* is analytic on the domain

$$|x_1^*| < q, |x_2^*| \leqslant 1, \cdots, |x_n^*| \leqslant 1.$$

It is interesting whether in general the function f^* is analytic or not on the domain

$$|x_1^*| = q, |x_2^*| = 1, \cdots, |x_n^*| = 1.$$

REFERENCES

[1] M Endo. Analytic properties of functions represented by p-adic integration. Lecture Notes in Pure ans Applied Mathematics, Marcel Dekker, 1996.

[2] M Endo, A Escassut. A counter example of p-adic Hartogs theorem on a product of closed disks. Comm Math Univ St Pauli 46, 1997.

[3] A Escassut. Analytic Elements in p-adic Analysis. World Scientific, 1995.

[4] N Koblitz. p-adic Numbers, p-adic Analysis, and Zeta-Functions. Springer Verlag, 1977.

[5] S Lang. Cyclotomic fields. I,II Springer 1978, 1980.

[6] M Lazard. Les zeros des fonctions analitiques d'une variable sur un corps value complet. Inst Hautes Etudes scient 14:223–251, 1962.

[7] B Ja Levin. Distribution of zeros of entire functions. Moscow 1956 (translated by AMS 1964).

[8] MS Stawski. Hartogs theorem in some non archimedean fields. Dokl Acad Nauk USSR 161:776–779, 1965.

[9] MS Stawski. Outer linear measure and Hartogs theorem in some valued fields. Mathematicheskii Sbornik 70:113–131, 1966.

[10] MS Stawski. On the theorem of Hartogs in non-archimedean valued fields. J of Number Therory 16:75–86, 1983.

[11] WH Schikhof. Ultrametric calculus. Cambridge Univ Press, 1984.

[12] J Tate. Rigid analytic spaces. Inventiones mathematicae 12:257–289, 1971.

The Fourier transform for p-adic tempered distributions

N DE GRANDE-DE KIMPE, Department of Mathematics, Vrije Universiteit Brussel, Pleinlaan 2, B-1050 Brussels, Belgium.

A KHRENNIKOV, Department of Mathematics, Statistics and Computer Sciences, University of Växjö, S-35195, Sweden.

L VAN HAMME, Faculty of Applied Sciences, Vrije Universiteit Brussel, Pleinlaan 2, B-1050 Brussels, Belgium.

Abstract. We introduce a Fourier transform for distributions defined on a subspace of the space $C^\infty(\mathbb{Z}_p)$. The values of these distributions are analytic functions satisfying certain growth conditions. This Fourier transform has all the properties one expects it to have. The main result is a p-adic variant of the classical Paley-Wiener theorem.

0 INTRODUCTION

In this paper we introduce a natural p-adic variant of the classical Fourier transform of a distribution, which has all the properties needed for applications. Our approach makes intensive use of the Mahler bases for the spaces of test functions and so it is completely different from the one in [3]. The main result is a p-adic version of the classical Paley-Wiener theorem. Its proof is based on an identification of our spaces with spaces of sequences.

NOTATION The field of scalars K is always the field $\mathbb{Q}_p$ (for some prime number p) and we write $|\cdot|$ for the p-adic valuation. A sequence a in K will always have the form $a = (\alpha_0, \alpha_1, \alpha_2, \ldots)$ and, unless specified otherwise, $\sum$ will always be a sum from 0 to ∞. We denote by E the domain of convergence of the exponental function e^x, i.e.

$$E = \{\alpha \in \mathbb{Q}_p \mid |\alpha| \leqslant \frac{1}{p}\} \text{ if } p \neq 2,$$

$$E = \{\alpha \in \mathbb{Q}_p \mid |\alpha| \leqslant \frac{1}{4}\} \text{ if } p = 2.$$

For all further information about the exponential function we refer to [5]. We also refer to [5] for the definition of the spaces $C^k(\mathbb{Z}_p)$, $k = 0, 1, \ldots$, and $C^\infty(\mathbb{Z}_p)$ and for the description of their elements in terms of the Mahler bases.

For elementary functions such as

$$\mathbb{Z}_p \to \mathbb{Q}_p : x \to x^m$$

$$\mathbb{Z}_p \to \mathbb{Q}_p : x \to \binom{x}{n} = \frac{x(x-1)\ldots(x-n+1)}{n!}, \quad n = 1, 2, \ldots, \binom{x}{0} = 1$$

we shall use the value (i.e. x^m, $\binom{x}{n}$) to denote the function itself. This simplifies the notation and will not cause any confusion for the reader.

1 SEQUENCE SPACES

For all the results in this section we refer to [4] 3.2. We recall however some of the main notations and facts that will be useful for us. First note that in [4] the sequences (α_n) are of the form $(\alpha_n) = (\alpha_1, \alpha_2, \alpha_3, \ldots)$ while for our applications the sequences start with α_0. This does of course not affect the results. So, let $B = (b_n^k)$, $n = 0, 1, 2, \ldots$; $k = 0, 1, 2, \ldots$ be an infinite matrix of strictly positive real numbers satisfying the condition $b_n^k \leqslant b_n^{k+1}$ for all k and n.

1.1 Define $c_0(\mathbb{N}, b^k) = \{(\alpha_n) | \lim_n |\alpha_n| b_n^k = 0\}$. This is ([4] p. 209) a Banach space for the norm

$$||(\alpha_n)||_k = \sup_n |\alpha_n| b_n^k.$$

Clearly $c_0(\mathbb{N}, b^{k+1}) \subset c_0(\mathbb{N}, b^k)$, for all k, and the canonical injections are continuous. Put $\Lambda^0(B) = \bigcap_k c_0(\mathbb{N}, b^k)$ and consider on $\Lambda^0(B)$ the projective limit topology τ_p^0. Then ([4] section 3.2):

i) $\Lambda^0(B)$ is a perfect sequence space.

ii) The topology τ_p^0 can be generated by the restriction to $\Lambda^0(B)$ of the increasing sequence of norms $(|| \cdot ||_k)$ defined as above.

iii) $\Lambda^0(B)$, τ_p^0 is a polar Fréchet space.

iv) If $\lim_n \dfrac{b_n^k}{b_n^{k+1}} = 0$ for all k then $\Lambda^0(B)$, τ_p^0 is nuclear (and hence reflexive).

1.2 Define $l^\infty(\mathbb{N}, \frac{1}{b^k}) = \{(\alpha_n) | \sup_n \dfrac{|\alpha_n|}{b_n^k} < \infty\}$. This is ([4] p. 209) a Banach space for the norm

$$||(\alpha_n)||_k = \sup_n \frac{|\alpha_n|}{b_n^k}.$$

Clearly $l^\infty(\mathbb{N}, \frac{1}{b^k}) \subset l^\infty(\mathbb{N}, \frac{1}{b^{k+1}})$, for all k, and the canonical injections are continuous. Put $\Lambda_\infty(B) = \bigcup_k l^\infty(\mathbb{N}, \frac{1}{b^k})$ and provide it with the inductive limit topology τ_i^∞. Then

i) $\Lambda_\infty(B)$ $(= \Lambda^0(B)^\times)$ is a perfect sequence space that is algebraically isomorphic to $(\Lambda^0(B), \tau_p^0)'$.

ii) $\Lambda_\infty(B)$, $\tau_\imath^\infty$ is complete.

iii) If $\lim_n \dfrac{b_n^k}{b_n^{k+1}} = 0$ for all k then $\Lambda_\infty(B)$, $\tau_\imath$ is nuclear and $\tau_\imath^\infty = n(\Lambda_\infty(B), \Lambda^0(B))$.

We are interested here in two special cases.

1.3 Take $b_n^k = n^k$ $(0^0 = 1)$. Then we denote by s_k the corresponding sequence space $c_0(\mathbb{N}, b^k)$, i.e. $s_k = \{(\alpha_n) |\ \lim_n |\alpha_n| n^k = 0\}$, and by s the space $\Lambda^0(B)$, i.e. $s = \{(\alpha_n) |\ \lim_n\ |\alpha_n| n^k = 0$ for all $k\}$. (This space is called the space of "rapidly decreasing sequences"). We write τ_p for the projective limit topology on s.

Note that the sequence $(n!)$ is an element of s. Indeed, by [5] p. 71, $\lim_n |n!|^{\frac{1}{n}} = p^{\frac{1}{1-p}} < 1$. Hence $\lim_n (|n!| n^k)^{\frac{1}{n}} < 1$ and so the real series $\sum_{n=1}^{\infty} |n!| n^k$ converges for all k and the conclusion follows.

We then obtain directly from 1.1:

i) s is a perfect sequence space.

ii) The topology τ_p can be generated by the increasing sequence of norms $(\|\cdot\|_k)$, where $\|(\alpha_n)\|_k = \sup_n |\alpha_n| n^k$.

iii) s, τ_p is a nuclear, polar, reflexive Fréchet space.

We denote by s^k the space $l^\infty(\mathbb{N}, \frac{1}{b^k})$, i.e. $s^k = \{(\alpha_n) |\ \sup_n \dfrac{|\alpha_n|}{n^k} < \infty\}$, and by $s^\times$ the space $\Lambda_\infty(B)$ (compare with 1.2 i)), i.e. $s^\times = \{(\alpha_n) |\ \sup_n \dfrac{|\alpha_n|}{n^k} < \infty$ for some $k \in \{0, 1, 2, \ldots\}\}$. Its inductive limit topology will be denoted by $\tau_\imath$.

Then we obtain directly form 1.2:

i) $s^\times$ is a perfect sequence space that, as a vector space, can be identified with $(s, \tau_p)'$.

ii) $s^\times, \tau_\imath$ is nuclear and complete.

iii) $\tau_\imath = n(s^\times, s)$.

1.4 Take $b_n^k = \frac{n^k}{|n!|}$ $(0^0 = 1)$. Now, exactly as in 1.3 we denote the corresponding sequence spaces by S_k, S, S^k, $S^\times$ and the topologies on S and $S^\times$ by τ_p and $\tau_\imath$. Then these spaces have the same properties as those mentioned in 1.3.

Also note that

PROPOSITION 1.5 *$s^\times$ is a dense subspace of $S^\times, \tau_\imath$.*

Proof. If $(\alpha_n) \in s^\times$ then there is a k such that $\sup_n \dfrac{|\alpha_n|}{n^k} < \infty$. Hence $\sup_n \dfrac{|\alpha_n||n!|}{n^k} < \infty$ and so $(\alpha_n) \in S^\times$.

Take now $\beta = (\beta_n) \in S^\times$. By [2] and 1.2 iii) we have $\beta = \sum_n \beta_n e_n$ where the series converges in $\tau_\imath = n(S^\times, S)$. Take $\alpha^{(n)} = (\beta_1, \beta_2, \ldots, \beta_n, 0, 0, \ldots)$. Then $\lim_n \alpha^{(n)} = \beta$ in $S^\times, \tau_\imath$ and we are done.

2 THE SPACE OF TEST FUNCTIONS AND THE SPACE OF DISTRIBUTIONS

DEFINITION 2.1 It is well known that the Mahler polynomials $\binom{x}{n}$, $n = 0, 1, 2, \ldots$, form a basis for the space $C^\infty(\mathbb{Z}_p)$. In fact ([5] p. 166) a function $f : \mathbb{Z}_p \to \mathbb{Q}_p$ belongs to $C^\infty(\mathbb{Z}_p)$ if and only if it can be written as

$$f(x) = \sum a_n \binom{x}{n}, \quad \text{with} \quad \lim_n |a_n| n^k = 0 \quad \text{for} \quad k = 1, 2, \ldots$$

On $C^\infty(\mathbb{Z}_p)$ there is a natural locally convex topology determined by the increasing sequence of norms $(||\cdot||_k)$, where

$$||f||_k = \sup_n |a_n| n^k, \quad k = 0, 1, \ldots \quad (0^0 = 1).$$

The space $C^\infty(\mathbb{Z}_p)$ equipped with this locally convex topology will be the space of test functions. It follows immediately that the space $C^\infty(\mathbb{Z}_p)$ can, as a locally convex space, be identified with the space s of the rapidly decreasing sequences (see 1.3). The identification being

$$f \leftrightarrow (a_n), \quad f(x) = \sum a_n \binom{x}{n} \quad (x \in \mathbb{Z}_p).$$

More concretely

PROPOSITION 2.2 *For $k = 0, 1, 2, \ldots$ each of the spaces $C^k(\mathbb{Z}_p)$* (see [5] p. 86) *can, as a Banach space, be identified with the space s_k defined in section 1.3. The identification being as above. Moreover $C^\infty(\mathbb{Z}_p) = \bigcap_k C^k(\mathbb{Z}_p)$* (see [5] p. 86).

Proof. By [5], 54 A p. 165, the norm on $C^k(\mathbb{Z}_p)$ is equivalent to the norm

$$||f||_k = \sup_n |a_n| n^k, \quad f \in C^k(\mathbb{Z}_p).$$

This gives (see 1.3)

PROPOSITION 2.3 *The locally convex space $C^\infty(\mathbb{Z}_p)$ is a nuclear, reflexive, Fréchet space.*

EXAMPLES 2.4

2.4.1 Every function $f : \mathbb{Z}_p \to \mathbb{Q}_p$ which is analytic on $\mathbb{Z}_p$ is an element of $C^\infty(\mathbb{Z}_p)$ (see [5] p. 91).

2.4.2 For $y \in E$ the function $\mathbb{Z}_p \to \mathbb{Q}_p : x \to e^{xy}$ is analytic in $\mathbb{Z}_p$. By 2.4.1 it is an element of $C^\infty(\mathbb{Z}_p)$. This can also be proved directly form the Mahler expansion

$$e^{xy} = \sum_{n=0}^{\infty} (e^y - 1)^n \binom{x}{n} \qquad ([5], 52.2 p.153)$$

2.4.3 If $f \in C^{\infty}(\mathbb{Z}_p)$ then $f^{(n)} \in C^{\infty}(\mathbb{Z}_p)$ and

$$||f^{(n)}||_k \leqslant ||f||_{k+n} \quad \text{for} \quad k = 0, 1, 2, \ldots$$

Indeed, observe that it is sufficient to prove this for $n = 1$.

Let $f \in C^{\infty}(\mathbb{Z}_p)$ be written as $f(x) = \sum_{n=0}^{\infty} a_n \binom{x}{n}$. Then it is not hard to see that

$$f'(x) = \sum_{j=1}^{\infty} b_j \binom{x}{j} \quad \text{with} \quad b_j = \sum_{i=1}^{\infty} (-1)^{i-1} \frac{a_{i+j}}{i}.$$

We first prove that $f' \in C^{\infty}(\mathbb{Z}_p)$, i.e. $\lim_j |b_j| j^k = 0$ for $k = 1, 2, \ldots$. Fix k and choose $\varepsilon > 0$. Then, since $\frac{1}{|i|} \leqslant i \leqslant i + j$, we have

$$|b_j| \leqslant \max_{i \geqslant 1} \frac{|a_{j+i}|}{|i|} \leqslant \max_{i \geqslant 1} \frac{|a_{j+i}|(i+j)^{k+1}}{(i+j)^k}. \tag{*}$$

Since $f \in C^{\infty}(\mathbb{Z}_p)$, $\lim_n |a_n| n^{k+1} = 0$. Hence, there exists a j_0 such that

$$|a_{i+j}|(i+j)^{k+1} < \varepsilon \quad \text{for} \quad j \geqslant j_0, \quad i = 1, 2, 3, \ldots.$$

So $|b_j| \leqslant \varepsilon \max_{i \geqslant 1} \frac{1}{(i+j)^k} \leqslant \frac{\varepsilon}{j^k}$ and we conclude that $\lim_j |b_j| j^k = 0$.

To show that $||f'||_k \leqslant ||f||_{k+1}$ we use (*) to get $||f'||_k = \sup_j |b_j| j^k \leqslant ||f||_{k+1}$.

2.4.4 Let $x \in \mathbb{Z}_p$. If $f \in C^{\infty}(\mathbb{Z}_p)$ then $x^m f \in C^{\infty}(\mathbb{Z}_p)$ for $m = 0, 1, \ldots$ and

$$||x^m f||_k \leqslant 2^{m+k} ||f||_k, \quad k = 0, 1, \ldots.$$

Indeed, observe that it is sufficient to prove this for $m = 1$.

Writing $f(x) = \sum_{n=0}^{\infty} a_n \binom{x}{n}$ and using $x\binom{x}{n} = n\binom{x}{n} + (n+1)\binom{x}{n+1}$ we see that $xf(x) = \sum_{n=1}^{\infty} n(a_n + a_{n-1}) \binom{x}{n}$. Hence we have to show that for $k = 1, 2, \ldots$

$$\lim_n |n||a_n + a_{n-1}| n^k = 0. \tag{**}$$

Now $|n||a_n| n^k \leqslant |a_n| n^k$ while

$$|n||a_{n-1}| n^k \leqslant \frac{|a_{n-1}| n^k (n-1)^k}{(n-1)^k} \leqslant 2^k |a_{n-1}|(n-1)^k, \quad n \neq 1.$$

Combining these two inequalities we get (**) since $\lim_n |a_n| n^k = 0$. We now show that

$$||xf||_k \leqslant 2^k ||f||_k \quad \text{for} \quad k = 0, 1, 2, \ldots .$$

This follows from

$$||xf||_k = \sup_{n\geqslant 1} |n||a_n + a_{n-1}|n^k \leqslant \sup_{n\geqslant 1} |a_n + a_{n-1}|n^k$$
$$\leqslant max\left(\sup_{n\geqslant 1} |a_n|n^k,\ \sup_{n\geqslant 2} |a_{n-1}|n^k,\ |a_0|\right)$$
$$\leqslant max\left(||f||_k,\ \sup_{n\geqslant 2} |a_{n-1}|(n-1)^k(\frac{n}{n-1})^k\right)$$
$$\leqslant max\left(||f||_k,\ 2^k||f||_k\right) \leqslant 2^k||f||_k.$$

NOTATION 2.5 Let X be a locally convex space and $f : X \to K$ a map. We denote by $< x, f >$ the value of f at x.

DEFINITION 2.6 The topological dual space $[C^\infty(\mathbb{Z}_p)]'$ of $C^\infty(\mathbb{Z}_p)$ will be called the space of distributions. It follows immediately from 1.3 that, as a vector space, the space $[C^\infty(\mathbb{Z}_p)]'$ can be identified with the space $s^\times$ and we transfer the inductive limit topology τ_ι on $s^\times$ to the space $[C^\infty(\mathbb{Z}_p)]'$. More concretely

PROPOSITION 2.7 *For each k the Banach spaces* $[C^k(\mathbb{Z}_p)]'$ *and* s^k (see 1.3) *are linearly homeomorphic. The identification being*

$$\varphi \in [C^k(\mathbb{Z}_p)]' \leftrightarrow (< \binom{x}{n}, \varphi >)_n.$$

Moreover $[C^\infty(\mathbb{Z}_p)]' = \bigcup_k [C^k(\mathbb{Z}_p)]'$.

Proof. For $(a_n) \in s^k$ define a map φ from the Mahler polynomials to K by

$$< \binom{x}{n}, \varphi >= a_n, \quad n = 0, 1, 2, \ldots$$

It is easy to see that then, for $f \in C^k(\mathbb{Z}_p)$, $f(x) = \sum \alpha_n \binom{x}{n}$, the series $\sum \alpha_n a_n$ is convergent. Now define the map $\varphi : C^k(\mathbb{Z}_p) \to K$ by $< f, \varphi >= \sum \alpha_n a_n$. Obviously φ is linear and continuous i.e. $\varphi \in [C^k(\mathbb{Z}_p)]'$.

Consider now the linear map

$$T : s^k \to [C^k(\mathbb{Z}_p)]' : (a_n) \to \varphi$$

where φ is defined as above. We show that T is the desired linear homeomorphism.

- T is injective: If $T((a_n)) = \bar{0}$ then in particular $< \binom{x}{n}, \varphi >= 0$ for all n.

- T is surjective: Take $\varphi \in [C^k(\mathbb{Z}_p)]'$. Then there exists a $K > 0$ such that $| < f, \varphi > | \leqslant K||f||_k$ for all f in $C^k(\mathbb{Z}_p)$. In particular, since $||\binom{x}{n}||_k = n^k$, $| < \binom{x}{n}, \varphi > | \leqslant Kn^k$ for all n. Putting $a_n =< \binom{x}{n}, \varphi >$, $n = 0, 1, 2, \ldots$ we find that $(a_n) \in s^k$ and $T((a_n)) = \varphi$.

- T is continuous: Indeed

$$||T((a_n))|| = ||\varphi|| = \sup_{||f||_k\leqslant 1} | < f, \varphi > |$$
$$\leqslant \sup_{||f||_k\leqslant 1} (\max \frac{|\alpha_n||a_n|n^k}{n^k})$$
$$\leqslant ||(a_n)||_k \sup_{||f||_k\leqslant 1} \max |\alpha_n|n^k \leqslant ||(a_n)||_k$$

Now apply the open mapping theorem.
The second part of the proposition follows directly from [4] Theorem 1.3.7.

As an immediate consequence of 1.3 we then obtain

PROPOSITION 2.8 *The locally convex space $[C^\infty(\mathbb{Z}_p)]', \tau_i$ is nuclear and complete.*

EXAMPLES 2.9

2.9.1 For $\varphi \in [C^\infty(\mathbb{Z}_p)]'$, $f \in C^\infty(\mathbb{Z}_p)$ and $x \in \mathbb{Z}_p$ we define $< f, x^m\varphi >= < x^m f, \varphi >$ (This definition makes sense by 2.4.4). Then $x^m\varphi \in [C^\infty(\mathbb{Z}_p)]'$.

Indeed, observe that it is sufficient to prove this for $m = 1$.

Obviously the map $x\varphi$ is linear. Further, by the continuity of φ, there exists $K > 0$ and a $k \in \{0, 1, 2, \ldots\}$ such that $| < f, x\varphi > | \leqslant K||xf||_k$. Now apply the second part of 2.4.4 to conclude that $x\varphi$ is continuous.

2.9.2 For $a \in \mathbb{Z}_p$ we denote by δ_a the linear map $\delta_a : C^\infty(\mathbb{Z}_p) \to K : f \to f(a)$. Then $\delta_a \in [C^\infty(\mathbb{Z}_p)]'$.

Indeed, for $f \in C^\infty(\mathbb{Z}_p)$, $f(x) = \sum \alpha_n \binom{x}{n}$ we have

$$| < f, \delta_a > | = |\sum \alpha_n \binom{a}{n}| \leqslant \max_n |\alpha_n| = ||f||_0.$$

Hence δ_a is continuous.

2.9.3 The n-th derivative $\varphi^{(n)}$ of a distribution $\varphi \in [C^\infty(\mathbb{Z}_p)]'$ is defined by

$$< f, \varphi^{(n)} >=< f^{(n)}, \varphi >, \quad f \in C^\infty(\mathbb{Z}_p).$$

This definition makes sense by 2.4.3 and it follows from 2.4.3 that then $\varphi^{(n)} \in [C^\infty(\mathbb{Z}_p)]'$.

2.10 The convolution $\varphi * \psi$ of two distributions φ and ψ.

Identify $\varphi \in [C^\infty(\mathbb{Z}_p)]'$ (respectively $\psi \in [C^\infty(\mathbb{Z}_p)]'$) as in 2.7 with the sequence $a_n =< \binom{x}{n}, \varphi >\in s^\times$ (resp. $b_n =< \binom{x}{n}, \psi >\in s^\times$).

Now consider the convolution $(a_n) * (b_n)$ of the sequences (a_n) and (b_n) i.e. (see [5] p. 106)

$$(a_n) * (b_n) = (c_n) \text{ with } c_n = \sum_{j=0}^{n} a_j b_{n-j}, \quad n = 0, 1, 2, \ldots$$

There exists a $k > 1$ such that (a_n) and (b_n) are in s^k. It is then easy to see that $(a_n) * (b_n) \in s^{2k} \subset s^\times$. Finally define $\varphi * \psi$ as the unique element of $[C^\infty(\mathbb{Z}_p)]'$ that corresponds to the sequence $(a_n) * (b_n)$.

3 THE FOURIER TRANSFORM

DEFINITION 3.1 For $y \in E$ we define the function

$$exp\ y\cdot\ :\ \mathbb{Z}_p \to \mathbb{Q}_p \quad \text{by} \quad (exp\ y\cdot)(x) = e^{xy}.$$

For a distribution $\varphi \in [C^\infty(\mathbb{Z}_p)]'$ we define the Fourier transform $\mathcal{F}\varphi$ of φ by

$$(\mathcal{F}\varphi)(y) = < exp\ y\cdot, \varphi >, \quad y \in E \quad \text{(compare 2.4.2)}.$$

This is the natural p-adic variant of the classical definition of the Fourier transform. Our first aim is to give some alternative expressions for $\mathcal{F}\varphi$.

THE STIRLING NUMBERS. Let $S(m,n)$ be the Stirling numbers of the second kind (see e.g. [1] for the definition of these numbers). Put $A_{nm} = n!S(m,n)$. Since $S(m,n)$ is an integer we see that A_{nm} is an integer divisible by $n!$.

LEMMA 3.2

i) $x^n = \displaystyle\sum_{k=0}^{n} A_{kn}\binom{x}{k}$.

ıı) $(e^y - 1)^n = \displaystyle\sum_{r=1}^{\infty} A_{nr}\frac{y^r}{r!},\ y \in E$.

Proof. These classical results are well-known (see e.g. [1] p. 40-41).

THEOREM 3.3   Let $y \in E, \varphi \in [C^\infty(\mathbb{Z}_p)]'$.

i) $(\mathcal{F}\varphi)(y) = \sum a_n(e^y - 1)^n$, where, as before, $a_n = < \binom{x}{n}, \varphi >,\ n = 0,1,2,\ldots$.

ıı) $(\mathcal{F}\varphi)(y) = a_0 + \displaystyle\sum_{r=1}^{\infty} < x^r, \varphi > \frac{y^r}{r!}$.

Proof. i) $e^{yx} = (e^y - 1 + 1)^x = \sum(e^y - 1)^n\binom{x}{n}$. Then apply φ.

ii) Using i) and lemma 3.2 we find, for $y \in E$:

$$(\mathcal{F}\varphi)(y) = a_0 + \sum_{n=1}^{\infty} a_n \left(\sum_{r=1}^{\infty} A_{nr}\frac{y^r}{r!} \right).$$

In order to prove ii) we need to change the order of summation. It is not hard to see that this is allowed (apply [5] p. 62).

COROLLARY 3.4   The function $\mathcal{F}\varphi : E \to \mathbb{Q}_p$ is analytic on E (see 3.3 ii)).

PROPOSITION 3.5   Let $y \in E, \varphi \in [C^\infty(\mathbb{Z}_p)]'$ then (compare 2.9)

i) $\mathcal{F}(x^m\varphi)(y) = (\mathcal{F}\varphi)^{(m)}(y),\ x \in \mathbb{Z}_p$.

ıı) $(\mathcal{F}\delta_a)(y) = e^{ay}$.

ııı) $(\mathcal{F}\varphi^{(n)})(y) = y^n(\mathcal{F}\varphi)(y)$.

Proof. i) We prove that $(\mathcal{F}(x\varphi))(y) = (\mathcal{F}\varphi)'(y)$. Clearly

$$(\mathcal{F}(x\varphi))(y) = < x exp\ y\cdot, \varphi > .$$

By [5] p. 155 we have

$$xe^{xy} = \sum_{n=1}^{\infty} n[(e^y-1)^n + (e^y-1)^{n-1}]\binom{x}{n}$$
$$= \sum_{n=1}^{\infty} n[e^y-1]^{n-1}e^y\binom{x}{n}.$$

Then apply φ and 3.3.i).

ii) Follows directly from the definition of $\mathcal{F}$.

iii) We prove that $(\mathcal{F}\varphi')(y) = y(\mathcal{F}\varphi)(y)$. This follows directly from $(\mathcal{F}\varphi')(y) = < (exp\ y\cdot)'_x, \varphi >$.

THEOREM 3.6 *The Fourier transform of the convolution of two distributions. For* $\varphi,\ \psi \in [C^\infty(\mathbb{Z}_p)]'$ *we have* $\mathcal{F}(\varphi * \psi) = (\mathcal{F}\varphi).(\mathcal{F}\psi)$.

Proof. Put as before $< \binom{x}{n}, \varphi >= a_n$, $< \binom{x}{n}, \psi >= b_n$, $n = 0, 1, 2, \ldots$. Then, by definition (see 2.10)

$$< \binom{x}{n}, \varphi * \psi >= c_n \quad \text{with} \quad c_n = \sum_{j=0}^{n} a_j b_{n-j}.$$

Hence, by (3.3. i))

$$(\mathcal{F}(\varphi * \psi))(y) = \sum (e^y-1)^n c_n.$$

On the other hand (again by 3.3. i))

$$(\mathcal{F}\varphi)(y).(\mathcal{F}\psi)(y) = (\sum (e^y-1)^n a_n).(\sum (e^y-1)^n b_n).$$

Multiplying the two power series we get the desired result.

REMARK 3.7 Denote by $\mathcal{A}_k(E)$, $k = 0, 1, 2, \ldots$, the vector space of the functions $f : E \to \mathbb{Q}_p$ which are analytic on E and which, when written as $f(y) = \sum_{r=0}^{\infty} b_r \frac{y^r}{r!}$, satisfy the growth condition $\sup_r \frac{|b_r|}{r^k} < \infty$. Then $\mathcal{A}_k(E)$ can, as a vector space, be identified with s^k and we transfer the norm $\|\cdot\|_k$ on s^k to the space $\mathcal{A}_k(E)$.

Put $\mathcal{A}(E) = \bigcup_k \mathcal{A}_k$ and consider on $\mathcal{A}(E)$ the inductive limit topology $\tau_\imath$. Then $\mathcal{A}(E), \tau_\imath$ is linearly homeomorphic to $s^\times, \tau_\imath$.

PROPOSITION 3.8 *If* $\varphi \in [C^\infty(\mathbb{Z}_p)]'$, *then* $\mathcal{F}\varphi \in \mathcal{A}(E)$.

Proof. Take $\varphi \in [C^\infty(\mathbb{Z}_p)]'$. For $y \in E$ we have (3.3.ii))

$$(\mathcal{F}\varphi)(y) = a_0 + \sum_{r=1}^{\infty} < x^r, \varphi > \frac{y^r}{r!}$$

Now identify this power series with the sequence of its coefficients, say (b_r) with $b_0 = a_0$. It is then sufficient to prove that $(b_r) \in s^\times$. We have, for r fixed ($r \geqslant 1$), with the notation of 3.3

$$|b_r| = |< x^r, \varphi >| \leqslant \max_{n=1,\ldots,r} |A_{nr}||a_n|.$$

Also, since $(a_n) \in s^\times$, there exists a $k \in \{1, 2, 3, \ldots\}$ and $K > 0$ such that $|a_n| \leqslant Kn^k$ for all n. This gives

$$|b_r| \leqslant K \left(\max_{n=1,\ldots,r} n^k \right) = Kr^k.$$

Hence $(b_r) \in s^k \subset s^\times$.

COROLLARY 3.9 *The Fourier transform $\mathcal{F}$ can be considered as a linear map*

$$\mathcal{F} : s^\times \to s^\times.$$

In fact we have shown that, for all k the restriction of $\mathcal{F}$ to s^k is a linear map from s^k to s^k.

The rest of this section is devoted to the study of this map $\mathcal{F}$. We first need a lemma.

LEMMA 3.10 *Let* $b_r = \sum_{n=1}^{r} A_{nr} a_n$, $r = 1, 2, \ldots$. *Then* $a_n = \frac{1}{n!} \sum_{r=1}^{n} \alpha_{nr} b_r$, $n = 1, 2, \ldots$, *with* $|\alpha_{nr}| \leqslant 1$ *for all* n *and all* r.

Proof. For $n \geqslant 1$ we define integers α_{nr} by

$$\binom{x}{n} = \frac{1}{n!} \sum_{r=1}^{n} \alpha_{nr} x^r \qquad (*)$$

Obviously $|\alpha_{nr}| \leqslant 1$. Now consider the linear map $L : \mathbb{Q}_p[x] \to \mathbb{Q}_p[x]$ defined by $L(\binom{x}{n}) = a_n$. Since $x^r = \sum_{n=1}^{r} A_{nr} \binom{x}{n}$ (lemma 3.2.i)) we see that $L(x^r) = b_r$. Now apply L to (*).

THEOREM 3.11 *The linear map $\mathcal{F} : s^\times, \tau_i \to s^\times, \tau_i$ is a continuous injection.*

Proof. $\mathcal{F}$ is injective by proposition 3.8 and lemma 3.10.

By [4] p. 166 and by the definition of the inductive limit topology, the continuity of $\mathcal{F}$ follows if we prove that, for each k, the restriction $\mathcal{F}|_{s^k} : s^k, ||\cdot||_k \to s^k, ||\cdot||_k$ is continuous. So fix k and take $(a_n) \in s^k$. Then $||\mathcal{F}(a_n)||_k = \sup_r \frac{|b_r|}{r^k}$. On the other hand (as in 3.8)

$$|b_r| \leqslant \max_{n=1,\ldots,r} \frac{|A_{nr}||a_n|n^k}{n^k} \leqslant r^k ||(a_n)||_k$$

and the conclusion follows.

REMARK 3.12 Unfortunately the map $\mathcal{F}$ is not surjective i.e. we don't have the p-adic variant of the classical Paley-Wiener theorem. Indeed, with the same notations as above we have (since A_{nr} is divisible by $n!$)

$$|b_r| \leqslant \max_{n=1,\ ,r} |n!||a_n| \leqslant \max_n |n!|n^k K.$$

It follows that $(b_r) \in l^\infty \subset s^\times, l^\infty \neq s^\times$.

In order to solve this problem we will change our space of test functions. This will be done in the next section.

4 THE SPACE OF RAPIDLY DECREASING C^∞-FUNCTIONS AND THE SPACE OF TEMPERED DISTRIBUTIONS

DEFINITION 4.1 An element f of $C^\infty(\mathbb{Z}_p)$, $f(x) = \sum \alpha_n \binom{x}{n}$, will be called "rapidly decreasing" if

$$\lim_n \frac{|\alpha_n|n^k}{|n!|} = 0, \quad k = 1, 2, \ldots.$$

We use the term "rapidly decreasing" in analogy to the classical situation. Obviously the rapidly decreasing elements of $C^\infty(\mathbb{Z}_p)$ form a vector subspace of the space $C^\infty(\mathbb{Z}_p)$. We denote this subspace by $RD[C^\infty(\mathbb{Z}_p)]$ or, since no confusion is possible, simply by RD.

A natural locally convex topology τ on RD is generated by the increasing sequence of norms $(||\cdot||_k)$ where, for $f \in RD$,

$$||f||_k = \sup_n \frac{|\alpha_n|n^k}{|n!|} = 0, \quad k = 0, 1, 2, \ldots, \quad 0^0 = 1.$$

Note that we use here the same notation $||\cdot||_k$ as in section 2 with a different meaning. But it will always be clear from the context which definition is meant. The locally convex space RD, τ will be our new space of test functions.

With the same type of identifications as those made in section 2 we find that the space RD, τ can, as a locally convex space, be identified with the sequence space S from 1.4.

This gives

PROPOSITION 4.2 *The space RD, τ is a nuclear, reflexive, Fréchet space.*

EXAMPLES 4.3

4.3.1 For $y \in E$ the function $exp\ y\cdot$. defined in 3.1 is an element of RD.

Indeed, to prove this we first consider the case where $p \geqslant 3$. For $x \in \mathbb{Z}_p$, $y \in E$ we have

$$(exp\ y\cdot)(x) = \sum_{n=0}^{\infty}(e^y - 1)^n \binom{x}{n}$$

and

$$|e^y - 1| = |y| \leqslant \frac{1}{p}.$$

Hence it is sufficient to prove that for $k = 1, 2, \ldots$ the real limit $\lim_n \dfrac{n^k}{p^n|n!|} = 0$.

Now, with the notations of [5] p. 70

$$|n!| = p^{-\lambda(n)} \text{ and } \lim_n \frac{\lambda(n)}{n} = \frac{1}{p-1} \leqslant \frac{1}{2}.$$

Choose n_0 such that $\frac{\lambda(n)}{n} < \frac{3}{4}$ for $n \geqslant n_0$. Then, for each k and $n \geqslant n_0$ we have

$$0 \leqslant \frac{n^k}{p^n|n!|} < \frac{n^k}{p^{n/4}}$$

and the conclusion follows.

The same type of proof works for $p = 2$, now taking into account= that for $p = 2$, $y \in E$ implies $|y| \leqslant \frac{1}{4}$.

4.3.2 If $f \in RD$ then $f^{(n)} \in RD$, $(n = 0, 1, 2, \ldots)$ and $||f^{(n)}||_k \leqslant ||f||_k$ for $k = 0, 1, 2, \ldots$.

Indeed, observe that it is sufficient to prove this for $n = 1$.
First note that for all $j \in \{0, 1, 2, \ldots\}$

$$\max_{i \geqslant 1} \frac{|(i+j)!|}{|i| \cdot |j!|} \leqslant 1 \qquad (*)$$

(Indeed $\frac{(i+j)!}{i\ j!} = (i-1)!\binom{i+j}{j}$).

With the notations of 2.4.3 we have to prove that $\lim_j \dfrac{|b_j|j^k}{|j!|} = 0$, $k = 1, 2, \ldots$, where $b_j = \sum_{i=1}^{\infty}(-1)^{i-1}\dfrac{a_{i+j}}{i}$. The proof is straightforward since

$$||(b_j)||_k = \sup_j \frac{|b_j|j^k}{|j!|}$$

while, for j fixed,

$$\frac{|b_j|j^k}{|j!|} \leqslant \max_{i \geqslant 1} \frac{|a_{i+j}|(i+j)^k|(i+j)!|}{|j!| \cdot |i| \cdot |(i+j)!|} \leqslant ||f||_k$$

by (*).

4.3.3 Let $x \in \mathbb{Z}_p$. If $f \in RD$ then $x^m f \in RD$ $(m = 0, 1, 2, \ldots)$ and $||x^m f||_k \leqslant 2^{mk} ||f||_k$.

The proof is analogous to the proof of 2.4.4 and is therefore omitted.

REMARK 4.4 It follows from [5], p. 166 that every $f \in RD$ is analytic= on $\mathbb{Z}_p$.

However, not every function $f : \mathbb{Z}_p \to \mathbb{Q}_p$ that is analytic on $= \mathbb{Z}_p$ is an element of RD as is shown by the following example.

EXAMPLE 4.5 A function $f \in C(\mathbb{Z}_p)$, written as $f(x) = \sum a_n \binom{x}{n}$, is analytic on $\mathbb{Z}_p$ iff $\lim_n \frac{|a_n|}{|n!|} = 0$ ([5], p. 166). On the other hand $f \in RD$ iff $\lim_n \frac{|a_n| n^k}{|n!|} = 0$, for $k = 1, 2, \ldots$. Hence it is sufficient to construct a sequence (b_n) such that $\lim_n b_n = 0$ while $n|b_n|$ does not tend to 0. Such a sequence is for instance

$$(b_n) = (p, p, \ldots, p, p^2, p^2, \ldots, p^2, p^3, p^3, \ldots, p^3, \ldots, p^r, p^r, \ldots, p^r, \ldots).$$

were we have p times p, $p^2 - p$ times p^2, ..., $p^r - p^{r-1}$ times p^r, ...

DEFINITION 4.6 We denote by T the topological dual space of the space RD and we call the elements of T *tempered distributions*.

Then, whith the same type of identification as in 2.1 and 2.2 the space T can, as a vector space, be identified with the sequence space $S^\times$ from 1.4. We transfer the inductive limit topology τ_ι on $S^\times$ to the space T. This gives (by 1.4 and 1.5).

PROPOSITION 4.7 i) T, τ_ι *is nuclear and complete.*

ii) $[C^\infty(\mathbb{Z}_p)]'$ *is a dense subspace of* T, τ_ι.

EXAMPLES 4.8

4.8.1 Let $x \in \mathbb{Z}_p$. If $\varphi \in T$ then $x^m \varphi \in T$. Here $x^m \varphi$ is defined as in 2.9.1.

The result then follows directly from 4.3.3.

4.8.2 For all $a \in \mathbb{Z}_p$ the distribution δ_a (as defined in 2.9.2) is an element of T.

Indeed, by 2.9.2 $\delta_a \in [C^\infty(\mathbb{Z}_p)]'$.

4.8.3 If $\varphi \in T$ then $\varphi^{(n)} \in T$, $n = 0, 1, 2, \ldots$, where $\varphi^{(n)}$ is defined as in 2.9.3.

This follows directly from 4.3.2.

4.9 The convolution of two tempered distributions.

Let $\varphi, \psi \in T$ and identify φ (respectively ψ) as in 2.10 with the sequence (a_n), $a_n = < \binom{x}{n}, \varphi >$, $n = 0, 1, 2, \ldots$ (respectively (b_n), $b_n = < \binom{x}{n}, \psi >$, $n = 0, 1, 2, \ldots$). Now consider the convolution $(c_n) = (a_n) * (b_n)$ of the sequences (a_n) and (b_n) as defined in 2.10. Then a proof analogous to the proof in 2.10 shows that $(c_n) \in S^\times$. Finally we define $\varphi * \psi$ as the unique element of T that corresponds to the sequence $(c_n) \in S^\times$.

5 THE FOURIER TRANSFORM FOR TEMPERED DISTRIBUTIONS

LEMMA 5.1 *If $\varphi \in T$ the series $\sum_{r=1}^{\infty} < x^r, \varphi > \frac{y^r}{r!}$ converges for all $x \in \mathbb{Z}_p$ and all $y \in E$.*

Proof. Take $\varphi \in T$, $x \in \mathbb{Z}_p$ and $y \in E$. We have $x^r \in RD$ and, for $k = 1, 2, 3, \ldots$,

$$||x^r||_k = ||\sum_{n=1}^{r} A_{nr}\binom{x}{n}||_k \leqslant \max_{n=1,\ldots,r} |A_{nr}|\frac{n^k}{n!} \leqslant r^k.$$

Since $\varphi \in T = (RD)'$ there exists a $k \in \{1, 2, 3, \ldots\}$ and $K > 0$ such that

$$| < x^r, \varphi > | \leqslant K||x^r||_k \leqslant Kr^k.$$

Hence it is sufficient to prove that $\lim_r \frac{r^k y^r}{r!} = 0$ and this has already been done in 4.3.1.

DEFINITION 5.2 (of the Fourier transform $\hat{\varphi}$ of a tempered distribution φ) Let $\varphi \in T$. Then we define for $y \in E$

$$\hat{\varphi}(y) = a_0 + \sum_{r=1}^{\infty} < x^r, \varphi > \frac{y^r}{r!} \quad (x \in \mathbb{Z}_p,\ a_0 =< \binom{x}{0}, \varphi >=< x^1, \varphi >).$$

Note that by 3.3 we have

$$\hat{\varphi} = \mathcal{F}\varphi \quad \text{whenever} \quad \varphi \in [C^\infty(\mathbb{Z}_p)]'.$$

LEMMA 5.3 *With the same notations as in 3.3 and 3.8 we have: If $(a_n) \in S^k$, then $(b_r) \in s^k$.*

Proof. We know that $\sup_n \frac{|a_n||n!|}{n^k} = K < \infty$. It follows that for $r \geqslant 1$ fixed

$$|b_r| \leqslant \max_{n=1,\ldots,r} |A_{nr}||a_n| \leqslant Kr^k \quad \text{(by 3.2)}.$$

Hence $(b_r) \in s^k$.

REMARK 5.4 Identifying the space of the tempered distributions T with the sequence space $S^\times$ as in 4.6 and the space $\mathcal{A}(E)$, defined in 3.7, with the sequence space $s^\times$ (as in 3.7) we see that the Fourier transform $\wedge$ can be considered as a map

$$\wedge : S^\times \to s^\times : (a_n) \to (b_r)$$

where $b_0 = a_0$ and for $r \geqslant 1$ $b_r = \sum_{n=1}^{r} A_{nr}a_n$ (see 5.3).

THEOREM 5.5 (Paley-Wiener) *The Fourier transform* $\wedge$ *is a linear homeomorphism from* T, τ_i *onto* $\mathcal{A}(E), \tau_i$.

Proof. We have to prove that the map $\wedge$ defined in 5.4 is a linear homeomorphism. Obviously this map is linear and its injectivity follows from 3.10.

For the surjectivity of $\wedge$ take a sequence $(b_r) \in s^{\times}$, i.e. there exists a $k \in \{0,1,2,\ldots\}$ with $\sup_r \frac{|b_r|}{r^k} < \infty$. Now define (a_n) as follows: $a_0 = b_0$ and for $n = 1,2,\ldots$ the a_n are the solutions of the system of equations

$$b_r = \sum_{n=1}^{r} A_{nr} a_n \quad r = 1,2,\ldots,$$

as done in 3.10. Then obviously $\wedge((a_n)) = (b_r)$. Moreover $(a_n) \in S^k$ i.e. $\sup_n \frac{|a_n||n!|}{n^k} < \infty$. This follows directly from the expression of the a_n given in 3.10.

In fact we have shown that $\wedge_k$, the restriction of $\wedge$ to S^k, is for each k a linear bijection $S^k \to s^k$.

As in the proof of 3.11 it is now sufficient to show that this map= is continuous for the norms on S^k and s^k.

So let $(a_n) \in S^k$. Then

$$\|\widehat{(a_n)}^k\|_k = \|(b_r)\|_k = \sup_r \frac{|b_r|}{r^k} = \sup_r \frac{|\sum_{n=1}^{r} A_{nr} a_n|}{r^k}$$
$$\leqslant \sup_r \frac{\max_{n=1,\ldots,r} |A_{nr}||a_n|}{r^k} \leqslant \sup_r \left(\max_{n=1,\ldots,r} \frac{|n!||a_n|}{n^k} \right).$$

On the other hand

$$\|(a_n)\|_k = \sup_n \frac{|a_n||n!|}{n^k}$$

and the desired conclusion follows.

We finally show that the Fourier transform $\wedge$ has all the desired properties.

PROPOSITION 5.6 *Let* $y \in E, \varphi \in T$ *then*

i) $\widehat{(x^m \varphi)}(y) = \hat{\varphi}^{(m)}(y)$, $x \in \mathbb{Z}_p$.

ii) $\hat{\delta}_a(y) = e^{ay}$ $(\hat{\delta} = \overline{1})$.

iii) $\hat{\varphi}^{(m)}(y) = y^m \hat{\varphi}(y)$, *hence* $\hat{\delta}_a^{(m)}(y) = y^m e^{ay}$.

iv) $\widehat{\varphi * \psi} = \hat{\varphi} \cdot \hat{\psi}$.

This follows immediately from definition 5.2.

REFERENCES

[1] Comtet. Analyse combinatoire, vol 2. Presses Universitaires de France, 1970.

[2] N De Grande-De Kimpe. Perfect locally K-convex sequence spaces. Proc Kon Ned Akad Sci A74:471–482, 1971.

[3] N De Grande-De Kimpe, A Khrennikov. The non-archimedean Laplace Transform. Bull Belg Math Soc 3:225–237, 1996.

[4] N De Grande-De Kimpe, J Kakol, C Perez-Garcia, WH Schikhof. p-Adic locally convex inductive limits. In : WH Schikhof, C Perez-Garcia, I Kakol, ed. p-Adic Functional Analysis. New York: Marcel Dekker, 1997, pp 159–222.

[5] WH Schikhof. Ultrametric Calculus. Cambridge: Cambridge University Press, 1984.

On the Mahler coefficients of the logarithmic derivative of the p-adic gamma function

L VAN HAMME Faculty of Applied Sciences, Vrije Universiteit Brussel, Pleinlaan 2, B-1050 Brussels, Belgium, lvhamme@vub.ac.be

Abstract. Let $\Gamma_p(x)$ denote the p-adic gamma function. The Mahler expansion of $\frac{d}{dx}[lg\Gamma_p(1+x)]$ can be written in the form

$$\frac{\Gamma_p'(1+x)}{\Gamma_p(1+x)} = \Gamma_p'(0) - \sum_{n=1}^{\infty} \frac{u_n}{n}\binom{x}{n}$$

We study the Mahler coefficients u_n and calculate the sum of several series involving these numbers u_n.

1 INTRODUCTION

Let p be a prime number and let $\mathbb{Z}_p$ denote, as usual, the ring of p-adic integers. The p-adic valuation will be denoted by $|\cdot|$. Let χ_p be the characteristic function of the set $p\mathbb{Z}_p$, i.e.

$$\chi_p(x) = 1 \ \text{ if } \ |x| < 1$$
$$\chi_p(x) = 0 \ \text{ if } \ |x| = 1.$$

Since χ_p is continuous it has a Mahler expansion of the form

$$\chi_p(x) = \sum_{n=0}^{\infty} u_n^{(p)} \binom{x}{n} \tag{1}$$

The first aim of this paper is to study the numbers $u_n^{(p)}$. These numbers depend on p but in order to simplify the notation we shall write u_n for $u_n^{(p)}$.
The numbers u_n are interesting because they are related to the p-adic gamma function Γ_p. (See [1] for the definition and properties of the p-adic gamma function Γ_p).

Indeed we have

PROPOSITION 1

$$\frac{\Gamma_p'(1+x)}{\Gamma_p(1+x)} = \Gamma_p'(0) - \sum_{n=1}^{\infty} \frac{u_n}{n}\binom{x}{n} \tag{2}$$

Proof. Since $\frac{\Gamma_p'(1+x)}{\Gamma_p(1+x)}$ is continuous on $\mathbb{Z}_p$ there exists a Mahler expansion of the form

$$\frac{\Gamma_p'(1+x)}{\Gamma_p(1+x)} = \sum_{n=0}^{\infty} \alpha_n \binom{x}{n} \tag{*}$$

Putting $x = 0$ we see that $\alpha_0 = \frac{\Gamma_p'(1)}{\Gamma_p(1)}$.
Proposition 35.1 from [1] says that

$$\begin{aligned} \Gamma_p(x+1) &= -x\Gamma_p(x) \quad if \ |x| = 1 \\ &= -\Gamma_p(x) \quad if \ |x| < 1. \end{aligned}$$

Hence

$$\begin{aligned} \frac{\Gamma_p'(1+x)}{\Gamma_p(1+x)} - \frac{\Gamma_p'(x)}{\Gamma_p(x)} &= \frac{1}{x} \quad if \ |x| = 1 \\ &= 0 \quad if \ |x| < 1. \end{aligned} \tag{**}$$

Taking $x = 0$ we get $\frac{\Gamma_p'(1)}{\Gamma_p(1)} = \frac{\Gamma_p'(0)}{\Gamma_p(0)} = \Gamma_p'(0)$ since $\Gamma_p(0) = 1$.
Hence $\alpha_0 = \Gamma_p'(1)$.
In order to calculate α_n for $n \geqslant 1$ we write (**) in the form

$$\frac{\Gamma_p'(1+x)}{\Gamma_p(1+x)} - \frac{\Gamma_p'(x)}{\Gamma_p(x)} = \frac{1-\chi_p(x)}{x}$$

Using (*) we get

$$\frac{1-\chi_p(x)}{x} = \sum_{n=0}^{\infty} \alpha_n \left\{ \binom{x}{n} - \binom{x-1}{n} \right\} = \sum_{n=0}^{\infty} \alpha_n \binom{x-1}{n-1}$$

$$\chi_p(x) = 1 - \sum_{n=0}^{\infty} \alpha_n x \binom{x-1}{n-1} = 1 - \sum_{n=0}^{\infty} n\alpha_n \binom{x}{n}$$

Comparing this with (1) we conclude that $-n\alpha_n = u_n$.
This proves (2).
This paper is organized as follows. In section 2 we study the numbers u_n. In section 3 we prove several expansions involving the numbers u_n. The main theorem of this paper is proved in section 4. It is a rather general theorem from which we deduce new series containing the numbers u_n.

2 THE NUMBERS u_n

The properties of the numbers u_n that we will prove in this section all follow from the following proposition.

PROPOSITION 2

$$u_n = \frac{1}{p}\sum_{\zeta}(\zeta - 1)^n \quad for \quad n \geqslant 1$$

where ζ runs through the primitive p-th roots of unity.

Proof. Let Δ denote the difference operator defined by $(\Delta f)(x) = f(x+1) - f(x)$. The coefficient of $\binom{x}{n}$ in the Mahler expansion of a continuous function f is equal to $(\Delta^n f)(0)$. Hence $u_n = (\Delta^n \chi_p)(0)$. To calculate this we use the well-known fact that

$$p\chi_p(m) = \sum_{\zeta}\zeta^m \tag{3}$$

where ζ runs through all the p-th roots of unity and m is a natural number. Applying the operator Δ n times to (3) we get the desired result.

COROLLARY 1

$$u_n = \sum_{k\equiv 0 \ mod \ p} (-1)^{n-k}\binom{n}{k}$$

This follows from

$$pu_n = \sum_{\zeta}(\zeta - 1)^n = \sum_{k=0}^{n}\binom{n}{k}(-1)^{n-k}\sum_{\zeta}\zeta^k$$

using (3).
Note that the corollary shows that the numbers u_n are integers.

COROLLARY 2

$$u_{(2n+1)p} = 0 \quad if \quad p \neq 2$$

$$u_{mp} = (-1)^{mp}\sum_{r=0}^{m}(-1)^{rp}\tbinom{mp}{rp} = (-1)^{m}\sum_{r=0}^{m}(-1)^{r}\tbinom{mp}{rp} \text{ since } p \neq 2.$$

$$= (-1)^m\sum_{r=0}^{m}(-1)^{m-r}\tbinom{mp}{(m-r)p} = \sum_{r=0}^{m}(-1)^{r}\tbinom{mp}{rp} = (-1)^m u_{mp}.$$

Hence $u_{mp} = 0$ if m is odd.

COROLLARY 3

$$\sum_{k=1}^{p}\binom{p}{k}u_{n+k-1} = 0 \quad for \quad n \geqslant 1.$$

Indeed $(\zeta - 1)$ is a root of the equation

$$(x+1)^p = 1 \quad i.e. \quad \sum_{k=1}^{p}\binom{p}{k}x^{k-1} = 0$$

Hence $\sum_{k=1}^{p} \binom{p}{k}(\zeta - 1)^{k-1+n} = 0.$

Summing over ζ and using the proposition we get the linear recursion of the corollary. Note that the corollary is not true for $n = 0$.

For each real number t let $[t]$ denote the greatest integer $\leqslant t$.

COROLLARY 4

$$u_n \text{ is divisible by } p^e \text{ with } e = [\frac{n-1}{p-1}]$$

In the recursion

$$u_{n+p-1} + \binom{p}{1} u_{n+p-2} + \binom{p}{2} u_{n+p-3} + \ldots + p u_n = 0 \qquad (4)$$

all the coefficients (except the first one) are divisible by p. If we take $n = 1$ in (4) and note that the numbers u_n are integers we conclude that p divides u_n. Taking $n = 2, 3, \ldots$ we see that u_n is divisible by p for $n \geqslant p$. In the same way we can deduce from (4) that u_n is divisible by p^2 for $n \geqslant 2p - 1$. Similarly we conclude that p^3 divides u_n for $n \geqslant 3p - 2$ etc. This proves the corollary.

EXAMPLES

From these corollaries one easily deduces the following :

$u_n = (-1)^n$ for $n = 0, 1, 2, \ldots, p-1$ (use cor. 1)

$u_p = 0$ for $p \neq 2$

$u_{2p} = 2 - \binom{2p}{p}$ (cor. 1).

If $p = 2$ one has $u_0 = 1$, $u_n = (-1)^n 2^{n-1}$ for $n \geqslant 1$ (prop. 2).

Finally we calculate u_n when $p = 3$.

Let ζ be a primitive 3-th root of unity. Then

$$\zeta^2 + \zeta + 1 = 0, \quad (\zeta - 1)^2 = -3\zeta,$$

$$(\zeta - 1)^6 = -27\zeta^3 = -27, \quad (\zeta - 1)^{6+n} = -27(\zeta - 1)^n$$

Hence $u_{6+n} = -27u_n$. Together with the starting values

$$u_0 = 1, \ u_1 = -1, \ u_2 = 1, \ u_3 = 0, \ u_4 = -3, \ u_5 = 9$$

this determines the values of u_n for $p = 3$.

3 SOME SERIES INVOLVING THE NUMBERS u_n

Let $\mathbb{C}_p$ be the completion of the algebraic closure of $\mathbb{Q}_p$ and let lgx denote the Iwasawa logarithm of $x \in \mathbb{C}_p{}^{\times}$.

PROPOSITION 3

$$\sum_{n=1}^{\infty} \frac{u_n}{n} x^n = -\frac{1}{p} lg[(1+x)^p - x^p] \qquad |x| \leqslant 1, \quad x \in \mathbb{C}_p \tag{5}$$

Proof. If $\zeta^p = 1, \zeta \neq 1$ then $(\zeta - 1)^{-1}$ is a root of the polynomial $(1+x)^p - x^p$. Hence

$$(1+x)^p - x^p = \prod_{\zeta} [1 - (\zeta - 1)x],$$

where ζ runs through the primitive p-th roots of unity. Now

$$lg[1 - (\zeta - 1)x] = -\sum_{n=1}^{\infty} (\zeta - 1)^n \frac{x^n}{n}$$

since $|\zeta - 1| < 1$ and $|x| \leqslant 1$. Summing over ζ we see that the proposition follows from proposition 2.

COROLLARY 1 We mention three special cases of the proposition :

$$\sum_{n=1}^{\infty} \frac{(-1)^n}{n} u_n = 0 \qquad (x = -1)$$

$$\sum_{n=1}^{\infty} \frac{1}{n} u_n = -\frac{1}{p} lg(2^p - 1) \qquad (x = 1)$$

$$\sum_{n=1}^{\infty} \frac{(-1)^n}{n 2^n} u_n = (1 - \frac{1}{p} lg2), \quad p \neq 2 \qquad (x = -1) \tag{6}$$

From this last equality we can deduce the following congruence

$$\frac{2^{p-1} - 1}{p} \equiv \sum_{n=1}^{p-1} \frac{1}{n 2^n} \ (mod\ p), \quad p \neq 2. \tag{7}$$

This is an old arithmetical result (see [2] p. 111).
To prove (7) we reduce both sides of (6) *mod p*.
Since $u_p = 0$ and p^2 divides u_n for $n \geqslant 2p - 1$ it follows from corollary 4 of proposition 2 that p divides $\frac{u_n}{n}$ for $n \geqslant p$.
Since $u_n = (-1)^n$ for $n = 1, 2, \ldots, p-1$ the L.H.S. of (6) is congruent to

$$\sum_{n=1}^{p-1} \frac{1}{n 2^n} \ (mod\ p)$$

Now $(1 - \frac{1}{p}) lg2 = \frac{1}{p} lg 2^{p-1} = \frac{1}{p} lg(1 + 2^{p-1} - 1) \equiv \frac{2^{p-1}-1}{p} \ (mod\ p)$
This proves (7).
For the remainder of this paper K will denote a non-archimedean non-trivially valued complete field containing $\mathbb{Q}_p$. We continue to denote the valuation by $|\cdot|$.

COROLLARY 2

$$\sum_{n=1}^{\infty} u_n x^n = \frac{(1+x)^{p-1}}{(1+x)^p - x^p} \qquad |x| \leqslant 1, \quad x \in K$$

For $x \in \mathbb{C}_p$ this is clear by taking the derivative of (5). This implies that the formula is true when we consider $\sum u_n x^n$ as a formal power series. Hence the result is valid for $x \in K$, $|x| \leqslant 1$.

COROLLARY 3 *For* $p \neq 2$

$$\sum_{n=1}^{\infty} \frac{u_n}{n} x^n = \sum_{n=1}^{\infty} \frac{(-1)^n}{n} u_n (1+x)^n \qquad |x| \leqslant 1, \; x \in \mathbb{C}_p.$$

Since $p \neq 2$ this follows from (5) by replacing x by $-1-x$.

PROPOSITION 4 *In the ring of formal power series we have*

$$\left\{\sum_{n=0}^{\infty} (-1)^{np} \frac{x^{np}}{(np)!}\right\} \left\{\sum_{n=0}^{\infty} \frac{x^{np}}{(np)!}\right\} = \sum_{n=0}^{\infty} u_{np} \frac{x^{np}}{(np)!} \qquad (a)$$

$$\left\{\sum_{n=0}^{\infty} (-1)^n u_n \frac{x^n}{n!}\right\} \left\{\sum_{n=0}^{\infty} u_n \frac{x^n}{n!}\right\} = \sum_{n=0}^{\infty} u_{np} \frac{x^{np}}{(np)!} \qquad (b)$$

Proof. We start from a well-known formula from the theory of finite differences. For a sequence $f(n)$, $n = 0, 1, 2, \ldots$, the following identity between formal power series is valid

$$\left\{\sum_{n=0}^{\infty} (-1)^n \frac{x^n}{n!}\right\} \left\{\sum_{n=0}^{\infty} f(n) \frac{x^n}{n!}\right\} = \sum_{n=0}^{\infty} (\Delta^n f)(0) \frac{x^n}{n!}.$$

If $f(n) = \chi_p(n)$ then $(\Delta^n \chi_p)(0) = u_n$ and hence

$$\left\{\sum_{n=0}^{\infty} (-1)^n \frac{x^n}{n!}\right\} \left\{\sum_{n=0}^{\infty} \frac{x^{np}}{(np)!}\right\} = \sum_{n=0}^{\infty} u_n \frac{x^n}{n!}. \qquad (8)$$

Replacing x by $x\zeta$ and summing over all primitive p-th roots of unity ζ we obtain formula (a). To prove (b) we replace x by $-x$ in (8) and multiply the result with (8). Observing that

$$\left\{\sum_{n=0}^{\infty} \frac{x^n}{n!}\right\} \left\{\sum_{n=0}^{\infty} (-1)^n \frac{x^n}{n!}\right\} = 1$$

and using (a) we obtain formula (b).

COROLLARY

$$\begin{aligned} \textstyle\sum_{k=0}^{n} \binom{n}{k} (-1)^k u_k u_{n-k} &= u_n \quad \textit{if } p \textit{ divides } n \\ &= 0 \quad \textit{if } n \textit{ is not divisible by } p. \end{aligned}$$

This follows from (b) by comparing the coefficient of x^n on both sides.

REMARK 1 If $p \neq 2$ the R.H.S. of (a) and (b) can be simplified since in that case $u_{(2n+1)p} = 0$.

REMARK 2 In the special case where $p = 3$ formula (a) is mentioned in [3] p. 62.

REMARK 3 When $p = 5$ the R.H.S. of (a) can be expressed by means of the Fibonacci numbers F_n and the Lucas numbers L_n. These numbers are defined as follows

$$\begin{array}{lll} F_0 = 0, & F_1 = 1, & F_{n+1} = F_n + F_{n-1} \\ L_0 = 2, & L_1 = 1, & L_{n+1} = L_n + L_{n-1}. \end{array}$$

The result is

$$\sum_{n=0}^{\infty} \frac{x^{5n}}{(5n)!} \sum_{n=0}^{\infty} (-1)^n \frac{x^{5n}}{(5n)!}$$

$$= 1 - 2\sum_{n=1}^{\infty} F_{(2n-1)5} 5^{5n-3} \frac{x^{(2n-1)10}}{((2n-1)10)!} + 2\sum_{n=1}^{\infty} L_{10n} 5^{5n-1} \frac{x^{20n}}{(20n)!}$$
$$= 1 - 2.5^2 F_5 \frac{x^{10}}{10!} + 2L_{10} 5^4 \frac{x^{20}}{(20)!} - 2F_{15} 5^7 \frac{x^{30}}{(30)!} + 2L_{20} 5^9 \frac{x^{40}}{(40)!} - \cdots.$$

We omit the details.

4 A GENERAL THEOREM

Recall that Δ is the difference operator. Let T denote the translation operator, i.e. $(Tf)(x) = f(x+1)$, and suppose, momentarily, that Δ and T act on the space of polynomials. From the theory of finite differences we know that a finite identity of the form

$$\sum_{n=0}^{N} a_n x^n = \sum_{n=0}^{N} b_n (1+x)^n$$

can be transformed in an equality between operators

$$\sum_{n=0}^{N} a_n \Delta^n = \sum_{n=0}^{N} b_n T^n,$$

i.e.

$$\sum_{n=0}^{N} a_n (\Delta^n f)(0) = \sum_{n=0}^{N} b_n f(n).$$

This is only true for polynomials !

With this in mind one can hope that corollary 3 of proposition 3 will give rise to a formula of the type

$$\sum_{n=1}^{N} \frac{u_n}{n} (\Delta^n f)(0) = \sum_{n=1}^{N} (-1)^n \frac{u_n}{n} f(n). \tag{9}$$

In real analysis there is not much hope that this will work since the series in (9) are usually divergent. In p-adic analysis this works as the following theorem shows.

THEOREM *Let (a_n), (b_n) be two null sequences with values in K. The following conditions are equivalent :*

i) $\sum_{n=0}^{\infty}(-1)^n a_n \binom{x+n}{n} = \sum_{n=0}^{\infty}(-1)^n b_n \binom{x}{n} \qquad x \in \mathbb{Z}_p.$

ii) *For every bounded sequence $f : \mathbb{N} \to K$ we have*

$$\sum_{n=0}^{\infty} a_n(\Delta^n f)(0) = \sum_{n=0}^{\infty} b_n f(n).$$

iii) $\sum_{n=0}^{\infty} a_n x^n = \sum_{n=0}^{\infty} b_n(1+x)^n \qquad |x| < 1, \quad x \in \mathbb{Z}_p.$

iv) $b_n = \sum_{r=0}^{\infty}(-1)^r a_{n+r}\binom{n+r}{r} \qquad n = 0, 1, 2, \ldots.$

v) $a_n = \sum_{r=0}^{\infty} b_{n+r}\binom{n+r}{r} \qquad n = 0, 1, 2, \ldots.$

Proof. We first need a result from the theory of finite differences. Consider the sequence $g(n) = (-1)^n\binom{x}{n}$, $n = 0, 1, 2, \ldots$, then $(\Delta^n g)(0) = (-1)^n\binom{x+n}{n}$. Here x is a parameter and Δ acts on n. This follows from the well-known identity (in the ring of formal power series)

$$\sum_{n=0}^{\infty} g(n)t^n = \sum_{n=0}^{\infty}(\Delta^n g)(0)\frac{t^n}{(1-t)^{n+1}}.$$

Put $g(n) = (-1)^n\binom{x}{n}$ and observe that

$$\sum_{n=0}^{\infty}(-1)^n\binom{x}{n}t^n = (1-t)^x = \frac{1}{1-t}\left(1+\frac{t}{1-t}\right)^{-x-1} = \sum_{n=0}^{\infty}\binom{-1-x}{n}\frac{t^n}{(1-t)^{n+1}}.$$

Hence $(\Delta^n g)(0) = \binom{-1-x}{n} = (-1)^n\binom{x+n}{n}$.
We are now ready to prove the theorem.

(i) implies (ii)

Since $f : \mathbb{N} \to K$ is bounded there exists a measure μ on $\mathbb{Z}_p$ such that

$$(-1)^n f(n) = \int_{\mathbb{Z}_p}\binom{x}{n}\mu(x).$$

From the preliminary result it follows that

$$(\Delta^n g)(0) = (-1)^n\int_{\mathbb{Z}_p}\binom{x+n}{n}\mu(x).$$

Since both series in (i) are uniformly convergent we may integrate termwise with respect to μ to obtain (ii).

(ii) implies (iii)
In (ii) take $f(x) = (1+u)^x$ with $|u| < 1$, $u \in \mathbb{Z}_p$ then $(\Delta^n g)(0) = u^n$.

(iii) implies (iv)

$$\begin{aligned}\sum_{n=0}^{\infty} b_n(1+x)^n &= \sum_{n=0}^{\infty} a_n(x+1-x)^n \\ &= \sum_{n=0}^{\infty} a_n \sum_{k=0}^{n} \tbinom{n}{k}(x+1)^k(-1)^{n-k} \\ &= \sum_{k=0}^{\infty}(-1)^k(x+1)^k \sum_{n\geqslant k}(-1)^n a_n \tbinom{n}{k}\end{aligned}$$

Comparing the coefficient of $(x+1)^n$ in the first and the last series we obtain (iv). The change in the order of summation of the double series is allowed by exercise 23.B of [1] and because $lim_{n\to\infty} a_n = 0$ and the general term of the double series is 0 for $k > n$.

(iv) implies (i)
Let c_n be the coefficient of $\binom{x}{n}$ in the Mahler expansion of the L.H.S. of (i) i.e.

$$\sum_{n=0}^{\infty}(-1)^n a_n \tbinom{x+n}{n} = \sum_{n=0}^{\infty} c_n \binom{x}{n}.$$

To calculate c_n we apply k times the operator Δ to the L.H.S. This gives $\sum_{n\geqslant k}(-1)^n a_n \binom{x+n}{n-k}$. Putting $x=0$ we obtain

$$c_k = \sum_{n\geqslant k}(-1)^n a_n \binom{n}{n-k} = (-1)^k b_k$$

by (iv). This proves (i).

(iii) implies (v)
The proof is similar to the proof of the implication (iii)⇒(iv). We compare the coefficient of x^n in

$$\sum_{n=0}^{\infty} a_n x^n = \sum_{n=0}^{\infty} b_n \sum_{k=0}^{n} \binom{n}{k} x^k = \sum_{k=0}^{\infty} x^k \sum_{n\geqslant k} b_n \binom{n}{k}.$$

(v) implies (i)
If we calculate the coefficient d_n in $\sum b_n \binom{x+n}{n} = \sum d_n \binom{x}{n}$ in the same way as in the proof of the implication (iv)⇒(i) we find that $d_n = a_n$.
Hence

$$\sum_{n=0}^{\infty} b_n \binom{x+n}{n} = \sum_{n=0}^{\infty} a_n \binom{x}{n}.$$

Changing x into $-1-x$ we get (i).

COROLLARY 1 If $f : \mathbb{N} \to K$ is bounded then

$$\sum_{n=0}^{\infty} \frac{u_n}{n} (\Delta^n f)(0) = \sum_{n=0}^{\infty} (-1)^n \frac{u_n}{n} f(n) \qquad p \neq 2.$$

Indeed corollary 3 of proposition 3 shows that we may take $a_n = \frac{u_n}{n}$ and $b_n = \frac{(-1)^n}{n} u_n$ (for $p \neq 2$).

EXAMPLE

$$\sum_{n=0}^{\infty} \frac{u_n^2}{n} = \sum_{n=0}^{\infty} \frac{u_{2np}}{2np} \qquad p \neq 2$$

Take $f = \chi_p$ in corollary 1 and use the fact that $u_{(2n+1)p} = 0$.

COROLLARY 2 If $f : \mathbb{N} \to K$ is bounded and $(\Delta^n f)(0) = (-1)^{n+1} f(n)$ then

$$\sum_{n=0}^{\infty} (-1)^n \frac{f(n) u_n}{n} = 0 \qquad \textit{for} \;\; p \neq 2$$

This is an immediate consequence of corollary 1.
As an example of a sequence satisfying the conditions of corollary 2 we have the following sequence of polynomials.

EXAMPLE Define $P_n(z)$ as follows

$$P_0(z) = 0, \quad P_1(z) = 1,$$

$$P_{n+1}(z) = P_n(z) + z P_{n-1}(z), \quad n \geqslant 1, z \in K, |z| \leqslant 1,$$

then $\sum_{n=1}^{\infty} (-1)^n P_n(z) \frac{u_n}{n} = 0 \; (p \neq 2)$.
For $z = 1$ $P_n(1) = F_n$ (Fibonacci number) and we get
$\sum_{n=0}^{\infty} (-1)^n F_n \frac{u_n}{n} = 0.$
Proof. We see by induction that the sequence $P_n(z), \quad n = 0, 1, 2, \ldots$, is bounded. Moreover $P_n(z) = \frac{\alpha^n - \beta^n}{\alpha - \beta}$, where α en β are the roots of $t^2 - t - z = 0$. Note that $\alpha + \beta = 1$. Putting $f(n) = P_n(z)$ we see that
$(\Delta^n f)(0) = \frac{(\alpha-1)^n - (\beta-1)^n}{\alpha - \beta} = \frac{(-\beta)^n - (-\alpha)^n}{\alpha - \beta} = (-1)^{n+1} f(n).$

REMARK 1 Let $g : \mathbb{Z}_p \to K$ be a continuous function and define the function h by $h(x) = g(-1 - x)$. Then the sequences $a_n = (\Delta^n g)(0)$ and $b_n = (-1)^n (\Delta^n h)(0)$ satisfy the conditions of the theorem. To see this we look at the Mahler expansions

$$g(x) = \sum_{n=0}^{\infty} (\Delta^n g)(0) \binom{x}{n} = \sum_{n=0}^{\infty} a_n \binom{x}{n},$$

$$g(-1-x) = \sum_{n=0}^{\infty}(-1)^n a_n \binom{x+n}{n},$$

$$h(x) = \sum_{n=0}^{\infty}(\Delta^n h)(0)\binom{x}{n} = \sum_{n=0}^{\infty}(-1)^n b_n \binom{x}{n}.$$

Hence the relation $h(x) = g(-1-x)$ is nothing else than condition (i) of the theorem. Conversely every couple of sequences (a_n), (b_n) satisfying (i) arises in this way. (Define $g(x)$ as $\sum_{n=0}^{\infty} a_n \binom{x}{n}$).

REMARK 2 If the sequences (a_n), (b_n) satisfy the conditions of the theorem then the sequences $((n+1)a_{n+1})$, $((n+1)b_{n+1})$ also satisfy these conditions. To see this it suffices to take the derivative of (iii). This is a source of new formulas such as e.g.

$$(-1)^{n+1}u_{n+1} = \sum_{r=0}^{\infty}(-1)^r \binom{n+r}{r} u_{n+1+r}$$

5 A FORMULA FOR C^1-FUNCTIONS

Let C^1 denote the set of continuously differentiable functions from $\mathbb{Z}_p$ into K. In [1] p. 159 it is shown that C^1 is a Banach space for the norm $||\cdot||_1$ defined as follows. If

$$f(x) = \sum_{n=0}^{\infty} a_n \binom{x}{n},$$

then we define

$$||f||_1 = max(|a_0|, sup_{n\geqslant 1} n|a_n|).$$

THEOREM 1 *For $f \in C^1$ we have*

$$\sum_{n=1}^{\infty}(-1)^n \frac{u_n}{n} f(x-n) = f'(x) - \frac{1}{p}\sum_{n=1}^{\infty}\frac{(-1)^{n-1}}{n}(\Delta^{np} f)(x) \quad p \neq 2.$$

Proof. We first prove that the theorem is true for polynomials. Note that it is sufficient to prove the formula for $x = 0$. Indeed, if

$$\sum_{n=1}^{\infty}(-1)^n \frac{u_n}{n} f(-n) = f'(0) - \frac{1}{p}\sum_{n=1}^{\infty}\frac{(-1)^{n-1}}{n}(\Delta^{np} f)(0) \qquad (10)$$

is true for all polynomial functions $t \to P(t)$, then (10) is true for the polynomial function $t \to P(t+x)$. To prove (10) for polynomials it suffices to consider the case $f(t) = \binom{t}{N}$ for $N = 0, 1, 2, \ldots$. In that case $(\Delta^{np} f)(0) = \binom{t}{N-np}$ and the series in the R.H.S. of (10) can contain at most one term that is non zero. This happens if N is divisible by p. Hence (10) takes the form

$$\sum_{n=1}^{\infty}(-1)^{n+N}\frac{u_n}{n}\frac{n(n+1)\ldots(n+N-1)}{N!} = \frac{(-1)^{N-1}}{N} + \frac{(-1)^N}{N}\chi_p(N).$$

Multiplying with $(-1)^N N$ and changing N into $N+1$ we get

$$\sum_{n=1}^{\infty}(-1)^n u_n\binom{n+N}{n} = -1+\chi_p(N+1)$$

or

$$\sum_{n=0}^{\infty}u_n\binom{-1-N}{n} = \chi_p(-N-1).$$

Since this is true because of (1) we have proved the theorem for polynomials.
Let $C(\mathbb{Z}_p)$ denote the space of continuous functions $\mathbb{Z}_p \to K$ equipped with the supremum norm $||\cdot||_\infty$. Consider three linear operators $T_i : C^1 \to C(\mathbb{Z}_p)$ $i = 1,2,3$ defined as follows. For $f \in C^1$

$$(T_1 f)(x) = \sum_{n=1}^{\infty}(-1)^n \frac{u_n}{n} f(x-n),$$

$$(T_2 f)(x) = f'(x),$$

$$(T_3 f)(x) = \sum_{n=1}^{\infty}\frac{(-1)^n}{n}(\Delta^{np} f)(x).$$

It is clear that both series are uniformly convergent and define continuous functions. From the first part of the proof it follows that the theorem is true for polynomials. Since the polynomials are dense in C^1 it is sufficient to prove that the operators T_1, T_2 and T_3 are continuous. Since $|\frac{u_n}{n}| \leqslant 1$ (see the proof of corollary 1 of proposition 3) we have

$$|(T_1 f)(x)| = \left|\sum_{n=1}^{\infty}(-1)^n \frac{u_n}{n} f(x-n)\right| \leqslant sup_{n\geqslant 0}\left|\frac{u_n}{n}\right| |f(x)| \leqslant ||f||_\infty.$$

In [1] it is shown that $||f||_\infty \leqslant ||f||_1$ hence $||T_1 f||_\infty \leqslant ||f||_1$ and T_1 is continuous. In order to prove that T_2 is continuous we use the Mahler expansions

$$f(x) = \sum_{n=0}^{\infty} a_n \binom{x}{n}, \qquad f'(x) = \sum_{n=0}^{\infty} b_n\binom{x}{n}.$$

It is known that $b_n = \sum_{i=1}^{\infty}\frac{(-1)^{i-1}}{i} a_{i+n}$. Using the fact that $\frac{1}{|i|} \leqslant i$ we see that

$$|b_n| \leqslant sup_{i\geqslant 1}\frac{|a_{i+n}|}{|i|} \leqslant sup_{i\geqslant 1}|a_{i+n}|(i+n).\frac{i}{i+n} \leqslant ||f||_1.$$

Hence $||f'||_\infty = sup_n|b_n| \leqslant ||f||_1$ and T_2 is continuous.
Using the same notations as before the continuity of T_3 follows from

$$(\Delta^i f)(x) = \sum_{n\geqslant i} a_n\binom{x}{n-i} = \sum_{n=0}^{\infty} a_{n+i}\binom{x}{n},$$

$$\left|\frac{(\Delta^i f)(x)}{i}\right| \leqslant \frac{1}{|i|} sup_n |a_{n+i}| \leqslant sup_n |a_{n+i}|(n+i) \leqslant ||f||_1.$$

We end this paper with a theorem that is not related to the numbers u_n.

THEOREM 2 *Let f and g be continuous functions from $\mathbb{Z}_p$ into K. Then*

$$\sum_{n=0}^{\infty} \binom{x}{n} f(x-n)(\Delta^n g)(0) = \sum_{n=0}^{\infty} \binom{x}{n} g(x-n)(\Delta^n f)(0).$$

Proof. Since f and g are continuous $lim_{n\to\infty}(\Delta^n f)(0) = 0$, $lim_{n\to\infty}(\Delta^n g)(0) = 0$. This implies that all series we consider are convergent. We start from the Mahler expansion $g(x) = \sum_{n=0}^{\infty} \binom{x}{n} (\Delta^n g)(0)$. Let m be a natural number. Replacing x by $x - m$ we get

$$\sum_{n=0}^{\infty} \binom{x}{m} \binom{x-m}{n} (\Delta^n g)(0) = \binom{x}{m} g(x-m).$$

Since $\binom{x}{m}\binom{x-m}{n} = \binom{x}{n}\binom{x-n}{m}$ we get

$$\sum_{n=0}^{\infty} \binom{x}{n} \binom{x-n}{m} (\Delta^n g)(0) = \binom{x}{m} g(x-m).$$

$$\begin{aligned}
&\sum_{m=0}^{\infty} (\Delta^m f)(0) \binom{x}{m} g(x-m) \\
&= \sum_{m=0}^{\infty} (\Delta^m f)(0) \sum_{n=0}^{\infty} \binom{x}{n} \binom{x-n}{m} (\Delta^n g)(0) \\
&= \sum_{n=0}^{\infty} \binom{x}{n} (\Delta^n g)(0) \sum_{m=0}^{\infty} (\Delta^m f)(0) \binom{x-n}{m} \\
&= \sum_{n=0}^{\infty} \binom{x}{n} (\Delta^n g)(0) f(x-n)
\end{aligned}$$

The change of the order of summation in the double is allowed since $lim_{n\to\infty}(\Delta^n f)(0) = 0$ and $lim_{n\to\infty}(\Delta^n f)(0) = 0$.

REFERENCES

[1] WH Schikhof. Ultrametric Calculus. Cambridge University Press, 1984.

[2] LE Dickson. History of the Theory of Numbers, 1. New York: Chelsea Publishing Company, 1952.

[3] B Berndt. Ramanujan's Notebooks, II. New York: Springer, 1989.

p-adic (dF)-spaces

AK KATSARAS, V BENEKAS Department of Mathematics, University of Ioannina, 45110 Ioannina, Greece.

Abstract. Non-Archimedean (dF)-spaces are introduced and some of their properties are investigated. It is shown that every (dF)-space is nuclear and that a space E is nuclear iff it is topologically isomorphic to a subspace of a product of (dF)-spaces.

INTRODUCTION

If E is a real or complex locally convex space, then the polar dual E^p of E is the topological dual space E' of E endowed with the topology of precompact convergence. The space E is called polar reflexive (in the terminology of Köthe [12]) if the canonical mapping $J: E \to E^{pp}$ is a topological isomorphism. A locally convex space E is called a (dF)-space (in the terminology of Brauner [2]) or a (DCF)-space (in the terminology of Hollstein [8]) if E is polar reflexive and it has a fundamental sequence of compact sets. Several authors have studied properties of such spaces. In this paper we introduce the non-Archimedean (dF)-spaces. The c-dual E^c, of a non-Archimedean locally convex space E, is the dual space E' of E endowed with the topology of compactoid convergence. E is said to be a (dF)-space if it has a fundamental sequence of compactoid sets and the canonical mapping $J_E : E \to E^{cc}$ is a topological isomorphism. We study some of the properties of (dF)-spaces. We show that every such space E is a complete nuclear space and its topology is the finest of all nuclear topologies τ on E for which $(E, \tau)' = E'$. In contrast to what happens in the classical case, the class of (dF)-spaces is not stable under the taking of closed subspaces and separated quotients. We prove that there is a relation between nuclear spaces and (dF)-spaces. More specifically we show that a locally convex space E is nuclear iff it is topologically isomorphic to a subspace of a cartesian product of (dF)-spaces or even to a subspace of D^I, for some index set I and some (dF)-space D.

1 PRELIMINARIES

Throughout this paper, $\mathbb{K}$ will stand for a complete non-Archimedean valued field whose valuation is non-trivial. By a seminorm, on a vector space E over $\mathbb{K}$, we will mean a non-Archimedean seminorm. Also, by a locally convex space, we will mean a non-Archimedean locally convex space over $\mathbb{K}$. Let now E be a locally convex space. We will denote by $cs(E)$ the collection of all continuous seminorms on E. For a subset S of E, we will denote by $co(S)$ the absolutely convex hull of S. In case $S = \{x_1, x_2, \dots, x_n\}$ is a finite set, we have

$$co(S) = \{\sum_{k=1}^{n} \lambda_k x_k : \lambda_k \in \mathbb{K}, |\lambda_k| \leqslant 1\}.$$

The topological dual space (or simply the dual space) of E will be denoted by E'. By $\sigma(E, E')$ and $\sigma(E', E)$ we will denote the weak topologies on E and E' respectively. For a subset A of E we will denote by A^o and A^{oo} the polar and the bipolar of A, respectively. If A is absolutely convex, then A^e is the edged hull of A (see [13]). If E is Hausdorff, then we will denote by $\widehat{E}$ its completion. A subset A of E is called compactoid if, for each neighborhood V of zero, there exists a finite subset S of E such that $A \subset co(S) + V$. In this case, if $|\lambda| > 1$, we may choose $S \subset \lambda A$. A linear map T, from a locally convex space E to another F, is called compact if there exists a neighborhood V of zero in E such that $T(V)$ is a compactoid subset of F. A Hausdorff locally convex space E is called nuclear if, for each $p \in cs(E)$, the quotient map $\pi_p : E \to E_p$ is compact, where E_p is the quotient space $E/kerp$, $kerp = \{x \in E : p(x) = 0\}$, and E_p is equipped with the norm $||\pi_p(x)|| = p(x)$. Also, if V is a neighborhood of zero in E, then E'_{V^o} is the vector subspace of E' spanned by V^o and on E'_{V^o} we use the Minkowski functional p_{V^o} of V^o (which is a norm). A sequence (a_n) in E' is locally null if it is a null sequence in E'_{V^o} for some neighborhood V of zero in E. The locally convex space E is a semi-Montel space ((SM)-space) if every bounded subset of E is compactoid. A (gDF)-space is a locally convex space E which has a fundamental sequence (B_n) of bounded sets and whose topology is the finest of all locally convex topologies τ on E which coincide with the topology of E on each B_n. As in the classical case, a (gDF)-space E has the countable neighborhood property, i.e. for each sequence (V_n) of neighborhoods of zero there exists a sequence (λ_n) of non-zero scalars, such that $\bigcap_n \lambda_n V_n$ is a neighborhood of zero. A metrizable locally convex space, with the countable neighborhood property, is normable. For all unexplained terms, concerning locally convex spaces, we refer to [13] and [17].

2 THE c-DUAL SPACE

For a locally convex E, we will denote by E^c the dual space E' of E endowed with the topology $\tau_c = \tau_c(E', E)$ of uniform convergence on the compactoid subsets of E. We will refer to E^c as the c-dual space of E. The space $E^{cc} = (E^c)^c$ will be called the c-bidual space of E. Let

$$J_E : E \to E^{cc}, \quad <J_E x, x'>=<x', x>, \quad x \in E,\, x' \in E'.$$

Clearly J_E is linear. If E' separates the points of E (in particular if E is a Hausdorff polar space), then J_E is one-to-one.

DEFINITION 2.1 A Hausdorff polar space E is called:
a) c-semireflexive if J_E is onto.
b) c-complete if every closed compactoid subset of E is complete.

In case E is c-semireflexive, we will usually write $(E^c)' = E$ (identifying E with $J_E(E)$).

PROPOSITION 2.2 ([10], Theorem 4.7) *If E is a polar space, then E is c-semireflexive iff it is c-complete.*

For a locally convex space E, we will denote by $\tau_\sigma = \tau_\sigma(E', E)$ the finest locally convex topology on E' which agrees with $\sigma(E', E)$ on equicontinuous sets. By $\tau_{\sigma p} = \tau_{\sigma p}(E', E)$ we will denote the finest polar topology on E' which is coarser than τ_σ.

PROPOSITION 2.3 ([10], Proposition 3.1) *If E is a Hausdorff polar space, then $(E', \tau_\sigma)' = (E', \tau_{\sigma p})' = \widehat{E}$.*

PROPOSITION 2.4 *Let E be a Hausdorff polar space. Then:*
(1) $\tau_{\sigma p}$ is the finest polar topology on E' which agrees with $\sigma(E', E)$ on equicontinuous sets.
(2) $\tau_{\sigma p}$ coincides with the topology of uniform convergence on the compactoid subsets of $\widehat{E}$.

Proof. (1) It follows easily from the fact that

$$\sigma(E', E) \leqslant \tau_{\sigma p} \leqslant \tau_\sigma.$$

(2) We will consider E' as the dual space of both E and $\widehat{E}$. We have that $\sigma(E', E) = \sigma(E', \widehat{E})$ on equicontinuous subsets of E' and so $\sigma(E', \widehat{E}) \leqslant \tau_{\sigma p}$. Let $G = (E', \tau_{\sigma p})$. For each $\tau_{\sigma p}$ -neighborhood of zero W, let W^o be the polar of W in $\widehat{E}$ and consider the family

$$A = \{W^o : W \quad \text{a polar neighborhood of zero in} \quad G\}.$$

Let γ be the topology on E' of uniform convergence on the compactoid subsets of $\widehat{E}$ and let ω be the topology on $\widehat{E}$ of uniform convergence on the compactoid subsets of G. By [11], Remark 3.1, every member of A is ω-compactoid.

Claim. If τ is the topology of E and $\widehat{\tau}$ the topology of $\widehat{E}$, then $\widehat{\tau} \leqslant \omega$. Indeed, let V be a polar $\widehat{\tau}$-neighborhood of zero and let $D = V \cap E$. Then the polars D^o and V^o, of D, V in E', coincide. Since D^o is a $\sigma(E', E)$-compactoid and since

$\sigma(E',E) = \tau_{\sigma p}$ on D^o, it follows that V^o is a compactoid subset of G and so $V = V^{oo}$ is an ω-neighborhood of zero, which proves our claim.

By the above claim, each member of A is $\widehat{\tau}$-compactoid and so $\tau_{\sigma p} \leqslant \gamma$. Since γ is a polar topology, in order to finish the proof, it suffices to show that $\gamma = \sigma(E',E)$ on equicontinuous subsets of E'. If W is a neighborhood of zero in E and if $\widehat{W}$ is the closure of W in $\widehat{E}$, then W and $\widehat{W}$ have the same polar in E'. Since $\gamma = \sigma(E',\widehat{E}) = \sigma(E',E)$ on $(\widehat{W})^o = W^o$, the result follows.

COROLLARY 2.5 *Let E be a Hausdorff polar space. Then:*

(a) *If E is complete, then $\tau_c(E',E) = \tau_{\sigma p}(E',E)$.*

(b) *If E is c-complete, then E is complete iff $\tau_c(E',E) = \tau_{\sigma p}(E',E)$.*

PROPOSITION 2.6 *Let E be a Hausdorff polar space and let*

$$G = \cup\{\overline{A} : A \quad \text{compactoid subset of} \quad E\},$$

where $\overline{A}$ denotes the closure of A in $\widehat{E}$. Then $(E',\tau_c)' = G$.

Proof. Since $\tau_c \leqslant \tau_\sigma$, it follows that

$$(E',\tau_c)' \subset (E',\tau_\sigma)' = \widehat{E}.$$

Each $x \in \widehat{E}$ defines a linear functional on E' by $<x,f>= \widehat{f}(x)$, where $f \in E'$ and $\widehat{f}$ is the unique continuous extension of f to $\widehat{E}$. Let now $x \in (E',\tau_c)'$, $x \in \widehat{E}$. There exists an absolutely convex compactoid subset A of E such that $|<x,f>| \leqslant 1$ for all f in the polar A^o of A in E'. Thus x belongs to the polar A^{oo} of A^o in $\widehat{E}$. Since $\overline{A}$ is an absolutely convex compactoid in $\widehat{E}$, it follows that $\overline{A}$ is weakly closed in $\widehat{E}$ by [13], Theorem 5.13. Thus

$$x \in A^{oo} = (\overline{A})^e \subset \lambda\overline{A},$$

if $|\lambda| > 1$, and so $x \in G$. Conversely, let $z \in \overline{A}$, for some absolutely convex compactoid subset A of E and let (x_δ) be a net in A converging to z. Then, for $f \in A^o$, we have $|<z,\widehat{f}>| = \lim|<x_\delta,f>| \leqslant 1$ and so $z \in (E',\tau_c)'$. This clearly completes the proof.

We also have the following easily established

PROPOSITION 2.7 *Let E be a polarly bornological space. For a subset D of E^c, the following are equivalent:*

(1) *D is compactoid.*

(2) *D is bounded.*

(3) *D is equicontinuous.*

PROPOSITION 2.8 *Let $E = \Pi E_\imath$ and $G = \oplus G_\jmath$, where $(E_\imath)_{\imath \in I}$ and $(G_\jmath)_{\jmath \in J}$ are families of Hausdorff locally convex spaces. Then*

$$E^c \cong \oplus E_\imath^c \quad \text{and} \quad G^c \cong \Pi G_\jmath^c.$$

Proof. As it is well known, E' is algebraically isomorphic to $\oplus E_i'$ via the isomorphism

$$f \mapsto (f_i), \quad f_i = f|_{E_i},$$

(we may consider E_i as a topological subspace of E). For each $i \in I$, the projection $E \to E_i$ is continuous and so the canonical mapping $g_i : E_i^c \to E^c$ is continuous, which implies that the topology of E^c is coarser than the direct sum topology τ of $M = \oplus E_i^c$. On the other hand, let W be a convex τ-neighborhood of zero. For each i, there exists an absolutely convex compactoid subset A_i of E_i, such that $A_i^o \subset W$, where A_i^o is the polar of A_i in E_i^c (we may consider E_i^c as a topological subspace of M). The set $A = \Pi A_i$ is a compactoid subset of E with $A^o \subset W$. This proves the result for E^c. For the G^c, we first observe that G' is algebraically isomorphic to $\Pi G_j'$. For each $j \in J$, the canonical map $G_j \to G$ is continuous and so the projection $G^c \to G_j^c$ is continuous, which implies that the topology of G^c is finer than the product topology on $F = \Pi G_j^c$. On the other hand, let A be an absolutely convex compactoid subset of G and let $W = A^o$ be its polar in G'. Since A is bounded, there exists (as in the classical case) a finite subset J_1 of J such that $x_j = 0$ for all $x = (x_j) \in A$ and all $j \notin J_1$. Let $\pi_j : G \to G_j$ be the projection map and let V_j be the polar A_j^o of $A_j = \pi_j(A)$ in G_j'. The set

$$D = \{f = (f_j) \in \Pi G_j^c : f_j \in V_j \quad \text{for} \quad j \in J_1\}$$

is a product neighborhood of zero with $D \subset W$. This clearly completes the proof.

COROLLARY 2.9 *If $E = \Pi E_i$ and $G = \oplus G_j$ and if each E_i and each G_j is c-semireflexive, then E and G are c-semireflexive.*

PROPOSITION 2.10 *Let $E = \varinjlim E_n$, where each E_n is a locally convex space. If the inductive limit is compactoid regular, then*

$$E^c \cong \varprojlim E_n^c \quad \textit{(topologically)}.$$

Proof. It is well known that $E' \cong \varprojlim E_n'$ algebraically. Since the inclusion maps $E_n \to E$ are continuous, it follows that each of the canonical maps

$$E^c \to E_n^c, \quad f \mapsto f|_{E_n},$$

is continuous and so the topology of E^c is finer than the projective topology. On the other hand, let A be an absolutely convex compactoid in E. Since the inductive limit is compactoid regular, there exists n such that A is compactoid in E_n. Let A^o and D_n be the polars of A in E' and E_n', respectively. If

$$\phi_n : E^c \to E_n^c, \quad f \mapsto f|_{E_n},$$

then $\phi_n^{-1}(D_n)$ is a neighborhood of zero for the projective topology. Since $\phi_n^{-1}(D_n) \subset A^o$, the results follows.

3 POLARLY c-BARRELLED SPACES

DEFINITION 3.1 A polar c-barrel, in a locally convex space E, is a polar subset W of E with the following property: For every compactoid subset A of E, there exists a zero-neighborhood V in E with $V \cap A \subset W$. The space E is polarly c-barrelled if every polar c-barrel is a neighborhood of zero.

We have the following easily established

PROPOSITION 3.2 a) *If F is a subspace of a polarly c-barrelled space E , then E/F is polarly c-barrelled.*

(b) *Every infrabarrelled (in particular every bornological space) is polarly c-barrelled.*

PROPOSITION 3.3 *Let E be a locally convex space.*

(1) *If E is polarly c-barrelled, then each compactoid subset of E^c is equicontinuous.*

(2) *If E is polar, then:*

a) $J_E^{-1} : J_E(E) \to E$ *is continuous.*

b) *The following are equivalent:*

(i) *E is polarly c-barrelled.*

(ii) *Every compactoid subset of E^c is equicontinuous.*

(iii) $J_E : E \to E^{cc}$ *is continuous.*

Proof. (1) Assume that E is polarly c-barrelled and let H be a compactoid subset of E^c. If A is a compactoid subset of E, then $H \subset co(S) + A^o$, for some finite subset S of E', and so $A \cap S^o \subset H^o$, which proves that H^o is a polar c-barrel and hence a neighborhood of zero.

(2) Suppose that E is polar. a) If V is a polar neighborhood of zero in E, then V^o is a compactoid subset of E^c, by [11], Remark 3.1, and so its polar D in E^{cc} is a neighborhood of zero in E^{cc}. Moreover,

$$J_E^{-1}(D \cap J_E(E)) = V^{oo} = V.$$

b) (ii) $\Rightarrow$ (iii) If H is a compactoid subset of E^c, then (by our hypothesis) $H \subset V^o$, for some neighborhood V of zero in E. If D is the polar of H in E^{cc}, then $J_E(V) \subset D$, which proves that J_E is continuous.

(iii) $\Rightarrow$ (i) Let W be a polar c-barrel in E. Let $|\lambda| > 1$ and let A be an absolutely convex compactoid in E. Without loss of generality, we may assume that A is a polar set in E. Since W is a polar c-barrel and since on compactoid sets the topology of E coincides with the weak topology, there exists a finite subset S of E' such that

$$[(\lambda A)^o + co(S)]^o = (\lambda A) \cap S^o \subset W$$

and so

$$W^o \subset [(\lambda A)^o + co(S)]^{oo}.$$

By Schikhof [14], Corollary 1.2, the set $[(\lambda A)^o + co(S)]^e$ is weakly closed. Thus

$$W^o \subset [(\lambda A)^o + co(S)]^e \subset A^o + co(\lambda S).$$

This proves that W^o is a compactoid subset of E^c and hence its polar D in E^{cc} is a neighborhood of zero. In view of our hypothesis $J_E^{-1}(D)$ is a neighborhood of zero in E. Since $J_E^{-1}(D) = W^{oo} = W$, the result follows.

PROPOSITION 3.4 *If $E = \oplus E_\imath$, where $(E_\imath)_{\imath \in I}$ is a collection of polarly c-barrelled spaces, then E is polarly c-barrelled.*

Proof. It follows from the fact that, if D is a polar c-barrel in E, then $D \cap E_\imath$ is a polar c-barrel in $E_\imath$, for each i.

THEOREM 3.5 *Let $E = \Pi E_\imath$, where each $E_\imath$ is a Hausdorff locally convex space. If each $E_\imath$ is barrelled (resp. polarly barrelled, resp. polarly c-barrelled), then E is barrelled (resp. polarly barrelled, resp. polarly c-barrelled).*

Proof. Let W be a barrel in E. By an argument analogous to the one used in [1], Lemma 3.7, we show that there exists a finite subset J of I such that the set $M = \bigoplus_{\imath \in I \setminus J} E_\imath$ is contained in W (we consider M as a subset of E). Since the closure of M in E is the space $\prod_{\imath \in I \setminus J} E_\imath = G$, it follows that $G \subset W$. Since $F = \prod_{\imath \in J} E_\imath = \bigoplus_{\imath \in J} E_\imath$, it follows that F is barrelled (resp. polarly barrelled, resp. polarly c-barrelled). Also F is a closed topological subspace of E. The set $A = W \cap F$ is a barrel in F. If W is a polar barrel or a polar c-barrel, then A is a polar barrel or a polar c-barrel respectively. Thus if each $E_\imath$ is barrelled (resp. polarly barrelled, resp. polarly c-barrelled) and if W is a barrel (resp. polar barrel, resp. polar c-barrel), then A is a neighborhood of zero in F. In this case, it follows easily that $B = A + \prod_{\imath \in I \setminus J} E_\imath$ is a neighborhood of zero in E. Since $B \subset W$, we get that W is a neighborhood of zero in E. This clearly completes the proof.

As the following Proposition shows, barrelledness is a three-space property.

PROPOSITION 3.6 *Let F be a closed subspace of a locally convex space E. If both F and E/F are barrelled, then E is barrelled.*

Proof. Let W be a barrel in E and let $\phi : E \to E/F$ be the quotient map. Since $W \cap F$ is a barrel in F, there exists a convex neighborhood V of zero in E with $V \cap F = W \cap F$. Next we observe that $M = \overline{\phi(W \cap V)}$ is a barrel in E/F. If $x \in E$ is such that $\phi(x) \in M$, then for each neighborhood D of zero in E we have that $\phi(x) \in \phi(W \cap V) + \phi(D)$, and so $x \in W \cap V + D + F$, which implies that $x \in \overline{W \cap V + F}$. Thus

$$\phi^{-1}(M) \subset \overline{W \cap V + F}$$

and so $\overline{W \cap V + F}$ is a neighborhood of zero in E. Now, for each convex neighborhood D of zero in E, we have

$$\begin{aligned} V \cap (\overline{W \cap V + F}) &\subset V \cap (W \cap V + F + D \cap V) \\ &\subset V \cap (W \cap V + F \cap V + D \cap V) \\ &\subset V \cap (W \cap V + W \cap F + D \cap V) \\ &\subset D + W, \end{aligned}$$

and so

$$V \cap (\overline{W \cap V + F}) \subset \overline{W} = W,$$

which proves that W is a neighborhood of zero.

PROBLEM 3.7 *Is the property of polar barrelldness or polar c-barrelldness a three-space property?*

PROPOSITION 3.8 *Let F be a dense subspace of a locally convex space E. If F is polarly c-barrelled, then E is polarly c-barrelled. In particular, a completion of a polarly c-barrelled space is polarly c-barrelled.*

Proof. Let W be a polar c-barrel in E. Then, $V = W \cap F$ is a polar c-barrel in F and so it is a neighborhood of zero in F. Thus, $\overline{W \cap F}$ is a neighborhood of zero in E and so W is a neighborhood of zero since $\overline{W \cap F} \subset W$.

PROPOSITION 3.9 *Let E be a Hausdorff locally convex space and let F be a subspace of E of finite codimension in E. Then:*

a) *If F is polarly c-barrelled, then E is polarly c-barrelled.*

b) *If F is closed and E is polarly c-barrelled, then F is polarly c-barrelled.*

Proof. a) Since the closure of F in E is polarly c-barrelled (by the preceeding Proposition), we may assume that F is closed. Let G be any algebraic complement of F. Then G is a topological complement. If n is the dimension of G, then G is topologically isomorphic to $\mathbb{K}^n$ and so G is barrelled (and hence polarly c-barrelled). Since $E \cong F \times G$, the result follows from Treorem 3.5.

b) Let G be any algebraic (and hence topological) complement of F and let $\pi : E \to F$, $x \mapsto x_2$, where $x = x_1 + x_2$ with $x_1 \in G$, $x_2 \in F$. Then π is continuous. If W is a polar c-barrel in F, then $D = \pi^{-1}(W)$ is a polar c-barrel in E and so D is a neighborhood of zero in E. Since $D \cap F = W$, the result follows.

PROPOSITION 3.10 *Let E be a locally convex space. Then:*

a) *If E is c-semireflexive, then E^c is polarly c-barrelled.*

b) *If E is polarly c-barrelled, then E^c is c-semireflexive.*

Proof. a) Let H be a polar c-barrel in E^c. Since E is c-semireflexive, there exists a subset A of E such that $H = A^o$. Let W be a polar neighborhood of zero in E. Since W^o is a compactoid subset of E^c and since $\tau_c = \sigma(E', E)$ on W^o, there exists a finite subset S of E such that

$$[W + co(S)]^o = W^o \cap S^o \subset H$$

and so $A \subset [W + co(S)]^{oo}$. Since W is $\sigma(E, E')$-closed, the set $[W + co(S)]^e$ is $\sigma(E, E')$-closed (by [14], Corollary 1.2) and so

$$A \subset [W + co(S)]^e \subset \lambda W + co(\lambda S),$$

if $|\lambda| > 1$. This clearly proves that A is compactoid in E and so $H = A^o$ is a neighborhood of zero in E^c.

b) Assume that E is polarly c-barrelled. Since E^c is a polar space, in order to show that E^c is c-semireflexive, it suffices to show that E^c is c-complete (by Proposition 2.2). So, let D be a closed absolutely convex compactoid in E^c. By Proposition 3.3.(1), D is equicontinuous. So, without loss of generality, we may assume that $D = V^o$ for some neighborhood V of zero in E. But then D is $\sigma(E', E)$-complete and hence D is complete in E^c since $\sigma(E', E) = \tau_c$ on V^o. This clearly completes the proof.

DEFINITION 3.11 A Hausdorff polar space E is called c-reflexive if $J_E : E \to E^{cc}$ is a topological isomorphism.

In view of Propositions 2.2 and 3.3, a Hausdorff polar space E is c-reflexive iff it is c-complete and polarly c-barrelled.

Using Corollary 2.9, Proposition 3.4 and Theorem 3.5, we get the following

THEOREM 3.12 *Let $E = \Pi E_i$ and $G = \oplus G_j$, where $(E_i)_{i \in I}$ $(G_j)_{j \in J}$ are families of c-reflexive spaces. Then, E and G are c-reflexive.*

COROLLARY 3.13 *Every minimal-Hausdorff locally convex space and every space with the finest locally convex topology is c-reflexive.*

Proof. It follows from the fact that in the former case the space is topologically isomorphic to some product $\mathbb{K}^I$ ([16], Theorem 7) while in the latter to some $\bigoplus_I \mathbb{K}$.

Since every barrelled space is polarly c-barrelled, we have the following

PROPOSITION 3.14 *Every polar Fréchet space is c-reflexive. Also the c-dual of a c-reflexive space is c-reflexive.*

PROPOSITION 3.15 *Every reflexive space E is c-reflexive.*

Proof. Since the strong dual of any locally convex space is Hausdorff and polar, it follows that E is a Hausdorff polar space. The strong topology $b = b(E', E)$ is polar and $(E', b)' = E''$. Thus any $\sigma(E', E'')$-bounded set is strongly bounded. The space E is c-semireflexive since $(E^c)' \subset E''$. If D is a compactoid subset of E^c, then D is $\sigma(E', E'')$-bounded and hence D is strongly bounded. If W is the polar of D in E'', then D is a neighborhood of zero in E'' and so $D^o = J_E^{-1}(W)$ is a neighborhood of zero in E. Now, Proposition 3.3.(2) implies that E is polarly c-barrelled. This clearly completes the proof.

DEFINITION 3.16 A locally convex space E is called a c-space if every function $f : E \to \mathbb{K}$, whose restriction to each compactoid subset of E is continuous, is continuous on E.

It is easy to see that every metrizable space is a c-space.

PROPOSITION 3.17 *Let E be a c-space and let X be a zero-dimensional topological space. If a function $f : E \to X$ is such that its restriction to each compactoid subset of E is continuous, then f is continuous on E.*

Proof. Let $x_0 \in E$ and let W be a clopen neighborhood of $f(x_0)$ in X. If ϕ is the $\mathbb{K}$-characteristic function on W and if $h = \phi \circ f$, then the restriction of h , to each compactoid subset of E, is continuous and so h is continuous. Let V be a neighborhood of x_0 in E such that $|h(x) - h(x_0)| = |h(x) - 1| < 1$ if $x \in V$. Now, for $x \in V$, we have $h(x) \neq 0$ and so $f(x) \in W$, which proves that f is continuous at x_o.

PROPOSITION 3.18 *Every c-space E is polarly c-barrelled.*

Proof. Let $\mathcal{U}$ be the family of all polar c-barrels. Then, $\mathcal{U}$ is a base at zero for a locally convex topology $\tau_{\mathcal{U}}$ on E. Let

$$f : E \to (E, \tau_{\mathcal{U}}), \quad f(x) = x.$$

It is clear that the restriction of f to each compactoid subset of E is continuous and so f is continuous by the preceeding Proposition. Hence, every polar c-barrel in E is a neighborhood of zero, i.e. E is polarly c-barrelled.

PROPOSITION 3.19 *If E is a c-space, then both E^c and the strong dual E'_b are complete.*

Proof. Let (f_δ) be a Cauchy net in E^c. Then (f_δ) is $\sigma(E', E)$-Cauchy and hence we get a linear function $f : E \to \mathbb{K}$, $f(x) = \lim f_\delta(x)$.

Let A be an absolutely convex compactoid subset of E and let μ be a non-zero element of $\mathbb{K}$. There exists δ_0 such that $f_\delta - f_{\delta_0} \in \mu A^o$, for $\delta \geqslant \delta_0$, and so $|f(x) - f_{\delta_0}(x)| \leqslant |\mu|$ for all $x \in A$. If

$$V = \{x \in E : |f_{\delta_0}(x)| \leqslant |\mu|\},$$

then $|f(x)| \leqslant |\mu|$ for all $x \in V \cap A$. This proves that the restriction of f to each compactoid subset of E is continuous and so $f \in E'$. Moreover $f_\delta \to f$ in E^c. Thus, E^c is complete. The completeness of the strong dual E'_b follows from the fact that the strong topology is stronger than the topology of E^c and from the fact that E'_b has a base at zero consisting of $\sigma(E', E)$-closed (and hence τ_c-closed) sets.

4 THE ASSOCIATED POLARLY c-BARRELLED TOPOLOGY

Let (E, τ) be a locally convex space. The collection of all polar c-barrels in E is a base at zero for a locally convex topology τ^β. We have the following two easily established Propositions.

PROPOSITION 4.1 a) *τ^β is a polar topology.*
b) *If (E, τ) is a polar space, then $\tau \leqslant \tau^\beta$.*
c) *τ is polarly c-barrelled iff $\tau^\beta \leqslant \tau$.*

PROPOSITION 4.2 *If $(E, \tau), (F, \gamma)$ are locally convex spaces and $f : (E, \tau) \to (F, \gamma)$ is a continuous linear map, then $f : (E, \tau^\beta) \to (F, \gamma^\beta)$ is continuous.*

Assume now that (E, τ) is a polar space and let γ be the locally convex topology on E generated by the family of all polar seminorms. Since γ^β is polar and finer than γ, we get that $\gamma = \gamma^\beta$ and so γ is polarly c-barrelled. Thus, there are polarly c-barrelled topologies finer than τ. We will denote by τ^{pcb} the intersection of all polarly c-barrelled topologies finer than τ. The topology $\omega = \tau^{pcb}$ is polarly c-barrelled. Indeed, let W be a polar c-barrel in (E, ω). If ϕ is a polarly c-barrelled topology finer than τ, then $\omega \leqslant \phi$ and so W is a polar c-barrel in (E, ϕ), which implies that W is a ϕ-neighborhood of zero. This proves that W is a neighborhood of zero in (E, ω) and so ω is polarly c-barrelled. Clearly τ^{pcb} is the weakest polarly c-barrelled topology finer than τ. We will refer to τ^{pcb} as the *polarly c-barrelled topology associated to τ*. The topology ω is polar. Indeed, let ω_1 be the corresponding polar topology. Then $\tau \leqslant \omega_1 \leqslant \omega$ and so $\tau \leqslant \tau^\beta \leqslant \omega_1^\beta \leqslant \omega^\beta \leqslant \omega$. Since $\omega_1 \leqslant \omega_1^\beta$ and since ω_1^β is polar, it follows that $\omega_1 = \omega_1^\beta$, i.e. ω_1 is polarly c-barrelled, which implies that $\omega_1 = \omega$.

PROPOSITION 4.3 *Let (E, τ), (F, γ) be polar locally convex spaces and let $f : E \to F$ be a linear map. If f is (τ, γ)-continuous, then f is also $(\tau^{pcb}, \gamma^{pcb})$-continuous.*

Proof. Let τ_1 be the supremum of all polar topologies ϕ on F for which $f : (E, \omega) \to (F, \phi)$ is continuous, where $\omega = \tau^{pcb}$. Then τ_1 is a polar topology. Since $\omega^\beta = \omega$, we get (using Proposition 4.2) that $\tau_1 = \tau_1^\beta$ and so τ_1 is polarly c-barrelled. Since $\tau_1 \geq \gamma$, we have that $\tau_1 \geq \gamma^{pcb}$ and so

$$f : (E, \omega) \to (F, \gamma^{pcb})$$

is continuous.

PROPOSITION 4.4 *Let (E, τ) be a polar space and let $\omega = \tau^{pcb}$. If (x_α) is an ω-Cauchy net in E which converges to x in (E, τ), then $x_\alpha \to x$ in (E, ω).*

Proof. Let Φ be the collection of all polar locally convex topologies ϕ on E, with $\tau \leqslant \phi \leqslant \omega$, which have the following property: If (x_α) is a ϕ-Cauchy net which converges to some x with respect to the topology τ, then $x_\alpha \to x$ in (E, ϕ). If $\phi = \sup \Phi$, then it is easy to see that $\phi \in \Phi$. Also, since ϕ^β has a base at zero consisting of ϕ-closed sets and since $\phi \leqslant \phi^\beta$, it follows that $\phi^\beta \in \Phi$ and so $\phi = \phi^\beta$, which implies that ϕ is polarly c-barrelled. Finally, since $\tau \leqslant \phi \leqslant \omega$, we get that $\phi = \omega$ and the result follows.

COROLLARY 4.5 *If (E, τ) is a polar complete space, then both (E, τ^β) and (E, τ^{pcb}) are complete.*

PROPOSITION 4.6 *If τ is a polar topology on E, then τ and τ^β have the same compactoid sets.*

Proof. It follows from the fact that $\tau = \tau^\beta$ on τ-compactoid sets.

PROPOSITION 4.7 *If (E, τ) is a c-complete space, then (E, τ^{pcb}) is c-complete and so (E, τ^{pcb}) is c-reflexive.*

Proof. Let $\omega = \tau^{pcb}$ and A be an ω-closed compactoid in (E, ω). Then A is τ-compactoid and so $\overline{A}^\tau$ is τ-compactoid. Let (x_δ) be an ω-Cauchy net in A. Then (x_δ) is τ-Cauchy and hence $x_\delta \to^\tau x$ for some $x \in \overline{A}^\tau$. By Proposition 4.4, $x_\delta \to x$ in (E, ω). Since A is ω-closed, we have that $x \in A$, and the result follows.

PROPOSITION 4.8 *If (E, τ) is a polar space, then $\tau = \tau^{pcb}$ on τ-compactoid sets.*

Proof. Let Ω be the collection of all locally convex topologies γ on E, with $\tau \leqslant \gamma \leqslant \tau^{pcb}$, which agree with τ on each τ-compactoid set. If $\phi_1, \phi_2 \in \Omega$, then $\phi = \sup\{\phi_1, \phi_2\} \in \Omega$. Using this, we get that, if $\phi = \sup \Omega$, then $\phi \in \Omega$. If now A is a τ-compactoid, then A is ϕ-compactoid and so $\phi^\beta|_A \leqslant \phi|_A = \tau|_A$, which implies that $\phi^\beta \in \Omega$ since $\tau \leqslant \tau^\beta \leqslant \phi^\beta \leqslant \tau^{pcb}$. Thus $\phi^\beta \leqslant \phi$ and hence ϕ is polarly c-barrelled. This implies that $\phi = \tau^{pcb}$ and the result follows.

COROLLARY 4.9 *If τ is a polar locally convex topology on E, then τ and τ^{pcb} have the same compactoid sets.*

PROPOSITION 4.10 *Let (E, τ) be a polar space and let $\omega = \tau^{pcb}$. If τ_1 is the topology on E of uniform convergence on the compactoid subsets of E^c, then:*

(1) $\omega \geqslant \tau_1$.

(2) *If every linear form on E, whose restriction to each τ-compactoid set is continuous, is τ-continuous on E, then $\omega = \tau_1$.*

Proof. (1) Let $G = (E, \omega)$. Then, the identity map $I : E^c \to G^c$ is continuous, and so every compactoid subset D of E^c is also compactoid in G^c. Since G is polarly c-barrelled, it follows that every such D is ω-equicontinuous (by Proposition 3.3.(1)). This clearly proves that $\tau_1 \leqslant \omega$.

(2) Assume that the condition is satisfied. Since the topologies ω and τ have the same compactoid sets and agree on each such set, our hypothesis implies that (E, τ) and $G = (E, \omega)$ have the same c-dual space. Let now W be a polar ω-neighborhood of zero and let $D = W^o$ be its polar in E'. Given a polar compactoid subset A of (E, τ), there exists a finite subset S of E' such that $A \cap S^o \subset W$. Now

$[A^o + co(S)]^o \subset W$ and so

$$W^o \subset [A^o + co(S)]^{oo} = [A^o + co(S)]^e \subset \lambda A^o + co(\lambda S),$$

if $|\lambda| > 1$. This proves that W^o is a compactoid subset of E^c and so $W = W^{oo}$ is a τ_1-neighborhood of zero. This clearly completes the proof.

5 (dF)-SPACES

DEFINITION 5.1 A locally convex space E is hemicompactoid if it has a fundamental sequence (A_n) of compactoid sets, i.e. every compactoid subset of E is contained in some A_n.

Motivated from Definition 1.1 in [2], we give the following

DEFINITION 5.2 A locally convex space E is a (dF)-space if it is a hemicompactoid c-reflexive space.

As in the classical case (see [2]), we have the following

PROPOSITION 5.3 *A locally convex space E is a (dF)-space iff E is topologically isomorphic to G^c, for some polar Fréchet space G.*

Proof. If E is a (dF)-space, then $G = E^c$ is metrizable and c-complete (since G is c-reflexive) and so G is complete, i.e. G is a Fréchet space. Moreover, E is topologically isomorphic to G^c. Conversely, assume that E is topologically isomorphic to the c-dual of a polar Fréchet space G. Then E is c-reflexive since G (and hence G^c) is c-reflexive. Also, E has a fundamental sequence of compactoid sets, by Proposition 2.7, and so E is a (dF)-space.

COROLLARY 5.4 *Every (dF)-space is complete.*

COROLLARY 5.5 *Every (dF)-space is an (SM)-space and hence E^c coincides with the strong dual E'_b of E.*

Proof. It follows from Propositions 2.7 and 5.3.

THEOREM 5.6 *Every (dF)-space E is nuclear. Moreover, the topology of E is the finest of all nuclear topologies τ on E with $(E,\tau)' = E'$.*

Proof. Let $G = E^c$. Then G is a Fréchet space and E is topologically isomorphic to G^c. Moreover, G coincides with the strong dual E'_b of E by Corollary 5.5. Now E is an (SM)-space with a fundamental sequence of bounded sets and so it is nuclear by Corollary 3.9 in [6]. To prove the last assertion, let τ be a nuclear topology on E with $(E,\tau)' = (E,\tau_1)'$, where τ_1 is the topology of E. Then, τ and τ_1 have the same bounded sets and so they yield the same strong dual. Since τ is nuclear, it is the topology of uniform convergence on the locally null sequences in $(E,\tau)'$. If

(f_n) is such a sequence, then (f_n) is strongly null and so (f_n) is a null sequence in $E'_b = G$. This clearly proves that $\tau \leqslant \tau_1$ and the result follows.

PROPOSITION 5.7 *The locally convex direct sum, of a sequence of (dF)-spaces, is a (dF)-space.*

Proof. Let $E = \bigoplus_n E_n$, where each E_n is a (dF)-space. Each E_n^c is a Fréchet polar space and so $F = \prod_n E_n^c$ is a Fréchet polar space. Moreover

$$F^c \cong \oplus E_n^{cc} \cong \oplus E_n = E$$

and so E is a (dF)-space by Proposition 5.3.

REMARK The cartesian product of a sequence of (dF)-spaces need not be a (dF)-space. In fact the space $\mathbb{K}^{\mathbb{N}}$, with the product topology, is an infinite dimensional metrizable space and so it is not a (dF)-space by Theorem 5.9 below.

PROPOSITION 5.8 *A Hausdorff polar space E is a (dF)-space iff it is a semi-Montel complete (gDF)-space.*

Proof. Assume that E is a (dF)-space and so E is topologically isomorphic to G^c for some polar Fréchet space G. In view of Proposition 2.7 and Corollary 5.4, E is a complete (SM)-space with a fundamental sequence of bounded sets. Moreover, since G is metrizable, the topology of G^c coincides with the finest locally convex topology which agrees with $\sigma(G', G)$ on equicontinuous subsets of G' (by [10], Theorem 4.5). It follows that E is a (gDF)-space. Conversely, suppose that E is a semi-Montel complete (gDF)-space. Then E is a hemicompactoid space and $(E^c)' = E$. If H is a compactoid subset of E^c, then, for each bounded subset A of E, there exists a finite subset S of E' such that $H \subset co(S) + A^o$ and so $S^o \cap A \subset H^o$. Since E is a (gDF)-space, it follows that H^o is a neighborhood of zero and so H is an equicontinuous subset of E'. In view of Proposition 3.3.(2), E is polarly c-barrelled and so E is c-reflexive. This clearly completes the proof.

By the next Theorem there are no infinite-dimensional metrizable (dF)-spaces.

THEOREM 5.9 *A metrizable locally convex space E is a (dF)-space iff it is finite-dimensional.*

Proof. The condition is clearly sufficient. For the necessity, assume that E is a metrizable (dF)-space. Then it is a (gDF)-space and hence it has the countable neighborhood property. This implies that E is normable. If $\|\cdot\|$ is a norm on E giving its topology, then the closed unit ball will be compactoid and so E is finite dimensional.

PROPOSITION 5.10 *Let f be a linear functional on a (dF)-space E. Then, f is continuous if $f|_A$ is continuous for every compactoid subset A of E.*

Proof. We may assume that $E = G^c$, for some polar Fréchet space G. Assume that the restriction of f, to every compactoid subset of E, is continuous. Since every equicontinuous subset of G' is compactoid in G^c and since on equicontinuous subsets of G' the topology $\tau_c(G', G)$ coincides with $\tau_\sigma = \tau_\sigma(G', G)$, it follows that f is τ_σ-continuous. Since $\tau_\sigma = \tau_c$ (by [10], Theorem 4.5), the result follows.

We will next look at the question of whether closed subspaces and quotient spaces of (dF)-spaces are (dF)-spaces.

PROPOSITION 5.11 *Let E be a (dF)-space and let (A_n) be a fundamental sequence of absolutely convex compactoid subsets of E such that $\lambda A_n \subset A_{n+1}$, for all n, where $|\lambda| > 1$. Let F be a closed subspace of E, $\pi : E \to E/F$ the quotient map and $B_n = \pi(A_n)$. Then:*

a) *(B_n) is a fundamental sequence of compactoid subsets of E/F.*

b) *E/F is an (SM)-space.*

c) *The topology of E/F coincides with the finest of all locally convex topologies τ on E/F which agree with the quotient topology on each B_n.*

d) *E/F is polarly c-barrelled.*

Proof. We first observe that each B_n is compactoid and that d) holds by Proposition 3.2.

c) Let W be an absolutely convex subset of E/F such that, for each n, there exists a convex neighborhood W_n of zero in E/F with $W_n \cap B_n \subset W$. Each $V_n = \pi^{-1}(W_n)$ is a neighborhood of zero in E and $A_n \cap V_n \subset \pi^{-1}(W)$. Since E is a (gDF)-space, it follows that $\pi^{-1}(W)$ is a neighborhood of zero in E and so W is a neighborhood of zero in E/F.

a) & b) Let B be a bounded subset of E/F and assume that B is not contained in any B_n. Since $\lambda B_n \subset B_{n+1}$, B is not absorbed by any B_n. Thus, there exists $x_n \in B$ with $x_n \notin \lambda^n B_n$. Since, for each convex neighborhood W in E/F, we have

$$\overline{\pi(A_n)} + W = \pi(A_n) + W,$$

we have that $\lambda^{-n} x_n \notin B_n + W_n$, for some convex neighborhood W_n of zero in E/F. Set

$$D = \bigcap_n (B_n + W_n).$$

By c) and by [9], Theorem 5.2, D is a neighborhood of zero in E/F with $\lambda^{-n} x_n \notin D$, for all n, which is a contradiction since $\lambda^{-n} x_n \to 0$ in E/F. This clearly completes the proof.

THEOREM 5.12 (compare with Proposition 1.9 in [2]) *Let F be a closed subspace of a (dF)-space E, and let F^o be its polar in $G = E^c = E'_b$. Then: (1) $(E/F)^c$ is topologically isomorphic to $M = F^o$.*

(2) F is a (dF)-space iff G/F^o is polar. In this case $F \cong (G/F^o)^c$ topologically. In particular this happens, if G is strongly polar.

(3) E/F is a (dF)-space iff M has the weak extension property (W.E.P.) in G. In this case, $E/F \cong M^c$ topologically. In particular, if G is strongly polar, then E/F is a (dF)-space.

Proof. (1) Let $\phi : E \to E/F$ be the quotient map. Then

$$\Phi = \phi' : (E/F)^c \to E^c = G$$

is continuous, one-to-one and $\Phi((E/F)^c) = M$. Since (by Proposition 5.11) a subset D of E/F is compactoid iff $D \subset \phi(A)$, for some compactoid subset A of E, and since the compactoid subsets of E coincide with the equicontinuous subsets of G', in order to show that Φ^{-1} is continuous, it suffices to prove that, for each convex neighborhood V of zero in G, we have

$$\Phi^{-1}(V \cap F^o) \subset [\overline{\phi(V^o)}]^o = [\phi(V^o)]^o.$$

But, if $x \in V \cap F^o$, then for $y \in V^o$ we have

$$|<\Phi^{-1}(x), \phi(y)>| = |<x, y>| \leqslant 1$$

and so $\Phi^{-1}(x) \in [\phi(V^o)]^o$. This clearly completes the proof of (1).

(2) Since E is nuclear (and hence strongly polar), the space F is weakly closed in E and so $F^{oo} = F$. The restriction map

$$\omega : E^c \to F^c, \quad f \mapsto f|_F,$$

is continuous and F^o coincides with the kernel of ω. Moreover, ω is onto since E is strongly polar. Thus, the induced map

$$T : E^c/F^o \to F^c$$

is continuous, one-to-one and onto. If F is a (dF)-space, then F^c is a Fréchet polar space. Since E^c/F^o is also a Fréchet space, it follows (from the open-mapping Theorem) that T is a topological isomorphism and so E^c/F^o must be polar (since F^c is polar). Conversely, assume that G/F^o is a polar space (this in particular is the case when G is strongly polar because then G/F^o is strongly polar). We claim that $S = T^{-1}$ is continuous. Indeed, let D be a polar neighborhood of zero in $Z = E^c/F^o$. Given $|\lambda| > 1$, there exists a convex neighborhood V of zero in G such that $\pi(V) \subset \lambda^{-1}D$, where $\pi : E^c \to Z$ is the quotient map. Set $W = V^o \cap F$. If W^o is the polar of W in E', then

$$W^o = (V^o \cap F^{oo})^o = (V + F^o)^{oo} = [\overline{V + F^o}^{\sigma(G,G')}]^e \subset \lambda(\overline{V + F^o}^{\sigma(G,G')}).$$

Since π is continuous with respect to the weak topologies and since $\pi(V + F^o) = \pi(V) \subset \lambda^{-1}D$, it follows that $\pi(W^o) \subset D$, since D is weakly closed in Z. Thus

$$\omega(W^o) = T(\pi(W^o)) \subset T(D).$$

Since W is a compactoid subset of F and since $\omega(W^o)$ coincides with the polar of W in F, our claim follows. To finish the proof of (2), we first observe that F is c-complete and hence $(F^c)' = F$ algebraically. Let now H be an absolutely convex compactoid subset of F^c. Then $B = T^{-1}(H)$ is compactoid in Z. By Proposition

2.5 in [3], there exists an absolutely convex compactoid subset A of G such that $B \subset \pi(A)$. Let U be a convex neighborhood of zero in E, such that $A \subset U^o$, and set $X = U \cap F$. Let X^o be the polar of X in E'. If $y \in H$, then $y = Tx$, for some $x \in B$, and so $y = T \circ \pi(z)$ for some $z \in U^o \subset X^o$. Thus

$$H \subset T \circ \pi(U^o) \subset \omega(X^o).$$

Since $\omega(X^o)$ is the polar of X in F', it follows that every compactoid subset of F^c is an equicontinuous subset of F' and thus F is polarly c-barrelled by Proposition 3.3.(2). It is now clear that F is a (dF)-space. Moreover

$$F \cong F^{cc} \cong (E^c/F^o)^c.$$

(3) Assume that $M = F^o$ has the W.E.P. in G. To prove that E/F is a (dF)-space, it suffices to show that E/F is c-semireflexive since the rest of the proof follows from Proposition 5.11. So, let h be in the dual space of $(E/F)^c$. If Φ is as in the proof of (1), then $g = h \circ \Phi^{-1} \in M'$. Since M has the W.E.P. in G, there exists $f \in G'$ with $g = f|_M$ (see [5]). Now, for $x \in M$,

$$<\phi(f), \Phi^{-1}(x)> = <f, x> = <g, x> = <h, \Phi^{-1}(x)>$$

and so $\phi(f) = h$. This proves that E/F is c-semireflexive and thus E/F is a (dF)-space. Moreover, since $M \cong (E/F)^c$, we have that $E/F \cong M^c$. Conversely, assume that E/F is a (dF)-space and let $g \in M'$. Then $h = g \circ \Phi \in (E/F)^{cc}$ and so $h = \phi(f)$ for some $f \in E = G'$. Now, for $x \in M$, we have

$$<f, x> = <\phi(f), \Phi^{-1}(x)> = <h, \Phi^{-1}(x)> = <g, x>, \quad \text{i.e. } g = f|_M$$

and so M has the W.E.P. in G. This completes the proof.

REMARK As Schikhof has shown in [15], there exists a polar Banach space G with a weakly closed subspace M not having the W.E.P. in G. For such a space, $E = G^c$ is a (dF)-space, $F = M^0$ a closed subspace of E and $F^o = M$ does not have the W.E.P. in G and so E/F is not a (dF)-space. Thus a separated quotient of a (dF)-space need not be a (dF)-space.

6 A UNIVERSAL NUCLEAR SPACE

For a Hausdorff polar locally convex space E, let us denote by $\widetilde{E}$ the space E endowed with the associated nuclear topology (see [4]). As it is shown in [4], the topology of $\widetilde{E}$ coincides with the topology of uniform convergence on the locally null sequences in E'. In case E is a normed space, a sequence in E' is locally null iff it is a null sequence with respect to the usual norm on E'.

THEOREM 6.1 *Every nuclear space is topologically isomorphic to a subspace of a cartesian product* $\widetilde{c}_0{}^I$, *for some index set* I.

Proof. Let $\{(f_k^{(\imath)})_{k\in\mathbb{N}} : i \in I\}$ be the family of all locally null sequences in E'. For each $i \in I$, let $p_\imath = \sup_k |f_k^{(\imath)}|$. Since E is nuclear, its topology is generated by the family of seminorms $(p_\imath)_{\imath\in I}$. Set $G_\imath = E_{p_\imath}$ and let $\pi_\imath : E \to G_\imath$ be the quotient map. Let

$$T_\imath : E \to c_o, \quad T_\imath x = (f_k^{(\imath)}(x))_{k\in\mathbb{N}}.$$

Then $T_\imath$ induces a linear isometry $S_\imath : G_\imath \to c_o$. Set $M_\imath = S_\imath(G_\imath)$. Since c_o is strongly polar and since $c_o' = \ell^\infty$, it follows that the adjoint map $S_\imath' : \ell^\infty \to G_\imath'$ is onto. Since the polar $M_\imath^o$ of $M_\imath$ in ℓ^∞ is the kernel of $S_\imath'$, we get an algebraic isomorphism

$$\Phi_\imath : \ell^\infty / M_\imath^o \to G_\imath'.$$

This map is an isometry. In fact, let $z \in \ell^\infty$ and let

$$\omega_\imath : \ell^\infty \to \ell^\infty / M_\imath^o$$

be the quotient map. The map

$$h : M_\imath \to \mathbb{K}, \quad h(y) =< z, y >= \sum_{k=1}^{\infty} z_k y_k,$$

belongs to $M_\imath'$ and $||h|| = ||S_\imath' z||$. Since c_o is strongly polar, given $\varepsilon > 0$ there exists $w \in \ell^\infty$ such that $< w, y >= h(y)$, for $y \in M_\imath$, and

$$||w|| \leqslant ||h|| + \varepsilon = ||S_\imath' z|| + \varepsilon.$$

Now

$$||\omega_\imath(z)|| = ||\omega_\imath(w)|| \leqslant ||w|| \leqslant ||S_\imath' z|| + \varepsilon,$$

which proves that

$$||\Phi_\imath(\omega_\imath(z))|| \geq ||\omega_\imath(z)||.$$

It is also easy to see that

$$||\Phi_\imath(\omega_\imath(z))|| \leq ||\omega_\imath(z)||$$

and so $\Phi_\imath$ is an isometry. We claim that

$$S_\imath^{-1} : M_\imath \to \widetilde{G}_\imath$$

is continuous if we consider $M_\imath$ as a subspace of $\widetilde{c}_o$. Indeed, let (ϕ_n) be a null sequence in $G_\imath'$. Since null sequences in $\ell^\infty / M_\imath^o$ are images under $\omega_\imath$ of null sequences in ℓ^∞, there exists a null sequence (α_n) in ℓ^∞ such that $\phi_n = \Phi_\imath(\omega_\imath(\alpha_n))$. Let q be defined on c_o by $q(y) = \sup_n |<y, \alpha_n>|$. Then q is continuous on $\widetilde{c}_o$ and, for $y = S_\imath x$, we have $\sup_n |\phi_n(x)| = q(y)$ which proves our claim. Since $S_\imath : \widetilde{G}_\imath \to \widetilde{c}_o$ is continuous, it follows that $S_\imath : \widetilde{G}_\imath \to M_\imath$ is a topological isomorphism if we consider $M_\imath$ as a subspace of $\widetilde{c}_o$. Let now τ be the topology of E and let τ_o be the projective topology with respect to the quotient maps $\pi_\imath : E \to \widetilde{G}_\imath$. Clearly $\tau \geqslant \tau_o$. We will prove that $\tau \leqslant \tau_o$ by showing that each $\pi_\imath : (E, \tau_o) \to G_\imath$ is continuous. Indeed, let

$W_\imath$ be a convex neighborhood of zero in E such that $(f_k^{(\imath)})_{k\in\mathbb{N}}$ is a null sequence in $E'_{W_\imath^o}$. There exists $j \in I$ such that

$$V_j = \{x \in E : p_j(x) \leqslant 1\} \subset W_\imath.$$

By [7], Lemma 2.5.1, G'_j is isometric to $E'_{V_j^o}$ via the adjoint map $\pi'_j : G'_j \to E'$. If $h_k = (\pi'_j)^{-1}(f_k^{(\imath)})$, then (h_k) is a null sequence in G'_j and

$$\sup_k |h_k(\pi_j(x))| = p_\imath(x), \quad x \in E.$$

Thus the canonical map $\pi_{\imath j} : \widetilde{G}_j \to G_\imath$ is continuous. Since $\pi_\imath = \pi_{\imath j} \circ \pi_j$ and since $\pi_j : (E, \tau_o) \to \widetilde{G}_j$ is continuous, it follows that $\pi_\imath : (E, \tau_o) \to G_\imath$ is continuous. This clearly proves that $\tau \leqslant \tau_o$ and so $\tau = \tau_o$. Finally, let $X = \widetilde{c}_o{}^I$ and let

$$T : E \to X, \quad x \mapsto (S_i(\pi_\imath x))_{\imath \in I}.$$

If $D = T(E)$, then $T : E \to D$ is a topological isomorphism, which completes the proof.

PROPOSITION 6.2 a) *The completion of $\widetilde{c}_o$ coincides with the c-dual of ℓ^∞.*
b) *$\widetilde{c}_o$ is a (dF)-space iff it is complete.*
c) *If $\mathbb{K}$ is not spherically complete, then $\widetilde{c}_o$ is a (dF)-space.*

Proof. a) Let G be the c-dual of ℓ^∞. In view of [13], Proposition 8.2, the topology of G coincides with the topology of uniform convergence on the null sequences of ℓ^∞. Also, a sequence, in the dual space $c'_o = \ell^\infty$ of c_o, is locally null iff it is norm-null. Thus $\widetilde{c}_o$ is a topological subspace of G. Since G is complete (by Proposition 3.19), it only remains to show that c_o is dense in G. Let $M = \widetilde{c}_o$ and let M^o be the polar of M in the dual space ℓ^∞ of G. Let $z \in M^o$. For each $x \in c_o$, we have $< x, z >= 0$. This clearly proves that $z = 0$. It follows that $M^o = \{0\}$ and so $M^{oo} = G$, which implies that $\overline{M}^{\sigma(G,\ell^\infty)} = G$. Since G is a (dF)-space, it is nuclear and hence strongly polar. Now, by [13], Corollary 4.9, we get that $G = \overline{M}$ and so M is dense in G.

b) If M is complete, then $M = \widehat{M} = G$ and so M is a (dF)-space. The converse follows from Corollary 5.4.

c) It follows from the well known fact that, for $\mathbb{K}$ non-spherically complete, we have $(\ell^\infty)' = c_o$ (see [17], Theorem 4.17).

Since every subspace of a nuclear space is nuclear and every cartesian product of a family of nuclear spaces is nuclear, combining Theorems 5.6, 6.1 and Proposition 6.2, we get the following

THEOREM 6.3 *For a locally convex space E, the following are equivalent:*
(1) *E is nuclear.*
(2) *E is topologically isomorphic to a subspace of $\widetilde{c}_o{}^I$, for some index set I.*
(3) *There exists a (dF)-space D and an index set I such that E is topologically isomorphic to a subspace of D^I.*
(4) *E is topologically isomorphic to a subspace of a product of (dF)-spaces.*

REFERENCES

[1] V Benekas, AK Katsaras. Topological vector spaces over valued fields. Glas Mat 28 (48):241–258, 1993.

[2] K Brauner. Duals of Fréchet spaces and a generalization of the Banach-Dieudonné Theorem. Duke Math J 40:845–855, 1974.

[3] N De Grande-De Kimpe. Projective locally K-convex spaces. Proc Kon Ned Akad Wet A87:247–254, 1984.

[4 N De Grande-De Kimpe. Nuclear topologies on non-Archimedean locally convex spaces. Proc Kon Ned Akad Wet A90:279–292, 1987.

[5] N De Grande-De Kimpe, C Perez-Garcia. Weakly closed subspaces and the Hahn-Banach extension property in p-adic Analysis. Proc Kon Ned Akad Wet A91:253–261, 1988.

[6] N De Grande-De Kimpe, C Perez-Garcia. p-adic semi-Montel spaces and polar inductive limits. Result Math 24:66–75, 1993.

[7] N De Grande-De Kimpe, J Kąkol, C Perez-Garcia, WH Schikhof. p-adic locally convex inductive limits. In: WH Schikhof, C Perez-Garcia, J Kąkol, ed. p-Adic Functional Analysis. New York: Marcel Dekker, 1997, pp 153–222.

[8] R Hollstein. (DCF)-Räume und lokalkonvexe tensorprodukte. Arch Math XXIX:
524–531, 1997.

[9] AK Katsaras. Spaces of non-archimedean valued functions. Boll Unione Math Ital (6) 5-B:603–621, 1986.

[10] AK Katsaras, A Beloyiannis. On the topology of compactoid convergence in non-archimedean spaces. Ann Math Blaise Pascal 3:135–153, 1996.

[11] AK Katsaras, A Beloyiannis. On non-archimedean weighted spaces of continuous functions. In: WH Schikhof, C Perez-Garcia, and J Kąkol, ed. p-Adic Functional Analysis. New York: Marcel Dekker, 1997, pp 232–252.

[12] G Köthe. Topological Vector Spaces I. New York: Springer-Verlag, 1969.

[13] WH Schikhof. Locally convex spaces over non-spherically complete valued fields I-II. Bul Soc Math Belg B 38:187–224, 1986.

[14] WH Schikhof. The continuous image of a p-adic compactoid. Proc Kon Ned Akad Wet A92:119–123, 1989.

[15] WH Schikhof. The complementation of ℓ^∞ in p-adic Banach spaces. In: F

Baldassarri, S Bosch, B Dwork, ed. p-Adic Analysis. Berlin: Springer-Verlag, 1990, pp 342–350.

[16] WH Schikhof. Minimal-Hausdorff p-adic locally convex spaces. Ann Math Blaise Pascal 2:259–266, 1995.

[17] ACM van Rooij. Non-Archimedean Functional Analysis. New York: Marcel Dekker, 1978.

On the weak basis theorems for p-adic locally convex spaces

JERZY KĄKOL Faculty of Mathematics and Informatics, A. Mickiewicz University, 60-769 Poznań, Matejki 48-49, Poland.

THOMAS GILSDORF Department of Mathematics, University of North Dakota, 58202 Grand Fork, USA.

Abstract. It is known that if E is a locally convex space over a spherically complete non-trivially valued complete field K, then the weak and the original topologies of E have the same convergent sequences. This implies in particular, that for such spaces the "*weak basis theorem holds*", i.e. if E is a locally convex space with a weak Schauder basis (x_n), then (x_n) is a Schauder basis in the original topology of E. If K is not spherically complete, then the sequence space ℓ_∞ provides a concrete example of a Banach space with a weak Schauder basis (the unit vectors) which is not a Schauder basis. In this paper the *weak basis theorem* and related concepts are studied for spaces over nonspherically complete K.

1 INTRODUCTION

Throughout K denotes a non-archimedean non-trivially valued complete field. For fundamentals on locally convex spaces E over K (lcs) we refer to [28] and [26].

Let E be a Hausdorff lcs over K. A sequence (x_n) in E is called a (topological) *basis* for E if every element $x \in E$ can be written uniquely as $x = \sum_n \alpha_n x_n$ with $\alpha_n \in K$. If the coefficient (linear) functionals $f_n : x \to \alpha_n$ are continuous, then (x_n) is called a *Schauder basis*. Recall that Banach [2] proved that for any real or complex Banach space any basis is a Schauder basis (the "continuity theorem"). This line of research for the *real or complex* spaces was continued by many of specialists, for instance Newns [19] extended Banach's result to Fréchet spaces. Arsove and

This paper was prepared when the first named author visited the Department of Mathematics of the University of North Dakota in Grand Forks as Visiting Professor, 1997.

Edwards, Floret proved the same for metrizable complete topological vector spaces and sequentially retractive (LB)-spaces, respectively, [1], [5].

The Schauder basis (x_n) is called *equicontinuous* if the sequence of partial sum operators (S_n) is equicontinuous, where $S_n(x) = \sum_{k=1}^{n} f_k(x)x_k$, $n \in \mathbb{N}$, $x \in E$.

Equicontinuous bases for real or complex lcs were studied by McArthur and Rethenford in [18]. They proved e.g. that *every Schauder basis in a barrelled space is equicontinuous*. For real or complex lcs the following problem is known as the "weak basis problem".

"Is every weak Schauder basis for E, i.e. Schauder basis for the weak topology $\sigma(E, E')$, *a Schauder basis for* E*?"*.

Arsove and Edwards proved [1] that the answer is positive if E is a barrelled space. Dubinsky and Retherford observed [4] that the answer is negative in general. In [8] De Grande-De Kimpe solved completely the weak basis problem for lcs E having a $\sigma(E', E)$-sequentially complete topological dual E'. We refer the reader to the monograph [12] for more detail.

Let E be a Hausdorff lcs over K with a weak Schauder basis. Then the weak topology $\sigma(E, E')$ is Hausdorff. We shall say that for a lcs E *the weak basis theorem* holds if every weak Schauder basis in E is a Schauder basis. If K is *spherically complete*, then every weakly convergent sequence is convergent; E is sometimes called an (*Orlicz-Pettis space, (O.P.)-space*), cf. [22], 1.2. Therefore, *the weak basis theorem holds for a lcs* E *over spherically complete* K , since the sequence $(x - S_n(x))$ is a null-sequence in the weak topology and hence in the original one, where $S_n(x)$, $n \in \mathbb{N}$, is a partial sum associated with a weak Schauder basis (x_n).

In [7], [9] De Grande-De Kimpe proved the following result.

$(*)$ *Every Schauder basis in a barrelled (even in a G-) space* E *is an orthogonal basis, when* K *is spherically complete.*

Recall [7] that a basis (x_n) in a lcs E is said to be an *orthogonal basis* if the topology of E can be determined by a family $\mathcal{P}$ of non-archimedean seminorms p satisfying the condition:

$$\text{If } x \in E,\ x = \sum_{n=1}^{\infty} f_n(x)x_n, \text{ then } p(x) = \max_n p(f_n(x)x_n) \text{ for all } p \in \mathcal{P}.$$

Clearly every orthogonal basis is equicontinuous since $p(S_n(x)) \leqslant \max_{1 \leqslant j \leqslant n} p(f_j(x)x_j) \leqslant p(x)$, $n \in \mathbb{N}$.

Note also that the assumption concerning the spherical completeness in $(*)$ can be removed. We extend this result, see Proposition 5. It is clear that every lcs with a Schauder basis is a space of countable type. By Monna and Springer [17] every non-archimedean Banach space of countable type is linearly isomorphic to the sequence space c_0; hence it has a Schauder basis, cf. [27], 3.16, [24]. It is still unknown if every Fréchet space of countable type has a Schauder basis. Concrete examples of lcs of countable type having no Schauder basis were constructed also by De Grande-De Kimpe [7]. See also [20] for more recent results on this subject.

It is known that every lcs of countable type is (O.P.), [22], 1.3. Therefore the weak basis problem in the p-adic functional analysis has the following simple solution.

For a lcs E *over* K *with a weak Schauder basis the weak basis theorem holds iff* E *is an (O.P.)-space.*

This fact shows how this "p-adic situation" differs from its real or complex counterpart. In [16] Martinez-Maurica and Perez-Garcia proved that the space ℓ_1 has a quotient ℓ_1/M which is dual-separating and of countable type but without a Schauder basis.

A "good" example of a lcs which is not (O.P.) is the space ℓ_∞ of all bounded sequences in K, provided K is nonspherically complete, with the topology defined by the norm $||x|| = \sup_n |x_n|$, $x = (x_n) \in \ell_\infty$. The unit vectors (e_n) compose a weak Schauder basis in ℓ_∞, cf. [13], but ℓ_∞ is even not of countable type. Nevertheless, (e_n) is a *basic sequence* for ℓ_∞, i.e. (e_n) is a Schauder basis of the closed linear span of (e_n) in ℓ_∞.

Therefore, in the non-archimedean counterpart the following questions arise.

(1) *For which locally convex spaces is every topological basis a Schauder basis?*

(2) *Is a weak Schauder basis (x_n) in a lcs E a basic sequence in E?.*

(3) *Let $E = (E, \tau)$ be a lcs with a Schauder basis (x_n). Does there exist the finest locally convex topology on E compatible with τ and having (x_n) as a Schauder basis?*

(4) *Let E be an A-Banach space with a weak Schauder basis. Assume that every weak Schauder basis in E is a basis (basic sequence) in E. Does there exist on E a non-archimedean norm giving the topology of E defined by the original norm of E?*

We prove a non-archimedean counterpart of a result of Floret: *In any sequentially retractive (LB)-space any basis is an equicontinuous basis.* Hence all strict (LB)-spaces have such property. It turns out (as we show) that the answer concerning (2) is positive if E is a polarly barrelled polar space. We provide a wide class of non-polar spaces E with a weak Schauder basis which is a basic sequence in the original topology of E. The problem (3) always has the answer "yes".

We show that (4) has a positive solution. We provide also some examples illustrating problems mentioned above.

A-normed spaces (A-Banach spaces) $E = (E, ||.||)$ are defined as vector spaces over K endowed with a norm satisfying the triangle inequality. A norm $||.||$ on a vector space E is said to be *non-archimedean* if it verifies the inequality $||x + y|| \leqslant \max\{||x||, ||y||\}$ for all $x, y \in E$. Clearly, the topology generated by a non-archimedean norm is locally convex in the sense of Monna.

An A-normed space E will be called *non-archimedean* if the original topology of E can be defined by a non-archimedean norm.

A lcs E is called of *countable type* if for every continuous seminorm p on E the normed space $F =: E/\ker p$ is of countable type, i.e. F contains a countable subset whose linear span is dense in F. A lcs E is called a *polar* space if its topology is generated by polar seminorms. Recall, that a seminorm p is *polar*, if $p = \sup\{|f| : f \in E^*, |f| \leqslant p\}$, where E^* denotes the algebraic dual of E.

A locally convex topology γ on a lcs (E, τ) is called *compatible* with τ, if τ and γ have the same continuous linear functionals on E, i.e. $(E, \tau)' = (E, \gamma)'$.

If G is a vector subspace of (E, τ), $\tau|G$, τ/G denote the topology τ restricted to G and the quotient topology on the quotient space E/G, respectively. If α is a finer locally convex topology on E/G, we denote by $\xi := \tau \vee \alpha$ the weakest locally convex topology on E such that $\tau \leqslant \xi$, $\xi/G = \alpha$, $\xi|G = \tau|G$, cf. [3]. The sets $U \cap q^{-1}(V)$ compose a basis of neighbourhoods of zero for ξ, where U, V run over neighbourhoods of zero for τ and α, respectively, where $q : E \to E/G$ denotes the quotient map.

A metrizable and complete lcs will be called a *Fréchet* space. A lcs E is called a *barrelled* (*polarly barrelled*) space [26] if every closed and absorbing absolutely convex (polar) subset of E is a neighbourhood of zero. By the Baire Category Theorem it follows that every Fréchet (hence Banach) space is barrelled. Every barreled space is polarly barrelled; the converse implication is not true in general, [10].

A locally convex space E is called *dual-separating* if its topological dual E' separates points of E from zero, i.e. the weak topology of E is Hausdorff. Every Hausdorff polar lcs is dual-separating, cf. [26]. Note also that every lcs E with a weak basis (x_n) and its associated sequence (f_n) of linear functionals (not necessarily continuous) is dual-separating. By $\mathcal{P}(E)$ or $\mathcal{P}$ we will denote the set of all continuous non-archimedean seminorms defining the topology of E.

We shall say that a lcs (E, τ) is an (LB)-space, see [11], if there exists an increasing sequence (E_n, τ_n) of Banach spaces (called a defining sequence for (E, τ)) covering E such that $\tau_{n+1}|E_n \leqslant \tau_n$ for all $n \in \mathbb{N}$, and τ is the finest locally convex topology on E such that $\tau|E_n \leqslant \tau_n$ for all $n \in \mathbb{N}$.

Let E be an A-normed space and let B be a bounded neighbourhood of zero in E. Let W be its absolutely convex envelope and let p be the Minkowski functional of W. Clearly p is a non-archimedean seminorm on E and generates a seminormed topology ν on E, called the *Mackey envelope* of the topology τ. Clearly $\nu \leqslant \tau$. The topologies τ and ν are *compatible*. Moreover ν is the finest locally convex topology on E weaker than τ. Hence $\sigma(E, E') \leqslant \nu \leqslant \tau$. If E is *dual-separating*, then p is a non-archimedean norm. The sequence spaces ℓ_p over K (the definition as in the real or complex "case") are A-Banach spaces for $1 \leqslant p \leqslant \infty$. Note also the following simple observation, see [15]. *Let E be a dual-separating A-normed space of countable type. Then:*

(a) *The completion of (E, ν) is isomorphic to the sequence space c_0.*

(b) *$(E, \sigma(E, E'))$ has a Schauder basis.*

(c) *E' endowed with its natural non-archimedean norm is isomorphic to ℓ_∞.*

2 RESULTS

In the sequel we will assume that K is NONSPHERICALLY COMPLETE, if nothing more is mentioned.

We know already that the weak basis theorem fails for spaces over nonspherically complete K. We know also that if the weak basis theorem holds for a lcs E with a weak Schauder basis, then E does not contain a subspace (linearly) isomorphic to ℓ_∞. Indeed, in that case E is of countable type.

Following van Rooij, see [27], a lcs E is said to have *property* $(*)$ if *for every subspace F of E of countable type, each $f \in F'$ has a continuous linear extension to the whole space E.*

It is easy to see that property $(*)$ is equivalent to the property which says that every closed subspace of E of countable type is weakly closed.

As we have already mentioned the weak basis theorem holds only for (O.P.)-spaces. The following simple observation is the starting point of this note.

PROPOSITION 1 *Let $E = (E, \tau)$ be a lcs with a weak Schauder basis (x_n). The following assertions are equivalent.*

(i) *Every weak Schauder basis in E is a Schauder basis in E.*

(ii) *(x_n) is a Schauder basis.*

(iii) *E is of countable type.*

(iv) *E has property $(*)$.*

(v) *E is an Orlicz-Pettis space.*

If additionally, E is weakly sequentially complete, then (iv) *is equivalent to*

(vi) *E is polar and does not contain a subspace linearly isomorphic to ℓ_∞.*

Proof. The implications (i)$\Rightarrow$(ii)$\Rightarrow$(iii) and (v) $\Rightarrow$(i) are obvious. The impplication (iii)$\Rightarrow$(iv) follows from [26], 4.2 and 4.4. The implication (iv)$\Rightarrow$(v) is in [23], 1.6. The equivalence "(vi) iff (v)" follows from [23], 4.7.

Combining Proposition 1 with 2.4, 2.5, 2.6, 2.7 of [22] one gets the following concrete examples:

If E is a Fréchet (O.P.)-space and X is a zerod imensional Hausdorff topological space, then the weak basis theorem holds for every space $PC(X, E)$, $C(X, E)$ and $C_b(X, E)$, $(C_b(X, E), \tau_u)$ (provided X is pseudocompact in the last case).

We proved in [14], 3.2, (see also [25], 2.1 and [9], 2.1) that:

For a lcs (E, τ) there exists the finest locally convex topology ν of countable type compatible with τ.

The topology ν is the supremum topology of the all locally convex topologies of countable type on E, compatible with τ. The following version of this fact refers to problem (3).

PROPOSITION 2 *Every lcs (E, τ) with a Schauder basis (x_n) admits the finest locally convex topology ν_0 compatible with τ and having (x_n) as a Schauder basis. The topology ν_0 is the supremum topology of all the locally convex topologies on E of countable type compatible with τ.*

Proof. Let ν be the supremum topology of all the locally convex topologies on E of countable type compatiblc with τ. As we proved in [14], 3.2, this topology is compatible with τ and of countable type. Therefore (x_n) is also a Schauder basis for ν by Proposition 1. Since every lcs with a Schauder basis is of countable type, we are done.

It is known [7] that if (x_n) is an orthogonal basis in a lcs E and (f_n) is the corresponding sequence of continuous linear functionals, then (f_n) is equicontinuous iff (x_n) is *regular*, i.e. away from a neighbourhood of zero.

We have another characterization of this property.

PROPOSITION 3 *Let (E, τ) be a lcs with an orthogonal basis (x_n) and the corresponding sequence (f_n) of continuous linear functionals. Then the sequence $(\alpha_n f_n)$ is equicontinuous for a suitable choice of scalars (α_n) in K, $\alpha_n \neq 0$, $n \in \mathbb{N}$, iff E admits a continuous polar norm.*

Proof. Suppose that (α_n) is such a sequence in K. Then $p(x) = \sup_n |\alpha_n f_n(x)|$ is a continuous norm on E and $p(x) = \sup\{|f(x)| : f \in E', |f| \leqslant p\}$. Hence p is polar.

Conversely, suppose that q is a continuous polar norm on E. Let $\mathcal{P}$ be a family of continuous seminorms defining the topology τ such that $p(x) = \max_n p(f_n(x)x_n)$ for all $x \in$E and $p \in \mathcal{P}$. Take non-zero $\alpha_n \in K$, $n \in \mathbb{N}$, such that $|\alpha_n| \leqslant q(x_n)$ for all $n \in \mathbb{N}$. Since q is a continuous norm, there exists $p \in \mathcal{P}$ such that $q \leqslant p$. Then

$$|\alpha_n f_n(x)| \leqslant (|\alpha_n|)^{-1}|\alpha_n f_n(x)|q(x_n) = q(\alpha_n f_n(x)x_n(\alpha_n)^{-1}) \leqslant \\ p(\alpha_n f_n(x)x_n(\alpha_n)^{-1}) \leqslant p(x),$$

for all $n \in \mathbb{N}$. Therefore the sequence $(\alpha_n f_n)$ is equicontinuous.

EXAMPLE 4 We know already that the weak basis theorem fails in general. Note also that problem (2) has a negative solution for A-Banach spaces. The following example (suggested by Schikhof) provides a weak Schauder basis in the sequence space ℓ_1 which is not a basic sequence in the original topology: Let E be the sequence space $(\ell^1, \|.\|)$ of all K-sequences $x = (x_n)$ such that $\|x\| = \sum_{n=1}^{\infty} |x_n| < \infty$. Let (e_n) be the sequence of the unit vectors in E. For every $k \in \mathbb{N}$ let

$$f_k = \sum_{i=2^n}^{2^{n+1}-1} e_i$$

if $n \in \mathbb{N} \cup \{0\}$ and $k = 2^n$
and

$$f_k = e_k$$

otherwise. It is easy to see that (f_n) is a weak Schauder basis in E. Indeed, the sequence (f_n) is a norm-orthogonal basis of the space c_0, hence a weak Schauder basis of the space $\ell^1 \subset c_0$. We show that (f_n) is not a basic sequence in E. Observe that the closed linear span of (f_n) is E. The vector $x = (\xi_n)$, where

$$\xi_k = t^n$$

if $n \in \mathbb{N} \cup \{0\}$ and $k = 2^{2^n}$
and

$$\xi_k = 0$$

otherwise, where $t \in K$, $0 < |t| < 1$, belongs to E. On the other hand, assume that $x = \sum_{n=1}^{\infty} a_n f_n$, where $a_n \in K$, converges in the original topology of E. Then $a_{2^{2^n}} = \xi_{2^{2^n}}$ for all $n \in \mathbb{N} \cup \{0\}$; hence $\|a_{2^{2^n}} f_{2^{2^n}}\| = |t^n| 2^{2^n} \to \infty$, a contradiction since $\|a_n f_n\| \to 0$.

We will see that problem (2) has a positive solution for polar Banach spaces. In [7] De Grande-De Kimpe proved that every Schauder basis in a barrelled space is orthogonal (note that the assumption in [7] that K is spherically complete can be removed). Our next proposition extends this result.

PROPOSITION 5 *Let E be a polarly barrelled polar lcs and let (x_n) be a weak Schauder basis in E with its associated sequence (f_n) in E' such that $\{f_n(x)x_n :$*

$n \in \mathbb{N}\}$ is bounded in E for every $x \in E$. Then the sequence (x_n) is an orthogonal basis of the closed linear span G of (x_n).

Proof. Observe that for each $x \in E$ the set $\{f_n(x)x_n : n \in \mathbb{N}\}$ is weakly bounded and hence bounded because E is polar, [26], 7.5.

Let $\mathcal{P}$ be a family of polar continuous seminorms on E defining the original topology of E. For $p \in \mathcal{P}$ put $p^*(x) = \sup_n p(f_n(x)x_n)$, $x \in E$. Let $\mathcal{P}^* = \{p^* : p \in \mathcal{P}\}$. Clearly

$$\{x \in E : p^*(x) \leqslant 1\} = \bigcap_n \{x \in E : p(f_n(x)x_n) \leqslant 1\}$$

and for every $n \in \mathbb{N}$ the seminorm $x \to p(f_n(x)x_n)$ is a continuous polar seminorm and p^* is a polar seminorm. Hence the set $\{x \in E : p^*(x) \leqslant 1\}$ is a polar barrel in E; so it is a neighbourhood of zero. Therefore p^* is continuous. Now we prove that (x_n) is a basic sequence in E. Take $p \in \mathcal{P}$ and $\epsilon > 0$. Since E is polarly barrelled and the sequence (p_n) of polar continuous seminorms $p_n(x) = p(f_n(x)x_n)$, $n \in \mathbb{N}$, $x \in E$, is pointwise bounded, then the sequence (p_n) is equicontinuous [26]. Then there exists an absolutely convex neighbourhood of zero U in E such that $p(f_n(x)x_n) < \epsilon$ and $p(x) < \epsilon$ for every $x \in U$ and $n \in \mathbb{N}$. Therefore

$$p(\sum_{k=1}^{n} f_k(x)x_k) < \epsilon, \quad x \in U, \quad n \in \mathbb{N}.$$

For $x \in G$ let y be a finite linear combination of the x_n such that $x - y \in U$. Let $m \in \mathbb{N}$ be such that $\sum_{k=1}^{n} f_k(y)x_k = y$ for all $n \geqslant m$. Then

$$p(x - \sum_{k=1}^{n} f_k(x)x_k) \leqslant \max\{p(x-y), p(\sum_{k=1}^{n} f_k(x-y)x_k)\} < \epsilon.$$

This proves that (x_n) is a basic sequence. On the other hand, $p(x) \leqslant p^*(x)$ on G. We proved that on G the family $\mathcal{P}^*$ is equivalent to the family $\mathcal{P}$ and finally this shows that (x_n) is a basic sequence in E and (x_n) is an orthogonal basis of the space G.

A sequence satisfying the conclusion of the last result will be called an *orthogonal basic sequence*. As a direct consequence of Proposition 5 we note the following

COROLLARY 6 *Every Schauder basis in a polarly barrelled space is an orthogonal basis. Every weak Schauder basis in a polarly barrelled polar lcs is an orthogonal basic sequence. In particular, every polar Banach space has such property.*

Note that there are "plenty" of non-polar normed spaces with a weak Schauder basis which "is only" a basic sequence.

EXAMPLE 7 Let (E, τ) be an infinite dimensional Banach space over a nonspherically complete K with a Schauder basis (x_n). By G we denote the (dense) linear span of the x_n, $n \in \mathbb{N}$. Then $\dim E/G = \operatorname{card} K$, compare the argument of 2.4 in

[14]. Note that $\dim \ell_\infty/c_0 = \operatorname{card} K$, see [14], 2.1. Now we consider the quotient space E/G with an isomorphic copy α of the original topology of ℓ_∞/c_0. Then the topology $\xi = \tau \vee \alpha$ is normed and strictly finer than τ. Since

$$\xi|G = \tau|G \text{ and } \xi/G = \alpha,$$

then G is ξ-closed in E.

Moreover τ and ξ are compatible normed topologies. Indeed, take $f \in (E,\xi)'$, $f \neq 0$. Consider the restriction $f|G$ of f to the space G. Note that $f|G$ is non-zero. Otherwise $h(q(x)) = f(x) \in (E/G, \alpha)'$, where $q : E \to E/G$ is the canonical quotient map, and h is non-zero, a contradiction since $(\ell_\infty/c_0)' = \{0\}$, [27], 4.15. There exists a τ-continuous linear extension g of f to E. Note that $f = g$, otherwise $h(q(x)) = f(x) - g(x)$ defines a non-zero α-continuous linear functional, again a contradiction. Finally, note that (E,ξ) is non-polar. Indeed, otherwise the topologies ξ and τ would have the same bounded sets (they are compatible and normed) and consequently $\xi = \tau$, a contradiction. We constructed a normed (non-complete and non-polar) space (E,ξ) with a weak Schauder basis (x_n). It is easy to see that (x_n) is a Schauder (and orthogonal) basis in the ξ-closed subspace G in (E,ξ).

We do not know if the above construction leads the conclusion that (E,ξ) is not barrelled. We note only that $(G, \tau|G)$ is not barrelled, cf. [10], 2.1.3.

As we know [7], [9], every Schauder basis in a barrelled space is orthogonal. Is the converse implication also true? *If E is a lcs with an orthogonal basis and every Schauder basis in E is orthogonal, then E is barrelled.* Clearly in a (polarly) barrelled space any (polar) seminorm p such that $p(x) = \sup_n p(f_n(x)x_n)$ is continuous. We note only the following simple fact.

PROPOSITION 8 *Let E be a lcs and let (x_n) be a Schauder basis in E and (f_n) its associated sequence in E'. Assume that every seminorm p on E such that $p(x) = \sup_n p(f_n(x)x_n)$ is continuous. Then E is quasibarrelled, i.e. every bornivorous barrel in E is a neighbourhood of zero.*

Proof. Let V be a bornivorous barrel in E and let p be its Minkowski functional. Define $p_1(x) = \sup_n p(f_n(x)x_n)$, $x \in E$. Clearly

$$p_1(f_n(x)x_n) = \sup_m p(f_m(f_n(x)x_n)x_m) = p(f_n(x)x_n).$$

Thus $p_1(x) = \sup_n p_1(f_n(x)x_n)$. By assumption p_1 is continuous; hence p is continuous (note that $p \leqslant p_1$ since V is closed). Consequently V is a neighbourhood of zero.

REMARKS 9 (a) It is known, see [9], Lemma, p. 20, that *a barrelled (even a G-) space with a Schauder basis (x_n) does not admit a strictly finer locally convex topology for which (x_n) is still a basis.*

Note that this Lemma was originally formulated for spherically complete K; this assumption can be removed. The proof of this useful fact needed however some extra sequence space techniques. The following simple observation (strictly connected with our Example 7) is much easier to prove, cf. also [26], 7.9, to get this fact using a different approach.

If (E,τ) is a barrelled space, then there is on E no strictly finer locally convex topology ξ compatible with τ such that (E,ξ) has a Schauder basis.

Indeed, assume (x_n) is a Schauder basis for (E,ξ), where ξ is a strictly finer locally convex topology on E compatible with τ, with its associated sequence (f_n) in $(E,\xi)' = (E,\tau)'$. For every $n \in \mathbb{N}$ let $S_n(x) = \sum_{k=1}^{n} f_k(x)x_k$. Then the maps $S_n : (E,\tau) \to (E,\xi)$ are continuous and the sequence $(S_n(x))$ converges to $I(x)$, where $I : (E,\tau) \to (E,\xi)$ is the identity map. By the Banach-Steinhaus theorem I is continuous, so $\tau = \xi$, a contradiction.

(b) Let E be a barrelled space and let (x_n) be a Schauder basis for E with the associated sequence (f_n) in E'. Let F be the completion of E and for every $n \in \mathbb{N}$ let g_n be the continuous linear extension of the f_n to the space F. Then (x_n) is a Schauder basis in F with (g_n) as the associated sequence. Indeed, let q be a continuous polar seminorm on F and p its restriction to E. Clearly for every $n \in \mathbb{N}$ the seminorm $p_n : x \to p(g_n(x)x_n)$ is continuous and polar and the sequence (p_n) is pointwise bounded. Since E is barrelled, (p_n) is equicontinuous. Then the seminorms $y \to p(g_n(y)x_n)$ form an equicontinuous sequence on F. Therefore for $\epsilon > 0$ there exists a neighbourhood of zero V in F such that

$$q(\sum_{k=1}^{n} g_k(y)x_k) = p(\sum_{k=1}^{n} g_k(y)x_k) < \epsilon$$

and $q(y) < \epsilon$ for every $n \in \mathbb{N}$ and $y \in V$. Take $y \in F$. Then there exists $x \in E$ such that $x - y \in V$. There exists $N \in \mathbb{N}$ such that

$$q(x - \sum_{k=1}^{n} g_k(x)x_k) < \epsilon \text{ for all } n \geqslant N.$$

Then

$$q(y - \sum_{k=1}^{n} g_k(y)x_k) \leqslant \max\{q(x-y), p(x - \sum_{k=1}^{n} g_k(x)x_k), q(\sum_{k=1}^{n} g_k(x-y)x_k)\} < \epsilon$$

for $n \geqslant N$.

This fact applies to deduce that *a sequentially complete barrelled space E with a Schauder basis (x_n) is complete*. Indeed, if F is the completion of E and $y \in F$, then $(\sum_{k=1}^{n} g_k(y)x_k)$ is a Cauchy sequence in E, where (g_k) is as above. Then this sequence converges to $z \in E$. Since (x_n) is a Schauder basis in F, then $z = y$. Hence $F = E$.

(c) It turns out that property from (a), without the assumption that E is barrelled, can be described in another way: Similarly as for the "real or complex case" one gets the following non-archimedean counterpart of Theorem 3 of [11]. Let E be a locally convex space with a Schauder basis (x_n) and its associated sequence (f_n) in E'. *The following assertions are equivalent*: (i) If T is a linear map of E into a locally convex space F such that $T(\sum_{k=1}^{n} f_k(x)x_k) \to T(x)$, $x \in E$, then T is continuous. (ii) If p is a seminorm on E such that $p(x - \sum_{k=1}^{n} f_k(x)x_k) \to 0$, $x \in E$, then p is continuous. (iii) There exists no locally convex topology on E strictly finer that the original one such that (x_n) is a Schauder basis.

Since (iii) holds for any barrelled space over K, cf. [9], Lemma, p.20, then *in a barrelled space E with a Schauder basis (x_n) and its associated sequence (f_n) in E', a seminorm p is continuous iff $p(x - \sum_{k=1}^{n} f_k(x)x_k) \to 0$ for any $x \in E$.* Note also that originally in [9] Lemma was formulated for spherically complete K; this assumption can be removed.

(d) Consider the following condition: (+) *If a linear functional f on E satisfies the condition $f(\sum_{k=1}^{n} f_k(x)x_k) \to f(x)$ for any $x \in E$, then f is continuous.*

As in [11], p.123, one shows that if E is a barrelled space and (+) is satisfied, then (ii) (from (c)) holds, too. This combined with Theorem 6 of [11] (which works also in the non-archimedean case) and (c) leads the following: *Let E be a barrelled lcs with a Schauder basis. (i) The strong dual of E is complete. (ii) Any sequentially continuous linear map of E into a lcs is continuous.* Note that (ii) follows from (a) and (c).

Recall [5], [10], that an (LB)-space E (with its defining sequence (E_n)) is *sequentially retractive* if every null-sequence in E is contained (as a null-sequence) in some E_n. Recall also that every strict (LB)-space is sequentially retractive [10]. Our next result corresponds to the problem (1). It is well-known (as in the "classical case") that in any Banach (or Fréchet) space over K every basis is an equicontinuous basis. Combining the arguments used by De Grande-De Kimpe in [7], p. 397 and Floret in [5] one gets the following.

PROPOSITION 10 *Let (E,τ) be a sequentially retractive (LB)-space with its defining sequence (E_n,τ_n) of Banach spaces. Then any basis in (E,τ) is an equicontinuous basis.*

Proof. Let (x_n) be a basis in $E = (E,\tau)$ with the associated sequence (f_n) of linear functionals. For every $n \in \mathbb{N}$ define the space

$$F_n = \{x \in E_n : x = \sum_k f_k(x)x_k \text{ with respect to } \tau_n\}.$$

Endow every F_n with the topology τ_n^* generated by the following norm

$$||(\sum_j f_j(x)x_j)||_n^* = \max_j ||(f_j(x)x_j)||_n,$$

where $||.||_n$ denotes the non-archimedean norm defining the topology τ_n. Clearly E is the union of the F_n, $n \in \mathbb{N}$, and the inclusion $F_n \hookrightarrow F_{n+1}$ is continuous for every $n \in \mathbb{N}$. Moreover $||.||_n \leqslant ||.||_n^*$ for every $n \in \mathbb{N}$. If τ^* is the inductive limit topology on E of the sequence (F_n,τ_n^*), then clearly $\tau \leqslant \tau^*$. Since every f_j is continuous in (F_n,τ_n^*), then it is continuous in (E,τ^*), [10], 1.1.6. Using the same argument as in [7], 397, one gets that every (F_n,τ_n^*) is a Banach space. Therefore the identity map I between (LB)-spaces (E,τ) and (E,τ^*) has closed graph. By the closed graph theorem, [6], 2.3, we deduce that $\tau = \tau^*$.

We do not know if every basis in a barrelled space is a Schauder basis.

Finally we consider problem (4), see Introduction. It is known that every null-sequence in a non-archimedean Banach space is summable, [27], p. 5. We already constructed on the space ℓ^1 a weak Schauder basis which is not a basic sequence in the original topology of ℓ^1. It is also clear that ℓ^p, $0 < p < \infty$, does not admit a non-archimedean norm equivalent to the original one. The following extension of the main result of [13] explains this "particular situation".

THEOREM 11 *Let $E = (E, \|.\|)$ be an A-Banach space with a weak Schauder basis (x_n). Consider the following conditions.*

(i) *E is a non-archimedean Banach space.*

(ii) *Every weak Schauder basis in E is a Schauder basis.*

(iii) *Every weak Schauder basis in E is a basic sequence in E which is strongly bounded.*

(iv) *Every weak Schauder basis (y_n) in E is strongly bounded, i.e. $\{f_n(x)y_n : n \in \mathbb{N}\}$ is bounded in $(E, \|.\|)$ for every $x \in E$, where (f_n) is the sequence of continuous linear functionals associated with the basis (y_n).*

Then (ii)⇒(iii)⇒(iv), (ii)⇒(i). *If K is spherically complete, then* (iv)⇒(i). *Moreover, if E is an (O.P.)-space, then* (i)⇒(ii).

We shall need the following useful perturbation theorem from [13].

LEMMA 12 *Let (x_n) be a Schauder basis in a non-archimedean Banach space $(E, \|.\|)$ with coefficient functionals (f_n). If (y_n) is a sequence in E such that $\sup_n \|f_n\|\|x_n - y_n\| = a < 1$, then (y_n) is also a Schauder basis in $(E, \|.\|)$.*

Hence, if a non-archimedean Banach space E has a Schauder (orthogonal) basis and F is a dense subspace of E, then E has a Schauder (orthogonal) basis whose elements belong to the space F.

Proof of Theorem 11. The implications (ii)⇒(iii)⇒(iv) are obvious. Also the implication (i)⇒(ii) holds provided E is an (O.P.)-space, see Proposition 1. Now we prove the implication (iv)⇒(i). We may assume that $x_n \to 0$ in $(E, \|.\|)$. Since K is spherically complete, then (E, ν) is an (O.P.)-space, so (x_n) is also a Schauder basis in the Mackey envelope (E, ν), see Proposition 1. Let (f_n) be the corresponding sequence of continuous linear functionals associated with (x_n). The space (E, ν) is barrelled. Indeed, let τ be a locally convex topology on E determined by the norm $\|.\|$. Let U be a barrel in (E, ν). Since U is a barrel in (E, τ), so it is a τ-neighbourhood of zero. Then $\epsilon B \subset U$ for some non-zero $\epsilon \in K$, where B is as in the introduction. Therefore $\epsilon \operatorname{conv} B \subset U$ and U is a ν-neighbourhood of zero. This proves that (E, ν) is barrelled. Since (E, ν) is barrelled, we deduce from (b), Remarks 9, that (x_n) is a Schauder basis in the completion Y of (E, ν). By ν we denote also the norm of Y and Y'.

Now assume that E is not non-archimedean. Then the topologies determined by $\|.\|$ and ν, respectively, are different. There exists a sequence (w_n) in E such that

$$\nu(w_n) \to 0 \text{ and } \inf_n \|w_n\| > 0 \ , \sup_n \nu(f_n)\nu(w_n) = a < 1,$$

where (f_n) now denotes the sequence of the continuous linear extensions of associated functionals to the completion Y.

Set $y_n = x_n + w_n$, $n \in \mathbb{N}$. By Lemma 12 the sequence (y_n) is a Schauder basis in Y, hence it is a Schauder basis in $(E, \sigma(E, E'))$, too. Let (g_n) be the sequence of linear functionals associated with the basis (y_n). The sequence (y_n) is a null-sequence in (E,ν) and also $\inf_n \|y_n\| > 0$.

By assumption (y_n) is strongly bounded in $(E, \|.\|)$. Hence for every $x \in E$ there exists $M(x) > 0$ such that $\|g_n(x)y_n\| \leqslant M(x)$ for all $n \in \mathbb{N}$. There exists $L > 0$ such that $\|y_n\| > L$, for all $n \in \mathbb{N}$. Therefore

$$L\,|g_n(x)| \leqslant \|g_n(x)y_n\| \leqslant M(x) \text{ for all } n \in \mathbb{N}.$$

This proves that the sequence (g_n) is pointwise bounded. Since (E,ν) is barrelled, the Banach-Steinhaus theorem applies to deduce that (g_n) is equicontinuous in (E,ν). But $y_n \to 0$ and $g_n(y_n) = 1$ for all $n \in \mathbb{N}$, a contradiction.

(ii)⇒(i): We may assume that (x_n) is a Schauder basis in $(E, \|.\|)$, otherwise we are done. Hence (x_n) is a Schauder basis in (E,ν). Next we proceed as in the case (iv)⇒(i).

REMARK 13 We do not know if the following result is true: *Let E be a Banach space with a weak Schauder basis. Then E is a polar space iff every weak Schauder basis in E is a basic sequence;* comp. Corollary 6.

Our next result supplements also Example 7.

PROPOSITION 14 *Let (E,τ) be an infinite-dimensional dual-separating Banach space. Then E admits two normed topologies τ_1 and τ_2 strictly finer than τ such that:*

(i) *The topologies τ_k are compatible with τ, $k = 1,2$.*

(ii) *The spaces (E,τ_k) are not of countable type, $k = 1,2$.*

(iii) *The completion of (E,τ_k) is not dual-separating, $k = 1,2$.*

(iv) *The space (E,τ_k) has a quotient isomorphic to the space ℓ_∞/c_0, $k = 1,2$.*

(v) *(E,τ) is isomorphic to the quotient $Y_k/\ker Y_k$, $k = 1,2$, where Y_k is the completion of (E,τ_k) and $\ker Y_k = \{y \in Y_k : f(y) = 0 \text{ for all } f \in Y_k'\}$.*

(vi) *$\inf\{\tau_1,\tau_2\} = \tau$, where $\inf\{\tau_1,\tau_2\}$ is the finest locally convex topology on E which is weaker than τ_1 and τ_2.*

In particular, if (E,τ) has a Schauder basis, then (E,τ_1) and (E,τ_2) are not of countable type but have a weak Schauder basis.

We shall need the following two lemmas; the proof of the first one proceeds as in the classical case, see [21], 9.2.4.

LEMMA 15 *Let E be a locally convex space and assume that F is a closed subspace of E. If the quotient E/F and F are suprabarrelled spaces, then E is suprabarrelled.*

A lcs E will be called *suprabarrelled* (called also a *(db)-space*) [21] if given any increasing sequence of subspaces of E covering E, one of them is both dense and barrelled in E. Clearly every Fréchet space is suprabarrelled and every suprabarrelled space is barrelled.

Note that the next lemma is also valid in the class of arbitrary locally convex spaces.

LEMMA 16 *Let (E,τ) be a normed space admitting a weaker normed topology γ. Let F be a subspace of E such that $\gamma|F = \tau|F$ and $\gamma/F = \tau/F$. Then $\tau = \gamma$.*

Proof. Let (x_n) be a null-sequence in (E,γ) and assume that U is a bounded neighbourhood of zero in τ. Let $q : E \to E/F$ be the quotient map. By our assumption there exist sequences (h_n) in $\mathbb{N}$ and (t_n) in K such that $t_n \to 0$ and $q(x_n) \in q(t_n U)$ for all $n \in \mathbb{N}$. For every $n \in \mathbb{N}$ there exists $u_n \in U$ and $y_n \in F$ such that $x_n = t_n u_n + y_n$ for all $n \in \mathbb{N}$. Since $y_n = x_n - t_n u_n \in F$ and (y_n) is a null-sequence in γ, then, by the assumption, it is a null-sequence in τ. Therefore $x_n \in U$ for n enough large, i.e. (x_n) is a null-sequence in (E,τ).

Now we are ready to prove Proposition 14.

Proof of Proposition 14. By [14], 2.2, $\dim E \geqslant \operatorname{card} K$. First we find a dense Baire subspace G_1 in (E,τ) such that $\operatorname{codim} G_1 = \operatorname{card} K$. Let $(x_t)_{t\in T}$ be a Hamel basis of E. Take a partition (T_n) of T such that $\bigcup T_n = T$, $\operatorname{card} T_n = \operatorname{card} T$, $n \in \mathbb{N}$. Then $E_n = \operatorname{lin}\{x_t : t \in \bigcup_{k=1}^{n} T_k\}$ compose an increasing sequence of subspaces of E covering E and $\dim E_n = \operatorname{codim} E_n$ for all $n \in \mathbb{N}$.

Since (E,τ) is a Baire space, there exists $n \in \mathbb{N}$ such that E_m is dense in E for all $m \geqslant n$. Now it is clear how to get the desired subspace G_1.

Let τ_1 be the topology on E obtained as in Example 7, i.e. τ_1 is strictly finer than τ and τ_1 is normed and compatible with τ. Moreover (E,τ_1) is not of countable type and $(E,\tau_1)/G_1$ is isomorphic to the quotient ℓ_∞/c_0. Since $(G_1,\tau_1|G_1)$ and $(E/G_1,\tau_1/G_1)$ are suprabarrelled spaces, then, by Lemma 15, the space (E,τ_1) is suprabarrelled.

Now we construct a dense subspace G_2 in (E,τ_1) such that $\operatorname{codim} G_2 = \operatorname{card} K$. Let H be a complemented (algebraically) subspace of G_1 in E. Clearly $\dim H = \operatorname{card} K$. Using the same argument as above we construct an increasing sequence $(G_1 + H_n)$ of subspaces of E covering E and such that $\operatorname{codim}(G_1 + H_n) = \operatorname{card} K$ for all $n \in \mathbb{N}$.

Since (E,τ_1) is suprabarrelled, then one of the $G_2 = G_1 + H_n$ is both barrelled and dense in (E,τ_1). Let τ_2 be the topology on E (associated with (E,τ) and the dense subspace G_2) constructed as in Example 7. Then τ_2 is strictly finer than τ, compatible with τ and normed. Also $\tau_2|G_2 = \tau|G_2$. The quotient space $(E/G_2,\tau_2/G_2)$ is isomorphic to the space ℓ_∞/c_0. This proves (i) and (iv).

Since the quotients of (E,τ_1) and (E,τ_2) are isomorphic to the space ℓ_∞/c_0, then the spaces (E,τ_1) and (E,τ_2) are not of countable type. This proves (ii).

Since $\tau|G_2 = \tau_2|G_2$, then $\tau|G_2 = \inf\{\tau_1,\tau_2\}|G_2$. On the other hand, G_2 is a dense subspace of (E,τ_1).

Therefore

$$\tau/G_2 = \inf\{\tau_1,\tau_2\}/G_2 = \tau_1/G_2.$$

Using Lemma 16 we derive that $\tau = \inf\{\tau_1,\tau_2\}$.

Now we prove (v). Let $q : Y_k \to Y_k/\ker Y_k$ be the quotient map and fix a number k, where $k = 1,2$. We may assume also that E is a subspace of Y_k. First observe that $Y_k = E + \ker Y_k$ (algebraically). Indeed, if $z \in E \cap \ker Y_k$, then $f(z) = 0$ for all $f \in Y_k'$. Since τ and τ_k are compatible and (E,τ) is dual-separating, then $z = 0$.

If $y \in Y_k$, then there exists a sequence (y_n) in E such that $y_n \to y$ in Y_k. Since τ is complete, then $y_n \to x \in E$ in τ. Hence

$$f(x-y) = f(x-y_n) + f(y_n - y) \text{ and } f(x-y_n) \to 0,\ f(y_n - y) \to 0,$$

for all $f \in Y_k'$. Hence $x - y \in Y_k$.

Finally we prove that $q|E : (E,\tau) \to Y_k/\ker Y_k$ is an isomorphism. Since $q|E$ is one-to-one, it is enough to prove that $q|E$ is continuous. We show that $q|E$ has a closed graph; then the continuity of $q|E$ follows from the closed graph theorem, cf. [6].

Let (x_n) be a null-sequence in (E,τ) such that $(q|E)(x_n) \to y \in Z_k$, where $Z_k = Y_k/\ker Y_k$. Choose $a \in Y_k$ such that $q(a) = y$. Let (U_n) be a decreasing basis of neighbourhoods of zero in Y_k. There exist sequences (k_n) in $\mathbb{N}$ and (u_n) such that $u_n \in U_n$ and $q(x_{k_n} - a) = q(u_n)$ for all $n \in \mathbb{N}$. Therefore $x_{k_n} - a - u_n \in \ker Y_k$. Hence $f(x_{k_n} - a - u_n) = 0$ for all $f \in Y_k'$.

Since $f(x_{k_n}) \to 0$ and $f(u_n) \to 0$, we derive that $f(a) = 0$ for all $f \in Y_k'$. Consequently, $a \in \ker Y_k$, so $y = q(a) = 0$. The proof of (v) is complete. Now the property (iii) follows directly from the proof of (v).

It is known that the space ℓ_∞ provides a concrete example of a polar Banach space which is not an (O.P.)-space provided K is nonspherically complete. By [22], 2.3, every metrizable O.P. space is polar. There are (O.P.)-spaces which are not polar, [23], 4.1: For a sufficiently large set I the space $c_0(I)$ admits a locally convex topology τ between the weak one of $c_0(I)$ and the original one such that $(c_0(I),\tau)$ is an (O.P.)-space but non-polar. This argument applies also to get the following

PROPOSITION 17 *Let (E,τ) be a polar lcs which contains a subspace isomorphic to ℓ_∞. Then E admits a locally convex topology α such that $\sigma(E,E') < \alpha < \tau$ and (E,α) is non-polar.*

Proof. By F we denote also a subspace of (E,τ) isomorphic to ℓ_∞. First we construct such a topology on the space F. Let τ_∞ be the original topology of ℓ_∞. Let $q : \ell_\infty \to \ell_\infty/c_0$ be the quotient map and define the seminorm p on ℓ_∞ by $p(x) = \|q(x)\|$, where $\|.\|$ is the original norm on ℓ_∞/c_0. Then the seminorm p satisfies the condition of 4.1 of [23], i.e. if f is a linear functional on ℓ_∞ such that $|f(x)| \leqslant p(x)$ on ℓ_∞, then $f = 0$. Indeed, put $h(q(x)) = f(x)$ and assume that f is non-zero. The functional h is well-defined and continuous, a contradiction since ℓ_∞/c_0 does not have non-zero continuous linear functionals.

Therefore, by [23], the proof of 4.1, there exists on ℓ_∞ a locally convex topology α_∞ such that

$$\sigma(\ell_\infty, c_0) < \alpha_\infty < \tau_\infty$$

and $(\ell_\infty, \alpha_\infty)$ is non-polar. The topology α_∞ is induced by $\sigma(\ell_\infty, c_0)$ and the single seminorm p. By α_∞ we denote also the corresponding topology on F with the desired conditions transformed from ℓ_∞.

Let α be the locally convex topology on E defined by the following basis of neighbourhoods of zero $U_\tau + V_\infty$, where sets U_τ and V_∞ run bases of neighbourhoods of zero from τ and α_∞, respectively.

Clearly $\alpha \leqslant \tau$. For an absolutely convex $\sigma(E, E')$-neighbourhood of zero U_σ there exists a τ-neighbourhood of zero V such that $V \subset U_\sigma$. Then the set $U_\sigma \cap F + V$ is a neighbourhood of zero in α and it is contained in the set $U_\sigma + V \subset U_\sigma + U_\sigma \subset U_\sigma$. Therefore

$$\sigma(E, E') \leqslant \alpha \leqslant \tau.$$

Note also that $\alpha | F = \alpha_\infty$ and $\alpha / F = \tau / F$. Since (F, α_∞) is non-polar, then (E, α) is non-polar. Therefore the inequalities between $\sigma(E, E')$, α and τ are proper.

Note that, according to [22], 3.2, every polar weakly sequentially complete Fréchet space which is not (O.P.) contains an isomorphic copy of ℓ_∞. Note also that, because of our assumption, the space E is not an (O.P.)-space; therefore it does not contain a basis.

REFERENCES

[1] MG Arsove, RE Edwards. Generalized bases in topological linear spaces. Stud Math 19:95–113, 1960.

[2] S Banach. Théorie des Operations Lineaires, Warszawa, 1932.

[3] S Dierolf. A note on the lifting of linear and locally convex topologies on a quotient space. Collect Math 31:193–198, 1980.

[4] E Dubinsky, JR Retherford. Schauder bases in compatible topologies. Stud Math 28:221–226, 1967.

[5] K Floret. Bases in sequentially retractive limit spaces. Stud Math 38:221–226, 1970.

[6] T Gilsdorf, J Kąkol. On some non-archimedean closed graph theorems. In: WH Schikhof, C Perez-Garcia, J Kąkol, ed. p-Adic Functional Analysis. New York: Marcel Dekker, 1997, pp 153–158.

[7] N De Grande-De Kimpe. On the structure of locally K-convex spaces with a Schauder basis. Proc Kon Ned Akad Wet 75:396–406, 1972.

[8] N De Grande-De Kimpe. Equicontinuous Schauder bases and compatible locally convex topologies. Proc Kon Ned Akad Wet 77:276–283, 1974.

[9] N De Grande-De Kimpe. Structure theorems for locally K-convex spaces. Proc Kon Ned Akad Wet 80:11–22, 1977.

[10] N De Grande-De Kimpe, J Kąkol, C Perez-Garcia, W Schikhof. p-adic locally convex inductive limits. In: WH Schikhof, C Perez-Garcia, J Kąkol, ed. p-Adic Functional Analysis. New York: Marcel Dekker, 1997, pp 159–222.

[11] O T Jones. Continuity on seminorms and linear mappings on a space with Schauder basis. Stud Math 34:121–126, 1970.

[12] PK Kamthan, M Gupta. The theory of bases and cones. Research Notes in Math, 1994.

[13] J Kąkol. The weak basis theorem for K-Banach spaces. Bull Soc Math Belg 45:1–4, 1993.

[14] J Kąkol, C Perez-Garcia, W Schikhof. Cardinality and Mackey topologies of non-archimedean Banach and Fréchet spaces. Bull Pol Acad Sci Math 44:131–141, 1996.

[15] J Kąkol, C Perez-Garcia, W Śliwa. Remarks on quasi-Banach spaces over valued fields. Funct Approximatio 20:23–24, 1997.

[16] J Martinez-Maurica, C Perez-Garcia. Thge three space problem for a class of normed spaces. Bull Soc Math Belg 39:209–214, 1987.

[17] AF Monna, TA Springer. Sur la structure des espaces de Banach non-archimédiends. Proc Kon Ned Akad Wet 68:602–614, 1965.

[18] W McArthur, JR Retherford. Uniform and equicontinuous Schauder bases of subspaces. Can J Math 17:207–212, 1965.

[19] WF News. On the representation of analytic functions by infinite series. Philos Trans R Soc Lond 245:429–468, 1953.

[20] H Ochsenius, W Schikhof. Banach spaces over fields with an infinite rank valuation, these Proceedings.

[21] P Perez Carreras, J Bonet. Barrelled Locally Convex Spaces. Amsterdam: North-Holland, 1987.

[22] C Perez-Garcia, W Schikhof. The Orlicz-Pettis property in p-adic analysis. Collect Math 43:225–233, 1992.

[23] C Perez-Garcia, W Schikhof. Compact operators and the Orlicz-Pettis property in p-adic analysis. Report 9101, Departement of Mathematics, Catholic University, Nijmegen, The Netherlands: 1–27, 1991.

[24] M Van der Put. Espaces de Banach non-archimédiens. Bull Soc Math Fr 97:309–320, 1969.

[25] W Schikhof. The continuous linear image of a p-adic compactoid. Proc Kon Ned Akad Wet 92:119–123, 1989.

[26] W Schikhof. Locally convex spaces over non-spherically complete fields I-II. Bull Soc Math Belg 38:197–224, 1986.

[27] ACM van Rooij. Non-Archimedean Functional Analysis. New York: Marcel Dekker, 1978.

[28] J van Tiel. Espaces localement K-convexes, I-III. Proc Kon Ned Akad Wet 27:249–289, 1965.

Fractional differentiation operator over an infinite extension of a local field

ANATOLY N KOCHUBEI Institute of Mathematics, Ukrainian National Academy of Sciences, Tereshchenkivska 3, Kiev, 252601 Ukraine

1 INTRODUCTION

Let k be a non-Archimedean local field of zero characteristic. Consider an increasing sequence of its finite extensions

$$k = K_1 \subset K_2 \subset \ldots \subset K_n \subset \ldots .$$

The infinite extension

$$K = \bigcup_{n=1}^{\infty} K_n$$

may be considered as a topological vector space over k with the inductive limit topology. Its strong dual $\overline{K}$ is the basic object of the non-Archimedean infinite-dimensional analysis initiated by the author [18]. Let us recall its main constructions and results.

Consider, for each n, a mapping $T_n : K \to K_n$ defined as follows. If $x \in K_\nu$, $\nu > n$, put

$$T_n(x) = \frac{m_n}{m_\nu} \mathrm{Tr}_{K_\nu/K_n}(x)$$

where m_n is the degree of the extension K_n/k, $\mathrm{Tr}_{K_\nu/K_n} : K_\nu \to K_n$ is the trace mapping. If $x \in K_n$ then, by definition, $T_n(x) = x$. The mapping T_n is well-defined and $T_n \circ T_\nu = T_n$ for $\nu > n$. Below we shall often write T instead of T_1.

The strong dual space $\overline{K}$ can be identified with the projective limit of the sequence $\{K_n\}$ with respect to the mappings $\{T_n\}$, that is with the subset of the direct product $\prod_{n=1}^{\infty} K_n$ consisting of those $x = (x_1, \ldots, x_n, \ldots)$, $x_n \in K_n$, for which $x_n = T_n(x_\nu)$ if $\nu > n$. The topology in $\overline{K}$ is defined by the seminorms

$$\|x\|_n = \|x_n\|, \quad n = 1, 2, \ldots,$$

where $\|\cdot\|$ is the extension to K of the absolute value $|\cdot|_1$ defined on k.

The pairing between K and $\overline{K}$ is defined as

$$< x, y >= T(xy_n)$$

where $x \in K_n \subset K$, $y = (y_1, \ldots, y_n, \ldots) \in \overline{K}$, $y_n \in K_n$. Both spaces are separable, complete, and reflexive. Identifying an element $x \in K$ with $(x_1, \ldots, x_n, \ldots) \in \overline{K}$ where $x_n = T_n(x)$, we can view K as a dense subset of $\overline{K}$. The mappings T_n can be extended to linear continuous mappings from $\overline{K}$ to K_n, by setting $T_n(x) = x_n$ for any $x = (x_1, \ldots, x_n, \ldots) \in \overline{K}$.

Let us consider a function on K of the form

$$\Omega(x) = \begin{cases} 1, & \text{if} \quad \|x\| \leq 1; \\ 0, & \text{if} \quad \|x\| > 1. \end{cases}$$

Ω is continuous and positive definite on K. That results in the existence of a probability measure μ on the Borel σ-algebra $B(\overline{K})$, such that

$$\Omega(a) = \int_{\overline{K}} \chi(< a, x >)\, d\mu(x), \quad a \in K,$$

where χ is a rank zero additive character on k. The measure μ is Gaussian in the sense of Evans [4]. It is concentrated on the compact additive subgroup

$$S = \left\{ x \in \overline{K} : \quad \|T_n(x)\| \leq q_n^{d_n/m_n} \|m_n\|, \; n = 1, 2, \ldots \right\}$$

where q_n is the residue field cardinality for the field K_n and d_n is the exponent of the different for the extension K_n/k. The restriction of μ to S coincides with the normalized Haar measure on S. On the other hand, μ is singular with respect to additive shifts by elements from $\overline{K} \setminus S$.

Having the measure μ, we can define a Fourier transform $\widehat{f} = Ff$ of a complex-valued function $f \in L_1(\overline{K})$ as

$$\widehat{f}(\xi) = \int_{\overline{K}} \chi(\langle \xi, x \rangle) f(x)\, d\mu(x).$$

Let $\mathcal{E}(\overline{K}) \subset L_1(\overline{K})$ be the set of "cylindrical" functions of the form $f(x) = \varphi(T_n(x))$ where $n \geq 1$, φ is a locally constant function. The fractional differentiation operator D^α, $\alpha > 0$, is defined on $\mathcal{E}(\overline{K})$ as $D^\alpha = F^{-1}\Delta^\alpha F$ where Δ^α is the operator of multiplication by the function

$$\Delta^\alpha(\xi) = \begin{cases} \|\xi\|^\alpha, & \text{if} \quad \|\xi\| > 1 \\ 0, & \text{if} \quad \|\xi\| \leq 1 \end{cases}, \qquad \xi \in K.$$

Correctness of this definition follows from the theorem of Paley-Wiener-Schwartz type for the transform F. The operator D^α is essentially self-adjoint on $L_2(\overline{K})$.

The operator D^α is an infinite-dimensional counterpart of the p-adic fractional differentiation operator introduced by Vladimirov [22] and studied extensively in [2], [7-9], [12-16], [21] and [23]. In some respects the infinite-dimensional D^α resembles its finite-dimensional analogue though it possesses some new features. For example, we shall show below that the structure of its spectrum depends on arithmetic properties of the extension K.

Both in the finite-dimensional and infinite-dimensional cases D^α admits a probabilistic interpretation. Namely, $-D^\alpha$ is a generator of a cadlag Markov process, which is a p-adic counterpart of the symmetric stable process. In analytic terms, that corresponds to a hyper-singular integral representation of $D^\alpha f$, for a suitable class of functions f. In [18] such a representation was obtained for $f \in \mathcal{E}(\overline{K})$, $f(x) = \varphi(T_n(x))$, φ locally constant:

$$(D^\alpha f)(x) = \psi(T_n(x))$$

where

$$\psi(z) = q_n^{d_n \alpha/m_n} \frac{1 - q_n^{\alpha/m_n}}{1 - q_n^{-1-\alpha/m_n}} \|m_n\|^{-m_n} \tag{1}$$

$$\times \int\limits_{x \in K_n \quad \|x\| \le q_n^{d_n/m_n} \|m_n\|} \left[\|x\|^{-m_n - \alpha} \|m_n\|^{m_n + \alpha} + 1 - q_n^{-1} q_n^{\alpha/m_n} - 1 q_n^{-d_n(1+\alpha/m_n)} \right]$$

$$\times \left[\varphi(z - x) - \varphi(z)\right] dx,$$

$z \in K_n$, $\|z\| \le q_n^{d_n/m_n} \|m_n\|$.

In this paper we shall show that there exists also a representation in terms of the function f itself:

$$(D^\alpha f)(y) = \int\limits_{\overline{K}} [f(y) - f(x + y)] \Pi(dx) \tag{2}$$

where Π is a measure on $B(\overline{K} \setminus \{0\})$ finite outside any neighbourhood of zero.

Another measure of interest in this context is the heat measure $\pi(t, dx)$ corresponding to the semigroup $\exp(-tD^\alpha)$, $t > 0$. We show that in contrast both to the finite-dimensional case and to the similar problem for the real infinite-dimensional torus [1], $\pi(t, dx)$ is not absolutely continuous with respect to μ, whatever is the sequence $\{K_n\}$.

2 SPECTRUM

We shall preserve the notation D^α for its closure in $L_2(\overline{K}) = L_2(S, d\mu)$. It is clear from the definition that the spectrum of D^α coincides with the closure in $\mathbb{R}$ of the range of the function $\Delta^\alpha(\xi)$, $\xi \in K$. In order to investigate the structure of the spectrum, we need some auxiliary results.

It follows from the duality theory for direct and inverse limits of locally compact groups [11] that the character group $\overline{K}^*$ of the additive group of $\overline{K}$ is isomorphic to K. The isomorphism is given by the relation

$$\psi(y) = \chi(< a_\psi, y >), \quad y \in \overline{K}, \tag{3}$$

where $a_\psi \in K$ is an element corresponding to the character ψ.

Denote $O = \{\xi \in K : \ \|\xi\| \leq 1\}, \quad O_n = O \cap K_n$.

LEMMA 1 *The dual group S^* of the subgroup $S \subset \overline{K}$ is isomorphic to the quotient group K/O.*

Proof. We may assume (without restricting generality) that $k = Q_p$. Let

$$S^{(n)} = \{z \in K_n : \ \|z\| \leq q_n^{d_n/m_n} \|m_n\|\}, \quad n = 1, 2, \ldots.$$

It is clear that $S^{(n)}$ is a compact (additive) group. If $z \in S^{(\nu)}$, $\nu > n$, then

$$\|T_n(z)\| = \|m_n\| \cdot \|m_\nu\|^{-1} \|\mathrm{Tr}_{K_\nu/K_n}(z)\|$$

where $|m_\nu^{-1} z|_\nu \leq q_\nu^{d_\nu}$, and $|\cdot|_\nu$ is the normalized absolute value on K_ν [3]. Therefore (see Chapter 8 in [24])

$$|m_\nu^{-1} \mathrm{Tr}_{K_\nu/K_n}(z)|_n \leq q_n^l$$

where $l \in \mathbb{Z}$, $e_{n\nu}(l-1) < d_\nu - d_{n\nu} \leq e_{n\nu} l$, $e_{n\nu}$ and $d_{n\nu}$ are the ramification index and the exponent of the different for the extension K_ν / K_n. On the other hand, $d_\nu = e_{n\nu} d_n + d_{n\nu}$, so that $l = d_n$, and we find that $T_n(z) \in S^{(n)}$.

Hence, $T_n : \ S^{(\nu)} \to S^{(n)}$. It is clear that T_n is a continuous homomorphism. As a result, S can be identified with an inverse limit of compact groups $S^{(n)}$ with respect to the sequence of homomorphisms T_n. Using the auto-duality of each field K_n, we can identify the group dual to $S^{(n)}$ with K_n/Φ_n where Φ_n is the annihilator of $S^{(n)}$ in K_n. On the other hand, $\Phi_n = O_n$.

Indeed,

$$\begin{aligned} \Phi_n &= \{\xi \in K_n : \ \chi(T(z\xi)) = 1 \quad \text{for all} \ \ z \in S^{(n)}\} \\ &= \{\xi \in K_n : \ |T(z\xi)|_1 \leq 1 \quad \text{for all} \ \ z \in S^{(n)}\}. \end{aligned}$$

If $\xi \in O_n$ then $|m_n^{-1} z \xi|_n \leq q_n^{d_n}$ for any $z \in S^{(n)}$, whence $\xi \in \Phi_n$ by the definition of the number d_n.

Thus $O_n \subset \Phi_n$. Conversely, suppose that $\xi \in \Phi_n \backslash O_n$, that is $\|\xi\| > 1$, $|T(z\xi)|_1 \leq 1$ for all $z \in S^{(n)}$. Let z be such that $\|z\| = \|m_n\| q_n^{d_n/m_n}$. Then

$$|m_n^{-1}|_n \cdot |\xi|_n \cdot |z|_n = |\xi|_n q_n^{d_n} > q_n^{d_n}.$$

It follows from the properties of the trace [24] that z can be chosen in such a way that $|\mathrm{Tr}_{K_n/k}(m_n^{-1} \xi z)|_1 > 1$, and we have a contradiction. So $\Phi_n = O_n$.

It follows from the identity

$$T(\xi T_n(z)) = T(\xi z), \quad \xi \in K_n, \ z \in K_\nu, \ \nu > n,$$

that the natural embeddings $K_n/O_n \to K_\nu/O_\nu$, $\nu > n$, are the dual mappings of the homomorphisms $T_n : \ S^{(\nu)} \to S^{(n)}$. Using the duality theorem [11], we find that

$$S^* = \lim_{\rightarrow} K_n/O_n \tag{4}$$

where the direct limit is taken with respect to the embeddings. The right-hand side of (4) equals K/O.

Note that the above isomorphism is given by the same formula (3) where this time a_ψ is an arbitrary representative of a coset from K/O.

Let $\varphi_a(x) = \chi(< a, x >)$, $x \in \overline{K}$, where $a \in K$, $\|a\| > 1$ or $a = 0$. It is known [18] that

$$(D^\alpha \varphi_a)(x) = \|a\|^\alpha \varphi_a(x) \tag{5}$$

for μ-almost all $x \in \overline{K}$. Note that the set of values of the function $a \mapsto \|a\|^\alpha$ with $\|a\| > 1$ coincides with the set

$$\left\{q_n^{\alpha N/m_n},\ n, N = 1, 2, \ldots\right\} = \left\{q_1^{\alpha N/e_n},\ n, N = 1, 2, \ldots\right\} \tag{6}$$

where e_n is the ramification index of the extension K_n/k. Denote its residue degree by f_n. It is well known that the sequences $\{f_n\}$, $\{e_n\}$ are non-decreasing and $e_n f_n = m_n$.

THEOREM 1 *Let $A \subset K$ be a complete system of representatives of cosets from K/O. Then $\{\varphi_a\}_{a\in A}$ is the orthonormal eigenbasis for the operator D^α in $L_2(S, d\mu)$. As a set, its spectrum equals the set (6) complemented with the point $\lambda = 0$. Each non-zero eigenvalue of D^α has an infinite multiplicity. The point $\lambda = 0$ is an accumulation point for eigenvalues if and only if $e_n \to \infty$.*

Proof. The first statement follows from (5) and Lemma 1. The assertion about the accumulation at zero is obvious from (6).

Let us construct A as the union of an increasing family $\{A_n\}$ of complete systems of representatives of cosets from K_n/O_n. Each A_n consists of elements of the form $a = \pi_n^{-N}\left(\sigma_1 + \sigma_2\pi_n + \cdots + \sigma_{N-1}\pi_n^{N-1}\right)$, $N \geq 1$, where $\sigma_1, \ldots, \sigma_{N-1}$ belong to the set F_n of representatives of the residue field corresponding to the field K_n, $\sigma_1 \neq 0$, π_n is a prime element of K_n. We have $|a|_n = q_n^N$ so that $\|a\| = q_1^{Nf_n/m_n} = q_1^{N/e_n}$. Meanwhile card $F_n = q_n = q_1^{f_n}$.

If $e_n \to \infty$ then the same value of $\|a\|$ corresponds to elements from infinitely many different sets A_n with different values of N (e_n is a multiple of e_{n-1} due to the chain rule for the ramification indices; see [5]). If the sequence $\{e_n\}$ is bounded, then it must stabilize, and we obtain the same value of $\|a\|$ for elements from infinitely many sets A_n with possibly the same N. However, in this case $f_n \to \infty$, the number of such elements (for a fixed N) is $Nq_1^{f_n} - 1$ ($\to \infty$ for $n \to \infty$). In both cases we see that all the non-zero eigenvalues have an infinite multiplicity.

Note that the cases where $e_n \to \infty$ or $e_n \leq$ const both appear in important examples of infinite extensions. Let K be the maximal unramified extension of k. Then one may take for K_n the unramified extension of k of the degree $n!$, $n = 2, 3, \ldots$. Here $e_n = 1$ for all n. On the other hand, if K is the maximal abelian extension of $k = Q_p$ then a possible choice of K_n is the cyclotomic extension $C_{n'} = Q_p(W_{n!})$ where W_l is the set of all roots of 1 of the degree l (see [10, 19]). Writing $n! = n'p^{l_n}$, $(n', p) = 1$, we see that $e_n = (p-1)p^{l_n - 1} \to \infty$ as $n \to \infty$.

3 HYPERSINGULAR INTEGRAL REPRESENTATION

The main aim of this section is the following result.

THEOREM 2 *There exists such a measure* Π *on* $B(\overline{K} \setminus \{0\})$ *finite outside any neighbourhood of the origin that such* D^α *has the representation (2) on all functions* $f \in \mathcal{E}(\overline{K})$.

In the course of the proof we shall also obtain some new information about the Markov process $X(t)$ generated by the operator $-D^\alpha$. In the seqvel we assume that $X(0) = 0$.

The projective limit topology on $\overline{K}$ coincides with the one given by the shift-invariant metric

$$r(x,y) = \sum_{n=1}^{\infty} 2^{-n} \frac{\|x-y\|_n}{1+\|x-y\|_n}, \quad x,y \in \overline{K}.$$

It is known [20] that the main notions and results regarding stochastic processes with independent increments carry over to the case of a general topological group with a shift-invariant metric.

Let $\nu(t,\Gamma)$ be a Poisson random measure corresponding to the process $X(t)$. Here $\Gamma \in B_0 = \bigcup_{\gamma>0} B_\gamma$,

$$B_\gamma = \{\Gamma \subset B(\overline{K}),\ \mathrm{dist}(\Gamma,0) \geq \gamma\}.$$

For any $\Gamma \in B_0$ we can define a stochastically continuous process $X_\Gamma(t)$, the sum of all jumps of the process $X(\tau)$ for $\tau \in [0,t)$ belonging to Γ. In a standard way [6] we find that

$$E\chi(<\lambda, X_\Gamma(t)>) = \exp\Big\{\int_\Gamma (\chi(<\lambda,x>)-1)\Pi(t,dx)\Big\}, \quad \lambda \in K, \tag{7}$$

where E denotes the expectation, $\Pi(t,\cdot) = E\nu(t,\cdot)$.

Let $\lambda \in K_n$. Consider the set

$$V_{\delta,n} = \{x \in \overline{K} :\ \|T_n(x)\| \geq \delta\}, \quad 0 < \delta \leq 1.$$

Let us look at the equality (7) with $\Gamma = V_{\delta,n}$, $\delta \leq \|\lambda\|^{-1}$. If $x \in V_{\delta,n}$ then $r(x,0) \geq 2^{-n}\frac{\|x\|_n}{1+\|x\|_n} \geq 2^{-n-1}\delta$, whence $V_{\delta,n} \in B_0$. If $\delta \leq \|\lambda\|^{-1}$, $x \notin V_{\delta,n}$, then

$$\| <\lambda, x> \| = \|T(\lambda T_n(x))\| \leq 1$$

which implies that the integral in the right-hand side of (7) coincides with the one taken over $\overline{K}$.

On the other hand, almost surely

$$\|X(t) - X_{V_{\delta,n}}(t)\|_n < \delta. \tag{8}$$

Indeed, let t_0 be the first exit time of the process $X(t) - X_{V_{\delta,n}}(t)$ from the set $\overline{K} \setminus V_{\delta,n}$. Suppose that $t_0 < \infty$ with a non-zero probability. Then

$$\|X(t_0 - 0) - X_{V_{\delta,n}}(t_0 - 0)\|_n < \delta, \tag{9}$$

$$\|X(t_0 + 0) - X_{V_{\delta,n}}(t_0 + 0)\|_n \geq \delta. \tag{10}$$

If $\|X(t_0 + 0) - X(t_0 - 0)\|_n < \delta$ then $X_{V_{\delta,n}}(t_0 + 0) = X_{V_{\delta,n}}(t_0 - 0)$, so that

$$\|[X(t_0 + 0) - X_{V_{\delta,n}}(t_0 + 0)] - [X(t_0 - 0) - X_{V_{\delta,n}}(t_0 - 0)]\|_n < \delta. \tag{11}$$

If, on the contrary, $\|X(t_0 + 0) - X(t_0 - 0)\|_n \geq \delta$, then

$$X_{V_{\delta,n}}(t_0 + 0) - X_{V_{\delta,n}}(t_0 - 0) = X(t_0 + 0) - X(t_0 - 0),$$

so the expression in the left-hand side of (11) equals zero, and the inequality (11) is valid too. In both cases it contradicts (9), (10). Thus $t_0 = \infty$ almost surely, and the inequality (8) has been proved. We arrive to the following formula of Lévy-Khinchin type.

LEMMA 2 *For any* $\lambda \in K,\ t > 0$

$$E\chi(<\lambda, X(t)>) = \exp\Big\{\int_{\overline{K}} [\chi(<\lambda, x>) - 1]\Pi(t, dx)\Big\}. \tag{12}$$

REMARK Lemma 2 can serve as a base to develope the theory of stochastic integrals and stochastic differential equations over $\overline{K}$. In fact, the techniques and results of [17] carry over to this case virtually unchanged.

Both sides of (12) can be calculated explicitly if we use the heat measure $\pi(t, dx)$ (see [18]). We have

$$E\chi(<\lambda, X(t)>) = \int_{\overline{K}} \chi(<\lambda, x>)\pi(t, dx) = \rho_\alpha(\|\lambda\|, t)$$

where

$$\rho_\alpha(s, t) = \begin{cases} e^{-ts^\alpha}, & \text{if} \quad s > 1 \\ 1, & \text{if} \quad s \leq 1, \end{cases} \tag{13}$$

$s \geq 0,\ t > 0$. It follows from the definitions that $\Pi(t, \cdot)$ is symmetric with respect to the reflection $x \mapsto -x$. Therefore

$$\int_{\overline{K}} [\chi(<\lambda, x>) - 1]\Pi(t, dx) = \begin{cases} -t\|\lambda\|^\alpha, & \text{if} \quad \|\lambda\| > 1 \\ 0, & \text{if} \quad \|\lambda\| \leq 1. \end{cases}$$

LEMMA 3 *Let* M_n *be a compact subset of* $K_n \setminus \{0\}$, $M = T_n^{-1}(M_n)$. *Then*

$$\Pi(t, M) = -t \int_{\eta \in K_n:\ \|\eta\| > 1} \|\eta\|^\alpha w_n(\eta)\, d\eta \tag{15}$$

where $w_n(\eta)$ is the inverse Fourier transform of the function $y \mapsto q_n^{-d_n/2}\omega_{M_n}^{(n)}(m_n y)$, $\omega_{M_n}^{(n)}$ is the indicator of the set M_n in K_n, and dx is the normalized additive Haar measure on K_n.

Proof. Let ω_M be the indicator of the set M in $\overline{K}$. Then $\omega_M(x) = \omega_{M_n}^{(n)}(T_n(x))$, $x \in \overline{K}$,

$$\omega_{M_n}^{(n)}(\xi) = \int_{K_n} \chi(\xi\eta) w_n(\eta)\, d\eta, \quad \xi \in K_n.$$

Since

$$\int_{K_n} w_n(\eta)\, d\eta = \omega_{M_n}^{(n)}(0) = 0,$$

we get

$$\omega_M(x) = \int_{K_n} [\chi(\eta T_n(x)) - 1] w_n(\eta)\, d\eta.$$

Integrating with respect to $\Pi(t, dx)$ and using (14) we come to (15).

It follows from Lemma 3 that $\Pi(t, dx) = t\Pi(1, dx)$. We shall write $\Pi(dx)$ instead of $\Pi(1, dx)$.

Proof of Theorem 2. Let $f(x) = \varphi(T_n(x))$, $x \in \overline{K}$, where φ is locally constant and in addition supp φ is compact, $0 \notin$ supp φ. It follows from Lemma 3 that

$$\int_{\overline{K}} f(x)\Pi(dx) = -\int_{K_n} \Delta^\alpha(\eta)\psi(\eta)\, d\eta \tag{16}$$

where ψ is the inverse Fourier transform of the function $y \mapsto q_n^{-d_n/2}\varphi(m_n y)$.

The right-hand side of (16) is an entire function with respect to α. Assuming temporarily Re $\alpha < -1$, we can use the Plancherel formula with subsequent analytic continuation (see [18]). As a result we find that for $\alpha > 0$

$$\int_{\overline{K}} f(x)\Pi(dx) = -q_n^{d_n\alpha/m_n} \frac{1 - q_n^{\alpha/m_n}}{1 - q_n^{-1-\alpha/m_n}} \int_{x \in K_n \cdot\ |x|_n \le q_n^{d_n}} \Big[|x|_n^{-1-\alpha/m_n} + \frac{1 - q_n^{-1}}{q_n^{\alpha/m_n} - 1} q_n^{-d_n(1+\alpha/m_n)}\Big] \varphi(m_n x)\, dx. \tag{17}$$

An obvious approximation argument shows that (17) is valid for any $f \in \mathcal{E}(\overline{K})$. Comparing (17) with (1) we obtain (2).

4 HEAT MEASURE

Recall that the heat measure $\pi(t, dx)$ corresponding to the operator $-D^\alpha$ is defined by the formula

$$\int_{\overline{K}} \chi(<\lambda, x>)\pi(t, dx) = \rho_\alpha(\|\lambda\|, t), \quad \lambda \in K,\ t > 0,$$

where ρ_α is given by (13).

THEOREM 3 *For each $t > 0$ the measure $\pi(t, \cdot)$ is not absolutely continuous with respect to μ.*

Proof. Let us fix $N \geq 1$ and consider the set

$$M = \left\{ x \in \overline{K} : \ \|T_n(x)\| \leq q_n^{d_n/m_n - N/f_n} \|m_n\|, \ n = 1, 2, \ldots \right\}$$

We shall show that $\mu(M) = 0$ whereas $\pi(t, M) \neq 0$.

Denote

$$M_n = \left\{ x \in \overline{K} : \ \|T_n(x)\| \leq q_n^{d_n/m_n - N/f_n} \|m_n\| \right\}, \quad n = 1, 2, \ldots.$$

It is clear that $M = \bigcap_{n=1}^{\infty} M_n$. Repeating the arguments from the proof of Lemma 1, we see that $M_\nu \subset M_n$ if $\nu > n$. Thus

$$\pi(t, M) = \lim_{n \to \infty} \pi(t, M_n), \quad \mu(M) = \lim_{n \to \infty} \mu(M_n).$$

It follows from the integration formula for cylindrical functions [18] that

$$\mu(M_n) = q_n^{-d_n} \|m_n\|^{-m_n} \int\limits_{z \in K_n:\ \|z\| \leq q_n^{d_n/m_n - N/f_n} \|m_n\|} dz$$

$$= q_n^{-d_n} |m_n|_n^{-1} \int\limits_{z \in K_n\ \ |z|_n \leq q_n^{d_n - Ne_n} |m_n|_n} dz = q_n^{-Ne_n} = q_1^{-Nf_n e_n}, \tag{18}$$

so that $\mu(M_n) = q_1^{-Nm_n} \to 0$ for $n \to \infty$. Thus $\mu(M) = 0$.

In a similar way (see [18])

$$\pi(t, M_n) = |m_n|_n^{-1} \int\limits_{z \in K_n:\ |z|_n \leq q_n^{d_n - Ne_n} |m_n|_n} \Gamma_\alpha^{(n)} \left(m_n^{-1} z, t\right) dz \tag{19}$$

where $\Gamma_\alpha^{(n)}$ is a fundamental solution of the Cauchy problem for the equation over K_n of the form $\frac{\partial u}{\partial t} + \partial_n^\alpha u = 0$. Here ∂_n^α is a pseudo-differential operator over K_n with the symbol $\Delta^\alpha(\xi)$. It is clear that

$$\Gamma_\alpha^{(n)}(\zeta, t) = q_n^{-d_n/2} \widetilde{\rho}_\alpha(\zeta, t)$$

where tilde means the local field Fourier transform:

$$\widetilde{u}(\zeta) = q_n^{-d_n/2} \int_{K_n} \chi \circ \mathrm{Tr}_{K_n/k}(z\zeta) u(z)\, dz\,,$$

for a complex-valued function u over K_n (sufficient conditions for the existence of $\widetilde{u}$ and the validity of the inversion formula

$$u(z) = q_n^{-d_n/2} \int_{K_n} \chi \circ \mathrm{Tr}_{K_n/k}(-z\zeta) \widetilde{u}(\zeta)\, d\zeta\,,$$

are well known).

Using the Plancherel formula we can rewrite (19) in the form

$$\pi(t, M_n) = \int\limits_{K_n} \widetilde{\Gamma}_\alpha^{(n)}(x,t) \widetilde{\beta}_n(x)\, dx$$

where

$$\beta_n(\zeta) = \begin{cases} 1, & \text{if } \ |\zeta|_n \le q_n^{d_n - Ne_n} \\ 0, & \text{if } \ |\zeta|_n > q_n^{d_n - Ne_n}. \end{cases}$$

We have $\widetilde{\Gamma}_\alpha^{(n)} = q_n^{-d_n/2} \rho_\alpha$,

$$\widetilde{\beta}_n(x) = \begin{cases} q_n^{\frac{d_n}{2} - Ne_n}, & \text{if } \ |x|_n \le q_n^{Ne_n} \\ 0, & \text{if } \ |\zeta|_n > q_n^{Ne_n} \end{cases}$$

(see e.g. [15]), so that

$$\begin{aligned} \pi(t, M_n) &= q_n^{-Ne_n} \int\limits_{|x|_n \le q_n^{Ne_n}} \rho_\alpha(\|x\|, t)\, dx \\ &\ge q_n^{-Ne_n} \int\limits_{|x|_n = q_n^{Ne_n}} \rho_\alpha\left(|x|_n^{1/m_n}, t\right) dx = \left(1 - q_n^{-1}\right) \exp\left(-t q_n^{\alpha N / f_n}\right) \\ &\ge \left(1 - q_1^{-1}\right) \exp\left(-t q_1^{\alpha N}\right). \end{aligned}$$

Hence, $\pi(t, M) > 0$.

ACKNOWLEDGEMENT This work was supported in part by the Ukrainian Fund for Fundamental Research (Grant 1.4/62).

REFERENCES

[1] AD Bendikov. Symmetric stable semigroups on the infinite-dimensional torus. Expo Math 13:39–79, 1995.

[2] AD Blair. Adelic path space integrals. Rev Math Phys 7:21–50, 1995.

[3] JWS Cassels, A Fröhlich, eds. Algebraic Number Theory. London and New York: Academic Press, 1967.

[4] SN Evans. Local field Gaussian measures. In: E Cinlar, KL Chung, RK Getoor, eds. Seminar on Stochastic Processes 1988. Boston: Birkhäuser, 1989, pp 121–160.

[5] IB Fesenko, SV Vostokov. Local Fields and Their Extensions. Providence: American Mathematical Society, 1993.

[6] II Gikhman, AV Skorohod. Theory of Stochastic Processes II. Berlin: Springer, 1975.

[7] S Haran. Riesz potentials and explicit sums in arithmetic. Invent Math 101:697–703, 1990.

[8] S Haran. Analytic potential theory over the p-adics. Ann Inst Fourier 43:905–944, 1993.

[9] RS Ismagilov. On the spectrum of the self-adjoint operator in $L_2(K)$ where K is a local field; an analogous of the Feynman-Kac formula. Theor Math Phys 89:1024–1028, 1991.

[10] K Iwasawa. Local Class Field Theory. Tokyo: Iwanami Shoten, 1980 (in Japanese; Russian translation, Moscow: Mir, 1983).

[11] S Kaplan. Extensions of the Pontrjagin duality. II: Direct and inverse sequences. Duke Math J 17:419–435, 1950.

[12] AN Kochubei. Schrödinger-type operator over p-adic number field. Theor Math Phys 86:221–228, 1991.

[13] AN Kochubei. Parabolic equations over the field of p-adic numbers. Math USSR Izvestiya 39:1263–1280, 1992.

[14] AN Kochubei. The differentiation operator on subsets of the field of p-adic numbers. Russ Acad Sci Izv Math 41:289–305, 1993.

[15] AN Kochubei. Gaussian integrals and spectral theory over a local field. Russ Acad Sci Izv Math 45:495–503, 1995.

[16] AN Kochubei. Heat equation in a p-adic ball. Methods Funct Anal Topol 2, No. 3–4:53–58, 1996.

[17] AN Kochubei. Stochastic integrals and stochastic differential equations over the field of p-adic numbers. Potential Anal 6:105–125, 1997.

[18] AN Kochubei. Analysis and probability over infinite extensions of a local field. Potential Anal. (to appear).

[19] JP Serre. Local Fields. New York: Springer, 1979.

[20] AV Skorohod. Random Processes with Independent Increments. Dordrecht: Kluwer, 1991.

[21] VS Varadarajan. Path integrals for a class of p-adic Schrödinger equations. Lett Math Phys 39:97–106, 1997.

[22] VS Vladimirov. Generalized functions over the field of p-adic numbers. Russ Math Surv 43, No. 5:19–64, 1988.

[23] VS Vladimirov, IV Volovich, EI Zelenov. p-Adic Analysis and Mathematical Physics. Singapore: World Scientific, 1994.

[24] A Weil. Basic Number Theory. Berlin: Springer, 1967.

Some remarks on duality of locally convex B_K-modules

A KUBZDELA Institute of Civil Engineering, University of Technology,
61-138 Poznan, Poland

Abstract. In this paper we study the duality for locally convex modules over the unit disk of a spherically complete valued field. We consider a dual pair for arbitrary locally convex B_K-modules and pairs consisting of a locally convex module and its dual. We prove that the Mackey topology for an arbitrary locally convex B_K-module exists. This extends some results of van Tiel [11] obtained for locally convex spaces.

1 INTRODUCTION

The locally convex modules over the valuation ring appear in non-archimedean analysis in a natural way, as absolutely convex subsets of locally convex spaces. Recently in some papers these objects were considered independently as a strictly larger class than the locally convex spaces. A new line of research concerning non-archimedean locally convex B_K-modules started with Schikhof in 1991 (see [10]). The openess of continuous homomorphisms from a metrizable complete compactoid to a locally convex B_K-module was studied. The theory of locally convex B_K-modules was deeply developped by Schikhof and Oortwijn in [5], [6]. The authors defined some several starting notions concerning duality of B_K-modules. A Hahn-Banach type theorem for B_K-modules over a valuation ring of spherically complete valued field K was proved.

In this note we continue this study, developping the duality theory for locally convex modules over the unit disk for spherically complete valued fields K. For a locally convex B_K-module A by A' we understand the topological dual of A, i.e. the set of all continuous homomorphisms $L(A, K_1)$, where $K_1 = K/B_K^-$. We obtain some basic facts concerning locally convex B_K-modules and their duals.

We prove that for an arbitrary locally convex B_K-module A the Mackey topology (the finest one among the all locally convex topologies compatible with the dual pair (A, A')) exists and it is a topology of the uniform convergence on the members of the collection all $\sigma(A', A)$-bounded and $\sigma(A', A)$-c-compact submodules of A'. Finally we include some remarks for compatible topologies on c-compact B_K-modules.

In [11] van Tiel proved that every locally convex space E over a spherically complete K admits the Mackey topology. For locally convex spaces over non-spherically complete valued fields the question of the existence of the Mackey topology has a negative answer. In [3] Kakol proved that the space l^∞ of all bounded sequences $x = (x_n)$ over a non-spherically complete K endowed with the norm $||x|| = \sup_n |x_n|$ does not have the Mackey topology. This problem has been studied also in [2], [4]. In [4] it was proved that an infinite metrizable and complete locally convex space E over K admits the Mackey topology if and only if K is spherically complete. Nevertheless, there exist locally convex spaces over non-spherically complete K (for instance polarly barreled spaces, see [9]) for which the Mackey topology exists.

2 TERMINOLOGY AND NOTATIONS

Throughout let K be a spherically complete non-archimedean, non-trivially valued field. We denote by B_K, B_K^- the unit disks $\{x \in K : |x| \leqslant 1\}, \{x \in K : |x| < 1\}$ respectively. A ***topological*** B_K***-module*** is a pair (A, τ), where A is a B_K-module and τ is a topology on A such that the addition $A \times A \to A$ and the scalar multiplication $B_K \times A \to A$ are continuous maps. A topological B_K-module (A, τ) is called *locally convex* if τ has a base of 0-neighbourhoods consisting of submodules of A. A submodule V of A is called τ^-*-open* if there exists a τ-open submodule U and a $\lambda \in K, |\lambda| > 1$, such that $\lambda U \subseteq V$; where for $|\lambda| > 1$ the set λU is defined as $\{a \in A : \lambda^{-1} a \in U\}$. The τ^--open submodules generate a locally convex module topology on A. A submodule V of A is called τ^+*-open* if for every $\lambda \in K, |\lambda| > 1$ the submodule λV is τ-open. The τ^+-open submodules generate a Hausdorff locally convex module topology on A.

A subset $B \subseteq A$ is called *convex* if $-x + B$ $(x \in B)$ is a submodule of A. A B_K-module is called ***simple*** if for every submodule B of A either $B = 0$ or $B = A$. A locally convex Hausdorff B_K-module (A, τ) is called *c-compact* if every collection of closed convex subsets of A with the finite intersection property has a non-empty intersection. A subset B of A is called *absorbing* if for every $a \in A$ there exists a $\lambda \in B_K$, $\lambda \neq 0$ such that $\lambda a \in B$. Let X be a subset of B_K-module. Then the set $co(X)$ means the submodule of A generated by X. A subset B of A is called *bounded* if for every 0-neighbourhood U of A there exists a non-zero $\lambda \in B_K$ with $\lambda B \subseteq U$. Let B be a submodule of a locally convex B_K-module A. Then the quotient topology on a B_K-module A/B is locally convex (if B is closed this topology is Hausdorff). We denote by $K_1 = K/B_K^-$ the locally convex B_K-module equipped with the discrete topology (it is the unique Hausdorff locally convex topology on K_1, see [6], 6.21).

Let A be a locally convex B_K-module. A *seminorm* on A is a map
$p : A \to [0, \infty)$ such that for all $x, y \in A$, $\lambda \in B_K$
(i) $p(\lambda x) \leqslant p(x)$,

(ii) $p(x+y) \leqslant \max(p(x), p(y))$,
(iii) if $\lambda_1, \lambda_2, \ldots \in B_K$, $\lambda_n \to 0$ then $p(\lambda_n x) \to 0$.
A seminorm p is called a *norm* if $p(x) = 0 \Longrightarrow x = 0$. A seminorm (norm) p is called *discrete* if $p(x) = 0$ for $x = 0$ and $p(x) = 1$ for $x \neq 0$.

Let Φ be a non-empty collection of seminorms on A. A subset $V \subseteq A$ is called Φ-*open* if for every $a \in V$ there exist $n \in N$, $p_1, \ldots, p_n \in \Phi$ and $\epsilon_1, \ldots, \epsilon_n > 0$ such that $\{x \in A : p_i(x-a) < \epsilon_i \ (i = 1, \ldots, n)\} \subseteq V$. The collection of all Φ-open subsets of A is called the Φ-*topology*. The Φ-topology is a locally convex topology on A. Every locally convex topology τ on a B_K-module A can be introduced as the Φ-topology for some collection of seminorms Φ. We say that Φ generates the topology τ if the Φ-topology equals τ.

Let A, B be B_K-modules. Let $T : (a,b) \to T(a,b)$ be a bihomomorphism $A \times B \to K_1$. We say that the bihomomorphism T puts the B_K-modules A and B into the *duality* (or (A,B) is a *dual pair*) if T is non-degenerate, i.e. $T(a,b) = 0$ for all $a \in A$ implies $b = 0$, $T(a,b) = 0$ for all $b \in B$ implies $a = 0$. We call T *separating* in A (in B) if for every $a \neq 0$, $a \in A$ ($b \neq 0$, $b \in B$) there exists $b \in B$ ($a \in A$) such that $T(a,b) \neq 0$. The set of all homomorphisms between A and K_1 is called the *algebraic dual* of A and denoted by $A^d = Hom(A, K_1)$. For a topological B_K-module A we denote by $A' = L(A, K_1)$ the *topological dual* of A. A' is a submodule of A^d consisting of all continuous homomorphisms.

Let (A,B) be a dual pair of locally convex B_K-modules. We define the *weak topology* on A (on B) $\sigma(A,B)$ (similarly $\sigma(B,A)$) as the topology induced by the product topology on K_1^B (K_1^A), where K_1^B denotes the product $\prod_{a \in B} K_1$ equipped with the product topology. The submodules $U = \{a \in A : f_i(a) = 0;\ i = 1, \ldots, n;\ f_i \in B\}$, $n \in N$, form a fundamental system of 0-neighborhoods for $\sigma(A,B)$ on A. We say that a topology ν on A is *compatible* with the duality (A,B) if $B = (A,\nu)'$. A Hausdorff topology ν on A is called *minimal* if there is no other locally convex Hausdorff topology on A strictly coarser then ν. Let A be a locally convex B_K-module and let Θ be a bornology on A consisting of bounded subsets of A. By a Θ-*topology* (or τ_Θ) we denote the topology on A of the uniform convergence on elements of Θ on A'. A fundamental system of 0-neighbourhood for τ_Θ is formed by submodules $\{f \in A' : f(x) = 0,\ x \in X\}$, $X \in \Theta$.

For another undefined notions and properties concerning B_K-modules we refer to [5], [6].

3 RESULTS

Note that if A is a torsion B_K-module then the set $Hom(A, K)$. This explains a reason why K should be replaced by K_1 It is proved ([6], 6.15, 6.22) that for a locally convex vector space E, the algebraical dual E^* and $Hom(E, K_1)$ (the topological dual E' and $L(E, K_1)$) are isomorphic as B_K-modules. We start with some elementary facts concerning the existence of continuous homomorphisms on locally convex B_K-modules.

PROPOSITION 1 *Let (A, τ) be a locally convex B_K-module and let Φ be a collection of seminorms on A generating the topology τ. Then a homomorphism $f : A \to K_1$ is continuous if and only if there exist seminorms $p_1, \ldots, p_n \in \Phi$ and $c > 0$, such* *that* $\| f(x) \| \leqslant c \cdot \max_{i=1,\ldots,n} p_i(x)$ *for every* $x \in A$, *and* $n \in N$, *where* $\| . \|$*denotes the discrete norm on* K_1.

Proof. Let $f : A \to K_1$ be a continuous homomorphism. Then there exist seminorms $p_1, ..., p_n \in \Phi$ and $\epsilon > 0$ such that $p_i(x) < \epsilon$ for $i = 1, 2, ..., n$ implies $f(x) = 0$. Let $p = \max_{i=1,\ldots,n} p_i$, and let $x \in A$ be such that $p(x) \geqslant \epsilon$. If $\| f(x) \| = 0$ we are done. Suppose that $\| f(x) \| = 1$. Taking $c = \epsilon^{-1}$ one gets $\| f(x) \| \leqslant c \cdot p(x)$.

PROPOSITION 2 (Hahn-Banach theorem)

(i) *Let* A *be a locally convex* B_K*-module, and* $B \subseteq A$ *a submodule. Then every* $g \in B'$ *can be extended to a* $f \in A'$.

(ii) *Let* A *be a* B_K*-module,* $B \subseteq A$ *a submodule, and* p *be a seminorm on* A. *If* $g : B \to K_1$ *is a homomorphism satisfying* $\|g(x)\| \leqslant p(x)$, *then there exists a homomorphism* $f : A \to K_1$, *extending* g *and* $c > 0$ *such that* $\|f(x)\| \leqslant c \cdot p(x)$, *where* $\|.\|$ *denotes the discrete norm on* K_1.

Proof. (i) See [6], 6.25. (ii) We equip the B_K-module A with the p-topology, the locally convex topology generated by the seminorm p. Since $\|g(x)\| \leqslant p(x)$, the homomorphism g is p-continuous and by (i) there exists a p-continuous homomorphism f, extending g to the whole B_K-module. From (1) we obtain that there exists $c > 0$ such that $\| f(x) \| \leqslant c \cdot p(x)$ for every $x \in A$.

REMARK. Let (E, τ) be a locally convex space (B_K-module) over K (B_K). We say that (E, τ) has the *extension property* if for every locally convex space (B_K-module) (G, ν), for every linear subspace (submodule) $H \subseteq G$ and for every continuous linear operator (homomorphism) $T : H \to E$ there exists a continuous linear operator (homomorphism) $U : G \to E$ extending T. It is known from [7], 4.18, that (E, τ) has the extension property if (E, τ) is spherically complete. For B_K-modules this fact fails. Let K be a spherically complete valued field. Put $E = B_K$, $G = K$, $H = B_K$ and the identity Id: $B_K \to B_K$. Then, as easily seen, there exists no extension $U : K \to B_K$ of Id being a homomorphism.

Note also the following fact, see [6], 6.26.

PROPOSITION 3 *Let* K *be a spherically complete valued field and let* A *be a locally convex Hausdorff* B_K*-module . Then* A' *separates points of* A.

In the sequel we will assume that K is spherically complete and (A, τ) is a Hausdorff locally convex B_K-module.

PROPOSITION 4 *Let* (A, τ) *be a locally convex* B_K*-module and let* B *be a closed submodule of* A. *Let* $a \in A \backslash B$. *Then there exists a homomorphism* $f \in A'$ *such that* $f(B) = 0$ *and* $f(a) = 1$.

Proof. Let $\pi : A \to A/B$ be the quotient homomorphism. Then $\pi(a) \neq 0$. By (3) there exists $g \in (A/B)'$ such that $g(\pi(a)) = 1$. Hence $f = g \circ \pi \in A'$ and $f(a) = 1$, $f(B) = 0$.

We prove the existence of the Mackey topology for locally convex B_K-modules over spherically complete valued field K by using some ideas of [2],[3],[4]. We shall need the following lemma.

LEMMA 5 *Let A be a B_K-module and let p, q be seminorms on A, $c_0 > 0$. Let $f : A \to K_1$ be a homomorphism such that $\| f(x) \| \leqslant c_0 \cdot \max\{p(x), q(x)\}$. Then there exists $c > 0$, and two homomorphisms $f_i : A \to K_1, i = 1, 2$, such that $f = f_1 + f_2$ and $\| f_1(x) \| \leqslant c \cdot p(x)$, $\| f_2(x) \| \leqslant c \cdot q(x)$, $x \in A$.*

Proof. We set $T(x, x) = f(x)$, $s(x, y) = \max\{p(x), q(y)\}$, $x, y \in A$. We endow the B_K-module $A \times A$ with the locally convex topology generated by the seminorm s. Since $\| T(x, x) \| = \| f(x) \| \leqslant c_0 \cdot \max\{p(x), q(x)\} = s(x, x)$, the homomorphism T is s-continuous and by (2) there exists an s-continuous homomorphism $T_0 : A \times A \to K_1$ extending T to the whole B_K-module, satisfying $\| T_0(x, y) \| \leqslant c \cdot s(x, y)$, where $c > 0, x, y \in A$. Put $f_1(x) = T_0(x, 0)$ and $f_2(x) = T_0(0, x)$. It is easy to see that $f_i, i = 1, 2$ are desired maps.

THEOREM 6 *Every locally convex B_K-module (A, τ) admits the Mackey topology.*

Proof. Let (A, τ) be a locally convex B_K-module and let Φ be the family of the all locally convex topologies on A compatible with the dual pair (A, A'). Let $\mu = \sup \Phi$ be the weakest locally convex topology on A which is finer that any topology from Φ. Let $f : (A, \mu) \to K_1$ be a continuous homomorphism. Then there exist on A seminorms $p_1, ..., p_n$ continuous in $\nu_1, ..., \nu_n$,respectively. Also there exists $c > 0$ such that $\| f(x) \| \leqslant c \cdot \max_{i=1,..,n} p_i(x)$ for every $x \in A$. By (5) we conclude that the homomorphism f can be given as a finite sum of τ-continuous homomorphisms. It follows that f is τ-continuous.

The following example provides the duality between modules B_K, K_1 and their finite products.

EXAMPLE (i) B_K and K_1 are topological duals of each other ([5], 4.5.22);

(ii) B_K^- and K/B_K are topological duals of each other (proof runs similarly as in (i));

(iii) $(K_1^n)' = B_K^n$. We have $K_1' = K_1^d$. Then for finite products one gets $(K_1^n)' = (K_1^n)^d$. Let $\nu = (\nu_1, ..., \nu_n) \in {B_K}^n$. We define a homomorphism $\varphi_\nu : K_1^n \to K_1$ by the formula $\varphi_\nu(\lambda_1 + B_K^-, ..., \lambda_n + B_K^-) = \lambda_1\nu_1 + ... + \lambda_n\nu_n + B_K^-$. The homomorphism $T : {B_K}^n \to (K_1^n)'$, $T(\nu) = \varphi_\nu$ is injective. If $\nu = (\nu_1, ..., \nu_n) \in {B_K}^n$, where $\varphi_\nu = 0$, then $\lambda_i\nu_i \in B_K^-$ for every $\lambda_i \in K$, $(i = 1, ..., n)$. This implies $\nu_i = 0$ for every $i = 1, ..., n$. Note that T is also surjective. Indeed, if $s \in (K_1^n)'$, then $s(\lambda_1 + B_K^-, ..., \lambda_n + B_K^-) = s(\lambda_1 + B_K^-, 0, ..., 0) + ... + s(0, ..., 0, \lambda_n + B_K^-)$. By (i) one gets that $s(0, ..., 0, \lambda_i + B_K^-, 0, ..., 0) = \lambda_i\nu_i + B_K^-$ for $\nu_i \in B_K$ and hence $s(\lambda_1 + B_K^-, ..., \lambda_n + B_K^-) = \lambda_1\nu_1 + ... + \lambda_n\nu_n + B_K^-$ for $(\nu_1, ..., \nu_n) \in {B_K}^n$.

PROPOSITION 7 *Let (A, B) be a dual pair of locally convex B_K-modules. Then $B = (A, \sigma(A, B))'$.*

Proof. Let $f \in (A, \sigma(A, B))'$. Let U be a 0-neighbourhood in $(A, \sigma(A, B))$ such that $f(U) = \{0\}$. Note that $U = \bigcap_{i=1,\dots,n} f_i^{-1}(\{0\})$ with $f_1, \dots, f_n \in B$. Then $\bigcap_{i=1,\dots,n} \ker f_i \subseteq \ker f$, where $\ker f = \{x \in A : f(x) = 0\}$. Let $T : A \to K_1^n$ be a homomorphism defined by the formula $T(a) = (f_1(a), \dots, f_n(a))$. Then $\ker T \subseteq \ker f$. This implies that there exists a homomorphism $s \in (K_1^n)' = B_K{}^n$ such that $s \circ T = f$, where $s((g_1, \dots, g_n)) = s_1 g_1 + \dots + s_n g_n$ for every $(g_1, \dots, g_n) \in K_1^n$. We obtain $f(x) = s(T(x)) = (s_1 f_1 + \dots + s_n f_n)(x)$ for every $x \in A$. It follows that $f = s_1 f_1 + \dots + s_n f_n$ and thus $f \in co(f_1, \dots, f_n) \subseteq B$. Obviously $B \subseteq (A, \sigma(A, B))'$ and finally $B = (A, \sigma(A, B))'$.

COROLLARY 8 *Let (A, τ) be a locally convex B_K-module. Then $A' = (A, \sigma(A, A'))'$.*

In the next part we characterize the Mackey topology on a B_K-module A, as a topology of the uniform convergence on the members of some covering of A'. We shall need the following technical properties.

PROPOSITION 9 *Let (A, τ) be a locally convex B_K-module and let B be a convex set in A. Then B is $\sigma(A, A')$-closed if and only if B is closed.*

Proof. Suppose B is a closed submodule of A. From (4) for every $a \in A \backslash B$ there exists a continuous homomorphism $f_a : A \to K_1$ such that $f_a(a) = 1$, and $f_a(B) = \{0\}$. Since $\ker f_a$ is weakly closed for every f_a and $B = \bigcap_{a \in A \backslash B} \ker f_a$ we conclude that B is weakly closed.

REMARK Let A be a locally convex B_K-module and A' its topological dual. Let $X \subseteq A$ be a subset of A. We define the *pseudopolar* of X as $X^p = \{f \in A' : f(X) = 0\}$ and *pseudobipolar* $X^{pp} = \{x \in A : X^p(x) = 0\}$. If $B \subseteq A$ is a closed submodule of A, then $B = B^{pp}$.

PROPOSITION 10 *Let X be a subset of a locally convex B_K-module (A, τ). Then*

(i) *X^p is a weakly closed submodule of A';*

(ii) *if X is absorbing in A, then X^p is a weakly bounded set in A';*

(iii) *if X is a 0-neighbourhood in A, then X^p is a weakly c-compact subset of A';*

(iv) *if X is bounded, then X^p is absorbing in A'.*

Proof. (i) Proof is straightforward.

(ii) Let $V = \{f \in A' : f(x_i) = 0,\ x_i \in A,\ i = 1, \dots, n\}$ be a 0-neighbourhood in $(A', \sigma(A', A))$. Since X is absorbing, there exists $\lambda \in B_K$ such that $\lambda x_i \in X$, $i = 1, \dots, n$. Take $f \in X^p$, then $\lambda f(x_i) = f(\lambda x_i) = 0$. Thus $\lambda X^p \subseteq V$.

(iii) Note that K_1^A is c-compact by [5], 5.3.13 and 5.3.3. Let $\overline{X^p}$ be the closure of X^p in K_1^A with respect to the product topology. Let $(f_i)_{i \in I}$ be a net in X^p converging to $f \in \overline{X^p}$ in K_1^A. It is easy to see that f is a homomorphism. Since $g(x) = 0$ for every $g \in X^p$ and $x \in X$, then X^p is an equicontinuous in $(A', \sigma(A', A))$. The map $p_a : K_1^A \to K_1$, $p_a((x_b)_{b \in A}) = x_a$ is continuous and $p_a(g) = g(a)$ for $g \in A'$, $a \in A$. Thus $p_a(\overline{X^p}) = \{0\}$ for $a \in X$ because $p_a(X^p) = \{0\}$ and $f(X) \subseteq \{0\}$ for each $f \in \overline{X^p}$. Hence $\overline{X^p} \subseteq A'$ and $\overline{X^p}$ is a closed submodule of K_1^A. By [5], 5.3.8 $X^p = \overline{X^p}$ is c-compact.

(iv) Take $f \in A'$. Then there exists a 0-neighbourhood $U \subseteq A$ such that $f(U) = 0$. We find $\lambda \in B_K$ such that $\lambda X \subseteq U$. Hence $f(\lambda X) = \lambda f(X) = 0$ and f is absorbed by X^p.

PROPOSITION 11 *Let A be a locally convex B_K-module and let τ be a locally convex topology compatible with respect to the dual pair (A, A'). Then τ is a Ω-topology, where Ω is a collection of $\sigma(A', A)$-bounded and $\sigma(A', A)$-c-compact submodules of A'.*

Proof. Let Ψ be a fundamental system of 0-neighborhoods on A consisting of closed submodules and $\Xi = \{U^p : U \in \Psi\}$. The collection Ξ covers A'. Let $f \in A'$, then there exists $U \in \Psi$ such that $f(U) = 0$. Hence $f \in U^p$. Since U is absorbing, then by (10) U^p is a $\sigma(A', A)$-bounded and $\sigma(A', A)$-c-compact submodule. By (9) we have $U = U^{pp}$ and then $\tau \subseteq \tau_\Xi$. On the other hand, if $W = \{a \in A : U^p(a) = 0\}$ is a 0-neighbourhood in (A, τ_Ξ), then $W \in \Psi$, and consequently $\tau_\Xi \subseteq \tau$. It follows that $\tau_\Xi = \tau$. Let $\Omega = \{co(\bigcup_{i=1,..,n} V_i),\ V_i \in \Xi,\ i = 1, ..., n\}, n \in N$. Take $X \in \Omega$. By [5], 4.3.8 the set X is $\sigma(A', A)$-bounded. Applying [11], 2.5 one gets that it is weakly c-compact.

Now we are ready to prove the theorem.

THEOREM 12 *Let A be a locally convex B_K-module. The (Mackey) Θ-topology, where Θ is formed by the collection of all $\sigma(A', A)$-bounded and $\sigma(A', A)$-c-compact submodules of A', is the finest locally convex topology which is compatible with (A, A').*

Proof. Applying the same argumentation as in the last part of the previous proof we obtain that if $V_1, ..., V_n \in \Theta$, then $co(\bigcup_{i=1,..,n} V_i)$ belongs to Θ for $n \in N$. Since every finite set $\{f_1, ..., f_n\}$ is contained in one of the members of the family Θ, we have $\sigma(A, A') \leqslant \tau_\Theta$ and $A' \subseteq (A, \tau_\Theta)'$. We denote by "$o$" a pseudopolarity for the dual pair $(A, (A, \tau_\Theta)')$ and "p" a pseudopolarity for the pair (A, A'). Let $f \in (A, \tau_\Theta)'$ be a homomorphism. By the continuity of f in τ_Θ there exists an open submodule $U \subseteq (A, \tau_\Theta)$ such that $f(U) = 0$. We may assume that $U = X^p$ for some $X \in \Theta$. Hence $f(X^p) = 0$ and $f \in X^{po} = X^{oo}$. Since X is weakly closed in A (as a weakly c-compact set) and the inclusion $I : (X, \sigma(A', A) \mid_X) \to ((A, \tau_\Theta)', \sigma((A, \tau_\Theta)', A))$ is continuous, X is closed in $(A, \tau_\Theta)'$ in the topology $\sigma((A, \tau_\Theta)', A)$ by [5], 5.3.10. Thus $X = X^{oo}$ and $f \in A'$. We conclude $A' = (A, \tau_\Theta)'$ and τ_Θ is compatible with (A, A'). As we noticed in (11), if τ is any locally convex topology on A which is compatible with the dual pair (A, A'), then τ is coarser than τ_Θ.

The following proposition provides some characterization of locally convex compatible topologies for c-compact B_K-modules. The topologies $\sigma(A, A')$ and $\mu(A, A')$ are in some sense minimal and maximal.

PROPOSITION 13 *Let (A, τ) be a c-compact locally convex B_K-module. Then:*

i) $\tau^- \leqslant \sigma(A, A') \leqslant \tau \leqslant \mu(A, A') \leqslant \tau^+$;

ii) *all locally convex (A, A')-compatible topologies on A have the same bounded sets.*

iii) *the weak topology $\sigma(A, A')$ is a minimal Hausdorff topology on A. If A contains no simple submodules, then the topologies τ^- and $\sigma(A, A')$ concide on A;*

iv) *the Mackey topology $\mu(A, A')$ is the finest locally convex Hausdorff topology on A for which A is still c-compact.*

In order to prove Proposition 13 we shall need the following simple observation.

LEMMA 14 *Let* (A, τ) *be a locally convex* B_K*-module and* $f : A \to K_1$ *be a homomorphism such that* $f^{-1}\{0\}$ *is closed. Then* f *is continuous.*

Proof. (of Lemma 14) If $f^{-1}\{0\}$ is closed then $A/\ker f$ is a locally convex Hausdorff B_K-module and there exists a bijective homomorphism $g : A/\ker f \to K_1$ such that $f = g \circ \pi$, where π is the quotient homomorphism $A \to A/\ker f$. Since $A/\ker f$ is isomorphic to K_1 and the discrete topology is the only locally convex Hausdorff topology on K_1, we conclude that f is continuous.

Proof. (of Proposition 13) i) Since $\sigma(A, A')$ is Hausdorff, by [6], 5.15 we get the inequality $\tau^- \leqslant \sigma(A, A')$. The identity Id:$(A, \mu(A, A')) \to (A, \tau)$ is a continuous homomorphism. From [6], 5.15, one obtains that Id:$(A, \mu(A, A')) \to (A, \tau^+)$ is open. Thus $\mu(A, A') \leqslant \tau^+$.

ii) By [6], 6.4, the topologies τ, τ^-, τ^+ have the same bounded sets. The conclusion follows from (i).

iii) Let ν be a Hausdorff locally convex topology on A weaker than $\sigma(A, A')$. Then (A, ν) is c-compact and $(A, \nu)' \subseteq (A, \tau)'$. Let f be a τ-continuous homomorphism on (A, τ). Then $f^{-1}\{0\}$ is closed in (A, τ). Since (A, ν) is Hausdorff, the map Id: $(A, \tau) \to (A, \nu)$ is continuous. By [5], 5.3.10 the image of a convex, closed subset of (A, τ) is closed in (A, ν). Then Id$(f^{-1}(0))$ is ν-closed. From (14) we conclude that f is ν-continuous, hence $\nu = \sigma(A, A')$. Since A contains no simple submodules, by [5], 4.4.12, one gets that τ^- is Hausdorff and finally $\tau^- = \sigma(A, A')$.

iv) By (9) all compatible topologies on A have the same closed convex sets. Hence $(A, \mu(A, A'))$ is c-compact. Let ν be a locally convex topology on A finer than $\mu(A, A')$ such that (A, ν) is c-compact. Applying the same argument as in iii) we obtain $(A, \nu)' = (A, \mu(A, A'))'$. Thus $\nu = \mu(A, A')$.

REMARK Let (E, τ) be a Hausdorff locally convex space over a spherically complete valued field. In [8] it is proved that on an absolutely convex, bounded and c-compact subset of E the initial and the weak topologies concide. For a c-compact B_K-module (A, τ) this is true if K is discrete, since the topologies τ^- and τ are the same. The following example shows that in general this property fails.

EXAMPLE Let the valuation on K be dense. Let A be the locally convex B_K-module K/B_K. Then A is c-compact. From [5], 4.1.5 we know that the discrete topology d and $||.||$−topology (generated by the norm $||\lambda + B_K|| = (|\lambda| - 1) \vee 0$, $\lambda \in K$) are the only locally convex Hausdorff topologies on A. It is easy to see that there does not exist a $\lambda \in K, |\lambda| > 1$, such that $\lambda\{0\} \subseteq \{0\}$ and d^- is not discrete. Since A contains no simple submodules, by [5], 4.4.12 one obtains that d^- is Hausdorff and equals the $||.||$-topology.

I would like to thank Prof. W. Schikhof and Prof. J. Kakol for their remarks concerning the final version of this paper.

REFERENCES

[1] S Bosch, U Güntzer, R Remmert. Non-archimedean analysis. No. 261. Springer Verlag, Berlin, Heidelberg, New York, Tokyo 1984.

[2] J Kakol. Remarks on spherical completeness of non-Archimedean valued fields. Indag Math 5:321–323, 1994.

[3] J Kakol. The Mackey-Arens property for spaces over valued fields, Bull Polon Acad Sci Math 42(2):97–101, 1994.

[4] J Kakol, C Perez-Garcia, WH Schikhof. On the cardinality of nonarchimedean Banach and Frechet spaces and the Mackey-type theorems. Bull Polon Acad Sci Math 44(2):131–141, 1996.

[5] S Oortwijn. Locally convex Modules over Valuation Rings. PhD thesis, Catholic University Nijmegen, 1995.

[6] S Oortwijn, WH Schikhof. Locally convex modules over the unit disk. In: WH Schikhof, C Perez-Garcia, J Kąkol, ed. p-Adic Functional Analysis. New York: Marcel Dekker, 1997, pp 305–326.

[7] JB Prolla. Topics in Functional Analysis over Valued Division Rings. North-Holland Publishing Company-Amsterdam, New York, Oxford, 1982.

[8] WH Schikhof. Topological Stability of p-adic compactoids under continuous injections. Report 8644, Catholic University Nijmegen, 1986.

[9] WH Schikhof. Locally convex spaces over non-spherically complete valued fields. Bull Soc Math Belg 38:187–224, 1986.

[10] WH Schikhof. Zero sequences in p-adic compactoids. In: JM Bayod, N De Grande De Kimpe, J Martinez-Maurica, ed. p-adic Functional Analysis. New York: Marcel Dekker, 1991, pp 227–236.

[11] J van Tiel. Espaces Localement K-convexes, I-III, Proc Kon Ned Akad Wet 68:249–289, 1965.

Spectral properties of p-adic Banach algebras

NICOLAS MAINETTI Laboratoire de Mathématiques Pures, Université Blaise Pascal, Clermont-Ferrand, 63177 Aubière Cedex, France.

Abstract. Let $\mathbb{K}$ be a complete ultrametric algebraically closed field, with respect to a non trivial absolute value, and let A be a commutative $\mathbb{K}$-Banach algebra with identity. Let $\mathrm{Mult}(A, \|\cdot\|)$ be the set of continuous multiplicative semi-norms of $\mathbb{K}$-algebra (with respect to the norm $\|\cdot\|$ of A) and let $\mathrm{Mult}_m(A, \|\cdot\|)$ be the set of the $\varphi \in \mathrm{Mult}(A, \|\cdot\|)$ whose kernel is a maximal ideal of A. If the norm of A is equal to its spectral semi-norm $\|\cdot\|_{si}$ defined as $\|x\|_{si} = \lim_{n\to+\infty} \| x^n \|^{\frac{1}{n}}$, we have proven that $\|t\|_{si} = \sup\{\psi(t) | \ \psi \in \mathrm{Mult}_m(A, \|\cdot\|)\}$. Moreover, if A has no divisors of zero, denoting by $s(t)$ the spectrum of any $t \in A$, we have shown that $\|t\|_{si} = \sup\{|\lambda| \mid \lambda \in s(t)\}$. Finally, we can obtain the same kind of results without the hypothesis that the norm of A is equal to its spectral semi-norm, by only assuming that they satisfy certain local conditions for an element t of A: the set $\{\frac{\|\frac{1}{t-a}\|}{\|\frac{1}{t-a}\|_{si}} \mid a \notin s(t)\}$ has to be bounded. We first begin by giving an example showing that these conditions — named Properties (r) and (r$'$) — are weaker than suppose the norm equivalent to its spectral norm. Finally, denoting by $sa(x)$ the set of $\lambda \in \mathbb{K}$ such that $x - \lambda$ belongs to a maximal ideal of codimension 1, and denoting by $\|x\|_{sa}$ the real number $\sup\{0, |\lambda| \mid \lambda \in sa(x)\}$, we give an example showing that we can have $\|x\|_{si} = \sup\{0, |\lambda| \mid \lambda \in s(x)\}$, $\forall x \in A$ but $\|\cdot\|_{si} \neq \|\cdot\|_{sa}$.

1 INTRODUCTION AND RESULTS

DEFINITIONS AND NOTATIONS Let $\mathbb{L}$ be a complete ultametric field, and let $\mathbb{K}$ be a complete ultrametric algebraically closed field with respect to a non trivial absolute value $|\cdot|$. $\mathbb{L}$ is said to be *strongly valued* if its residue class field, or if its valuation group, is not countable. In order to avoid any confusion, we denote by $|\cdot|_\infty$ the classical archimedean absolute value on $\mathbb{R}$. As usual, given $a \in \mathbb{K}$ and $r > 0$, we put $d(a,r) = \{x \in \mathbb{K} \mid |x - a| \leqslant r\}$, $d(a, r^-) = \{x \in \mathbb{K} \mid |x - a| < r\}$, $C(a,r) = \{x \in \mathbb{K} \mid |x - a| = r\}$. Besides, given $s > r$, we put $\Gamma(a, r, s) = d(a, s^-) \setminus d(a, r)$.

Given a ring $\mathrm{Max}(R)$ denotes the set of maximal ideals of R. Let $\mathbb{F}$ be an algebraically closed field and let B be a $\mathbb{F}$-algebra with identity. Given $t \in B$, we denote by $s(t)$ the *spectrum* of t (i.e. the set of the $\lambda \in \mathbb{F}$ such that $t - \lambda$ is not invertible). We denote by $\Upsilon_a(B)$ the set of $\mathbb{F}$-algebras homomorphisms from B into $\mathbb{F}$. Then we call *algebraic spectrum* of an element t of B the set $sa(t) = \{\varphi(t) \mid \varphi \in \Upsilon_a(B)\}$. Such a set is also the set of $\lambda \in \mathbb{F}$ such that $t - \lambda$ belongs to a maximal ideal of codimension 1.

REMARK When there will be any risk of confusion with these notations (essentially in the Section 4), we shall have to index them by the algebra on which they are defined. For example, if B is a subalgebra of A and t an element of B, we will note $s_B(t)$ the spectrum of t considered as an element of B to distinguish it from $s_A(t)$, spectrum of t as an element of A.

DEFINITIONS AND NOTATIONS Let A be a commutative $\mathbb{L}$-normed algebra with identity, whose norm is denoted by $\|\cdot\|$. A semi-norm φ of $\mathbb{L}$-algebra on A will be said to be *multiplicative* if it satisfies $\varphi(xy) = \varphi(x)\varphi(y)$, $\forall x, y \in A$ and to be *semi-multiplicative* if $\varphi(x^n) = \varphi(x)^n$, $\forall x \in A$, $\forall n \in \mathbb{N}$.

The map $\|\cdot\|_{si}$ defined on A as $\|x\|_{si} = \lim_{n\to\infty} \|x^n\|^{\frac{1}{n}}$ is an ultrametric semi-norm of $\mathbb{L}$-algebra called *spectral semi-norm* of A and which is obviously seen to be semi-multiplicative. We will denote by $\|\cdot\|_{sa}$ the map defined as $\|x\|_{sa} = \sup\{0, |\lambda| \mid \lambda \in sa(x)\} = \{\sup\{0, |\varphi(x)| \mid \varphi \in \Upsilon_a(A)\}$. It is clearly seen that $\|\cdot\|_{sa}$ is a semi-norm if and only if $\|x\|_{sa} < +\infty$ for all $x \in A$.

Following Guennebaud's notations [8], we denote by $\mathrm{Mult}(A, \|\cdot\|)$ the set of continuous multiplicative semi-norms of $\mathbb{L}$-algebra on A (with respect to the norm $\|\cdot\|$ of A). So, given $\varphi \in \mathrm{Mult}(A, \|\cdot\|)$, the set of $t \in A$ such that $\varphi(t) = 0$ is a closed prime ideal of A called *kernel* of φ, and denoted by $\mathrm{Ker}(\varphi)$. Then, we denote by $\mathrm{Mult}_m(A, \|\cdot\|)$ the set of $\varphi \in \mathrm{Mult}(A, \|\cdot\|)$ such that $\mathrm{Ker}(\varphi)$ belongs to $\mathrm{Max}(A)$.

Given a subset Σ of $\mathrm{Mult}(A, \|\cdot\|)$, the mapping $\|\cdot\|_\Sigma$ defined as $\|t\|_\Sigma = \sup\{\psi(t) \mid \psi \in \Sigma\}$ is obviously seen to be a semi-multiplicative semi-norm on A. In particular, when $\Sigma = \mathrm{Mult}_m(A, \|\cdot\|)$, we denote by $\|\cdot\|_m$ this semi-norm.

During the sixties, T.A. Springer proved that given a normed commutative $\mathbb{L}$-algebra A, for all $x \in A$, we have $\|x\|_{si} = \sup\{\psi(x) \mid \psi \in \mathrm{Mult}(A, \|\cdot\|)\}$, [13, Cor. 6.25].

A commutative $\mathbb{K}$-Banach algebra with identity whose norm is equivalent to its spectral semi-norm is said to be a *normal algebra*.

We will denote by (o), (q) and (s) the following properties:

(o) $\sup\{0, |\lambda| \mid \lambda \in s(t)\} = \|t\|_{sa}$

(q) $\sup\{0, |\lambda| \mid \lambda \in s(t)\} = \|t\|_{si}$ for every $t \in A$.

(s) $\|\cdot\|_{si} = \|\cdot\|_m$.

Given any $t \in A$, we denote by (r(t)) the local property:

$(r(t))$ There exists $M \in \mathbb{R}^+$ such that for all $a \in d(0, \|t\|_{si}) \setminus s(t)$, we have $\|\frac{1}{t-a}\| \leqslant M\|\frac{1}{t-a}\|_{si}$.

Given an element $t \in A$ such that $\sup\{0, |\lambda| \mid \lambda \in s(t)\} < \|t\|_{si}$, we denote by (r'(t)) the following property :

$(r'(t))$ There exists $\gamma \in]\sup\{0, |\lambda| \mid \lambda \in s(t)\}, \|t\|_{si}[$ and there exists $M \in \mathbb{R}^+$ such that for all $a \in \Gamma(b, \sup\{0, |\lambda| \mid \lambda \in s(t)\}, \|t\|_{si})$ where $b \in s(t)$, we have $\|\frac{1}{t-a}\| \leqslant M\|\frac{1}{t-a}\|_{si}$.

Finally, we say that A satisfies Property (r) if for all $t \in A$, Property (r(t)) is satisfied. In the same way, we say that A satisfies Property (r′) if for all $t \in A$ such that $\sup\{0, |\lambda| \mid \lambda \in s(t)\} < \|t\|_{si}$, Property (r′(t)) is satisfied.

REMARKS Properties (r(t)) and (r′(t)) are logically independent. However, Property (r) implies Property (r′). Moreover, Property (q) implies Property (r′).

Let A be a commutative $\mathbb{K}$-Banach algebra with identity whose norm is equivalent to its spectral semi-norm. Then A satisfies Property (r). The reciprocal implication is wrong, as it is shown in [3, Theorem VII.7], where it is constructed a Banach algebra whose norm is not spectral but in which the set $d(0, \|t\|_{si}) \setminus s(t)$ is empty for any element t.

Now, let A be a commutative $\mathbb{K}$-Banach algebra with identity. In 1976, Escassut showed that if every maximal ideal of A has codimension 1, then Property (s) holds in A [4, Corollary 4.4]. (In particular, this applies to Tate's algebras, whose maximal ideals are of codimension 1, on an algebraically closed field [14]). Next, using the holomorphic functional calculus, in [4, Theorem 7.5] he showed that if $\mathbb{K}$ is strongly valued, the equality holds in any commutative $\mathbb{K}$-Banach algebra with identity. But if K is not strongly valued, by [4, Theorem 7.5], counter examples show that Property (s) does not hold in the general case. In particular, there exists local commutative $\mathbb{K}$-Banach algebra whose spectral semi-norm is a norm.

When $\mathbb{K}$ is strongly valued, property (q) was proven in [4, Theorem 7.9], in assuming another additional hypothesis, like the integrity of A, but without assuming the norm to be the spectral norm. But counter examples given in [4] show this last equality does not hold when $\mathbb{K}$ is not strongly valued.

However, in [6] we have obtained such equalities, without assuming $\mathbb{K}$ to be strongly valued, provided the spectral semi-norm $\| \,.\, \|_{si}$ of A is a norm equivalent to its $\mathbb{K}$-Banach algebra norm. This has been made possible thanks to a recent basic result concerning a partition of any annulus by a family of disks [11]. Here we will show that we may adapt the proof to $\mathbb{K}$-Banach algebras satisfying Property (r) or (r′), by using Mittag-Leffler's Theorem [5,10].

In Lemma 1.1, we recall previous results given in [5,13].

LEMMA 1.1 *Let A be a commutative normed $\mathbb{L}$-algebra with identity, and let $x \in A$. Then $\| \cdot \|_{si}$ is an ultrametric semi-multiplicative semi-norm satisfying $\|x\|_{si} = \sup\{\varphi(x) \mid \varphi \in \mathrm{Mult}(A, \| \cdot \|)\}$ and there exists $\varphi \in \mathrm{Mult}(A, \| \cdot \|)$ such that $\varphi(x) = \|x\|_{si}$. Further, if A is complete, for every $\mathcal{M} \in \mathrm{Max}(A)$, there exists $\psi \in \mathrm{Mult}_m(A, \| \cdot \|)$ such that $\mathrm{Ker}(\psi) = \mathcal{M}$, and if $\mathcal{M}$ has finite codimension, such a ψ is unique.*

We now work over $\mathbb{K}$ and we first show that Property (r′) is more general than Property (r).

THEOREM 1.1 *There exists a commutative $\mathbb{K}$-Banach algebra with identity, which satisfies Property (q) (hence also Property (r′)), which does not satisfy Property (r), but which contains non scalar elements t satisfying Property (r(t)).*

We will also show that Properties (r(t)) and (r′(t)) locally imply Properties (s) and (q).

THEOREM 1.2 *Let A be a commutative* $\mathbb{K}$*-Banach algebra with identity.*

1. *Let* $t \in A$ *such that* $\sup\{0, |\lambda| \mid \lambda \in s(t)\} < \|t\|_{si}$*, satisfying Property* $(r'(t))$*. Then* $\|t\|_{si} = \|t\|_m$.
2. *Let* $t \in A$ *such that* $s(t) \neq \emptyset$*, satisfying Property* $(r'(t))$*. If A has no divisor of zero, then* $\|t\|_{si} = \sup\{0, |\lambda| \mid \lambda \in s(t)\}$.

REMARKS If t is an element of A such that $\|t\|_{si} = \sup\{0, |\lambda| \mid \lambda \in s(t)\}$, then $\|t\|_m = \|t\|_{si}$. So, we immediatly have the following corollary. We also remark that by Theorem 1.1, Properties (q) and (r) are not equivalent.

COROLLARY 1.1 *Let A be a commutative* $\mathbb{K}$*-Banach algebra with identity.*

1. *If A satisfies Property* (r')*, then Property (s) is satisfied.*
2. *If A has no divisor of zero, does not contain any proper extension of* $\mathbb{K}$ *and satisfies Property (r), then Property (q) is satisfied.*

COROLLARY 1.2 *Let A be a normal commutative* $\mathbb{K}$*-Banach algebra with identity. Then Property (s) is satisfied. Furthermore, if A has no divisor of zero and does not contain any proper extension of* $\mathbb{K}$*, then Property (q) is satisfied.*

COROLLARY 1.3 *Let A be a commutative* $\mathbb{K}$*-Banach algebra with identity which satisfies Property* (r') *and whose spectral semi-norm is a norm. Then the Jacobson radical of A is null.*

DEFINITIONS AND NOTATIONS Given a closed and bounded set D in $\mathbb{K}$, we denote by $R(D)$ the $\mathbb{K}$-algebra of rational functions of $\mathbb{K}[x]$ with no pole in D, by $\|\cdot\|_D$ the norm of uniform convergence on D and by $H(D)$ the completion of $R(D)$ for this norm, which is called the $\mathbb{K}$-Banach algebra of analytic elements on D [5,9].

Let R be a commutative ring with unity and without divisor of zero, provided with an ultrametric absolute value $|\cdot|$. For any $h = \sum_{i=0}^{n} a_i T^i \in R[T]$, we put $\|h\| = \sup_{0 \leqslant i \leqslant n} |a_i|$. It is known that $\|\cdot\|$ is a norm of R-algebra on $R[T]$ which is called the *Gauss norm* [5]. We denote by $R\{T\}$ the algebra of formal restricted power series in T with coefficients in R. The Gauss norm on $R[T]$ extends to $R\{T\}$ and if R is an ultrametric $\mathbb{L}$-Banach algebra, then $R\{T\}$ provided with the Gauss norm is also an ultrametric $\mathbb{L}$-Banach algebra.

Let $\mathcal{D} = d(0, 1^-)$ and let $E = H(\mathcal{D})\{T\}$. Since $H(D)$ is a commutative $\mathbb{K}$-Banach algebra with identity and without divisor of zero, and since the norm $\|\cdot\|_{\mathcal{D}}$ is multiplicative, we may provide $H(\mathcal{D})[T]$ with the Gauss norm. This norm extends to E and we will denote it by $\|\cdot\|$.

Let $\rho \in]0, 1[$, let $X = 1 - xT$ and let S be the set multiplicatively generated by the polynomials $X - \alpha$ with $\alpha \in d(0, \rho^-)$. We denote by F the $\mathbb{K}$-algebra $S^{-1}E$ and by W the set of $\psi \in \mathrm{Mult}(E, \|\cdot\|)$ satisfying $\psi(X) \geqslant \rho$.

LEMMA 1.2 *Let* $\psi \in W$ *and let* $h = \frac{f}{g} \in F$*, with* $f \in E$ *and* $g \in S$*. Then* $0 \notin \psi(S)$*. The mapping* $\overline{\psi}$*, defined as* $\overline{\psi}(\frac{f}{g}) = \frac{\psi(f)}{\psi(g)}$ *belongs to* $\mathrm{Mult}(F)$*. Furthermore, the mapping* ϕ *from F to* $\overline{\mathbb{R}}^+$ *defined as* $\phi(h) = \sup_{\psi \in W} \overline{\psi}(h)$ *is a semi-multiplicative norm of* $\mathbb{K}$*-algebra.*

NOTATIONS We will denote by $\|\cdot\|$ the norm on F obtained in Lemma 1.2, and by $\widehat{F}$ the completion of F with respect to this norm. Thus, $\widehat{F}$ is a normal commutative $\mathbb{K}$-Banach algebra with identity.

Given a $\mathbb{K}$-Banach algebra A, we will denote by $A[Y]^0$ (resp. $A\{Y\}^0$) the set of polynomials (resp. restricted power series) $f(Y)$ with coefficients in A satisfying $f(0) = 0$. $A\{\{Y\}\}$ will denote the set of Laurent series of the form $\sum_{-\infty}^{+\infty} a_n Y^n$ such that $a_n \in A$ for all $n \in \mathbb{Z}$, $\lim_{n\to+\infty} a_n = 0$ and $\lim_{n\to-\infty} \|a_n\|\rho^n = 0$. We put $V = H(\mathcal{D}) + \mathbb{K}\{T\}^0$.

THEOREM 1.3 *$\widehat{F}$ is a normal commutative entire $\mathbb{K}$-Banach algebra with identity satisfying Property (q) but not Property (o). Moreover, $\widehat{F}$ is equal to the set of Laurent series $\sum_{-\infty}^{+\infty} \theta_n X^n$ with $\theta_n \in V$, $\forall n \in \mathbb{Z}$, $\lim_{n\to+\infty} \|\theta_n\| = 0$, $\lim_{n\to-\infty} \|\theta_n\|\rho^n = 0$ and*

$$\|\sum_{-\infty}^{+\infty} \theta_n X^n\| = \max(\sup_{n\geqslant 0} \|\theta_n\|, \sup_{n<0} \|\theta_n\|\rho^n).$$

2 PROOF OF THEOREM 1.1

Proof of Theorem 1.1. Let $D = C(0,1)$. We provide $\mathbb{K}[x, \frac{1}{x}]$ with the norm $\|\cdot\|$ defined as: for $f = \sum_{n=s}^{t} a_n x^n$, we put

$$\|f\| = \sup_{s\leqslant n\leqslant t} (|a_n|(|n|_\infty + 1)).$$

We first begin to show that $\|\cdot\|$ is actually a norm of $\mathbb{K}$-algebra. It is immediatly seen that it is a norm of $\mathbb{K}$-vector space; let us show that $\|fg\| \leqslant \|f\|\|g\|$, $\forall f, g \in \mathbb{K}[x, \frac{1}{x}]$. Let $f = \sum_{i=s_1}^{t_1} a_i x^i$ and $g = \sum_{j=s_2}^{t_2} b_j x^j$. For any i and j such that $s_1 \leqslant i \leqslant t_1$ and $s_2 \leqslant j \leqslant t_2$, we have

$$\begin{aligned} \|f\|\|g\| &= |a_{i_0}|(|i_o|_\infty + 1)|b_{j_0}|(|j_0|_\infty + 1) \\ &\geqslant |a_i|(|i|_\infty + 1)|b_j|(|j|_\infty + 1) \\ &\geqslant |a_j b_j|(|i|_\infty + |j|_\infty + 1) \\ &\geqslant |a_i b_j|(|i + j|_\infty + 1). \end{aligned}$$

Let $fg = \sum_{h=s_3}^{t_3} c_h x^h$. It is clear that

$$|c_h| \leqslant \sup_{i+j=h} |a_i b_j|,$$

hence

$$|c_h|(|h|_\infty + 1) \leqslant \|f\|\|g\| \quad \text{for every } s_3 \leqslant h \leqslant t_3,$$

and

$$\|fg\| \leqslant \|f\|\|g\|.$$

It follows that $\|\cdot\|$ is a norm of $\mathbb{K}$-algebra on $\mathbb{K}[x, \frac{1}{x}]$.

We now consider $\mathbb{K}[x, \frac{1}{x}]$ as a $\mathbb{K}$-subalgebra of $R(D)$. It is clearly seen that for any $f = \sum_{i=s}^{t} a_i x^i \in \mathbb{K}[x, \frac{1}{x}]$, we have

$$\|f\|_D = \sup_{s \leqslant i \leqslant t} |a_i| \leqslant \sup_{s \leqslant i \leqslant t} (|a_i|(|i|_\infty + 1)) = \|f\|.$$

We denote by A the completion of $\mathbb{K}[x, \frac{1}{x}]$ for its norm $\|\cdot\|$ and we are going to show that A is algebraically isomorphic to a $\mathbb{K}$-subalgebra of $H(D)$. For this, it suffices to prove that given a sequence $(f_n)_{n\in\mathbb{N}}$ of $\mathbb{K}[x, \frac{1}{x}]$ converging for the norm $\|\cdot\|$, it also converges to the same limit for $\|\cdot\|_D$. For every $n \in \mathbb{N}$, we put $f_n = \sum_{i=s_n}^{t_n} a_{i,n} x^i$. If $i \geqslant t_n$ or $i \leqslant s_n$, we put $a_{i,n} = 0$. This sequence converges for the norm $\|\cdot\|$. It means that for all $\varepsilon > 0$, there exists $N \in \mathbb{N}$ such that for all $n \geqslant N$ we have

$$\|f_{n+1} - f_n\| = \sup_{i \in \mathbb{Z}} (|a_{i,n+1} - a_{i,n}|(|i|_\infty + 1)) \leqslant \varepsilon.$$

Hence, for all $i \in \mathbb{Z}$ we have

$$|a_{i,n+1} - a_{i,n}| \leqslant |a_{i,n+1} - a_{i,n}|(|i|_\infty + 1) \leqslant \varepsilon. \qquad (1)$$

Thus, the sequence $(a_{i,n})_{n\in\mathbb{N}}$ is a Cauchy sequence for every $i \in \mathbb{Z}$ and converges to a limit denoted by a_i. So the sequence $(f_n)_{n\in\mathbb{N}}$ converges in A to an element f of the form $f = \sum_{-\infty}^{+\infty} a_i x^i$. But by (1), the limit of the sequence of the f_n for the norm $\|f\|_D$, may also be written $\sum_{-\infty}^{+\infty} a_i x^i$. In particular, if $\lim_{n\to\infty} \|f_n\|_D = 0$, then $\lim_{n\to\infty} \|f_n\| = 0$. Therefore, A is algebraically isomorphic to a $\mathbb{K}$-subalgebra of $H(D)$.

The norm $\|\cdot\|_D$ is clearly multiplicative. Let us show that it is the spectral semi-norm (say the spectral norm) of $\|\cdot\|$.

Let $f = \sum_{-\infty}^{+\infty} a_n x^n \in A$. We may suppose, without loss of generality, that $\|f\|_D = 1$; it means that $\sup_{n\in\mathbb{Z}} |a_n| = 1$.

Let t be the largest integer in absolute value such that $\|f\| = |a_t|(|t|_\infty + 1)$. We have

$$|a_n| = \frac{|a_t|(|t|_\infty + 1)}{|n|_\infty + 1} \varepsilon_n, \ \forall n \in \mathbb{Z} \text{ with } \varepsilon_n \leqslant 1 \text{ and } \lim_{|n|_\infty \to +\infty} \varepsilon_n = 0.$$

Let $k \in \mathbb{N}$ such that

$$\varepsilon_n \leqslant \frac{1}{|t|_\infty + 1}, \ \forall |n|_\infty \geqslant k.$$

Let us consider a product $P = |a_{j_1} \cdots a_{j_q}|$, where the sequence $(j_h)_h \subset \mathbb{Z}$ satisfies $|j_h|_\infty \leqslant |j_{h+1}|_\infty$.

Let r be the largest integer such that $|j_r|_\infty \leqslant |t|_\infty$ and let s be the largest integer such that $|j_s|_\infty < k$. We may choose k arbitrarily large to suppose $r \leqslant s$.

Better, since we are only interested by the large values of q, we may suppose q large enough to have $r < s$. Then

$$P \leqslant \prod_{h=r+1}^{q} \frac{\varepsilon_{j_h}(|t|_\infty + 1)}{|j_h|_\infty + 1},$$

using $|a_{j_h}| \leqslant 1$ for $h \leqslant r$. Now, when $h > s$, we have $\varepsilon_{j_h}(|t|_\infty + 1) \leqslant 1$, hence

$$P \leqslant \frac{\prod_{h=r+1}^{s}(\varepsilon_{j_h}(|t|_\infty + 1))}{\prod_{h=r+1}^{q}(|j_h|_\infty + 1)} \leqslant \frac{\prod_{h=r+1}^{s}\varepsilon_{j_h}}{\prod_{h=s+1}^{q}(|j_h|_\infty + 1)}$$

since $|t|_\infty + 1 \leqslant |j_h|_\infty + 1$, $\forall h > r$.

Consider now $\|f^q\|$. We may obviously find $j_1, \cdots, j_q \in \mathbb{Z}$ such that

$$\|f^q\| \leqslant |a_{j_1} \cdots a_{j_q}|(|j_1|_\infty + \cdots + |j_q|_\infty + 1).$$

Hence, for q large enough we have

$$\|f^q\| \leqslant \left(\prod_{h=r+1}^{s} \varepsilon_{j_h}\right) \frac{(|j_1|_\infty + \cdots + |j_q|_\infty + 1)}{\prod_{h=s+1}^{q}(|j_h|_\infty + 1)} \leqslant \frac{|j_1|_\infty + \cdots + |j_q|_\infty + 1}{\prod_{h=s+1}^{q}(|j_h|_\infty + 1)}.$$

We remark that

$$|j_1|_\infty + \cdots + |j_q|_\infty + 1 \leqslant sk + \sum_{h=s+1}^{q} |j_h|_\infty,$$

hence that

$$\begin{aligned} \|f^q\| &\leqslant \frac{sk}{k^{q-s}} + \frac{1 + \sum_{h=s+1}^{q} |j_h|_\infty}{\prod_{h=s+1}^{q}(|j_h|_\infty + 1)} \\ &\leqslant 1 + \frac{sk}{k^{q-s}} \\ &\leqslant 1 + sk \leqslant 1 + qk, \end{aligned}$$

and finally that

$$\|f^q\|^{\frac{1}{q}} \leqslant \sqrt[q]{1 + qk}.$$

Consequently,

$$\lim_{q \to +\infty} \|f^q\|^{\frac{1}{q}} = 1.$$

So, we have proven that $\|\cdot\|_{si} = \|\cdot\|_D$ on A.

Moreover, comparing the sequences $(\|x^n\|)_{n\in\mathbb{N}}$ and $(\|x^n\|_{si})_{n\in\mathbb{N}}$, we immediatly see that the norm $\|\cdot\|$ is not equivalent to its spectral norm.

We now will show that A satisfies Property (q). We put $B = H(D)$. We have shown that we may identify A to a $\mathbb{K}$-subalgebra of B. Hence, for any element $f \in A$, we have $s_B(f) \subset s_A(f)$. But Krasner algebras, as our considered $H(D)$, satisfy Property (q) by definition of their norm, [2]. In particular, on A we have

$$\forall f \in A, \quad \|f\|_D = \sup\{|\lambda| \mid \lambda \in s_B(f)\} \leqslant \sup\{|\lambda| \mid \lambda \in s_A(f)\} \leqslant \|f\|_{si}.$$

But we have shown that on A, $\|\cdot\|_{si} = \|\cdot\|_D$, and then A satisfies Property (q) (hence also Property (r′)).

Let $f = \sum_{-\infty}^{+\infty} a_n x^n \in A$, be invertible in A. Then f is invertible in $H(D)$ and in particular, does not vanish on $C(0,1)$. Let q be the largest integer such that $\|f\|_D = |a_q|$ and let s be the smallest integer such that $\|f\|_D = |a_s|$. By classical results [1,5,12], here we have $q = s$. Hence, f is of the form $x^q \sum_{-\infty}^{+\infty} a_n x^n$, with $|a_0| > \sup_{n\in\mathbb{Z}^*} |a_n|$. We may suppose that $a_0 = 1$. Then, we have $\|f\|_D = \|f\|_{si} = 1$. We are going to show that if $q \neq 0$, then f satisfies Property (r(f)).

$f(x)$ is of the form $x^q(1+\phi(x))$ with $q \neq 0$, $\phi \in A$ and $\|\phi\|_{si} < 1$. Let $\lambda \in d(0,1)$ be such that $f - \lambda$ is invertible in A. Then $f - \lambda$ is invertible in $H(D)$. Remark that $|\lambda| < 1$, because if not, we will have $|a_{-q} - \lambda| = 1 = a_0$, hence $f - \lambda$ will have $|q|_\infty$ zeros in $C(0,1)$ which will contradict its invertibility, [1,5,12]. We have

$$f - \lambda = x^q(1+\phi(x)) - \lambda = x^q(1+\phi(x) - \lambda x^{-q}).$$

We put $\theta(x) = \phi(x) - \lambda x^{-q}$. Hence we have

$$\|\theta(x)\|_{si} \leqslant \max(\|\phi\|_{si}, |\lambda|).$$

We put $\rho = \max(\|\phi\|_{si}, |\lambda|)$, then $\rho < 1$. Let us consider $\frac{1}{f-\lambda} = x^{-q}(1+\theta(x))^{-1}$. We see that

$$(1+\theta(x))^{-1} = \sum_{n=0}^{+\infty} (-1)^n (\theta(x))^n,$$

and it follows that,

$$\|(1+\theta(x))^{-1}\| \leqslant \sup_{n\in\mathbb{N}} \|\theta(x)^n\|.$$

Now

$$\|\theta(x)^n\| \leqslant \max_{0\leqslant j\leqslant n} (\|\phi^{n-j}\| |\lambda|^j).$$

But since $\|\phi\|_{si} < 1$, the sequence $(\|\phi^m\|)_m$ is bounded (and has limit 0). Let $M \leqslant \sup_m \|\phi^m\|$. We deduce that

$$\|\frac{1}{f-\lambda}\| \leqslant |q|_\infty M \quad \text{which is independant from } \lambda.$$

On the other hand, since

$$\|\theta(x)\|_{si} \leqslant \max(\|\phi\|_{si}, |\lambda|) < 1,$$

we see that

$$\|(1+\theta(x))^{-1}\|_{si} = \|\sum_{n=0}^{+\infty} (-1)^n (\theta(x))^n\|_{si} = 1.$$

Hence,

$$\|\frac{1}{f-\lambda}\|_{si} = 1, \quad \forall \lambda \in d(0,1) \setminus s(f).$$

Finally, Property (r(f)) is actually satisfied.

We now will show that there exist elements $f \in A$ which do not satisfy Property (r(f)). Let $b \in \mathbb{K}$ be such that $|b| < 1$, we put $f = 1 + bx$. Let $a \in \mathbb{K}$ be such that $|b| < |a| < 1$, we put $\lambda = a + 1$. Then $\lambda \in d(0, \|f\|_{si}) \setminus s(f)$, and in particular, we have

$$\frac{1}{f - \lambda} = \frac{1}{a(1 + \frac{b}{a}x)} = \frac{1}{a}\sum_{n=0}^{+\infty}(-1)^n\left(\frac{b}{a}\right)^n x^n.$$

Then, we see that

$$\|\frac{1}{f-\lambda}\|_{si} = |\frac{1}{a}|,$$

and that

$$\|\frac{1}{f-\lambda}\|_{si} = |\frac{1}{a}| \sup_{n\in\mathbb{N}} |\left(\frac{b}{a}\right)^n|(n+1).$$

But the family $\{\sup_{n\in\mathbb{N}} \rho^n(n+1) \mid \rho < 1\}$ is not bounded. Hence, the family $\left\{ \frac{\|\frac{1}{f-(a+1)}\|}{\|\frac{1}{f-(a+1)}\|_{si}} \middle| a \in \Gamma(0, b, 1) \right\}$ is not bounded. Therefore, f does not satisfy Property (r(f)), and finally, A does not satisfy Property (r).

3 PROOF OF THEOREM 1.2

DEFINITIONS AND NOTATIONS A set D in $\mathbb{K}$ is said to be *infraconnected* if for every $a \in D$, the mapping I_a from D to $\mathbb{R}^+$ defined by $I_a(x) = |x - a|$ has an image whose closure in $\mathbb{R}^+$ is an interval. An annulus $\Gamma(a, r, l)$ is called an *empty annulus of D* if it satisfies $\Gamma(a, r, l) \cap D = \emptyset$, $r = \sup\{|\lambda| \mid \lambda \in D \cap d(a, r)\}$ and $l = \inf\{|\lambda| \mid \lambda \in D \setminus d(a, l^-)\}$. In other words, a set D is not infraconnected if and only if it admits an empty annulus.

Circular filters are defined in [3,5,7]. A circular filter is said to be *large* if its diameter is different from zero. Large circular filters are known to characterize the absolute values on $\mathbb{K}(X)$ in this way:

for each large circular filter $\mathcal{F}$ on $\mathbb{K}$, for each $h \in \mathbb{K}(X)$, $|h(x)|$ admits a limit along $\mathcal{F}$ denoted by $\varphi_{\mathcal{F}}(h)$, and then, $\varphi_{\mathcal{F}}$ defines an absolute value on $\mathbb{K}(X)$, extending this of $\mathbb{K}$, i.e. a multiplicative norm of $\mathbb{K}(X)$ [3,5,7]. Then, the mapping that associates a multiplicative norm on $\mathbb{K}(X)$ to a large circular filter $\mathcal{F}$ on $\mathbb{K}$, in this way, is a bijection from the set of large circular filters on $\mathbb{K}$.

Given such a multiplicative norm ψ on $\mathbb{K}(X)$, we will denote by $\mathcal{G}_\psi$ the large circular filter that defines ψ. Then each multiplicative semi-norm ψ on $R(D)$,

either it is a norm, and then it has a continuation to $\mathbb{K}(X)$, and is defined by a large circular filter on $\mathbb{K}$ that we will denote again by $\mathcal{G}_\psi$ [5],

or it is not a norm, and then, there exists $a \in D$ such that $\psi(h) = |h(a)|$ for every $h \in R(D)$ [3,5,7] and we will denote by $\mathcal{G}_\psi$ the filter of neighborhoods of the point a.

Given a in $\mathbb{K}$ and $r > 0$, we call a *classic partition of* $d(a, r)$ a partition of the form $(d(b_j, r_j^-))_{j\in I}$. The disks $d(b_j, r_j^-)$ are called the *holes* of the partition.

Let $\mathcal{P} = (d(b_j, r_j^-))_{j\in I}$ be a classic partition of $d(a, r)$. A annulus $\Gamma(b, r', r'')$ included in $d(a, r)$ and such that $d(b, r^-)$ contains at least a hole of $\mathcal{P}$ will be said to be $\mathcal{P}$-*minorated* if there exists $\delta > 0$ such that $r_j \geqslant \delta$ for every $j \in I$ such that $d(b_j, r_j^-) \subset \Gamma(b, r', r'')$.

Given a closed bounded set E in $\mathbb{K}$, we denote by $\widetilde{E}$ the smallest disk of the form $d(\alpha, \rho)$ that contains E (i.e. ρ is the diameter of E and α may be taken in E). Besides, $\widetilde{E} \setminus E$ admits a unique partition of the form $(d(\alpha_j, \rho_j^-))_{j \in J}$ such that for each $j \in J$, ρ_j is the distance from α_j to E. Then each disk $d(\alpha_j, \rho_j^-)$ is called a *hole* of E. A closed infraconnected set E included in $d(a, r)$ will be said to be a $\mathcal{P}$-*set* if $\widetilde{E} = d(a, r)$ and if every hole of E is a hole of $\mathcal{P}$.

For each $j \in I$, we denote by $\mathcal{F}_j$ the circular filter of center b_j and diameter r_j, and for every $h \in \mathbb{K}(X)$ we put $\|h\|_{\mathcal{P}} = \sup_{j \in I} \varphi_{\mathcal{F}_j}(h)$. Then, by [11] we know that $\|\cdot\|_P$ is a semi-multiplicative norm of $\mathbb{K}$-algebra on $\mathbb{K}(X)$.

Next, $H(\mathcal{P})$ will denote the completion of $\mathbb{K}(X)$ for this norm. Hence, $H(\mathcal{P})$ is a $\mathbb{K}$-Banach algebra provided with a semi-multiplicative norm.

Let $\mathbb{F}$ be an algebraically closed field, let A be a $\mathbb{F}$-algebra, let $t \in A$ and let $\mathcal{I}$ be the ideal of the $G(X) \in \mathbb{F}[X]$ such that $G(t) = 0$. If $\mathcal{I} = \{0\}$, we call 0 the *minimal polynomial of t*. If $\mathcal{I} \neq \{0\}$, we call *minimal polynomial of t* the unique monic polynomial that generates $\mathcal{I}$.

Lemma 3.1 is given in [11].

LEMMA 3.1 *Let $\mathcal{P}$ be a classic partition of a disk $d(a, r)$ and let E be a $\mathcal{P}$-set. Then we have $\|h\|_E = \|h\|_{\mathcal{P}}$ for every $h \in R(E)$.*

COROLLARY 3.1 *Let $\mathcal{P}$ be a classic partition of a disk $d(a, r)$ and let E be a $\mathcal{P}$-set. Then $H(E)$ is isometrically isomorphic to a $\mathbb{K}$-subalgebra of $H(\mathcal{P})$.*

Henceforth, given a classic partition $\mathcal{P}$ of a disk $d(a, r)$ and a $\mathcal{P}$-set E, we will consider $H(E)$ as a $\mathbb{K}$-subalgebra of $H(\mathcal{P})$.

Lemma 3.2 is proven in [6].

LEMMA 3.2 *Let A be a $\mathbb{K}$-algebra with identity and let $t \in A$. There exists a $\mathbb{K}$-algebras homomorphism Θ from $R(s(t))$ into A such that $\Theta(P) = P(t)$ for all $P \in \mathbb{K}[X]$. Moreover, Θ is injective if and only if t has a null minimal polynomial. Besides, for every $h \in R(s(t))$, we have $s(h(t)) = h(s(t))$.*

REMARKS When the homomorphism Θ in Lemma 3.2 is injective, the $\mathbb{K}$-subalgebra $B = \Theta(R(s(t)))$ is isomorphic to $R(s(t))$, and in fact is the whole subalgebra generated by t in A. So, in such a case, we may consider $R(s(t))$ as a $\mathbb{K}$-subalgebra of A.

Let $\mathcal{F}$ be the circular filter of center a and diameter r. This filter is secant with $C(a, r)$ if and only if $r \in |\mathbb{K}|$. This leads us to introduce the following definition.

DEFINITION A circular filter $\mathcal{F}$ will be said to be *approaching a circle* $C(a, r)$ if it is secant with this circle, or if it is the circular filter of center a and diameter r.

By results of [3,7], also given in [5], we have Lemma 3.3.

LEMMA 3.3 *Let $\mathcal{F}$ be a circular filter on a set $S \subset \mathbb{K}$, and let $a \in \mathbb{K}$. There exists a unique $r > 0$ such that $\mathcal{F}$ is approaching $C(a, r)$.*

As a consequence, we have Lemma 3.4.

LEMMA 3.4 *Let A be a commutative $\mathbb{K}$-Banach algebra with identity and let $t \in A$ have a null minimal polynomial. Let $a \in \mathbb{K}$, let $\psi \in Mult(A, \|\cdot\|)$, let $\widetilde{\psi}$ be the restriction of ψ to $R(s(t))$, and let $r = \psi(t-a)$. Then $\mathcal{G}_{\widetilde{\psi}}$ is approaching $C(a, r)$.*

PROPOSITION 3.1 *Let A be a commutative $\mathbb{K}$-Banach algebra with identity. Let $t \in A$ be such that the mapping Θ from $\mathbb{K}[x]$ into A defined as $\Theta(P) = P(t)$ is injective. Let $a \in \mathbb{K} \setminus s(t)$, and let $r = \|(t-a)^{-1}\|_{si}^{-1}$. There exists $\theta \in$ Mult$(A, \|.\|)$ whose restriction to $R(s(t))$ has a circular filter approaching $C(a, r)$.*

Proof. We consider $R(s(t))$ as a $\mathbb{K}$-subalgebra of A. For all $\phi \in$ Mult$(A, \|.\|)$ we denote by $\widetilde{\phi}$ the restriction of ϕ to $R(s(t))$. Let $\psi \in$ Mult$(A, \|.\|)$. If $\mathcal{G}_{\widetilde{\psi}}$ is secant with a disk $d(a, \rho)$ for some $\rho \in]0, r[$, then clearly we have $\psi(t-a) \leqslant \rho$ hence $\psi((t-a)^{-1}) > \frac{1}{r}$ and therefore $\|(t-a)^{-1}\|_{si} > \frac{1}{r}$ which contradicts the hypothesis. So $\mathcal{G}_{\widetilde{\psi}}$ is secant with $\mathbb{K} \setminus d(a, r^-)$.

Suppose that there exists $\rho > r$ such that, for every $\phi \in$ Mult$(A, \|.\|)$, $\mathcal{G}_\phi$ is not secant with $d(a, \rho)$. Clearly we have $\phi(t-a) \geqslant \rho$ for all $\phi \in$ Mult$(A, \|.\|)$ and therefore $\|(t-a)^{-1}\|_{si} < \frac{1}{r}$. As a consequence, for each $n \in \mathbb{N}^*$ we can find $\psi_n \in$ Mult$(A, \|.\|)$ such that $\mathcal{G}_{\widetilde{\psi}_n}$ is secant with $d(a, r + \frac{1}{n})$, and since it is also secant with $\mathbb{K} \setminus d(a, r^-)$, finally, it is secant with $\Gamma(a, r, r + \frac{1}{n})$. Since Mult$(A, \|.\|)$ is compact [5,8], we can extract from the sequence $(\psi_n)_{n \in \mathbb{N}}$ a subsequence $(\psi_{n_q})_{q \in \mathbb{N}})$ which converges in Mult$(A, \|.\|)$. So, without loss of generality, we may directly assume that the sequence is convergent. Let θ be its limit. For each $n \in \mathbb{N}^*$, $\mathcal{G}_{\widetilde{\psi}_n}$ is approaching a circle $C(a, r_n)$ with $r \leqslant r_n \leqslant r + \frac{1}{n}$. But putting $s_n = \widetilde{\psi}_n(t-a)$, by Lemma 3.4, it is secant with $C(a, s_n)$. Suppose $s_n \neq r_n$. But by Lemma 3.3, $\mathcal{G}_{\widetilde{\psi}_n}$ may not be approaching both $C(a, r_n)$ and $C(a, s_n)$. Hence we have $\widetilde{\psi}_n(t-a) = r_n$. Since $\lim_{n \to \infty} r_n = r$, we have $\widetilde{\theta}(t-a) = r$, hence by Lemma 3.3, $\mathcal{G}_{\widetilde{\theta}}$ is secant with $C(a, r)$. This completes the proof.

NOTATIONS AND DEFINITIONS Let A be a $\mathbb{K}$-normed algebra, and suppose that an element $x \in A$ has a null minimal polynomial and is such that $s(x)$ admits an empty annulus $\Gamma(a, r, l)$. Such an empty annulus is said to be *x-cleaved* if for every r', $r'' \in]r, l[$, with $r' < r''$, there exists $\psi \in$ Mult$(A, \|\cdot\|)$, such that the circular filter of the restriction of ψ to $R(s(x))$ is secant with $\Gamma(a, r', r'')$.

Let A be a commutative $\mathbb{K}$-Banach algebra with identity. Let $t \in A$ be such that the mapping Θ from $\mathbb{K}[x]$ into A defined as $\Theta(P) = P(t)$ is injective. Let $a \in s(t)$, and let $r = \|(t-a)\|_{si}$. For each $b \in d(a, r) \setminus s(t)$ we put $r_b = \frac{1}{\|\frac{1}{x-b}\|}$, $\Lambda_b = d(b, r_b^-)$. By Lemma 3.1 of [4], we know that if $c \in \Lambda_b$, then $\Lambda_c = \Lambda_b$. For every $b \in$

$d(a,r) \setminus s(t)$, we denote by ψ_b the element of $\text{Mult}(R(D))$ whose circular filter has center b and diameter r_b, so ψ_b satisfies $\psi_b(h) = \lim_{\substack{|x-b| \to r_b \\ |x-b| \neq r_b}} |h(x)| \; \forall h \in R(D)$.

For every $h \in R(D))$, we put $\|h\|_t = \max(\|h\|_D, \sup\{\psi_b(h) | \; b \in d(a,r) \setminus s(t)\})$.

As the Λ_b form a partition of $d(a,r) \setminus s(t)$, by [11], and Proposition 3.3 of [4], we have Proposition 3.2:

PROPOSITION 3.2 *Let A be a commutative $\mathbb{K}$-Banach algebra with identity. Let $t \in A$ be such that the mapping Θ from $\mathbb{K}[x]$ into A defined as $\Theta(P) = P(t)$ is injective. Then $\| \; . \; \|_t$ defines on $R(s(t))$ a semi-multiplicative norm satisfying $\|h\|_t \geqslant \|h(t)\|_{si}$ for every $h \in R(s(t))$.*

T-sequences are defined in [3,5]. They define filters — named T-filters — among them can be vanishing non trivially an analytic element. The following proposition makes a link between $\mathcal{P}$-minorated annulus, T-sequences and analytic elements vanished by T-filters.

PROPOSITION 3.3 *Let $\mathcal{P}$ be a classic partition of a disk $d(a,r)$, let $\Gamma(b,r',r'')$ be a $\mathcal{P}$-minorated annulus included in $d(a,r)$ and let $l \in]r',r''[$. There exists an increasing idempotent T-sequence $(T_n, 1)_{n\in\mathbb{N}} \subset \mathcal{P}$ of center b and diameter l together with a decreasing idempotent T-sequence $(T'_n, 1)_{n\in\mathbb{N}} \subset \mathcal{P}$ of same center and diameter. We put $T_n = d(a_n, r_n^-)$ and $T'_n = d(a'_n, r'_n)$ for all $n \in \mathbb{N}$. Moreover the set $E = d(a,r) \setminus \bigcup_{n\in\mathbb{N}} (T_n \cup T'n)$ forms a $\mathcal{P}$-set and there exist elements $f, g \in H(E)$ such that*

i) *$f(x) = 0$ for all $x \in E \setminus d(b,l)$ (resp. $g(x) = 0$ for all $x \in d(b,l) \cap E$).*

ii) *For all circular filter $\mathcal{G}$ secant with E but not secant with $E \setminus d(b,l)$ (resp. with $d(b,l) \cap E$) we have $\varphi_{\mathcal{G}}(f) \neq 0$ (resp. $\varphi_{\mathcal{G}}(g) \neq 0$).*

iii) *For each $n \in \mathbb{N}$, f (resp. g) is meromorphic in T_n (resp. T'_n), admits a_n (resp. a'_n) as a pole of order 1 and does not have any other pole in T_n (resp. T'_n).*

Proof. The existence of idempotent T-sequences is shown by [11, Proposition 2.5]. Of course, by definition, the set $E = d(a,r) \setminus \bigcup_{n\in\mathbb{N}} (T_n \cup T'n)$ is a $\mathcal{P}$-set because such T-sequences will not form an empty annulus. The existence of f and g satisfying the conditions of the proposition is immediatly given by [5, Theorem 37.2].

PROPOSITION 3.4 *Let A be a commutative $\mathbb{K}$-Banach algebra with identity and let $t \in A$ whose minimal polynomial is not null. Then we have $\|t\|_m = \|t\|_{si}$.*

Proof. We put $D = s(t)$. By Lemma 3.2, there exists a $\mathbb{K}$-algebras homomorphism Θ of $R(D)$ into A such that $\Theta(P) = P(t)$ for all $P \in \mathbb{K}[X]$.

The hypothesis that the minimal polynomial of t is not null implies that $\text{Ker}(\Theta) \neq \{0\}$, and so is an ideal of $R(D)$ generated by a monic polynomial $G(x) = \prod_{i=1}^{q}(x - a_i)$. Since $G(t) = 0$, for any $\psi \in \text{Mult}(A, \| \cdot \|)$ we have $\psi(G(t)) = 0$. Hence, there exists $l(\psi) \in \{1, \cdots, q\}$ such that $\psi(t - a_{l(\psi)}) = 0$. Then we have $\psi(t) = |a_{l(\psi)}|$. It means that $t - a_{l(\psi)}$ belongs to $\text{Ker}(\psi)$ and belongs to a maximal ideal $\mathcal{M}$ of

A. But, by Lemma 1.1 there exists $\theta_\psi \in \mathrm{Mult}_m(A, \|\cdot\|)$ such that $\mathrm{Ker}(\theta_\psi) = \mathcal{M}$. Hence, we have

$$\theta_\psi(t) = |a_{l(\psi)}| = \psi(t).$$

From this, we deduce that $\psi(t) \leqslant \|t\|_m$. But this is true for every $\psi \in \mathrm{Mult}(A, \|\cdot\|)$. Hence, we have $\|t\|_{si} \leqslant \|t\|_m$ and so, $\|t\|_{si} = \|t\|_m$.

We now are able to prove Theorem 1.2.

Proof of Theorem 1.2 We suppose that $\|t\|_m < \|t\|_{si}$, (resp. that $\sup\{|x| \mid x \in s(t)\} < \|t\|_{si}$). We put $D = s(t)$. In order to simplify the writing of the proof, we will suppose that $0 \in s(t)$, replacing eventually t by $t - \lambda$ where $\lambda \in \mathbb{K}$ with $|\lambda| < \|t\|_{si}$ and $0 \in s(t - \lambda)$.

By Lemma 3.2, there exists a $\mathbb{K}$-algebras homomorphism Θ of $R(D)$ into A such that $\Theta(P) = P(t)$ for all $P \in \mathbb{K}[X]$. We put $B = \Theta(R(D))$.

If $\mathrm{Ker}(\Theta) \neq \{0\}$, i.e. if the minimal polynomial of t is not null, then the hypothesis $\|t\|_m < \|t\|_{si}$ contradicts Proposition 3.4. Moreover, we remark that $\mathrm{Ker}(\Theta)$ admits a generator $G(X) \in \mathbb{K}[X]$ whose zeros lie in $s(t)$. If $deg(G) = 1$, then t lies in $\mathbb{K}$ (considered as a $\mathbb{K}$-subalgebra of A) and obviously we have $\psi(t) = |t|$, $\forall \psi \in \mathrm{Mult}(A, \|\cdot\|)$. This contradicts the hypothesis that there exists $t \in A$ such that $\sup\{|x| \mid x \in s(t)\} < \|t\|_{si}$. Next, if $deg(G) > 1$, then $\mathrm{Ker}(\Theta)$ is not prime, hence A contains divisors of zero, so this case does not concern the second statement.

Now we suppose $\mathrm{Ker}(\Theta) = \{0\}$. Hence B is isomorphic to $R(D)$. Furthermore, by Proposition 3.2, once $R(D)$ is provided with the norm $\|\cdot\|_t$, Θ is continuous. For all $\psi \in \mathrm{Mult}(A, \|\cdot\|)$, we denote by $\widetilde{\psi}$ the restriction of ψ to B and $\mathcal{G}_{\widetilde{\psi}}$ the circular filter associated to $\widetilde{\psi}$. We put $r = \|t\|_{si}$, $r'' = \sup\{|x| \mid x \in s(t)\}$ and $r' = \|t\|_m$. So we will suppose $r' < r$ (resp. $r'' < r$).

We will denote by γ the real number belonging to $]r'', r[$ whose existence is given by Property (r'(t)). Since we suppose that $r' < r$, we may choose γ such that $\gamma \in]r', r[$. Let $W = d(0, r)$ and let $s' \in]\gamma, r[$ (resp. and let $s'' \in]r'', r[$). Let $W' = d(0, s')$ (resp. $W'' = d(0, s'')$). For each $\alpha \in W \setminus W'$ (resp. $\alpha \in W \setminus W''$), we put $r_\alpha = \dfrac{1}{\|\frac{1}{t-\alpha}\|}$ and $\Lambda_\alpha = d(\alpha, r_\alpha^-)$. So, $(\Lambda_\alpha)_{\alpha \in W \setminus W'}$ (resp. $(\Lambda_\alpha)_{\alpha \in W \setminus W''}$) is a partition $\mathcal{T}'$ (resp. $\mathcal{T}''$) of $W \setminus W'$ (resp. $W \setminus W''$). So, we denote by $\mathcal{P}$ the classical partition of W defined by the classes of W' (resp. W'') union $\mathcal{T}$ (resp. union $\mathcal{T}'$). We clearly have

$$\|h\|_{\mathcal{P}} \geqslant \|h\|_t, \quad \forall h \in \mathbb{R}(s(t)).$$

Let $a \in s(t)$. In particular, the annulus $\Gamma(a, s', r)$ (resp. $\Gamma(a, s'', r)$) admits a partition by a subfamily $\mathcal{S}$ of $\mathcal{T}$ (resp. of $\mathcal{T}'$). Hence, by [11, Proposition 1.2], $\Gamma(a, s', r)$ (resp. $\Gamma(a, s'', r)$) contains a $\mathcal{P}$-minorated annulus $\Gamma(b, \rho, \sigma)$. Of course, we may choose σ as closed as we want to ρ. Then, if $|a - b| > \rho$, we take $\sigma \in]\rho, |a-b|[$. Next, we take $\lambda \in]\rho, \sigma[$. Clearly, b does not lie in $s(t)$, hence we may apply Proposition 3.1 to b and to the circle $C(b, r_b)$. So, there exists $\varphi_1 \in \mathrm{Mult}(A, \|\cdot\|)$ such that $\mathcal{G}_{\widetilde{\varphi_1}}$ approaches $C(b, r_b)$. In fact, by definition, we have $r_b \leqslant \rho$, hence $C(b, r_b)$ is included in $d(b, \rho)$ hence $\mathcal{G}_{\widetilde{\varphi_1}}$ is secant with $d(b, \rho)$. On the other hand, there certainly exists $\varphi_2 \in \mathrm{Mult}(A, \|\cdot\|)$ such that $\mathcal{G}_{\widetilde{\varphi_2}}$ approaches $C(a, r)$. Then, by Proposition 3.3, there exists an increasing idempotent T-sequence $(T_n, 1)_{n \in \mathbb{N}}$, of center b and diameter $\lambda \in]\rho, \sigma[$, together with a decreasing idempotent T-sequence $(T'_n, 1)_{n\in\mathbb{N}}$, of center b and diameter λ. For all $n \in \mathbb{N}$, we fix $a_n \in T_n$ and $a'_n \in T'_n$. We put

$$E = d(a, r) \setminus \bigcup_{n\in\mathbb{N}} (T_n \cup T'_n).$$

Always by Proposition 3.3, E is a $\mathcal{P}$-set and there exists f and g in $H(E)$ such that:

i) $f(x) = 0$ for all $x \in E \setminus d(b, l)$ (resp. $g(x) = 0$ for all $x \in d(b, l) \cap E$).

ii) For all circular filter $\mathcal{G}$ secant with E but not secant with $E \setminus d(b, l)$ (resp. with $d(b, l) \cap E$) we have $\varphi_{\mathcal{G}}(f) \neq 0$ (resp. $\varphi_{\mathcal{G}}(g) \neq 0$).

iii) For each $n \in \mathbb{N}$, f (resp. g) is meromorphic in T_n (resp. T'_n), admits a_n (resp. a'_n) as a pole of order 1 and does not have any other pole in T_n (resp. T'_n).

We will show that we may extend Θ by continuity to such elements f and g. Let us examine, for example, the case of f. Since f only admits poles of order 1 in the holes T_n, the Mittag-Leffler term f_n of f associated to the hole T_n is of the form

$$f_n = \frac{\lambda_n}{x - a_n}, \qquad \lambda_n \in \mathbb{K},$$

and the Mittag-Leffler term of f associated to $d(0, r)$ is identically zero [5]. Moreover, for all $n \in \mathbb{N}^*$, we clearly have

$$\|f_n\|_{\mathcal{P}} = \|f_n\|_E = \|f_n\|_t = \|f_n(t)\|_{si}.$$

Moreover, since t satisfies Property (r'(t)) (resp. Property (r(t))), there exists $M \in \mathbb{R}^+$ such that

$$\|\frac{1}{t - a_n}\| \leqslant M \|\frac{1}{t - a_n}\|_{si}, \quad \forall n \in \mathbb{N}^*.$$

For $n \in \mathbb{N}$, we put

$$h_n = \sum_{i=1}^{n} f_i.$$

The sequence $(h_n)_{n\in\mathbb{N}}$ converges to f in $H(E)$. By the inequalities above, the sequence $(\Theta(h_n))_{n\in\mathbb{N}}$ converges in A to an element denoted by $\widehat{f}$. Moreover, $\mathcal{G}_{\widetilde{\varphi}_1}$ and $\mathcal{G}_{\widetilde{\varphi}_2}$ being secant with E, we have

$$\widetilde{\varphi}_i \in \mathrm{Mult}(H(E), \|\cdot\|_{\mathcal{P}}), \quad \text{for } i = 1, 2.$$

Then, we have

$$\varphi_i(\widehat{f}) = \widetilde{\varphi}_i(f), \quad \forall i \in \{1, 2\}.$$

Similar arguments apply to g, and so, we may associate to g an element $\widehat{g}$ of A such that

$$\varphi_i(\widehat{g}) = \widetilde{\varphi}_i(g), \qquad \forall i \in \{1, 2\}.$$

Finally, we consider the element fg of $H(E)$. The holes $(T_n)_{n\in\mathbb{N}^*}$ and $(T'_n)_{n\in\mathbb{N}^*}$ are disjoint. Then, fg is meromorphic in any of this holes, admits only one pole by hole and this pole is of order 1. Hence we may consider an element $\widehat{fg}$ of A which clearly satisfies

$$\widehat{fg} = \widehat{f}\widehat{g},$$

and

$$\varphi_i(\widehat{fg}) = \varphi_i(\widehat{f})\varphi_i(\widehat{g}) = \widetilde{\varphi}_i(fg) = \widetilde{\varphi}_i(f)\widetilde{\varphi}_i(g), \quad \forall i \in \{1,2\}.$$

In $H(E)$, we have $fg = 0$, so we have $\widehat{f}\widehat{g} = 0$. Since $\varphi_1(\widehat{f})\varphi_2(\widehat{g}) \neq 0$, $\widehat{f}$ and $\widehat{g}$ are divisors of zero in A. Hence, if A does not have divisors of zero, we have

$$r'' = r,$$

i.e. $\sup\{0, |\lambda| \mid \lambda \in s(t)\} = \|t\|_{si}$.

Now, suppose $r' < r$. We first assume $|a - b| \leqslant \rho$, hence $d(a, r')$ is included in $\mathbb{K} \setminus d(b, \lambda)$. It is seen that for every $\phi \in \mathrm{Mult}_m(A, \|\cdot\|)$, we have $\phi(\widehat{g}) \neq 0$ because $\phi(t - a) \leqslant r'$ and therefore $\widehat{g}$ is invertible in A. But, as we saw, $\mathcal{G}_{\widetilde{\varphi}_2}$ approaches $C(a, r)$, and then $\varphi_2(\widehat{g}) = 0$, which contradicts the property $\widehat{g}$ invertible.

Finally, we assume $|a - b| > \rho$. Then we have $\widehat{d(a, r')} \subset \mathbb{K} \setminus d(b, \lambda)$, and therefore $\widehat{f}$ satisfies $\psi(\widehat{f}) \neq 0$ for all $\psi \in \mathrm{Mult}_m(A, \|\cdot\|)$. So $\widehat{f}$ is invertible in A. But we have seen that the set of $\psi \in \mathrm{Mult}(A, \|\cdot\|)$ such that $\mathcal{G}_{\widetilde{\psi}}$ is secant with $d(b, \lambda)$ is not empty. Such ψ satisfying $\psi(\widehat{f}) = 0$, this contradicts the property $\widehat{f}$ invertible. So, we have

$$r' = r,$$

i.e. $\|t\|_m = \|t\|_{si}$. This ends the proof of Theorem 1.2.

4 PROOF OF THEOREM 1.3

Proof of Lemma 1.2. First, we will check that $\phi(h) < +\infty$ for every $h \in F$. Let $h = \frac{f}{g} \in F$ with $f \in E$ and $g \in S$. We can write g in the form

$$g = \prod_{j=1}^{n} (X - \alpha_j)^{q_j}, \quad \alpha_j \in d(0, \rho^-), \quad q_j \in \mathbb{N}^*.$$

We put $q = \sum_{j=1}^{n} q_j$.

Let $\psi \in W$. Since $\psi(X) \geqslant \rho$, clearly we have $\psi(X - a_i) = \psi(X)$ and therefore $\psi(g) \geqslant \rho^q$. So, we have

$$\overline{\psi}(\frac{f}{g}) \leqslant \frac{\psi(f)}{\rho^q} \leqslant \frac{\|f\|}{\rho^q},$$

hence

$$\phi(h) \leqslant \frac{\|f\|}{\rho^q}.$$

As a consequence, ϕ is a semi-multiplicative semi-norm on F. Next, it is easily seen that W contains absolute values, because the classical norm of E belongs to W. Therefore, ϕ is a norm.

NOTATION Let A be a commutative ring with unity without divisor of zero, provided with an ultrametric absolute value $|\cdot|$. On $A[X]$, we will denote by $|||\cdot|||$ the Gauss norm associated to $|\cdot|$.

LEMMA 4.1 *Let A be a commutative ring with unity provided with an ultrametric absolute value $|\cdot|$. Let $P \in A[X]$ and let $Q(X) = P(1-X)$. Then $|||P||| = |||Q|||$.*

Proof. Let $P(X) = \sum_{j=0}^{n} a_j X^j$ and let $Q(X) = \sum_{j=0}^{n} b_j X^j$. Thus, we have

$$b_j = (-1)^j \sum_{i=j}^{n} \binom{i}{j} a_i,$$

and therefore $|b_j| \leqslant |||P|||$. Hence, $|||Q||| \leqslant |||P|||$. But we also have $P(X) = Q(1-X)$ hence $|||P||| \leqslant |||Q|||$. This ends the proof.

LEMMA 4.2 *Let A be a commutative ring with unity, provided with an ultrametric absolute value $|\cdot|$. Let $\rho \in]0,1[$ and let B be the set of series $\sum_{-\infty}^{+\infty} \theta_n X^n$ such that $\theta_n \in A$ for all $n \in \mathbb{Z}$, $\lim_{n\to+\infty} \theta_n = 0$ and $\lim_{n\to-\infty} |\theta_n|\rho^n = 0$. Then B is a A-subalgebra of $A\{X, \frac{1}{X}\}$.*

Proof. If A is a complete algebraically closed field, B is just the set of analytic elements in the set $\{x \in A \mid \rho \leqslant |x| \leqslant 1\}$, so the claim is trivial. In the general case the absolute value extends to an algebraically closed complete field, hence the claim holds in the same way.

Lemma 4.3 is given in [4].

LEMMA 4.3 *For every $\beta \in C(0,1)$, $(\beta - xT)E$ is a maximal ideal of infinite codimension of E. Furthermore, we have $sa_E(xT) = \mathcal{D}$ and $s_E(xT) = d(0,1)$.*

Proof of Theorem 1.3. First, we will show that $\widehat{F}$ does not satisfy Property (o). By Lemma 4.3, we have $s_E(xT) = d(0,1)$, and $sa_E(xT) = \mathcal{D}$. Let $\chi \in \Upsilon_a(E)$. For every $g \in S$, we check that $\chi(g) \neq 0$. Hence, χ has an extension to a $\mathbb{K}$-algebras homomorphism $\widetilde{\chi}$ from F to $\mathbb{K}$, defined as

$$\widetilde{\chi}(\frac{f}{g}) = \frac{\chi(f)}{\chi(g)}, \; \forall f \in E, \forall g \in S.$$

In particular, by Lemma 4.3, we check that $sa_E(X) = d(1, 1^-)$. So, $|\chi|$ lies in W and therefore $\widetilde{\chi}$ is continuous with respect to the norm $\|\cdot\|$ of F. Hence $\widetilde{\chi}$ has an extension to a $\mathbb{K}$-algebras homomorphism of $\widehat{F}$. As a consequence, we have $sa_{\widehat{F}}(xT) \supset sa_E(xT)$. But as $E \subset \widehat{F}$, trivially we have $sa_{\widehat{F}}(xT) \subset sa_E(xT)$. Therefore, we obtain $sa_{\widehat{F}}(xT) = \mathcal{D}$ and therefore

$$\|\frac{1}{X}\|_{sa}^{\widehat{F}} = 1.$$

Now by construction we check

$$s_F(X) = d(0,1) \setminus d(0, \rho^-). \tag{1}$$

Let $\chi \in \Upsilon_a(E)$ satisfy

$$|\chi(X)| \geqslant \rho, \tag{2}$$

and let $\Lambda = \chi(E)$ be provided with an ultrametric absolute value extending this of $\mathbb{K}$. Then by (2), χ satisfies $\chi(g) \neq 0$ for all $g \in S$. Thereby, χ has an extension to a $\mathbb{K}$-algebras homomorphism χ from F to Λ such that $|\widetilde{\chi}|$ belongs to W. Thus, $\widetilde{\chi}$ is continuous with respect to the norm $\|\cdot\|$ of F and therefore it has an extension to a $\mathbb{K}$-algebras homomorphism from $\widehat{F}$ to Λ. Then, we check that

$$s_{\widehat{F}}(X) \supset d(0,1) \setminus d(0,\rho^-).$$

But by (1) we have $d(0,1) \setminus d(0,\rho^-) \supset s_{\widehat{F}}(X)$. So, we obtain the equality

$$s_{\widehat{F}}(X) = d(0,1) \setminus d(0,\rho^-).$$

As a consequence, we have

$$s_{\widehat{F}}(\frac{1}{X}) = d(0,\frac{1}{\rho}) \setminus d(0,1^-),$$

and therefore

$$\sup\{0, |\lambda| \mid \lambda \in s_{\widehat{F}}(\frac{1}{X})\} = \frac{1}{\rho},$$

although by (1), $\|\frac{1}{X}\|_{sa}^{\widehat{F}} = 1$. This show that $\widehat{F}$ does not satisfy Property (o).

Now, we are going to show that $\widehat{F}$ is an entire ring, and first, we show that E is the $\mathbb{K}$-vector space of series f of the form $\sum_{j=0}^{+\infty} \theta_j X^j$, with $\theta_j \in V$,

$$\lim_{n\to+\infty} \|\theta_n\| = 0 \quad \text{and} \quad \|f\| = \sup_{j\geqslant 0} \|\theta_j\|. \tag{4}$$

Let $l \in H(\mathcal{D})$ and let $j \in \mathbb{N}^*$ be fixed. According to the classical properties of analytic elements and power series [5], we can write l in the form

$$l = \sum_{i=0}^{j-1} a_i x^i + x^j \alpha$$

with $\alpha \in H(\mathcal{D})$ and

$$\|l\|_{\mathcal{D}} = \sup(|a_0|, \cdots, |a_{j-1}|, \|\alpha\|_{\mathcal{D}}).$$

So, we have

$$l(x)T^j = \sum_{i=0}^{j-1} a_i x^i T^j + (xT)^j \alpha = \sum_{i=0}^{j-1} a_i (xT)^i T^{j-i} + (xT)^j \alpha.$$

Finally, we have

$$lT^j = \sum_{i=0}^{j-1} a_i (1-X)^i T^{j-i} + (1-X)^j \alpha. \tag{5}$$

On $H(\mathcal{D})[Y]$, we denote by $|||\cdot|||$ the Gauss norm.

Consider the polynomial

$$P(Y)=\sum_{i=0}^{j-1}(a_iT^{j-i})Y^i\in\mathbb{K}[T]^0[Y],$$

and let $Q(Y)=P(1-Y)$. By Lemma 4.1 we have

$$|||Q|||=|||P|||=\sup_{0\leqslant i\leqslant j-1}|a_i|.$$

We put $Q(Y)=\sum_{i=0}^{j-1}b_i(T)Y^i$, and then we have

$$|||Q(Y)|||=\sup_{1\leqslant i\leqslant j-1}\|b_i(T)\|=\sup_{1\leqslant i\leqslant j-1}|a_i|. \tag{6}$$

As a consequence, we obtain

$$\|\sum_{i=0}^{j-1}a_i(1-X)^iT^{j-i}\|=\|\sum_{i=0}^{j-1}b_i(T)X^i\|=|||Q|||=|||P|||=\sup_{0\leqslant i\leqslant j-1}|a_i|.$$

Finally, we obtain

$$\|Q(X)\|=\max(|a_0|,\cdots,|a_{j-1}|). \tag{7}$$

On the other hand, we have

$$|||\alpha(1-Y)^j|||=|||\alpha Y^j|||=\|\alpha\|.$$

But as $\|1-X\|=1$, in E, we obtain

$$\|\alpha(1-X)^j\|=\|\alpha\|. \tag{8}$$

By (5), we obtain

$$lT^j=Q(X)+(1-X)^j\alpha,\quad \alpha\in H(\mathcal{D}),\ Q(Y)\in\mathbb{K}[T]^0[Y],$$

and by (6), (7) et (8), we obtain

$$\|l\|=\max(\|Q(X)\|,\|(1-X)^j\alpha\|). \tag{9}$$

Now, consider

$$f=\sum_{j=0}^{+\infty}f_jT^j\in E.$$

For each $j\in\mathbb{N}$, f_jT^j has the form $Q_j(X)+\alpha_jS_j(X)$ with $Q_j(Y)\in\mathbb{K}[T]^0[Y]$, $\alpha_j\in H(\mathcal{D})$ and $S_j(Y)\in\mathbb{K}[Y]$, satisfying further

$$|||Q_j|||\leqslant\|f_j\|, \tag{10}$$

and

$$\|\alpha_j\| \leqslant \|f_j\|. \tag{11}$$

As a consequence, we have

$$|||Q_j(Y) + \alpha_j S_j||| \leqslant \|f_j\|. \tag{12}$$

Thus, by (10), the series $\sum_{j=0}^{+\infty} Q_j(Y)$ is converging in $\mathbb{K}\{T\}^0\{Y\}$ to a limit $\sum_{j=0}^{+\infty} \lambda_j(T)Y^j$. And by (11), the series $\sum_{j=0}^{+\infty} \alpha_j S_j(Y)$ is converging in $H(\mathcal{D})\{Y\}$ to a limit $\sum_{j=0}^{+\infty} \mu_j Y^j$. Furthermore, by (10) and (11), we check that

$$\sup_{j\in\mathbb{N}} \|\lambda_j\| \leqslant \|f\| \text{ and } \sup_{j\in\mathbb{N}} \|\mu_j\| \leqslant \|f\|.$$

On the other hand, obviously we have

$$\|f\| \leqslant \max(\sup_{j\geqslant 0} \|\lambda_j\|, \sup_{j\geqslant 0} \|\mu_j\|),$$

hence finally we obtain

$$\|f\| = \max(\sup_{j\geqslant 0} \|\lambda_j\|, \sup_{j\geqslant 0} \|\mu_j\|).$$

Putting $\theta_j = \lambda_j + \mu_j$, we have $\|\theta_j\| = \max(\|\lambda_j\|, \|\mu_j\|)$. This shows us that E is the set of series of the form (4) with $\theta_j \in V$ and $\|f\| = \sup_{j\in\mathbb{N}} \|\theta_j\|$.

Next, we will show that $\widehat{F}$ is the set G of series f of the form $\sum_{-\infty}^{+\infty} \theta_j X^j$ with $\theta_j \in V$, satisfying $\lim_{n\to+\infty} \|\theta_n\| = 0$ and $\lim_{n\to-\infty} \|\theta_n\|\rho^n = 0$. Since $\|\frac{1}{X}\| = \frac{1}{\rho}$, it is seen that G is included in $\widehat{F}$. Moreover, by Lemma 4.2, G is a $\mathbb{K}$-subalgebra of $\widehat{F}$.

Given $\alpha \in d(0, \rho^-)$, we have

$$\frac{1}{X-\alpha} = \sum_{n=0}^{+\infty} \frac{\alpha^n}{X^{n+1}} \in G.$$

So F is included in G, and therefore G is dense in $\widehat{F}$.

On the $\mathbb{K}$-algebra $H(\mathcal{D})[Y]$, by Proposition 1.17 of [5], we can obviously consider the following two absolute values ψ_0, ψ_1 defined as

$$\psi_0(\sum_{j=0}^{q} a_j Y^j) = \sup_{0\leqslant j\leqslant q} \|a_j\| \quad (a_j \in H(\mathcal{D})),$$

and

$$\psi_1(\sum_{j=0}^{q} a_j Y^j) = \sup_{0\leqslant j\leqslant q} \|a_j\|\rho^j \quad (a_j \in H(\mathcal{D})).$$

Both have an extension to $H(\mathcal{D})[Y, \frac{1}{Y}]$, and therefore, here, we can apply this to $H(\mathcal{D})[X, \frac{1}{X}]$ which is clearly isomorphic to $H(\mathcal{D})[Y, \frac{1}{Y}]$. Of course, we may also extend ψ_0 and ψ_1 to $H(\mathcal{D})[X, T, \frac{1}{X}]$ because in the field of fractions of $H(\mathcal{D})[X, \frac{1}{X}]$, we have $T = \frac{1-X}{x}$. In particular, ψ_0 (resp. ψ_1) has a restriction ψ_0' (resp. ψ_1') to $H(\mathcal{D})[T]$ satisfying

$$\psi_i'(T) = \frac{\psi_i'(1-X)}{\psi_i'(x)} = \psi_i'(1-X) = 1. \quad (i = 0, 1)$$

Thereby, we have

$$\psi_i'(\sum_{k=0}^{q} b_k T^k) \leqslant \sup_{0 \leqslant k \leqslant q} \psi_i'(b_k) = \sup_{0 \leqslant k \leqslant q} \|b_k\|. \quad (i = 0, 1)$$

This shows that $\psi_i'(f) \leqslant \|f\|$ for all $f \in H(\mathcal{D})[T]$, and therefore ψ_i' has an extension ψ_i'' to $H(\mathcal{D})\{T\}$, with $\psi_i'' \in \text{Mult}(E, \|\cdot\|)$ $(i = 0, 1)$. On the other hand, we check that both ψ_0'' and ψ_1'' belong to W. Thus, ψ_i'' has an extension ψ_i''' to F and to $\widehat{F}$ $(i = 0, 1)$. In particular, we have

$$\psi_i(f) \leqslant \|f\|, \ \forall f \in \widehat{F}. \quad (i = 0, 1) \quad (13)$$

Now, let $f = \sum_{-\infty}^{+\infty} \theta_n X^n \in G$. Of course, we have

$$\|f\| \leqslant \max(\sup_{n \geqslant 0} \|\theta_n\|, \sup_{n < 0} \|\theta_n\| \rho^n).$$

But by (13), we check that

$$\|f\| \geqslant \psi_0(f) \geqslant \sup_{n \geqslant 0} \|\theta_n\|,$$

and

$$\|f\| \geqslant \psi_1(f) \geqslant \sup_{n < 0} \|\theta_n\| \rho^n.$$

Thus, we have

$$\|f\| = \max(\sup_{n \geqslant 0} \|\theta_n\|, \sup_{n < 0} \|\theta_n\| \rho^n).$$

This shows that G is complete with respect to the norm $\|\cdot\|$, and therefore we have $G = \widehat{F}$. Clearly, $H(\mathcal{D})\{T\}\{\{X\}\}$ is a $\mathbb{K}$-subalgebra of $\widehat{F}$. But $H(\mathcal{D})\{T\}\{\{X\}\}$ contains V. Hence, we have

$$\widehat{F} = H(\mathcal{D})\{T\}\{\{X\}\}.$$

Let Ω be the field of fractions of $H(\mathcal{D})$ and let $\widehat{\Omega}$ be its completion with respect to the extension of $\|\cdot\|$ to Ω.

Now, we consider $(\widehat{\Omega}+\mathbb{K}\{T\}^0)\{\{X\}\}$ as a $\mathbb{K}$-vector space provided with the norm $\|\cdot\|$ extending this of $\widehat{F}$ as:
given $g \in \widehat{\Omega}$ et $h \in \mathbb{K}\{T\}^0$, we put

$$\|g+h\| = \max(\|g\|, \|h\|),$$

and then, given $\sum_{-\infty}^{+\infty} \omega_k X^k \in (\widehat{\Omega}+\mathbb{K}\{T\}^0)\{\{X\}\}$, we put

$$\|\sum_{-\infty}^{+\infty} \omega_k X^k\| = \max(\sup_{k\geqslant 0} \|\omega_k\|, \sup_{k<0} \|\omega_k\|\rho^k).$$

Thus, $\widehat{F}$ is a $\mathbb{K}$-subvector space of $(\widehat{\Omega}+\mathbb{K}\{T\}^0)\{\{X\}\}$. But, as $T = \frac{1-X}{x}$, and as $\|\frac{1}{x}\| = 1$ in $\widehat{\Omega}$, the algebra $\mathbb{K}\{T\}$ is clearly included in $\widehat{\Omega}\{\{X\}\}$, hence $\widehat{F}$ is included in $\widehat{\Omega}\{\{X\}\}$.

Of course, the multiplication in E is induced by this in $\widehat{\Omega}\{\{X\}\}$, and so is this of F. Furthermore, as the norm on $\widehat{F}$ is induced by this on $\widehat{\Omega}\{\{X\}\}$, the multiplication in $\widehat{F}$ is also induced by this in $\widehat{\Omega}\{\{X\}\}$. This shows that $\widehat{F}$ is a $\mathbb{K}$-subalgebra of $\widehat{\Omega}\{\{X\}\}$, and then $\widehat{F}$ is an entire $\mathbb{K}$-algebra. Hence, by Corollary 1.2, $\widehat{F}$ satisfies Property (q).

Acknowledgements. The author gratefully acknowledges the many helpful suggestions of Professor Alain Escassut during the preparation of the paper.

REFERENCES

[1] Y Amice. Les nombres p-adiques, PUF, 1975.

[2] A Escassut. Algèbres de Banach ultramétriques et algèbres de Krasner-Tate, Asterisque n. 10:1–107, 1973.

[3] A Escassut. Elements analytiques et filtres percés sur un ensemble infraconnexe, Ann Mat Pura Appl 110:335–352, 1976.

[4] A Escassut. The ultrametric spectral theory, Periodica Mathematica Hungarica 11(1):7–60, 1980.

[5] A Escassut. Analytic elements in p-adic analysis, World Scientific Publishing, Singapore, 1995.

[6] A Escassut, N Maïnetti. Spectral semi-norm of p-adic Banach algebra, Bulletin of the Belgian Mathematical Society, Simon Stevin, 5:79–91, 1998.

[7] G Garandel. Les semi-normes multiplicatives sur les algèbres d'éléments analytiques au sens de Krasner, Indag Math 37(4):327–341, 1975.

[8] B Guennebaud. Sur une notion de spectre pour les algèbres normées ultramétriques, thèse Université de Poitiers, 1973.

[9] M Krasner. Prolongement analytique uniforme et multiforme dans les corps valués complets: éléments analytiques, préliminaires du théorème d'unicité. C.R.A.S. Paris, A 239:468–470, 1954.

[10] M Krasner. Prolongement analytique uniforme et multiforme dans les corps valués complets: préservation de l'analycité par la convergence uniforme, Théorème de Mittag-Leffler généralisé pour les éléments analytiques, C.R.A.S. Paris, A 244:2570–2573, 1957.

[11] N Maïnetti. Algebras of abstract analytic elements, p-adic functional analysis, Lecture Notes in Pure and Applied Mathematics, Marcel Dekker 192:181–195, 1997.

[12] M Lazard. Les zéros des fonctions analytiques sur un corps valué complet, IHES, Publications Mathématiques 14:47–75, 1962.

[13] ACM Van Rooij. Non-Archimedean Functional Analysis, Marcel Decker, inc. 1978.

[14] J Tate. Rigid Analytic Spaces. Inventiones mathematicae 12(4):257–289, 1971.

ective isometries of spaces of continuous :tions

ENCE NARICI St. John's University, Jamaica, NY 11439
: naricil@stjohns.edu

RD BECKENSTEIN St. John's University Staten Island, NY 10301
: beckense@stjohns.edu

Abstract. Isometries $X \to X$ certainly do not have to be surjective, generally. When they are surjective, there can be powerful algebraic consequences—a surjective isometry between real normed spaces, for example, must be an affine map Isometries must be surjective sometimes. If X is finite, for example, any isometry $X \to X$ must be surjective; if X is a finite-dimensional normed space then any *linear* isometry must be surjective. Schikhof proved that for spherically complete fields F, all isometries $F \to F$ are surjective if (and only if) F has a finite residue class field. Hence, for example, any isometry $Q_p \to Q_p$ must be surjective. What about isometries of spaces of functions? There is a strong connection between *linear* isometries H of spaces of functions—even with very different types of metrics—and the property $fg = 0 \Rightarrow HfHg = 0$. We call a map H with this property *separating*. (Other aliases include *Lamperti operators, separation-preserving operators, disjoint operators, disjointness-preserving operators* and *d-homomorphisms*.) For Banach spaces $C(X)$ and $C(Y)$ of real-valued continuous functions on the compact spaces X and Y, a surjective linear isometry $H : C(X) \to C(Y)$ must be separating. And even though the norm is quite different, a linear isometry of $L_p[0,1]$ (or ℓ_p), $1 \leqslant p \leqslant \infty$, $p \neq 2$, onto itself must also be separating ([3], pp. 170-175). Now suppose that X and Y are compact 0-dimensional Hausdorff spaces and the functions in $C(X)$ and $C(Y)$ take values in a non-Archimedean valued field F. We show in Sec. 4 that a *linear* separating isometry $H : C(X) \to C(Y)$ is surjective iff it satisfies a 'functional separation' condition that we call 'detaching' (Def. 3.1). What happens if we weaken linear to additive? We lose the 'detaching iff surjective' equivalence but Schikhof's theorem (the one cited above), comes to the rescue when F is spherically complete and has a finite residue class field, then (Theorems 6 1 and 6 2) 'detaching' additive separating isometries $H : C(X) \to C(Y)$ must be surjective (and conversely). Some examples demonstrate the necessity of certain hypotheses

1 BACKGROUND

By *separating map* we mean a map H defined on some space Λ of functions, for all $f, g \in \Lambda$, $fg = 0 \Rightarrow HfHg = 0$. Any multiplicative map such as any ring homomorphism between rings of functions is obviously separating. Here are some other examples.

EXAMPLE 1.1 EXAMPLES OF SEPARATING MAPS

(a) COMPOSITION *Let K be any ring. Let X and Y be sets and let Λ and Γ be any spaces of K-valued functions closed under multiplication on X and Y, respectively, with multiplication defined pointwise in Λ and Γ. Let $h : Y \to X$ be any function. The composition map $f \mapsto f \circ h$ is separating.*

(b) WEIGHTED COMPOSITION *Let K be any topological ring. Let $C(X)$ and $C(Y)$ denote the spaces of K-valued continuous functions on the Tihonov spaces X and Y. Let $h : Y \to X$ be continuous and let $w \in C(Y)$. The weighted composition map $H : C(X) \to C(Y)$, $f \mapsto w \cdot (f \circ h)$ is separating; we call w the weight function. If $C(X)$ and $C(Y)$ carry their compact-open topologies, weighted compositions are continuous.*

(c) ISOMETRIES *Let K be any valued field and let $X = \{x_0\}$ be any point. Then $C(\{x_0\})$ "=" K (identify $a \in K$ with the map $\mathbf{a}(x_0) = a$). Any field isometry (i.e., additive and multiplicative) $u : K \to K$ is separating.*

(d) DIFFERENTIATION *Let $D(\mathbf{R}) \subset C(\mathbf{R})$ be the subspace of continuously differentiable functions with the compact-open topology. The differentiation operator is separating and discontinuous.*

(e) INTEGRATION IS NOT SEPARATING *The map $Hf(t) = \int_a^t f(x)\,dx$, on any space of integrable functions (closed under multiplication) is not separating: it maps triangles (i.e., hat functions) into functions that are eventually constant; thus, even though f and g may be disjoint 'triangles,' there will be points t such that $Hf(t)\,Hg(t) \neq 0$.*

Let $C(X)$ and $C(Y)$ denote the sup-normed Banach spaces of real- or complex-valued continuous functions on the compact Hausdorff spaces X and Y. If $H : C(X) \to C(Y)$ is a surjective linear isometry then not only must H be separating, it must be the following weighted composition map:

$$Hf(y) = H\mathbf{1}(y) \cdot f(h(y)) \quad \text{for any} \quad f \in C(X) \quad \text{and } y \in Y \tag{1.1}$$

where $h : Y \to X$ is a surjective homeomorphism, $\mathbf{1} \in C(X)$ maps every element in X into 1 and $|H\mathbf{1}(y)| \equiv 1$. The continuous function $H\mathbf{1}$ is the weight function in this case. Thus, the Banach space structure of $C(X)$ is enough to characterize a compact Hausdorff space X (i.e., $C(X)$ norm-isomorphic to $C(Y) \Rightarrow X \cong Y$; moreover, for example, $C[0,1]$ is not norm-isomorphic to $C\left([0,1]^2\right)$ because $[0,1] \ncong [0,1]^2$). Now drop compactness and assume only that X and Y are realcompact; instead of assuming that H is a linear isometry, assume only that H is a linear *biseparating* map (H a bijection with H and H^{-1} separating). Then H still must be a weighted composition (Eq. (1.1)) with a homeomorphism h from Y onto X ([1], Prop. 3). This latter result generalizes Hewitt's well-known result that the ring structure of $C(X)$ characterizes the realcompactification υX of X: if $C(X)$ is ring-isomorphic to $C(Y)$ then the realcompactification υX of X is homeomorphic to the realcompactification υY of Y.

What happens if we take F-valued functions $C(X,F)$ and $C(Y,F)$ instead of real - or complex - valued functions where F is a complete nonarchimedean nontrivially valued field? By [4], if X and Y are compact 0-dimensional Hausdorff spaces then:

- There are surjective linear isometries H other than weighted compositions but
- A surjective linear isometry $H : C(X,F) \to C(Y,F)$ is separating if and only if it is a weighted composition.

THEOREM 1.1 ([9], Cor. 2) *If X and Y are compact 0-dimensional Hausdorff spaces then a linear separating bijection $H : C(X,F) \to C(Y,F)$ is a bicontinuous weighted composition and X and Y are homeomorphic.*

The **N**-compactification $\upsilon_0 X$ ([12], p. 42) serves in place of the realcompactification for F-valued functions.

THEOREM 1.2 ([2], Prop. 6) *Let X and Y be 0-dimensional Hausdorff spaces. If the residue class field of F is of nonmeasurable cardinal and $H : C(X,F) \to C(Y,F)$ is a biseparating map then $\upsilon_0 X$ is homeomorphic to $\upsilon_0 Y$.*

Results like these are what motivated our interest in separating maps. We discuss the mechanism by which they work (namely, the support map) in Sec. 3. First, we list our notational conventions.

2 NOTATION

- $(F, |\ |)$ denotes a complete non-Archimedean, nontrivially valued field.
- X and Y are compact 0-dimensional Hausdorff spaces.
- $\mathrm{cl}\, U$ denotes the topological closure of the set U.
- $C(X)$ and $C(Y)$ denote the sup-normed Banach spaces of F-valued continuous functions on X and Y.
- $\mathbf{1}$ denotes the function that maps every $x \in X$ into $1 \in F$.
- $H : C(X) \to C(Y)$ is an additive separating map (for all $f, g \in C(X)$, $H(f+g) = (f+g)$ and $fg = 0 \Rightarrow HfHg = 0$) such that $H\mathbf{1}(y) \neq 0$ for all $y \in Y$.
- For any function f, $\mathrm{coz} f$ denotes the cozero set of f.
- $\boldsymbol{C}U$ denotes the complement of the set U.
- For $U \subset X$ or Y, $k_U$denotes the F-valued characteristic function of U.
- For $y \in Y$, the function $y^\wedge \circ H : C(X) \to F$, denotes the map $f \mapsto Hf(y)$, i.e., $y^\wedge \circ H$ is the evaluation of Hf at y.

3 MECHANISM: THE SUPPORT MAP

Any additive separating map $H : C(X) \to C(Y)$ induces a continuous map $h : Y \to X$. The mechanism is as follows. For each $y \in Y$ the set

$$\bigcap_{\substack{y^\wedge \circ Hf \neq 0 \\ f \in C(X)}} \operatorname{cl} \operatorname{coz} f = \operatorname{supp} y^\wedge \circ H$$

is a singleton (see Th. 3.1(a) below) called the support of $y^\wedge \circ H$. The map $h : Y \to X$, $y \mapsto \operatorname{supp} y^\wedge \circ H$, is called the support map of H and is continuous; we discuss its basic properties in Th. 3.1. *We reserve the letter h for the support map of H in the sequel.* The support map makes for the following duality—every separating 'connection' between $C(X)$ and $C(Y)$ establishes a continuous 'connection' between Y and X. For the sake of characterizing the strength of the connection established by h (a homeomorphism, for example) we must know when h is injective. This happens if and only if H satisfies the following functional separation condition (see Th. 3.1(d)).

DEFINITION 3.1 The separating map H is *detaching* if for any twodistinct points $y, y' \in Y$ there exist disjoint clopen subsets U and V of X such that $Hk_U(y)Hk_V(y') \neq 0$.

When the field F in which the continuous functions take values is spherically complete and has a finite residue class field, detaching is the key to characterize the surjectivity of additive separating isometries (Theorema 6.1 and 6.2).

We list some elementary properties of separating maps in Th. 3.1. Proofs can be found in [9] (3.2, 3.3, and 4.1) and [8].

THEOREM 3.1 *Let $H : C(X) \to C(Y)$ be (as throughout) an additive separating map such that $H\mathbf{1}(y) \neq 0$ for all $y \in Y$..*

(a) *For each $y \in Y$, $\operatorname{supp} y^\wedge \circ H = \cap_{y^\wedge \circ Hf \neq 0} \operatorname{cl} \operatorname{coz} f$ is a singleton. Also, the support map of H, $h : Y \to X$, $y \mapsto \operatorname{supp} y^\wedge \circ H$ is continuous.*

(b) *If $f = 0$ on an open subset U of X, then $Hf = 0$ on $h^{-1}(U)$; equivalently, if $f = g$ on U, then $Hf = Hg$ on $h^{-1}(U)$ $(f, g \in C(X))$.*

(c) *For any $f \in C(X)$, $h(\operatorname{coz} Hf) \subset \operatorname{cl} \operatorname{coz} f$. Consequently, if H is injective then $h(Y)$ is dense in X.*

(d) *The support map h is 1-1 if and only if H is detaching.* (Proof. If H is detaching and $y \neq y'$ there disjoint clopen subsets U and V of X such that $Hk_U(y)Hk_V(y') \neq 0$. By the way h is defined, $h(y) \in U$ and $h(y') \in V$. The converse is clear.)

(e) ([8], Theorem 2.2) *For any $y \in Y$, if $y^\wedge \circ H$ is continuous, then $[y^\wedge \circ H](f) = Hf(y) = H[f(h(y))\mathbf{1}](y)$ for all $f \in C(X)$. [$y^\wedge \circ H$ could be of this form even if H is not continuous* ([6], Example 23c).

4 THE QUESTION

First, consider the connection between detaching and surjectivity for linear isometries, namely,

THEOREM 4.1 *Let $H : C(X) \to C(Y)$ be an additive separating map such that $H\mathbf{1}(y) \neq 0$ for all $y \in Y$. If H is a linear isometry, then H is surjective iff H is detaching.*

Proof. If H is an isometry, then H is continuous so (Th. 3.1(e))

$$Hf(y) = H[f(h(y))\mathbf{1}](y) = f(h(y))\,H\mathbf{1}(y)$$

for all $f \in C(X)$, $y, y' \in Y$. If h is not injective, then there exist distinct $y, y' \in Y$ such that $h(y) = h(y')$. Thus $Hf(y) = f(h(y))\,H\mathbf{1}(y)$ and $Hf(y') = f(h(y'))\,H\mathbf{1}(y')$ so

$$Hf(y') = Hf(y)\frac{H\mathbf{1}(y')}{H\mathbf{1}(y)}.$$

If H is surjective, any $g \in C(Y)$ may be written as Hf for some $f \in C(X)$ so this means that $g'(y) = \alpha g(y)$ (α constant) for all $g \in C(Y)$ which is absurd. Therefore, if H is onto, h is injective. Conversely, suppose that H is a detaching linear isometry. Since Y is compact, $h(Y) = X$ by Th. 3.1(c) so h is a surjective homeomorphism by Th. 3.1(a,d). Hence, for any $g \in C(Y)$, $g = g \circ h^{-1} \circ h$. By Th. 3.1(e), for any $y \in Y$,

$$H\left(\frac{g \circ h^{-1}}{H\mathbf{1} \circ h^{-1}}\right)(y) = \frac{g(y)}{H\mathbf{1}(y)}H\mathbf{1}(y) = g(y)$$

and H is seen to be surjective.

What happens if we weaken linear to additive?

It fails: We show in Example 6.1, there exist detaching additive separating isometries that are not surjective. If F is spherically complete and has a finite residue class field, however, it is resuscitated: In this case (Theorems 6.1 and 6.2) an additive separating isometry H is surjective if and only if it is detaching.

5 SEPARATING ISOMETRIES

In this section we establish the first results relating detaching and surjectivity of additive separating isometries $H : C(X) \to C(Y)$.

For spaces of real-valued functions, a continuous and additive map $H : C(X) \to C(Y)$ must be linear by the density of rationals. This is not true if the functions are F-valued, $F \neq \boldsymbol{Q}_p$, even when H is an isometry.

EXAMPLE 5.1 *Nonlinear additive separating isometries of $C(X)$ onto $C(Y)$.*

Proof. For any complete nontrivially valued field F other than $\boldsymbol{Q}_p$ there exist nonlinear additive isometries of F onto itself ([5], Theorems 3.4-3.8). Let F be such a field and $k : F \to F$ be a nonlinear additive surjective isometry. Let X and Y be spaces which are homeomorphic via a homeomorphism $g : Y \to X$. Then $H_k : C(X) \to C(Y)$ with $f \mapsto k \circ f \circ g$ is an additive separating isometry which is not linear. Since $H_k(g)$ is a composite of continuous functions, $H_k(C(X)) \subset C(Y)$. Since k is an additive isometry, so also is H_k. Since for some $a \in F$ $H_k[a\mathbf{1}](y) = k(a) \neq ak(1)$, $H_k a\mathbf{1} \neq aH_k\mathbf{1}$ and H_k is not linear. To see that H_k is onto, let $y \in C(Y)$. Since k is an additive surjective isometry, so also is k^{-1}. Then $y = H_k(k^{-1} \circ y \circ g^{-1})$.

The collection of such maps can be broadened as follows. Let $U_1, \ldots, U_n$ be a clopen partition of X. Let $k_1, \ldots, k_n$ be distinct additive surjective isometries of F which are additive and not linear. Now define for $f \in C(X)$, $H_{k_1, \ ,k_n} f = k_i \circ f \circ g$ on $g^{-1}U_i$ for $i = 1, \ldots, n$. Since the k_i are not linear, $H_{k_1, \ ,k_n}$ is not linear.

The next example shows that additive separating isometries H need not be detaching.

EXAMPLE 5.2 *A nondetaching additive separating isometry* $H : C(X) \to C(Y)$.

Proof. Let Z be any compact Hausdorff 0-dimensional space. Let $X = Z \cup \{u\}$ and $Y = Z \cup \{u, v\}$ where u and v are isolated points of X and Y, respectively. Define $H : C(X) \to C(Y)$ by taking

$$Hf(y) = \begin{cases} f(y), & y \in X \\ f(u), & y = u \text{ or } v \end{cases}$$

It is clear that H is an additive separating isometry and that $H\mathbf{1} = 1$ never vanishes. Since $Hf(y) = f(u)$ for $y = u$ or v, $h(u) = h(v) = u$, h is not 1-1 and so H is not detaching.

Next we investigate the effect of separating isometries on 'truncated' functions.

PROPOSITION 5.1 *If* $H : C(X) \to C(Y)$ *is an additive separating isometry such that* $H\mathbf{1}(y) \neq 0$ *for all* $y \in Y$, *then for any* $f \in C(X)$ *and clopen subset* $U \subset X$, $H(fk_U) = (Hf)k_{h^{-1}(U)}$.

Proof. By Th. 3.1(c),

$$h[\text{coz}H(fk_U)] \subset \text{cl}\,\text{coz}fk_U \subset U, \quad \text{and} \quad h[\text{coz}H(fk_{\boldsymbol{C}U})] \subset \text{cl}\,\text{coz}fk_{\boldsymbol{C}U} \subset \boldsymbol{C}U.$$

Thus

$$\text{coz}H(fk_U) \subset h^{-1}(\text{cl}\,\text{coz}fk_u) \subset h^{-1}(U)$$

and

$$\text{coz}H(fk_{\boldsymbol{C}U}) \subset h^{-1}(\text{cl}\,\text{coz}fk_{\boldsymbol{C}U}) \subset h^{-1}(\boldsymbol{C}U).$$

Therefore $H(fk_U)k_{h^{-1}(\boldsymbol{C}U)} = H(fk_{\boldsymbol{C}U})k_{h^{-1}(U)} = 0$, so

$$\begin{aligned} Hf &= H(fk_U) + H(fk_{\boldsymbol{C}U}) \\ &= (H(fk_U) + H(fk_{\boldsymbol{C}U}))(k_{h^{-1}(U)} + k_{h^{-1}(\boldsymbol{C}U)}) \\ &= H(fk_U)k_{h^{-1}(U)} + H(fk_{\boldsymbol{C}U})k_{h^{-1}(\boldsymbol{C}U)}. \end{aligned}$$

Hence $Hf = H(fk_U)$ on $h^{-1}(U)$, and therefore $H(fk_U) = (Hf)k_{h^{-1}(U)}$.

DEFINITION 5.1 For an additive separating map $H : C(X) \to C(Y)$ and $y \in Y$ define

$$\begin{array}{rccc} g_y : & F & \longrightarrow & F \\ & a & \longmapsto & H[a\mathbf{1}](y) \end{array}$$

Note that if $y^\wedge \circ H$ is continuous then, by Th. 3.1(e), for any $f \in C(X)$, $g_y(f(h(y))) = Hf(y)$.

THEOREM 5.2 *Let $H : C(X) \to C(Y)$ be an additive separating isometry such that $H\mathbf{1}(y) \neq 0$ for all $y \in Y$. If $H : C(X) \to C(Y)$ is detaching, then,*
(a) $|H(\mathbf{1}y)| = 1$ *for all* $y \in Y$,
(b) *for all $y \in Y$, the map $g_y : F \to F$ is an additive isometry.*
(c) *H is surjective if and only if g_y is surjective for all $y \in Y$.*

Proof. (a) Since H is an isometry, $|H\mathbf{1}(y)| \leqslant 1$ for every $y \in Y$. If, for some $y \in Y$, $|H\mathbf{1}(y)| < 1$, there exists $\epsilon > 0$, for which we can choose a clopen neighborhood U of y such that $|H\mathbf{1}(u)| \leqslant 1 - \epsilon$ for all $u \in U$. Since h is 1-1 (Th. 3.1(d)), it follows that $h^{-1}h(U) = U$. Therefore, by Prop. 5.1, $H(\mathbf{1}k_{h(U)}) = (H\mathbf{1})k_U$. It follows that $\|H(\mathbf{1}k_{h(U)})\| \leqslant 1 - \epsilon$. Since $\|\mathbf{1}k_{h(U)}\| = 1$, this is a contradiction.

(b) We use the argument of (a) with $\mathbf{1}$ replaced by $a\mathbf{1}$ to deduce that g_y is an isometry.

(c) Surjectivity of H implies that each g_y is surjective. Conversely, suppose that g_y is surjective for all $y \in Y$. As H is an isometry, by Th. 3.1(c) and the fact that Y is compact, it follows that $h(Y) = X$. Since h is injective, h is a homeomorphism between X and Y. Let $g \in C(Y)$. For each $y \in Y$, let $f_y \in C(X)$ be such that $Hf_y(y) = H[f_y(h(y))\mathbf{1}](y) = g(y)$. For each $y \in Y$, choose clopen sets $U_y \subset Y$ such that $|g(y'') - g(y')| < \epsilon$ and $|H[f_y(h(y))\mathbf{1}](y'') - H[f_y(h(y))\mathbf{1}](y')| < \epsilon$ for all $y', y'' \in U_y$. By the compactness of Y, the sets U_y yield a finite clopen cover $U_{y_1}, \ldots, U_{y_n}$ of Y. We may assume the U's to be disjoint and we reindex them as $U_1, \ldots, U_n$. For each i, note that y_i satisfies the relation $H[f_i(h(y_i))\mathbf{1}](y_i) = g(y_i)$. By Prop. 5.1, $H(f_i(h(y_i))k_{h(U_i)}) = H(f_i(h(y_i))\mathbf{1})k_{U_i}$. Since $|g(y'') - g(y')| < \epsilon$, and $|H[f_i(h(y_i))\mathbf{1}](y'') - H[f_i(h(y_i))\mathbf{1}](y')| < \epsilon$ for all $y', y'' \in U_i$, the ultrametric inequality implies that

$$\left\| g - H\left(\sum_{i=1}^{n} f_i(h(y_i))k_{h(U_i)}\right) \right\| = \left\| g - \sum_{i=1}^{n} H(f_i(h(y_i))\mathbf{1})k_{U_i} \right\| < \epsilon$$

Thus, $H(C(X))$ is dense in $C(Y)$. As H is an isometry and $C(X)$ is complete (here is where the completeness of F is used), $H(C(X)) = C(Y)$.

The import of Th. 5.3 is that when the maps g_y of Def. 5.1 are surjective for all $y \in Y$ then H is detaching if H is biseparating, i.e., H is bijective and H and H^{-1} are separating. The following terminology makes Th. 5.3 easier to state.

DEFINITION 5.2 A map $A : C(X) \to C(Y)$ separates points of Y very strongly if, for any two distinct points $y_1, y_2 \in Y$ and $a \in F$, there exists $f \in C(X)$ such that $Af(y_1) = a$, and $Af = 0$ on a clopen neighborhood of y_2.

Theorem 5.3 provides a partial converse to Theorem 5.2.

THEOREM 5.3 *Let $H : C(X) \to C(Y)$ be an additive separating isometry such that $H\mathbf{1}(y) \neq 0$ for all $y \in Y$. If the g_y of Def. 5.1 are surjective for all $y \in Y$, then the following are equivalent:*

(a) *H is detaching.*

(b) *H separates points of Y very strongly and $H(a\mathbf{1})$ never vanishes if $a \neq 0$.*

(c) *The support map h is a homeomorphism and H is biseparating.*

Proof. $(c) \Rightarrow (a)$ By Th. 3.1(d), h injective implies that H is detaching.

$(c) \Rightarrow (b)$ Since H is biseparating, H is surjective, $H(C(X)) = C(Y)$, and so H separates the points of Y very strongly. Also, by Th. 5.2(b), $|Ha\mathbf{1}(y)| = |a| \neq 0$. for all $y \in Y$.

$(a) \Rightarrow (b)$ If (a) holds then, by Th. 3.1(c) and the fact that h is 1-1, it follows that h is a homeomorphism onto X. Let $y_1 \neq y_2 \in Y$, and let U be a clopen neighborhood of y_1 with $y_2 \notin U$. Choose $f \in C(X)$ such that $Hf(y_1) = a$. By Prop. 5.1, $(Hf)k_{h(U)}) = Hfk_{h^{-1}(h(U))} = (Hf)k_U$. The function $(Hf)k_U = 0$ on the clopen neighborhood $\mathcal{C}U$ of y_2 while $(Hf)(k_U(y_1)) = a$. The fact that $Ha\mathbf{1}$ never vanishes follows from the fact that g_y is an isometry (Th. 5.2(b)).

$(b) \Rightarrow (a)$ Suppose that H is not detaching so there exists $y_1 \neq y_2 \in Y$ with $h(y_1) = h(y_2)$. By hypothesis, there exists $f \in C(X)$ such that $Hf(y_1) = 1$ and $Hf(y_2) = 0$. Hence $Hf(y_1) = H[f(h(y_1))\mathbf{1}](y_1) = 1$. This means that $f(h(y_1)) \neq 0$; hence $Hf(y_2) = H[f(h(y_2))\mathbf{1}](y_2) = H[f(h(y_1))\mathbf{1}](y_2) = 0$. Since $f(h(y_1)) \neq 0$, this negates (b).

$(a) \Rightarrow (c)$ As noted in the proof of $(a) \Rightarrow (b)$, h is a homeomorphism from Y onto X. Since H is detaching, it is surjective by Th. 5.2(c). We show that $HfHg = 0$ implies that $fg = 0$. If $HfHg = 0$ then $H[f(h(y))\mathbf{1}](y)\,H[g(h(y))\mathbf{1}](y) = 0$ for all $y \in Y$. Hence $f(h(y)) = 0$ or $g(h(y)) = 0$ for each y. Since h is surjective, $f(x) = 0$ or $g(x) = 0$ for every $x \in X$.

EXAMPLE 5.3 *An additive separating isometry $H : C(X) \to C(Y)$ for which g_y is surjective for all $y \in Y$ that is neither surjective or detaching.*

Proof. Let $g : F \to F$ be any surjective additive isometry of F. Let X be any 0-dimensional compact Hausdorff space Let $Y = X \cup \{y_0\}$ where X carries its own topology in Y and y_0 is an isolated point. Let x_0 be any point in X and let $H : C(X) \to C(Y)$ be defined as follows; if $f \in C(X)$, $Hf(x) = f(x)$ for $x \in X$ and $Hf(y_0) = g(f(x_0))$. Hence $\hat{y_0} \circ H$ is surjective; however H is clearly a non-surjective isometry because $Hf(y_0) = g(f(x_0))$ is determined by f and cannot assume an arbitrary value for a fixed $f \in C(X)$. H is additive because g is additive. H is linear if and only if g is linear. Since $h(y_0) = h(x_0) = x_0$, H is not detaching.

6 MAIN RESULTS

If $H : C(X) \to C(Y)$ is a linear separating isometry, it follows from Th. 4.1 that H is detaching if and only if it is surjective. We show in Th. 6.2 that if H is only additive then detaching is still equivalent to surjective for all locally compact fields F, among others; if F does not have a finite residue class field (Example 6.1) then there exist detaching additive separating isometries that are not surjective. Theorems 6.1 and 6.2 prove the equivalence of H being surjective and detaching for spaces of continuous functions that take values in a spherically complete field with finite residue class field.

THEOREM 6.1 *Let $H : C(X) \to C(Y)$ be an additive separating isometry such that $H\mathbf{1}(y) \neq 0$ for all $y \in Y$. If F has a finite residue class field and H is surjective, then H is detaching.*

Proof. Suppose that H is surjective and let $c_1, \ldots, c_n$ be representatives of the residue class field of F. If $y_1 \neq y_2$ are such that $h(y_1) = h(y_2) = x$, let U and V be disjoint clopen neighborhoods of y_1 and y_2, respectively. We may choose $f_1, \ldots, f_n$ and $g_1, \ldots, g_n$ in $C(X)$ such that $Hf_i = c_i k_U$ and $Hg_i = c_i k_V$ for all i by the following argument: For each i, if U_i is a clopen neighborhood of x, by Prop. 5.1, $H(f_i k_{U_i}) = (Hf_i)k_{h^{-1}(U_i)}$. Since H is an isometry and U_i is arbitrary the fact that f_i is continuous implies that $|f_i(x)| = |c_i| = 1$. Similarly, for each i, $|g_i(x)| = |c_i| = 1$. For all i, choose clopen neighborhoods U_i of x such that $|f_i k_{U_i}(z)| = |f_i(x)|$ for all $z \in U_i$, and $|g_i k_{U_i}(z)| = |g_i(x)|$ for all $z \in V_i$. If $W = \cap_{i=1}^n U_i \cap V_i$, $|g_i k_{U_i}| = |g_i(x)|$ and $|f_i k_{U_i}| = |f_i(x)|$ for all $z \in W$ and $i = 1, \ldots, n$. By Prop. 5.1, $H(f_i k_W) = (Hf_i)k_{h^{-1}(W)} = c_i k_U k_{h^{-1}(W)}$ and $H(g_i k_W) = (Hg_i)k_{h^{-1}(W)} = c_i k_V k_{h^{-1}(W)}$. Since H is an isometry, it follows that

$$\|H(f_i k_W)\| = \|H(g_i k_W)\| = |f_i(x)| = |g_i(x)| = |c_i| = 1$$

for all i. Since $\left\|c_i k_V k_{h^{-1}(W)}\right\| = \left\|c_i k_U k_{h^{-1}(W)}\right\| = 1$ for all i and $\left\|c_i k_U k_{h^{-1}(W)} - c_j k_V k_{h^{-1}(W)}\right\| = 1$ for all i and j, it follows that

$$\begin{aligned}\left\|c_i k_U k_{h^{-1}(W)} - c_j k_V k_{h^{-1}(W)}\right\| &= \|f_i k_W - g_j k_W\| \\ &= \sup_{x' \in W} |f_i(x') - g_j(x')| \\ &= |f_i(x) - g_j(x)| = 1\end{aligned}$$

for all i and j. Similarly,

$$\begin{aligned}\left\|c_i k_U k_{h^{-1}(W)} - c_j k_U k_{h^{-1}(W)}\right\| &= \|f_i k_W - f_j k_W\| \\ &= \sup_{x' \in W} |f_i(x) - f_j(x)| \\ &= |f_i(x) - f_j(x)| = 1\end{aligned}$$

and

$$\begin{aligned}\left\|c_i k_V k_{h^{-1}(W)} - c_j k_V k_{h^{-1}(W)}\right\| &= \|g_i k_W - g_j k_W\| \\ &= \sup_{x' \in W} |g_i(x) - g_j(x)| \\ &= |g_i(x) - g_j(x)| = 1\end{aligned}$$

for all i, j. But this implies that the scalars $f_i(x)$, $g_j(x)$, and $f_i(x) - g_j(x)$ for all i and j are a set of $n^2 + 2n$ distinct representatives of the residue class field of F which is a contradiction. We conclude that $h(y_1) \neq h(y_2)$ and therefore by Th. 3.1(d) that H is detaching.

If F is spherically complete, the converse holds as well.

THEOREM 6.2 *Let $H : C(X) \to C(Y)$ be an additive separating isometry such that $H\mathbf{1}(y) \neq 0$ for all $y \in Y$. If the spherically complete field F has a finite residue class field and H is detaching, then H is surjective.*

Proof. Since H is a detaching isometry, it follows that $g_y(a) = H[a\mathbf{1}](y)$ is an isometry for all $y \in Y$ (Th. 5.2(b)). In [11], (Theorem 2), Schikhof proved that, for spherically complete fields F, all isometries $F \to F$ are surjective if and only if F has a finite residue class field. It therefore follows that each g_y is surjective. It follows from Th. 5.2(c) that H is surjective.

By Th. 5.3(c) the maps H of the previous two theorems are biseparating. In Example 6.1 the field F has an infinite residue class field and Th. 6.2 fails.

EXAMPLE 6.1 *A detaching additive separating isometry $H : C(X) \to C(Y)$ that is not surjective.*

Proof. Let L be a field and $\{x_a : a \in A\}$ an infinite transcendency basis over L. Let $K = L(x_a : a \in A)$ be the field L with $\{x_a : a \in A\}$ adjoined and assume that K carries the trivial valuation, with valuation of $\sum_{i \geqslant k} a_i x^i \in K(x)$ given by $\left|\sum_{i \geqslant k} a_i x^i\right| = r^k$ for a fixed r with $1 > r > 0$. Define a map g on K as follows: first consider a map g taking $\{x_a : a \in A\}$ onto a proper subset of itself, then apply g first to terms in K of the form

$$\sum c_{a_1, \ldots, a_n} (x_{a_1})^{n_1} (x_{a_2})^{n_2} \ldots (x_{n_k})^{n_k}, c_{a_1, \ldots, a_n} \in L$$

via the relation

$$g\left(\sum c_{a_1, \ldots, a_n} (x_{a_1})^{n_1} (x_{a_2})^{n_2} \ldots (x_{n_k})^{n_k}\right) = \sum c_{a_1, \ldots, a_n} g(x_{a_1})^{n_1} g(x_{a_2})^{n_2} \ldots g(x_{n_k})^{n_k}$$

As the terms in K are quotients of these terms, g can be extended to all of K by taking

$$g\left(\frac{\sum c_{a_1,\ldots,a_n} (x_{a_1})^{n_1} (x_{a_2})^{n_2} \cdots (x_{n_k})^{n_k}}{\sum c_{b_1,\ldots,b_m} (x_{b_1})^{m_1} (x_{b_2})^{m_2} \cdots (x_{b_{m_k}})^{m_k}}\right)$$
$$= \frac{g\left(\sum c_{a_1,\ldots,a_n} (x_{a_1})^{n_1} (x_{a_2})^{n_2} \ldots (x_{n_k})^{n_k}\right)}{g\left(\sum c_{b_1,\ldots,b_m} (x_{b_1})^{m_1} (x_{b_2})^{m_2} \ldots (x_{b_{m_k}})^{m_k}\right)}$$

Clearly g is a non-surjective additive injective map of K into K. Let $K(x)$ be the field of Laurent series in x with coefficients in K (cf. [5]). Extend g to $K(x)$ by letting $g(\sum_{i \geqslant k} a_i x^i) = \sum_{i \geqslant k} g(a_i) x^i$, where $\sum_{i \geqslant k} a_i x^i \in K(x)$. As in [5], Theorem 2.2, the map g induces a non-surjective, additive isometry of $K(x)$. As in Example 5.1, this leads to an additive, detaching isometry $H : C(X) \to C(Y)$ which is not surjective. The residue class field E of $K(x)$ is isomorphic to K. Since K is infinite, so is E.

7 STAR FIELDS

We close with a discussion of a condition on F which guarantees that there is an additive surjective separating isometry $H : C(X) \to C(Y)$ which is not detaching. For any singleton $\{x_0\}$, $C(\{x_0\})$ consists entirely of constant functions $\boldsymbol{a} = a\mathbf{1}$, $a \in F$, i.e., $x_0 \mapsto a$. Thus, $C(\{x_0\})$ is isometric to F.

DEFINITION 7.1 If Y is not a singleton and there exists an additive separating surjective isometry $J : F \to C(Y)$ with $J(\mathbf{1})(y) \neq 0$ for all $y \in Y$ then we say that F is a STAR FIELD.

For a star field F the map

$$\begin{array}{cccc} J^* : & C(\{x_0\}) & \longrightarrow & C(Y) \\ & \boldsymbol{a} & \longmapsto & (Ja)\,\mathbf{1} \end{array}$$

is an additive nondetaching separating surjective isometry with $J^*\mathbf{1}(y) \neq 0$ for all $y \in Y$. If $a, b \in F$ and $ab = 0$ then $a = 0$ or $b = 0$. Since J is additive, it follows that $J^*a = 0$ or $J^*b = 0$ and J^* is separating. Since the support map j of J^* is such that $j(y) = x_0$ for all $y \in Y$ and Y is not a singleton, J^* is not detaching by Th. 3.1(d). Thus, the existence of a star field F implies the existence of a nondetaching surjective additive isometry from $C(\{x_0\})$ onto $C(Y)$. The completion C_p of the algebraic closure of $\boldsymbol{Q}_p$ is a star field since there are linear isometries of C_p onto C_p^2 ([10], p. 229, Example 75.G).

We now show that, for complete fields F, the existence of compact 0-dimensional spaces X and Y and a additive separating surjective isometry $H : C(X) \to C(Y)$ is equivalent to F being a star field.

THEOREM 7.1 *There exist compact 0-dimensional Hausdorff spaces X and Y and a nondetaching additive separating surjective isometry $H : C(X) \to C(Y)$ with $H\mathbf{1}(y) \neq 0$ for all $y \in Y$ if and only if F is a star field.*

Proof. In view of the discussion preceding the theorem, we need only prove necessity. Suppose $H : C(X) \to C(Y)$ is a nondetaching additive separating surjective isometry. Since H is not detaching there exists $x_0 \in X$ such that $h^{-1}(\{x_0\}) = Z \subset Y$ is not a singleton by Th. 3.1(d). Since Z is closed, it is compact. We claim that

$$\begin{array}{cccc} J^* : & C(\{x_0\}) & \longrightarrow & C(Z) \\ & \boldsymbol{a} & \longmapsto & H\boldsymbol{a}|_Z \end{array}$$

is an additive separating nondetaching surjective isometry.

As $J^*\boldsymbol{a}$ is the restriction of the continuous function $H\boldsymbol{a}$, $J^*(C(\{x_0\})) \subset C(Z)$. As H is additive and separating, so is J^*. Since Z is not a singleton, J^* is not detaching. To show that J^* is surjective, let $g \in C(Z)$. As Z is closed in Y, by [5], Theorem 1.1, there exists a continuous extension $g^* \in C(Y)$ of g. Since H is surjective, there exists $f \in C(X)$ such that $Hf = g^*$. By Th. 3.1(e), $H[f(x_0)\mathbf{1}]|_Z = Hf|_Z = g^*|_Z = g$.

Next, we show that J^* is an isometry. For $g \in C(Z)$ choose $f \in C(X)$ such that $Hf|_Z = H[f(x_0)\mathbf{1}]|_Z = g$. To show that J^* is an isometry, we show that $\|g\| = |f(x_0)|$. Let $\{V_a : a \in A\}$ denote the clopen neighborhoods of x_0. Since $\bigcap_{a \in A} V_a =$

$\{x_0\}$, $h^{-1}\left(\bigcap_{a\in A} V_a\right) = \bigcap_{a\in A} h^{-1}(V_a) = h^{-1}(\{x_0\}) = Z$ and $\bigcup_{a\in A} h^{-1}(\boldsymbol{C}V_a) = \boldsymbol{C}Z$. By Th. 3.1(b), for each V_a, $H(k_{V_a}f) = Hf$ on $h^{-1}(V_a)$ and $H(k_{V_a}f) = 0$ on $\boldsymbol{C}\left(h^{-1}(V_a)\right) = h^{-1}(\boldsymbol{C}V_a)$. View (k_{V_a}) as a net indexed by (V_a) ordered by reverse inclusion. It is clear from the continuity of f that $\lim_{V_a} \|k_{V_a}f\| = |f(x_0)|$. As H is an isometry, $\lim_{V_a} \|Hk_{V_a}f\| = |f(x_0)|$, so it only remains to show that $\lim_{V_a} \|Hk_{V_a}f\| = \|g\|$. We show below that, for all $\epsilon < \|g\|/2$, there exists a clopen neighborhood V_ϵ of x_0 such that $\|H(k_{V_\epsilon}f)\| = \|g\|$.

Choose $\epsilon > 0$ such that $\epsilon < \|g\|/2$. For $z \in Z$, let $S_\epsilon(z) = \{y \in Y : |Hf(z) - Hf(y)| < \epsilon\}$. As Z is compact, there is a finite subcover $S_{\epsilon_i}(z_i)$, $i = 1, 2, \ldots, n$, of the open cover $\{S_\epsilon(z) : z \in Z\}$ of Z. Hence $Z \subset W_\epsilon = \bigcup_{i=1}^n S_\epsilon(z_i)$ and W_ϵ is clopen. As already noted, $\bigcup_{a\in A} h^{-1}(\boldsymbol{C}V_a) = \boldsymbol{C}Z$. Since the clopen set $\boldsymbol{C}W_\epsilon$ is disjoint from Z, it follows that $\boldsymbol{C}W_\epsilon \subset \bigcup_{i=1}^n h^{-1}(\boldsymbol{C}V_{a_i})$. Therefore there exists a clopen neighborhood V_ϵ of x_0 such that $h^{-1}(V_\epsilon) \subset W_\epsilon$. and $\boldsymbol{C}W_\epsilon \subset h^{-1}(\boldsymbol{C}V_\epsilon)$.

Since $H(k_{V\epsilon}f) = Hf$ on $h^{-1}(V_\epsilon)$, and $H(k_{V_\epsilon}f) = 0$ on $h^{-1}(\boldsymbol{C}V_\epsilon)$, it follows that $\|H(k_{V_\epsilon}f)\| = \max_{y\in h^{-1}(V_\epsilon)} |H(k_{V_\epsilon}f)(y)|$. For $y \in h^{-1}(V_\epsilon)$, then $y \in S_\epsilon(z_i)$ for some i. If $|H(k_{V_\epsilon}f)(y)| \geqslant \|g\|/2 > \epsilon$, it follows from the ultrametric inequality that $|H(k_{V_\epsilon}f)(y)| = H[f(x_0)\mathbf{1}](z_i) = |H(k_{V\epsilon}f)(z_i)| = |g(z_i)|$. Thus, $\|H(k_{V_\epsilon}f)\| \leqslant \|g\|$. As %$cozg \subset \mathrm{coz}H(k_{V_\epsilon}f)$, $\|H(k_{V_\epsilon}f)\| \geqslant \|g\|$.

THEOREM 7.2 *If F is a star field, then its residue class field is infinite.*

Proof. Let Y and J be as in Def. 7.1. If Y is finite and contains n points, $C(Y) = F^n$. Thus, the map J^* can be viewed as an additive, surjective isometry $J^* : F \to F^n$. Assuming that $A = \{a_1, \ldots, a_m\}$ is a complete set of representatives for the residue class field K of F, this property of J^* together with the fact that $|a_i - a_j| = 1$ $(i \neq j)$ can be used to show that the inverse image under J^* of the set $B = \{(b_1, \ldots, b_n) : b_i \in A, \text{ if } (b_1, \ldots, b_n) \neq \left(b_1', \ldots, b_n'\right) \text{ then } b_i \neq b_i' \text{ some } i\}$ produces a set of representatives for K with more than $2m(m-1)$ elements which is a contradiction.

If Y is infinite and U and V are disjoint clopen subsets of Y the inverse image under J^* of the functions bk_U and $b'k_V$ where $b, b' \in A = \{a_1, \ldots, a_m\}$ produces m^2 distinct representatives of K which is a contradiction.

If Y consists of n points, $C(Y) = F^n$. Then, as observed in the proof of Th. 7.2, the map $J^* : F \to F^n$ is an additive, surjective isometry. We may also consider a map $J_2^* : F^n \to F^{n^2}$ where $J_2^*(a_1, a_2, \ldots, a_n) = (J_2^*a_1, J_2^*a_2, \ldots, J_2^*a_n)$. J_2^* is also an additive, surjective isometry. Then $J_n^* \circ J^*$ produces an additive, surjective isometry from F to F^{n^2}. In this way we can manufacture additive surjective isometries J_m between F and F^{n^m} for all $m > 0$.

When Y is infinite, families of pairwise disjoint clopen subsets of Y can be used to show that F^n is additively isometric to an additive subgroup of F for all $n > 0$.

REFERENCES

[1] J Araújo, E Beckenstein, L Narici. Biseparating maps and homeomorphic realcompactifications. Journal of Mathematical Analysis and Applications 192:258–265, 1995.

[2] J Araújo, E Beckenstein, L Narici Separating maps and the non-Archimedean Hewitt theorem. Ann Math Blaise Pascal 2:19–27, 1995.

[3] S Banach Théorie des opérations linéaires. Chelsea, New York, 1932.

[4] E Beckenstein, L Narici A non-Archimedean Stone-Banach theorem. Proc AMS 100:242–246, 1987.

[5] E Beckenstein, L Narici. On continuous extensions, Georgian Math J 3:565–570, 1996.

[6] E Beckenstein, L Narici. Additive bijections of C(X), preprint.

[7] E Beckenstein, L Narici, W Schikhof *Isometries of valued fields.* In: WH Schikhof, C Perez-Garcia, J Kakol ed. p-Adic Functional Analysis. New York: Marcel Dekker, 1997, pp 29–38.

[8] E Beckenstein, L Narici, A Todd. Automatic continuity of linear maps on spaces of continuous functions. Manuscripta Math 62:257–275, 1988.

[9] L Narici, E Beckenstein, J Araújo Separating maps on rings of continuous functions . In: N de Grande-de Kimpe, S Navarro, WH Schikhof ed. Editorial de la Universidad de Santiago de Chile. Av Lib Bernardo O'Higgins 3363 Santiago, 1994, pp 69-82.

[10] W Schikhof. *Ultrametric calculus: An introduction to p-adic analysis.* Cambridge University Press, Cambridge, 1984.

[11] W Schikhof, *Isometrical embeddings of ultrametric spaces into non-Archimedean valued fields.* Indag Math 46:51–53, 1984.

[12] A van Rooij. *Non-archimedean functional analysis.* Marcel Dekker, New York, 1978.

On the algebras (c, c) and $(\ell_\alpha, \ell_\alpha)$ in non-archimedean fields

PN NATARAJAN Department of Mathematics, Ramakrishna Mission Vivekananda College, Chennai - 600 004, India

Throughout this paper, K denotes a complete non-trivially valued non-archimedean field and infinite matrices and sequences have entries in K. If $A = (a_{nk})$, $a_{nk} \in K$, $n, k = 0, 1, 2, \ldots$ is an infinite matrix and $x = \{x_k\}$, $x_k \in K$, $k = 0, 1, 2, \ldots$ is a sequence, by the A-transform of x, we mean the sequence $Ax = \{(Ax)_n\}$,

$$(Ax)_n = \sum_{k=0}^{\infty} a_{nk} x_k, \quad n = 0, 1, 2, \ldots,$$

where it is assumed that the series on the right converge. If X, Y are classes of sequences, we write $A = (a_{nk}) \in (X, Y)$ if $\{(Ax)_n\} \in Y$ whenever $x = \{x_k\} \in X$. We need the following classes of sequences:

$$c = \{x = \{x_k\} : x_k \to \ell, k \to \infty, \ell \in K\};$$

$$\ell_\alpha = \{x = \{x_k\} : \sum_{k=0}^{\infty} |x_k|^\alpha \text{ converges}\},$$

where $\alpha > 0$. It is known that c, ℓ_α are linear spaces with respect to coordinatewise addition and scalar multiplication; c is a non-archimedean Banach space with respect to the norm $||x|| = \sup_{k \geq 0} |x_k|$, $x = \{x_k\} \in c$ and if $\alpha \geqslant 1$, ℓ_α is a Banach space with respect to the norm $||x|| = (\sum_{k=0}^{\infty} |x_k|^\alpha)^{1/\alpha}$, $x = \{x_k\} \in \ell_\alpha$. $(c, c; P)$ denotes the class of all infinite matrices in (c, c) such that $\lim_{n\to\infty}(Ax)_n = \lim_{k\to\infty} x_k$, where $x = \{x_k\} \in c$, while $(\ell_\alpha, \ell_\alpha; P)$ denotes the class of all infinite matrices in $(\ell_\alpha, \ell_\alpha)$ for which $\sum_{n=0}^{\infty}(Ax)_n = \sum_{k=0}^{\infty} x_k$, where $x = \{x_k\} \in l_a$. In this context it is worthwile noting that $\sum_{k=0}^{\infty} x_k$ converges if and only if $\lim_{k\to\infty} x_k = 0$. The following theorems are well-known.

THEOREM A (see [4], [6]). $A = (a_{nk}) \in (c, c)$ *if and only if*

(i) $\sup_{n,k} |a_{nk}| < \infty$;

(ii) $\lim_{n\to\infty} a_{nk} = \delta_k$ *exists*, $k = 0, 1, 2, \ldots$;

and

(iii) $\lim_{n\to\infty} \sum_{k=0}^{\infty} a_{nk} = \delta$ *exists.*

Further $A \in (c, c; P)$ *if and only if* (i) *holds and* (ii) *and* (iii) *hold with* $\delta_k = 0$, $k = 0, 1, 2, \ldots$ *and* $\delta = 1$.

THEOREM B (see [5]). $A \in (\ell_\alpha, \ell_\alpha)$, $\alpha > 0$, *if and only if*

(i) $\sup_{k\geq 0} \sum_{n=0}^{\infty} |a_{nk}|^\alpha < \infty$.

Further $A \in (\ell_\alpha, \ell_\alpha; P)$ *if and only if* (i) *holds and*

(ii) $\sum_{n=0}^{\infty} a_{nk} = 1$, $k = 0, 1, 2, \ldots$.

THEOREM C (see [5]). *The class* $(\ell_\alpha, \ell_\alpha)$, $\alpha \geq 1$, *is a Banach algebra under the norm*

$$||A|| = \sup_{k\geq 0} (\sum_{n=0}^{\infty} |a_{nk}|^\alpha)^{1/\alpha}, \tag{1}$$

$A = (a_{nk}) \in (\ell_\alpha, \ell_\alpha)$, *with the usual matrix addition, scalar multiplication and multiplication.*

It was proved in [7] that the class (c, c) is a non-archimedean Banach algebra under the norm

$$||A|| = \sup_{n,k} |a_{nk}|, \tag{2}$$

$A = (a_{nk}) \in (c, c)$ with the usual matrix addition, scalar multiplication and multiplication.

For analysis in non-archimedean fields, see [1], [2].

We now state below some more results relating to the Banach algebra (c, c) omitting the proofs which are modelled on those for the real or complex case.

THEOREM 1 *The class* $(c, c; P)$, *as a subset of* (c, c), *is a closed* K*-convex semigroup with identity.*

REMARK 1 Note that $(c, c; P)$ is not an algebra since the sum of two elements in $(c, c; P)$ may not be in $(c, c; P)$.

We now introduce a convolution product following [3], p. 179.

DEFINITION 1 For $A = (a_{nk})$, $B = (b_{nk})$, define

$$(A * B)_{nk} = \sum_{i=0}^{k} a_{ni} b_{n,k-i}, \quad n, k = 0, 1, 2, \ldots. \tag{3}$$

$A * B = \big((A * B)_{nk}\big)$ is called the convolution product I of A and B.

Keeping the usual norm structure in the class (c, c) as defined by (2) and replacing the matrix product by the convolution product I as defined in (3) we have the following.

THEOREM 2 *(c, c) is a commutative Banach algebra with identity under the convolution product I. Further $(c, c; P)$, as a subset of (c, c), is a closed K-convex semigroup without identity.*

We now prove theorems analogous to Theorem 1 and Theorem 2 for the class $(\ell_\alpha, \ell_\alpha)$.

THEOREM 3 *The class $(\ell_\alpha, \ell_\alpha; P)$, as a subset of $(\ell_\alpha, \ell_\alpha)$, where $\alpha \geq 1$, is a closed K-convex semigroup with identity, the multiplication being the usual matrix multiplication.*

Proof. Let $\lambda, \mu, \gamma \in K$ such that $\lambda + \mu + \gamma = 1$ and $|\lambda|, |\mu|, |\gamma| \leq 1$. Let $A = (a_{nk})$, $B = (b_{nk})$, $C = (c_{nk}) \in (\ell_\alpha, \ell_\alpha; P)$. By Theorem C we know that $\lambda A + \mu B + \gamma C \in (\ell_\alpha, \ell_\alpha)$. Also,

$$\sum_{n=0}^{\infty} (\lambda a_{nk} + \mu b_{nk} + \gamma c_{nk}) = \lambda + \mu + \gamma = 1, \quad k = 0, 1, 2, \ldots,$$

so that, by Theorem B, $\lambda A + \mu B + \gamma C \in (\ell_\alpha, \ell_\alpha; P)$ and so $(\ell_\alpha, \ell_\alpha; P)$ is a K-convex subset of $(\ell_\alpha, \ell_\alpha)$.

Let now $A = (a_{nk}) \in \overline{(\ell_\alpha, \ell_\alpha; P)}$. Then there exist $A^{(m)} = (a_{nk}^{(m)}) \in (\ell_\alpha, \ell_\alpha; P)$, $m = 0, 1, 2, \ldots$ such that

$$\|A^{(m)} - A\| \to 0, \quad m \to \infty.$$

So given $\varepsilon > 0$, there exists a positive integer N such that

$$\|A^{(m)} - A\| < \varepsilon \quad \text{for all} \quad m \geq N$$

i.e.,

$$\sup_{k \geq 0} \Big(\sum_{n=0}^{\infty} |a_{nk}^{(m)} - a_{nk}|^\alpha\Big)^{1/\alpha} < \varepsilon \quad \text{for all} \quad m \geq N. \tag{4}$$

Again,

$$\begin{aligned} \Big|\sum_{n=0}^{\infty} a_{nk} - 1\Big|^\alpha &= \Big|\sum_{n=0}^{\infty} a_{nk} - \sum_{n=0}^{\infty} a_{nk}^{(N)}\Big|^\alpha, \quad \text{since} \quad A^{(N)} \in (\ell_\alpha, \ell_\alpha; P) \\ &\leq \sum_{n=0}^{\infty} |a_{nk} - a_{nk}^{(N)}|^\alpha, \quad \text{since the valuation on } K \text{ is non-archimedean} \\ &< \varepsilon^\alpha, \quad k = 0, 1, 2, \ldots, \quad \text{using (4).} \end{aligned}$$

So

$$\Big|\sum_{n=0}^{\infty} a_{nk} - 1\Big| < \varepsilon \text{ for all } \varepsilon > 0$$

and consequently

$$\sum_{n=0}^{\infty} a_{nk} = 1, \ k = 0, 1, 2, \ldots.$$

Thus $A \in (\ell_\alpha, \ell_\alpha; P)$ and $(\ell_\alpha, \ell_\alpha; P)$ is a closed subset of $(\ell_\alpha, \ell_\alpha)$.

It is clear that the unit matrix is in $(\ell_\alpha, \ell_\alpha; P)$ and it is the identity element of $(\ell_\alpha, \ell_\alpha; P)$.

To complete the proof of the theorem, it suffices to check closure under matrix product. If $A = (a_{nk})$, $B = (b_{nk}) \in (\ell_\alpha, \ell_\alpha; P)$, by Theorem C, $AB \in (\ell_\alpha, \ell_\alpha)$. It is in $(\ell_\alpha, \ell_\alpha; P)$ too since

$$\begin{aligned}\sum_{n=0}^{\infty} c_{nk} &= \sum_{n=0}^{\infty}\Big(\sum_{i=0}^{\infty} a_{ni} b_{ik}\Big)\\ &= \sum_{i=0}^{\infty} b_{ik}\Big(\sum_{n=0}^{\infty} a_{ni}\Big), \quad \text{since convergence is equivalent}\end{aligned}$$

to unconditional convergence (see [8], p. 133)

$$\begin{aligned}&= \sum_{i=0}^{\infty} b_{ik}, \quad \text{since} \quad \sum_{n=0}^{\infty} a_{ni} = 1, \ i = 0, 1, 2, \ldots\\ &= 1,\end{aligned}$$

since $\sum_{i=0}^{\infty} b_{ik} = 1, \ k = 0, 1, 2, \ldots$.

The proof of the theorem is now complete.

REMARK 2 $(\ell_\alpha, \ell_\alpha; P)$ is not an algebra since the sum of two elements of $(\ell_\alpha, \ell_\alpha; P)$ may not be in $(\ell_\alpha, \ell_\alpha; P)$.

We now introduce another convolution product.

DEFINITION 2 For $A = (a_{nk})$, $B = (b_{nk})$, define

$$(A \circ B)_{nk} = \sum_{i=0}^{n} a_{i,k} b_{n-i,k}, \quad n, k = 0, 1, 2, \ldots. \tag{5}$$

$A \circ B = \big((A \circ B)_{nk}\big)$ is called the convolution product II of A and B.

We keep the usual norm structure in $(\ell_\alpha, \ell_\alpha)$ as defined by (1) and replace matrix product by the convolution product II as defined by (5) and establish the following.

THEOREM 4 $(\ell_\alpha, \ell_\alpha), \alpha \geqslant 1$, *is a commutative Banach algebra with identity under the convolution product II and* $(\ell_\alpha, \ell_\alpha; P)$, *as a subset of* $(\ell_\alpha, \ell_\alpha)$, *is a closed* K-*convex semigroup with identity.*

Proof. We will prove closure under convolution product II. Let $A = (a_{nk})$, $B = (b_{nk}) \in (\ell_\alpha,\ell_\alpha)$ and $A \circ B = (c_{nk})$. Then

$$\begin{aligned}\sum_{n=0}^{\infty}|c_{nk}|^\alpha &= \sum_{n=0}^{\infty}|\sum_{i=0}^{n} a_{ik}b_{n-i,k}|^\alpha \\ &\leq \sum_{n=0}^{\infty}\sum_{i=0}^{n}|a_{ik}|^\alpha|b_{n-i,k}|^\alpha, \quad \text{since the valuation on } K \text{ is non-archimedean} \\ &= (\sum_{n=0}^{\infty}|a_{nk}|^\alpha)(\sum_{n=0}^{\infty}|b_{nk}|^\alpha) \\ &\leq \|A\|^\alpha\|B\|^\alpha, \quad k=0,1,2,\ldots\end{aligned}$$

so that $\sup_{k\geq 0}(\sum_{n=0}^{\infty}|c_{nk}|^\alpha) < \infty$ and so $A \circ B \in (\ell_\alpha,\ell_\alpha)$. Also $\|A \circ B\|^\alpha \leq \|A\|^\alpha\|B\|^\alpha$ which implies $\|A \circ B\| \leq \|A\|\,\|B\|$. It is clear that the convolution product II is commutative. The identity element is the matrix $E = (e_{nk})$ whose first row consists of 1's and which has 0's elsewhere i.e., $e_{0k} = 1$, $k = 0,1,2,\ldots$, $e_{nk} = 0$, $n = 1,2,\ldots$, $k = 0,1,2,\ldots$. Note that $\|E\| = 1$. We also note that $E \in (\ell_\alpha,\ell_\alpha;P)$ since $\sum_{n=0}^{\infty} e_{nk} = 1$, $k = 0,1,2,\ldots$. It suffices to prove that $(\ell_\alpha,\ell_\alpha;P)$ is closed under the convolution product II. Now

$$\begin{aligned}\sum_{n=0}^{\infty} c_{nk} &= \sum_{n=0}^{\infty}(\sum_{i=0}^{n} a_{ik}b_{n-i,k}) \\ &= (\sum_{n=0}^{\infty} b_{nk})(\sum_{n=0}^{\infty} a_{nk}) \\ &= 1, \quad k=0,1,2,\ldots,\end{aligned}$$

if $A, B \in (\ell_\alpha,\ell_\alpha;P)$ since in this case $\sum_{n=0}^{\infty} a_{nk} = \sum_{n=0}^{\infty} b_{nk} = 1$, $k = 0,1,2,\ldots$. The proof of the theorem is now complete.

We now obtain a Mercerian theorem supplementing an earlier Mercerian theorem of the author (see [5], Theorem 3.1).

THEOREM 5 *When $K = \mathbb{Q}_p$, the p-adic field for a prime p, if $y_n = x_n + \lambda(p^n x_0 + p^{n-1}x_1 + \cdots + px_{n-1} + x_n)$ and $\{y_n\} \in \ell_\alpha$, then $\{x_n\} \in \ell_\alpha$, $a \geq 1$, provided $|\lambda|_p < (1-\rho^\alpha)^{1/\alpha}$ where $\rho = |p|_p < 1$.*

Proof. Since $(\ell_\alpha,\ell_\alpha)$, $\alpha \geq 1$, is a Banach algebra under convolution product II, if $\lambda \in \mathbb{Q}_p$ is such that $|\lambda|_p < \frac{1}{\|A\|}$, $A \in (\ell_\alpha,\ell_\alpha)$, then $E - \lambda A$, where E is the identity element of $(\ell_\alpha,\ell_\alpha)$ under convolution product II, has an inverse in $(\ell_\alpha,\ell_\alpha)$. In this context, we recall that

$$E \equiv (e_{nk}) = \begin{bmatrix} 1 & 1 & 1 & \ldots \\ 0 & 0 & 0 & \ldots \\ 0 & 0 & 0 & \ldots \\ \ldots & \ldots & \ldots & \ldots \end{bmatrix}$$

We note that the equations

$$y_n = x_n + \lambda(p^n x_0 + p^{n-1} x_1 + \ldots + p x_{n-1} + x_n),\ n = 0, 1, 2, \ldots$$

are given by

$$(E + \lambda A) \circ x' = y',$$

where

$$A = \begin{bmatrix} 1 & 0 & 0 & \ldots \\ p & 0 & 0 & \ldots \\ p^2 & 0 & 0 & \ldots \\ \ldots & \ldots & \ldots & \ldots \end{bmatrix},$$

$$x' = \begin{bmatrix} x_0 & 0 & 0 & \ldots \\ x_1 & 0 & 0 & \ldots \\ x_2 & 0 & 0 & \ldots \\ \ldots & \ldots & \ldots & \ldots \end{bmatrix}$$

and

$$y' = \begin{bmatrix} y_0 & 0 & 0 & \ldots \\ y_1 & 0 & 0 & \ldots \\ y_2 & 0 & 0 & \ldots \\ \ldots & \ldots & \ldots & \ldots \end{bmatrix}.$$

It is clear that $A \in (\ell_\alpha, \ell_\alpha)$ with $\|A\| = \frac{1}{(1-\rho^\alpha)^{1/\alpha}}$, where $\rho = |p|_p < 1$. So, if $|\lambda|_p < (1 - \rho^\alpha)^{1/\alpha}$, $E + \lambda A$ has an inverse in $(\ell_\alpha, \ell_\alpha)$. Consequently, it follows that

$$x' = (E + \lambda A)^{-1} \circ y'.$$

Since $y' \in (\ell_\alpha, \ell_\alpha)$ and $(E + \lambda A)^{-1} \in (\ell_\alpha, \ell_\alpha)$, $x' \in (\ell_\alpha, \ell_\alpha)$ and so $\{x_n\} \in \ell_\alpha$. The proof of the theorem is now complete.

REFERENCES

[1] G Bachman. Introduction to p-Adic Numbers and Valuation Theory. New York: Academic Press, 1964.

[2] G Bachman, E Beckenstein and L Narici Functional Analysis and Valuation Theory. New York: Marcel Dekker, 1972.

[3] IJ Maddox. Elements of Functional Analysis. Cambridge: Cambridge University Press, 1977.

[4] AF Monna. Sur le théorème de Banach-Steinhaus. Indag Math 25:121–131, 1963.

[5] PN Natarajan. Characterization of a class of infinite matrices with applications. Bull Austral Math Soc 34:161–175, 1986.

[6] PN Natarajan. Criterion for regular matrices in non-archimedean fields. J Ramanujan Math Soc 6:185–195, 1991.

[7] MS Rangachari, VK Srinivasan. Matrix transformations in non-archimedean fields. Indag Math 26:422–429,, 1964.

[8] ACM van Rooij, WH Schikhof. Non-archimedean analysis. Nieuw Arch Wisk 29:120–160, 1971.

Banach spaces over fields with an infinite rank valuation

H OCHSENIUS* Facultad de Matemáticas, Universidad Católica de Chile, Casilla 306 - Correo 22, Santiago, Chile.

WH SCHIKHOF** Department of Mathematics, University of Nijmegen, Toernooiveld 6525 ED Nijmegen, The Netherlands

Abstract. For a field K, complete with respect to a valuation $|\ |$ of infinite rank, a basic theory of normed and Banach spaces is being developed. A crucial part is played by the G-modules introduced in 1.5. The results are applied to the class of the Norm Hilbert Spaces (NHS) i.e. Banach spaces for which closed subspaces admit projections of norm ≤ 1 We characterize NHS in several ways (Theorem 4.3.7). Bounded orthogonal sequences tend to 0 implying that every ball is a compactoid, a property that in rank 1 theory is shared only by the finite-dimensional spaces. Finally we describe in Section 5 those NHS for which there exists a Hermitean form $(\ ,\ )$, satisfying $|(x,x)| = \|x\|^2$ for all x. The first such so-called Form Hilbert Space (FHS) was discovered by Keller in 1980 [5] and its class was studied in several papers ([1], [3], [11], [18]).

INTRODUCTION

In real and complex Functional Analysis the basic and most elegant examples of Banach spaces are Hilbert spaces; therefore one might wonder why they are so absent in p-adic theory. In fact, attempts in the past, in Banach spaces over a complete field K with a rank 1 valuation $|\ |$ (i.e. $|\ |$ is real-valued), to introduce 'inner products' $(\ ,\ )$ that were compatible with the norm never resulted, for K not $\mathbb{R}$ or $\mathbb{C}$, into Hilbert-like spaces. It was a consequence of a beautiful theorem of M.P. Soler [18] (see below) that enabled one to understand that this feature was not accidental. In fact, we will show in the present paper (by independent means) that infinite-dimensional Banach spaces over finite rank valued fields are never 'form Hilbert' i.e. there is no inner product such that every closed subspace

* Supported by Fondecyt No. 1971131

** Supported by DIPUC, Universidad Católica de Chile and by the Netherlands Organization for Scientific Research (NWO)

has an orthogonal complement, see Corollary 4.4.6 for a precise formulation. We may summarize roughly by the slogan

'There are no p-adic Hilbert Spaces'.

One can generalize the perspective by removing in the above topology and norms obtaining a purely algebraic setting as follows. Let K be a field with an involution $a \mapsto a^*$ (that is allowed to be the identity). A vector space E over K with a Hermitean form $(\ ,\ )$: $E \times E \to K$ (linearity in the first variable and $(x,y) = (y,x)^*$ for all x and y) is called *orthomodular* if the projection theorem

$$X = X^{\perp\perp} \to E = X \bigoplus X^{\perp}$$

holds (where, as usual, for a subset X of E, $X^{\perp} := \{y \in E : (x,y) = 0 \quad \text{for all} \quad x \in X\}$).

QUESTION Do there exist infinite-dimensional orthomodular spaces, different from the Hilbert spaces over $\mathbb{R}$ and $\mathbb{C}$?

It has been open for quite some years until Keller [5] in 1980 gave an affirmative answer by constructing an example. The base field he employed turned out to have a valuation $|\ \ |$ of infinite rank and $x \mapsto \sqrt{|(x,x)|}$ behaves like a norm. In Keller's space every topologically closed subspace S is also orthogonally closed (i.e. $S^{\perp\perp} = S$) and it has a countable orthogonal base, but surprisingly enough, no base is ever orthonormal. In other examples that were found later one meets the same state of affairs. A systematic study of orthomodular spaces was made in [3], and in [10] one finds connections with non-archimedean analysis. A breakthrough came in 1995 with the following result.

THEOREM (Soler) [18] *If E is an orthomodular space over K admitting an orthonormal sequence then $K = \mathbb{R}$ or $\mathbb{C}$ and E is (linearly homeomorphic to) a Hilbert space.*

Thus, we have a contrasting catch-phrase.

'There do exist Hilbert spaces over valued fields with infinite rank'.

However one is tempted to add that such Hilbert spaces have peculiar properties. This is confirmed not only by Soler's Theorem, but by study of operators on such spaces, see [6], [7] where it is shown that self-adjoint operators behave in a strange way. From 4.3.7 of the present paper it even follows that every operator is compact!

We feel that these interesting phenomena more than justify our paper, which aims at setting up a theory of Banach spaces over fields with an infinite rank valuation, to form a new branch of Non-Archimedean Functional Analysis, encompassing Hilbert spaces as a special class.

Basics can be found in Sections 1-3. As one might expect, several results of rank 1 theory can easily be carried over. However there is one crucial point of difference making this part interesting; it lies in the choice of the range of the norm function. In fact, to include Hilbert space we must admit, by Soler's Theorem, the range X of the norm function to be strictly bigger than the set of values $|K|$ (otherwise, orthogonal bases could be transformed into orthonormal ones by suitable scalar multiplication). In addition, to be able to define the norm of operators, norms on quotient spaces, etc. one wants to take infima and suprema of bounded sets in X, so it is convenient to require the ordered set X to be Dedekind complete. All these considerations have lead to the introduction of so-called G-modules (see 1.5) as a natural 'home' for norm values. The notions of (algebraic and topological) types and the type condition (see 1.6) have no counterparts in rank 1 theory and play a crucial role in Sections 4-5 in which we apply the general theory to so-called norm Hilbert spaces i.e. spaces for which every closed subspace admits a projection of norm ≤ 1. We characterize them in various ways and describe the subclass of the 'form Hilbert spaces' of above.

Needless to say that this work needs continuation. The whole world of operators in Norm and Form Hilbert Spaces is still unexplored.

1 THE RANGE OF THE NORM FUNCTION

Section 1 deals mainly with the range sets of arbitrary rank valuations on a field K (linearly ordered groups G, 1.3 and 1.4) and of norms on K-vector spaces (the so-called G-modules, 1.5). Ultrametrics are being generalized to so-called scales (1.2) that have values in linearly ordered sets (1.1). These concepts are fundamental for the theory of normed and Hilbert-like spaces in Sections 2, 3 and 4.
A number of basic facts of Section 1 can be found in standard books such as [2], [13], [19]. The concepts of *topological type* and *type condition* of [3], Def. 21, 31 have been generalized in 1.6 to arbitrary G-modules so as to make them useful for general normed spaces.
We haven't found in the literature the G-modules, the scaled spaces, and the antipode in $G^{\#}$ (1.3.1).

1.2 Linearly ordered sets

Let X be an ordered set. A subset A of X is called *cofinal* (*coinitial*) in X if for every $x \in X$ there exists an $a \in A$ such that $a \geq x$ $(a \leq x)$. In the same spirit we define *cofinal* (*coinitial*) *sequences* and, more generally, *nets*.
Let $f : X \to Y$ where X, Y are linearly ordered sets. We say that f is *increasing* (*strictly increasing*) if $x, y \in X$, $x < y$ implies $f(x) \leq f(y)$ $(f(x) < f(y))$. In the same spirit we define *decreasing* (*strictly decreasing*) maps. Let $A \subset B \subset X$. We say that $s = \sup_B A$ if $s \in B$, $s \geq a$ for all $a \in A$, and if $t \in B$, $t \geq a$ for all $a \in A$ then $t \geq s$. In the same spirit we define $\inf_B A$. If $\sup_B A$ and $\sup_X A$ both exist then $\sup_B A \geq \sup_X A$, but we do not always have equality. If it is clear with respect to which set the supremum (infimum) is taken we sometimes omit the subscript X in $\sup_X A$ $(\inf_X A)$.

LEMMA 1.1.1 *Let X be a linearly ordered set, let $a \in X$. Then either* $\min\{x \in X : x > a\}$ *exists or* $\inf\{x \in X : x > a\} = a$.
Similarly, either $\max\{x \in X : x < a\}$ *exists or* $\sup\{x \in X : x < a\} = a$.

Proof. It suffices to prove the first assertion. Let $V := \{x \in X : x > a\}$ and suppose $\min V$ does not exist. Clearly a is a lower bound of V. If $b \in X$, $b > a$, then $b \in V$ and by assumption there is a $v \in V$, $v < b$. Hence, b is no lower bound of V i.e. a is the greatest lower bound (Remark. If $V = \varnothing$ then a is the largest element of X and $a = \inf V$.)

1.1.2. Continuity at 0

In the sequel it will be useful to extend a given linearly ordered set by adjoining one element called 0, for which $0 < x$ for all $x \in X$. (See, e.g. 1.2, 1.4, 2.1) Then the *extended set* $X \cup \{0\}$ is again linearly ordered, and has 0 as smallest element. Let Y be a second linearly ordered set, let $f : X \to Y$. The (*natural*) *extension* $f_0 : X \cup \{0\} \to Y \cup \{0\}$ extends f and maps 0 into 0. We will say that f (or f_0) is *continuous at* 0 if for each $\varepsilon \in Y$ there is a $\delta \in X$ such that $x \in X$, $x < \delta$ implies $f(x) < \varepsilon$. It is called *bicontinuous at* 0 if in addition to each $\delta \in X$ there is an $\varepsilon \in Y$ such that $f(x) < \varepsilon$ implies $x < \delta$.
For a net $i \mapsto x_i$ $(i \in I)$ in $X \cup \{0\}$ we say that $\lim_i x_i = 0$ if for each $\varepsilon \in X$ there exists an $i_0 \in I$ such that $x_i < \varepsilon$ for $i \geq i_0$. Then $f : X \to Y$ (or its extension f_0) is continuous at 0 if and only if, for each net $i \mapsto x_i$ in X, $\lim_i x_i = 0$ implies $\lim_i f(x_i) = 0$. It is bicontinuous at 0 if and only if, for each net $i \mapsto x_i$ in X, $\lim_i x_i = 0 \iff \lim_i f(x_i) = 0$.

For our purpose it is not necessary to enrich X with a largest element ∞. We only define the following. For a net $i \mapsto x_i$ in $X \cup \{0\}$ we say that $\lim_i x_i = \infty$ if for each $s \in X$ there is an i_0 such that $x_i > s$ for $i \geq i_0$.

LEMMA 1.1.3 *Let X, Y be linearly ordered sets without a smallest element and let $f : X \to Y$ be increasing. If $f(X)$ is coinitial in Y then f is bicontinuous at* 0.

Proof. Let $i \mapsto x_i$ be a net in X and suppose that not $\lim_i f(x_i) = 0$. Then there is an $\varepsilon \in Y$ such that $J := \{i \in I : f(x_i) \geq \varepsilon\}$ is cofinal. Since ε is not the smallest element of Y, and $f(X)$ is coinitial there is a $\delta \in X$ such that $f(x_i) > f(\delta)$ for all $i \in J$. Then $x_i > \delta$ for all $i \in J$. Hence not $\lim_i x_i = 0$.
Now suppose $\lim_i f(x_i) = 0$. To show that $\lim_i x_i = 0$, let $\varepsilon \in X$. Then $f(\varepsilon) \in Y$ so there is an $i_0 \in I$ such that $f(x_i) < f(\varepsilon)$ for $i \geq i_0$. Then $x_i < \varepsilon$ for $i \geq i_0$ and we are done.

A linearly ordered set X is called (*Dedekind*) *complete* if each nonempty subset of X that is bounded above has a supremum. Then also each nonempty subset of X that is bounded below has an infimum. (Proof. Let $V \neq \varnothing$ be bounded below. Then the set W consisting of all lower bounds of V is nonempty, and bounded above since $V \neq \varnothing$, so $s = \sup W$ exists; one verifies easily that $s \in W$.)

We now describe the construction of the completion of a linearly ordered set X. A subset S of X is called a *cut* if

(i) $S \neq \varnothing$, S is bounded above,
(ii) if $x \in S$, $y < x$ then $y \in S$,
(iii) if $\sup_X S$ exists then $\sup_X S \in S$.

Let $X^\#$ be the collection of all cuts of X. With the ordering by inclusion $X^\#$ becomes a linearly ordered set. To prove that $X^\#$ is complete, let $A \subset X^\#$ be nonempty and bounded above. There is a cut T such that $S \subset T$ for all $S \in A$. Then $V := \bigcup_{S \in A} S$ is nonempty and bounded above by T, and by adding $\sup_X V$ (if it exists) to V we obtain a cut that is easily seen to be $\sup_{X^\#} A$. We have the natural embedding $\varphi : X \to X^\#$ given by

$$\varphi(a) = \{x \in X : x \leq a\}.$$

φ is strictly increasing. Often we shall identify X and $\varphi(X)$, in other words we shall view φ as an inclusion. $X^\#$ is called *the completion* of X.

For reasons of quoting the next Proposition contains some redundancy.

PROPOSITION 1.1.4 *Let X be a linearly ordered set with completion $X^\#$. Then we have the following.*

(i) *If X is complete then $X = X^\#$.*

(ii) *X is cofinal and coinitial in $X^\#$.*

(iii) *For every $s \in X^\#$, $\{x \in X : x \leq s\}$ is a cut in X; every cut in X has this form.*

(iv) *If $s, t \in X^\#$, $s < t$ then there exist $x, y \in X$ with $s \leq x < t$, $s < y \leq t$.*

(v) *For each $s \in X^\#$,*

$$s = \sup_{X^\#}\{x \in X : x \leq s\} = \inf_{X^\#}\{x \in X : x \geq s\}.$$

(vi) *If $A \subset X$, $s = \sup_X A$ then $s = \sup_{X^\#} A$. If $t = \inf_X A$ then $t = \inf_{X^\#} A$.*

Proof. (i) Suppose X is complete, let S be a cut. Then, letting $s := \sup S$, we have $S = \{x \in X : x \leq s\}$, so $S = \varphi(s)$ where φ is as above, i.e. $X^\# = \varphi(X)$ or $X^\# = X$ by identification.
(ii) Let $s \in X^\#$. Then s is a cut, hence bounded above in X, so there is an $x \in X$ with $v \leq x$ for all $v \in s$. Hence $s \leq x$. It follows that X is cofinal in $X^\#$. Coinitiality follows from the fact that cuts are nonempty.
(iii) Obvious.
(iv) Let $C_1 := \{x \in X : x \leq s\}$, $C_2 := \{x \in X : x \leq t\}$. Then $C_1 \subset C_2$, but $C_1 \neq C_2$ so there exists a $y \in C_2 \setminus C_1$ i.e. $s < y \leq t$. To find an $x \in X$ with $s \leq x < t$, take $x := s$ if $s \in X$ and $x := y$ if $t \notin X$. If $s \notin X$ and $t \in X$ and there is no $z \in X$ with $s \leq z < t$ then $t = \sup_X C_1$, so $t \in C_1$ and $C_1 = C_2$, a contradiction.
(v) Let $V = \{x \in X : x \leq s\}$. Then clearly s is an upper bound of V in $X^\#$. If $v \in X^\#$, $v < s$ then by (iv) there is a $y \in X$ with $v < y \leq s$, so v is no upper bound of V. Hence $s = \sup_{X^\#} V$. Let $W = \{x \in X : x \geq s\}$. Then s is a lower bound of W in $X^\#$. If $t \in X^\#$, $t > s$ then by (iv) there is an $x \in X$ with $s \leq x < t$, so t is no lower bound of W. Hence $s = \inf_{X^\#} W$.
(vi) A is nonempty and bounded above in $X^\#$ so, by completeness, $t := \sup_{X^\#} A$ exists and $t \leq s$. If $t < s$ we would have a $y \in X$ with $t < y \leq s$ by (iv), so t is not an upper bound of A, contradiction. We leave the 'inf' part of the proof to the reader.

PROPOSITION 1.1.5 *Let X be a linearly ordered set, let $Y \subset X$ be a subset that is complete as a linearly ordered set. If Y is both cofinal and coinitial in X then there exists an increasing projection $P : X \to Y$ (i.e. $Py = y$ for each $y \in Y$).*

Proof. Set

$$Px := \sup_Y \{y \in Y : y \leq x\} \qquad (x \in X).$$

(By cofinality the set $\{y \in Y : y \leq x\}$ is bounded above in Y, by coinitiality it is not empty.) One checks easily that P satisfies the requirements.

REMARK In the above proof the map $x \mapsto \inf_Y\{y \in Y : y \geq x\}$ would also have solved the problem.

1.2 Scaled spaces

Let M be a set, let X be a linearly ordered set enriched with a smallest element, called 0. An *(X-valued) scale on M* is a map $d : M \times M \longrightarrow X \cup \{0\}$ satisfying

(i) $d(x, y) = 0 \Longleftrightarrow x = y$
(ii) $d(x, y) = d(y, x)$
(iii) $d(x, z) \leq \max\big(d(x, y), d(y, z)\big)$

for all $x, y, z \in M$. The set $M = (M, X, d)$ is called a *scaled space* (*ultrametric space* if $X = (0, \infty)$). For a nonempty subset S of M for which $\{d(x, y) : x, y \in S\}$ is bounded above in $X \cup \{0\}$ we define its diameter as diam $S = \sup_{X^\# \cup \{0\}}\{d(x, y) : x, y \in S\}$. For $a \in M, \varepsilon \in X$, we define, as usual, $B_M(a, \varepsilon) := B(a, \varepsilon) := \{x \in M : d(x, a) \leq \varepsilon\}$ (the 'closed' ball) and $B_M(a, \varepsilon^-) := B(a, \varepsilon^-) := \{x \in M : d(x, a) < \varepsilon\}$ (the 'open' ball). A subset U of M is called *open* if for each $a \in U$ there exists an $\varepsilon \in X$, such that $B(a, \varepsilon^-) \subset U$. The collection of those open sets form a topology, *the topology induced by d*. Each ball is clopen (= closed and open), two balls are either disjoint or ordered by inclusion, every point of a ball is a center. The induced topology is zerodimensional.

A *nest* of balls in a scaled space is a nonempty collection of balls that is linearly ordered by inclusion. If $\mathcal{C}$ is a nonempty collection of balls such that any two members have a nonempty intersection then $\mathcal{C}$ is a nest.

DEFINITION 1.2.1 A scaled space is *spherically complete* if each nest of balls has a nonempty intersection.

If $B_1 \subset B_2$ are balls in a scaled space and $B_1 \neq B_2$ then there exist an 'open' ball S and a 'closed' ball T such that $B_1 \subset S \subset B_2$, $B_1 \subset T \subset B_2$. Therefore, a scaled space is spherically complete if and only if each nest of 'open' ('closed') balls has a nonempty intersection. We use this fact to show that spherical completeness 'does not depend on the range space of d' in the following sense.

PROPOSITION 1.2.2 *Let (M, X, d) be a scaled space, let $Y := \{d(x, y) : x, y \in M,\ x \neq y\}$. Then (M, X, d) is spherically complete if and only if (M, Y, d) is spherically complete.*

Proof. It suffices to prove that spherical completeness of (M, Y, d) implies that of (M, X, d). Thus, let $\mathcal{C}$ be a nest of 'open' balls in (M, X, d). To show that $\bigcap \mathcal{C} \neq \varnothing$ we may assume that $\mathcal{C}$ has no smallest element. Let $B \in \mathcal{C}$. Then there exists a $B' \in \mathcal{C}$, $B' \subset B$, $B' \neq B$. Then $B' = \{x \in M : d(x,a) < r'\}$, $B = \{x \in M : d(x,a) < r\}$ for some $r, r' \in X$ with $r' < r$ and some $a \in B'$. If $\{s \in X : r' \leq s < r\} \cap Y = \varnothing$ then it would follow that $B = B'$, a contradiction. Thus, there is an $s \in Y$, $r' \leq s < r$ and $\tau(B) := \{x \in M : d(x,a) < s\}$ is a ball in (M, Y, d), between B' and B. It is easily seen that $\mathcal{D} := \{\tau(B) : B \in \mathcal{C}\}$ is a nest of balls in (M, Y, d), so $\varnothing \neq \bigcap \mathcal{D} \subset \bigcap \mathcal{C}$.

PROPOSITION 1.2.3 *Let (M, X, d) be a scaled space, let $V \subset M$ be a spherically complete subspace. Then each $x \in M$ has a best approximation in V i.e.* $\min\{d(x,v) : v \in V\}$ *exists.*

Proof. The collection $\{B(x,r) \cap V : r \in X,\ B(x,r) \cap V \neq \varnothing\}$ is a nest of balls in (V, X, d). By spherical completeness of V (thanks to Proposition 1.2.2 we do not have to specify the range space of d) it has a nonempty intersection, so there exists a $v \in V$ such that $d(x,v) \leq d(x,w)$ for all $w \in V$ and we are done.

PROPOSITION 1.2.4 *Let (M, X, d) be a scaled space. The following are equivalent.*
(α) *M is ultrametrizable.*
(β) *M is discrete or there exist $s_1 > s_2 > \ldots$ in X such that* $\lim_n s_n = 0$.

Proof. To prove $(\beta) \Rightarrow (\alpha)$, suppose we have $s_1 > s_2 > \ldots$ in X with $\lim_n s_n = 0$. For each $r \in X$, let $n_r := \min\{m \in \mathbb{N} : s_m \leq r\}$ and set $\phi(r) := 2^{-n_r}$. By adding the requirement $\phi(0) := 0$ we obtain an increasing map $\phi : X \cup \{0\} \to \{0, \frac{1}{2}, \frac{1}{4}, \frac{1}{8}, \ldots\}$ and it is easily checked that $\phi \circ d$ is an ultrametric on M yielding the same topology as d.
Conversely suppose (α) and let M be not discrete. Then it has a non-isolated point a. Let $U_1 \supset U_2 \supset \ldots$ be a neighbourhood base at a. There exist $u_n \in U_n$ $(n \in \mathbb{N})$ with $u_n \neq a$ for each n. As $\lim_{n\to\infty} u_n = a$ we have $\lim_n d(a, u_n) = 0$, so (β) is proved, for $n \mapsto s_n$ a suitably chosen subsequence of $n \mapsto \min\{d(a, u_j) : 1 \leq j \leq n\}$.

1.3 Linearly ordered groups

Throughout this paper G is an abelian multiplicatively written group with unit element 1. If G is linearly ordered such that $x, y, z \in G$, $x \leq y$ implies $xz \leq yz$ we call $G = (G, \leq)$ a *linearly ordered group.* Then $x, y, z \in G$, $x < y$ implies $xz < yz$ (if $xz \geq yz$ then $x = xzz^{-1} \geq yzz^{-1} = y$, a contradiction). It follows easily that G is torsion free and that, if $G \neq \{1\}$, G has no smallest or largest element.

A subset H of a linearly ordered group G is *convex* if $x, y \in H$, $z \in G$, $x \leq z \leq y$ implies $z \in H$. Each proper convex subgroup is bounded from below and from above. The set of convex subgroups is linearly ordered by inclusion. A convex subgroup H is called *principal* if there is an $a \in G$ such that H is the smallest convex subgroup of G containing a. The order type of the set of all principal subgroups $\neq \{1\}$ is called the *rank* of G. G has rank 1 if and only if it is, as an ordered group, isomorphic to a subgroup of $(0, \infty)$. For a proof, see [13]. The group $\bigoplus_{i \in \mathbb{N}} \mathbb{Z}_i$, where $\mathbb{Z}_i = \mathbb{Z}$ for each i, with the antilexicographic ordering, is an example of a group with infinite rank.

If H is a convex subgroup then G/H is in a natural way a linearly ordered group and the canonical quotient map $G \to G/H$ is an increasing homomorphism. If rank $G > 1$ then G is not complete. In fact, let H be a proper convex subgroup, $H \neq \{1\}$. If $s := \sup_G H$ would exist then $s^2 = s$, so $s = 1$, a contradiction.

Let G be a linearly ordered group. We extend the multiplication to its completion (see 1.1) $G^{\#}$ as follows. For $s, t \in G^{\#}$ set

$$s\,t = \sup\{g_1 g_2 : g_1, g_2 \in G,\ g_1 \leq s,\ g_2 \leq t\}.$$

Clearly this multiplication extends the one of G, is associative, commutative and has a unit 1. But the semigroup $G^{\#}$ is in general not a group. In fact, if rank $G > 1$ then we have $s^2 = s$ if $s = \sup H$ when H is a convex subgroup, $H \neq \{1\}$, $H \neq G$, but $s \neq 1$, so $G^{\#}$ is not a group. The order on $G^{\#}$ satisfies the following. If $s, t, s', t' \in G^{\#}$, $s \leq t$, $s' \leq t'$ then $ss' \leq tt'$.

REMARK The extension of the multiplication from G to $G^{\#}$ is, in general, not unique. In fact, the formula $s * t = \inf\{g_1 g_2 : g_1, g_2 \in G,\ g_1 \geq s,\ g_2 \geq t\}$ defines an extension $*$ to $G^{\#}$ of the multiplication of G that is also associative, commutative, for which 1 is a unit and such that $s, t, s', t' \in G^{\#}$, $s \leq t$, $s' \leq t'$ implies $s * s' \leq t * t'$. We have $s * t \geq st$ for all $s, t \in G^{\#}$. To see that $*$ differs from the multiplication of above, let $H \neq \{1\}$ be a proper convex subgroup, $s := \sup H$, $t := \inf H$. Then one proves easily that $s * t = s$ but $st = t$. (However, we will see in 1.5.4 that for any two extensions $*$ and $\cdot$ of the multiplication that are increasing in both variables we have $s * g = s \cdot g$ for each $s \in G^{\#}$, $g \in G$.) In contrast to this we will show now that the inversion map $g \mapsto g^{-1}$ extends uniquely to a decreasing map $G^{\#} \to G^{\#}$.

PROPOSITION 1.3.1 *Let G be a linearly ordered group. Then there is a unique decreasing map $\omega : G^{\#} \to G^{\#}$ extending $g \mapsto g^{-1}$ $(g \in G)$. It has the following properties.*

(i) $\omega(s) = \sup\{g \in G : sg \leq 1\} = \inf\{g \in G : 1 \leq sg\} \quad (s \in G^{\#})$.
(ii) $\omega(gs) = g^{-1}\omega(s) \quad (s \in G^{\#}, g \in G)$.
(iii) $\omega(st) \geq \omega(s)\omega(t) \quad (s, t \in G^{\#})$.
(iv) *ω^2 is the identity.*
(v) *For each net $i \mapsto s_i$ in $G^{\#}$,* $\lim_i s_i = 0 \iff \lim_i \omega(s_i) = \infty$.

Proof. Let ω be any decreasing extension of $g \mapsto g^{-1}$. Then if $s \in G^{\#}$ and $g \in G$ is such that $1 \leq sg$, then $g^{-1} \leq s$ so $g \geq \omega(s)$. We see that $\omega(s) \leq \inf\{g \in G : 1 \leq sg\}$. In the same vein we have $\omega(s) \geq \sup\{g \in G : sg \leq 1\}$. So, to prove existence, uniqueness and (i) it suffices to prove that $\sup A = \inf B$ where $A = \{g \in G : sg \leq 1\}$, $B = \{g \in G : sg \geq 1\}$. This is clear if $s \in G$ (then $\sup A = \inf B = s^{-1}$), so assume that $s \notin G$. Because G is coinitial and cofinal in $G^{\#}$ the sets A, B are nonempty. We have $A \cap B = \varnothing$, $A \cup B = G$, and A is a cut. (In fact, suppose $h := \sup_G A$ exists, but $h \notin A$. Then $sh > 1$ so $h^{-1} < s$. By 1.1.4 (iv) there exists a $b \in G$ with $h^{-1} < b \leq s$. We have $b \neq s$ so $b < s$ i.e. $1 < b^{-1}s$. But $b^{-1} < h$ so $b^{-1} \in A$, a contradiction.) So $t := \sup_{G^{\#}} A$ exists and $t \notin B$, so $B = \{x \in G : x \geq t\}$. By 1.1.4 (v), $t = \inf_{G^{\#}} B$. To prove (iii), let $V := \{g \in G : sg \leq 1\}$, $W := \{g \in G : tg \leq 1\}$, $X := \{g \in G : stg \leq 1\}$. Then $VW \subset X$, so $\omega(s)\omega(t) = \sup VW \leq \sup X = \omega(st)$. From (iii) it follows that $g^{-1}\omega(s) = \omega(g)\omega(s) \leq \omega(gs)$. But also $g\omega(gs) = \omega(g^{-1})\omega(gs) \leq \omega(s)$, so $\omega(gs) \leq g^{-1}\omega(s)$ and (ii) is proved. To prove (iv), let $s \in G^\#$. If $g_1, g_2 \in G$, $g_1 \leq s \leq g_2$ then $g_1 \leq \omega^2(s) \leq g_2$. It follows that $s \leq \omega^2(s) \leq s$ i.e. $\omega^2(s) = s$. Finally (v) follows from bijectivity and $x < \omega(\varepsilon) \iff \omega(x) > \varepsilon$ for each $\varepsilon \in G^\#$.

We will call the map $\omega : G^\# \to G^\#$ of 1.3.1 the *antipode*.

REMARK It is easy to prove that, for a proper convex subgroup H, $\omega(\sup H) = \inf H$.

1.3.2 Example. The completion of $\bigoplus \mathbb{Z}$

Let $G = \bigoplus_{i \in \mathbb{N}} G_i$ be the direct sum of the groups G_i, $i \in \mathbb{N}$, where G_i is the infinite cyclic group generated by g_i. Hence if $g \in G$, then $g = (g_i^{n_i})_{i \in \mathbb{N}}$, $n_i \in \mathbb{Z}$, $n_i \neq 0$ only for a finite number of indexes i. We order each G_i by $g_i^n < g_i^m$ if and only if $n < m$; with the antilexicographical order, G becomes a linearly ordered group. Below we shall describe $G^\#$, the completion of G (as in 1.1.4 and its preamble), as the set obtained by adjoining to G all the symbols of the form $(g_i^{n_i})_{i \in \mathbb{N}}$, such that for a given $m \in \mathbb{N}$, $m \geq 2$, $n_i = \infty$ for all $i < m$, $n_i \in \mathbb{Z}$ if $i \geq m$, $n_i \neq 0$ only for a finite number of indexes i. Such a symbol will denote the supremum of the set $\{(a_1, a_2, \ldots, a_{m-1}, g_m^{n_m}, \ldots, g_r^{n_r}, 1, \ldots) : a_i \in G_i \text{ for } i < m\}$. $G^\#$ is ordered antilexicographically, taking into account that for all i, $g_i^\infty > g_i^r$ for any $r \in \mathbb{Z}$. The infimum of the set $\{(a_1, a_2, \ldots, a_{m-1}, g_m^{n_m}, \ldots, g_r^{n_r}, 1, \ldots) : a_i \in G_i \text{ for } i < m\}$ corresponds to the symbol $(g_1^\infty, g_2^\infty, \ldots, g_{m-1}^\infty, g_m^{-1+n_m}, g_{m+1}^{n_{m+1}}, \ldots, g_r^{n_r}, 1, \ldots)$. We will prove that the set of all proper convex subgroups of G is the set of all H_n, $n \geq 0$, where $H_0 = \{1\}$, and $H_k = \{(g_i^{n_i})_{i \in \mathbb{N}} : n_i = 0 \text{ if } i > k\}$. We will denote by s_k the supremum of H_k, and by t_k its infimum. Therefore $s_k = (g_1^\infty, g_2^\infty, \ldots, g_k^\infty, g_{k+1}^0, 1, \ldots)$ and $t_k = (g_1^\infty, g_2^\infty, \ldots, g_k^\infty, g_{k+1}^{-1}, 1, \ldots)$. We will extend the multiplication of G to its completion, according to the formulas in 1.3. Then we have that for two elements $g = (g_i^{n_i})_{i \in \mathbb{N}}$, $h = (g_i^{t_i})_{i \in \mathbb{N}}$, where $n_i, t_i \in \mathbb{Z} \cup \{\infty\}$, $gh = (g_i^{n_i + t_i})_{i \in \mathbb{N}}$, with $\infty + m = m + \infty = \infty + \infty = \infty$ for all $m \in \mathbb{Z}$. This shows that any element w in the completion of G can be written as $w = s_n w'$ for some $n \in \mathbb{N}$ and some $w' \in G$, the number n is completely determined by w, but there are many possible choices for w'. As to the second extension $*$ of the multiplication in G, we shall prove that $g * h = gh$ in all cases except when $g = s_n g'$, $h = s_n h'$ with $g', h' \in G$. In that case $g * h = (1, \ldots, 1, g_{n+1}, 1, \ldots)gh$.

The construction of $G^\#$. Let $n_i \in \mathbb{Z}$ for $i = m, m+1, \ldots, r$, define $A = \{(a_1, a_2, \ldots, a_{m-1}, g_m^{n_m}, \ldots, g_r^{n_r}, 1, \ldots) : a_i \in G_i \text{ for } i < m\}$ and $C_A = \{g \in G : g \geq x \text{ for some } x \in A\}$. We will prove first that $\sup C_A$ does not exist in G. In fact, if $(g_i^{t_i})_{i \in \mathbb{N}}$ was the supremum of that set, then we would have the following inequalities. $(1, \ldots, 1, g_{m-1}, g_m^{n_m}, \ldots, g_r^{n_r}, 1, \ldots) \leq (g_i^{t_i})_{i \in \mathbb{N}} \leq (1, \ldots, 1, g_{m-1}, g_m^{1+n_m}, g_{m+1}^{n_{m+1}}, \ldots, g_r^{n_r}, 1, \ldots)$, so $t_i = n_i$ for $i \geq m+1$. Therefore $t_m = n_m$ or $t_m = 1 + n_m$; but in the first case $(1, \ldots, 1, g_{m-1}^{1+t_{m-1}}, g_m^{n_m}, \ldots, g_r^{n_r}, 1, \ldots)$ is an element of C_A bigger than $\sup C_A$, and in the second case $(1, \ldots, 1, g_{m-1}^{-1+t_{m-1}}, g_m^{1+n_m}, g_{m+1}^{n_{m+1}}, \ldots, g_r^{n_r}, 1, \ldots)$ is an upper bound of C_A that is smaller than $\sup C_A$. Therefore $\sup C_A$ does not exist in G. We shall denote the cut C_A by the symbol $(g_1^\infty, g_2^\infty, \ldots, g_{m-1}^\infty, g_m^{n_m}, \ldots, g_r^{n_r}, 1, \ldots)$. Now let F be a cut of G such that $\sup F$ does not exist in G, we will show that F is equal to a cut C_A for some set A as described before. Let t be an upper bound of F, pick some $f \in F$ and let $F' = \{x \in F : f \leq x < t\}$, it is clear that $F = \{g \in G : g \leq x \text{ for some } x \in F'\}$. By the definition of F' there is an $m \in \mathbb{N}$ such that for all elements $x = (g_i^{x_i})_{i \in \mathbb{N}} \in F'$ we have that $x_i = 0$ if $i > m$, while for some $x \in F'$, $x_m \neq 0$. We examine the m'th coordinate of the elements of F', if that set has no maximum, then define $A := \{(a_1, a_2, \ldots, a_{m-1}, a_m, 1, \ldots) : a_i \in G_i \text{ for } i < m+1\}$ and by a direct argument we see that F is the cut C_A, therefore $F = (g_1^\infty, g_2^\infty, \ldots, g_m^\infty, 1, 1, \ldots)$. If the set has a maximum, say ξ_m, we look at the set of the $(m-1)$th coordinates of the elements of F' whose mth coordinate is ξ_m. If that set has no maximum, then $F = (g_1^\infty, \ldots, g_{m-1}^\infty, \xi_m, 1, 1, \ldots)$, but if ξ_{m-1} is a maximum, then we continue with the elements of F' of the form $(*, \ldots, *, \xi_m, \xi_{m-1}, 1, \ldots)$. It is not possible to find in such a way elements $\xi_1, \xi_2, \ldots, \xi_m$ as indicated above, because in that case $(\xi_1, \xi_2, \ldots, \xi_m, 1, \ldots) \in G$ would be the supremum of F, contrary to our hypothesis. Therefore there has to be an $r \in \mathbb{N}$, $r > 1$, such that ξ_r exists, but ξ_{r-1} does not exist. Then $F = (g_1^\infty, \ldots, g_{r-1}^\infty, \xi_r, \ldots, \xi_m, 1, \ldots)$. Then $G^\#$, the collection of all cuts of G, is the union of the set of all cuts designed by the symbols of the form $(g_i^{n_i})_{i \in \mathbb{N}}$, such that for a given $m \in \mathbb{N}$, $m \geq 2$, $n_i = \infty$ for all $i < m$, $n_i \in \mathbb{Z}$ if $i \geq m$, $n_i \neq 0$ only for a finite number of indexes i, and the set of all the cuts that have a supremum in G. As in the preamble to 1.1.4 we identify those cuts with the elements of G. The order in $G^\#$ given by inclusion corresponds to the antilexicographic ordering, requiring that for all i, $g_i^\infty > g_i^r$ for any $r \in \mathbb{Z}$. By 1.1.4 (v), every element $s \in G^\#$ is the supremum of $\{x \in G : x \leq s\}$. Hence if $A = \{(a_1, a_2, \ldots, a_{m-1}, g_m^{n_m}, \ldots, g_r^{n_r}, 1, \ldots) : a_i \in G_i \text{ for } i < m\}$ for some particular choice of $n_m, n_{m+1}, \ldots, n_r$, then $\sup A = (g_1^\infty, g_2^\infty, \ldots, g_{m-1}^\infty, g_m^{n_m}, \ldots, g_r^{n_r}, 1, \ldots)$.

LEMMA 1.3.3 *Let A be as in the previous paragraph. Then the infimum of A is the element $g = (g_1^\infty, g_2^\infty, \ldots, g_{m-1}^\infty, g_m^{-1+n_m}, g_{m+1}^{n_{m+1}}, \ldots, g_r^{n_r}, 1, \ldots)$.*

Proof. Clearly $g \leq x$ for all $x \in A$. Suppose there exists a $b = (g_i^{b_i})_{i \in \mathbb{N}}, b_i \in \mathbb{Z} \cup \{\infty\}$ such that $g < b$, we shall prove that there is an $a \in A$ such that $g < a \leq b$. Without loss of generality we can assume that $g < b < (1, \ldots, 1, g_m^{n_m}, \ldots, g_r^{n_r}, 1, \ldots)$, so $b_i = 0$ if $i > r$, $b_i = n_i$ for $m < i \leq r$. Any element of the form $(*, \ldots, *, g_m^{-1+n_m}, g_{m+1}^{n_{m+1}}, \ldots, g_r^{n_r}, 1, \ldots)$ is smaller than or equal to g, therefore we must have $b_m = n_m$. Then the second inequality implies that $b_{m-1} \leq 0$. Then $a = (1, \ldots, 1, g_{m-1}^{-1+b_{m-1}}, g_m^{n_m}, \ldots, g_r^{n_r}, 1, \ldots)$ is an element in A that satisfies $g < a \leq b$. Hence $g = \inf A$.

Convex subgroups of G It is readily seen that the set of all proper convex subgroups of G is the set of all H_n, $n \geq 0$, where $H_0 = \{1\}$ and $H_k = \{(g_i^{n_i})_{i \in \mathbb{N}} : n_i = 0 \text{ if } i > k\}$. In what follows we denote by s_n the supremum of H_n, and by t_n its infimum. It is clear now that $s_0 = t_0 = 1$, and for $k > 0$,

$$s_k = (g_1^\infty, g_2^\infty, \ldots, g_k^\infty, 1, \ldots)$$
$$t_k = (g_1^\infty, g_2^\infty, \ldots, g_k^\infty, g_{k+1}^{-1}, 1, \ldots)$$

Multiplication in $G^\#$ For $s, t \in G^\#$ we have defined in 1.3 two products, both of them extending the multiplication in G, in fact,

$$st = \sup\{xy : x, y \in G, x \leq s, y \leq t\} \quad \text{and}$$
$$s * t = \inf\{xy : x, y \in G, x \geq s, y \geq t\}.$$

We look now for a simple formula for these products in $G^{\#}$, the crucial point is the following fact.

LEMMA 1.3.4 *Let $g \in G^{\#} \backslash G$ be the element $g = (g_i^{n_i})_{i \in \mathbb{N}}$ with $n_i = \infty$ if and only if $i \leq r$. Then g can be written as $g = s_r t = s_r * t$ for $t = (g_i^{t_i})_{i \in \mathbb{N}}$ with $t_i = 0$ if $i \leq r$, $t_i = n_i$ if $i > r$.*

Proof. Since $t \in G$ we have that $s_r t = \sup\{gt : g \in G, g \leq s_r\}$, and since $s_r = \sup(H_r \cup \{g \in G : g < h \text{ for any } h \in H_r\})$, we have that

$$s_r t = \sup\{gt : g \in H_r\}. \tag{$*$}$$

Then $s_r t = \sup\{(g_i^{k_i})_{i \in \mathbb{N}} : k_i = n_i \text{ if } i > r\} = (g_1^{\infty}, g_2^{\infty}, \ldots, g_r^{\infty}, g_{r+1}^{n_{r+1}}, \ldots) = g$. On the other hand $s_r * t = \inf\{gt : g \in G, g \geq s_r\}$, but as $s_r = (g_1^{\infty}, g_2^{\infty}, \ldots, g_r^{\infty}, 1, \ldots)$, by Lemma 1.3.3 we see that $s_r = \inf B$ with $B = \{(b_1, \ldots, b_r, g_{r+1}, 1, \ldots) : b_i \in G_i \text{ for } i \leq r\}$. Hence

$$\begin{aligned} s_r * t &= \inf\{gt : g \in B\} \\ &= \inf\{(g_i^{k_i})_{i \in \mathbb{N}} : k_i \in \mathbb{Z}, k_{r+1} = 1 + n_{r+1}, k_i = n_i, \quad \text{if } i > r+1\} \qquad (**) \\ &= g, \quad \text{by Lemma 1.} \end{aligned}$$

LEMMA 1.3.5 *If $t \in H_r$, then $s_r t = s_r * t = s_r$.*

Proof. By $(*)$ we have $s_r t = \sup\{gt : g \in H_r\} = \sup H_r$, and by $(**)$ $s_r * t = \inf\{gt : g \in B\} = \inf\{(g_i^{k_i})_{i \in \mathbb{N}} : k_i \in \mathbb{Z}, k_{r+1} = 1, \ k_i = 0 \text{ if } i > r+1\} = s_r$.

LEMMA 1.3.6 *Let $n, r \in \mathbb{N}$. If $r < n$ then $s_r s_n = s_r * s_n = s_n$. If $r = n$ then $s_n s_n = s_n$ but $s_n * s_n = \delta_{n+1} s_n$ with $\delta_{n+1} = (1, \ldots, 1, g_{n+1}, 1, \ldots)$.*

Proof. i) $s_r s_n = \sup\{xy : x, y \in G, x \leq s_r, y \leq s_n\} = \sup\{xy : x \in H_r, y \in H_n\}$. If $r \leq n$ then $H_r \subseteq H_n$ and $xy \in H_n$. Therefore the set above is contained in H_n, and since it clearly contains H_n, we have that $s_r s_n = s_n = s_n s_n$.
ii) $s_r * s_n = \inf\{xy : x, y \in G, x \geq s_r, y \geq s_n\}$. As in $(**)$ in the proof of Lemma 1.3.4, $s_r * s_n = \inf\{xy : x, y \in G, x \in B_r, y \in B_n\}$, where
$B_r = \{(b_1, \ldots, b_r, g_{r+1}, 1, \ldots) : b_i \in G_i \text{ for } i \leq r\}$
and $B_n = \{(c_1, \ldots, c_n, g_{n+1}, 1, \ldots) : c_i \in G_i \text{ for } i \leq n\}$.
If $r < n$ then $s_r * s_n = \inf\{(d_1, \ldots, d_n, g_{n+1}, 1, \ldots) : d_i \in G_i \text{ for } i \leq n\} = s_n$, but if $r = n$, then $s_r * s_n = \inf\{(d_1, \ldots, d_n, g_{n+1}^2, 1, \ldots) : d_i \in G_i \text{ for } i \leq n\} = (g_1^{\infty}, g_2^{\infty}, \ldots, g_n^{\infty}, g_{n+1}, 1, \ldots) = \delta_{n+1} s_n (> s_n)$.

COROLLARY 1.3.7 *Let $s, t \in G^{\#}$. Then $st = s * t$ for all cases, except when $s = s_n g$ and $t = s_n h$ for some g, $h \in G$. In that case $s * t = \delta_{n+1} st$ with $\delta_{n+1} = (1, \ldots, 1, g_{n+1}, 1, \ldots)$.*

Finally, from lemmas 1.3.4, 1.3.5, 1.3.6 we obtain

COROLLARY 1.3.8 *Let* $g, h \in G^{\#}$, $g = (g_i^{n_i})_{i \in \mathbb{N}}$, $h = (g_i^{k_i})_{i \in \mathbb{N}}$, $n_i, k_i \in \mathbb{Z} \cup \{\infty\}$. *Then* $gh = (g_i^{n_i + k_i})_{i \in \mathbb{N}}$ *with* $\infty + m = \infty + \infty = \infty$ *for all* $m \in \mathbb{Z}$.

We will leave the description of the antipode in $G^{\#}$ (see 1.3.1) to the reader.

1.4 Valued fields

Let G be a linearly ordered group. Like in 1.1.2 we add an element 0 to G, extend the ordering and the multiplication by declaring that $0 < g$ and $0.g = 0.0 = 0$ for all $g \in G$. A *valuation* on a field K (with *value group* G) is a surjective map $| \ | : K \to G \cup \{0\}$ such that for all $x, y \in K$

(i) $|x| = 0$ if and only if $x = 0$
(ii) $|x + y| \leq \max(|x|, |y|)$
(iii) $|xy| = |x|\,|y|$.

REMARK In this paper we prefer the multiplicative notation over the more commonly used additive one, to link up with the conventions in classical Functional Analysis.

The *rank* of the *valued field* $K = (K, | \ |)$ is the rank of G. We shall exclude the trivial valuation i.e. we assume $G \neq \{1\}$. The map $(x, y) \mapsto |x - y|$ $(x, y \in K)$ is a scale in the sense of 1.2 and its topology is a non-discrete field topology. The *valuation ring* $B_K := \{\lambda \in K : |\lambda| \leq 1\}$ has a unique maximal ideal $B_K^- := \{\lambda \in K : |\lambda| < 1\}$. The *residue class field* of K is $k := B_K / B_K^-$. The following theorem concerns metrizability of K.

THEOREM 1.4.1 *Let* $(K, | \ |)$ *be a valued field. The following are equivalent.*

(α) $(K, | \ |)$ *is (ultra) metrizable.*
(β) G *has a coinitial (cofinal) sequence.*
(γ) $K^\times := K \setminus \{0\}$ *contains a countable set* C *for which* $0 \in \overline{C}$.
(δ) K *contains a countable subset that is not closed.*

Proof. $(\alpha) \iff (\beta)$ follows from 1.2.4. The implications $(\beta) \Rightarrow (\gamma) \Rightarrow (\delta)$ are trivial. To prove $(\delta) \Rightarrow (\beta)$, let $\{\alpha_1, \alpha_2, \ldots\} \subset K$ be not closed, let α be in the closure, $\alpha \neq \alpha_n$ for each n. Then $n \mapsto \min(|\alpha - \alpha_1|, |\alpha - \alpha_2|, \ldots, |\alpha - \alpha_n|)$ is a sequence in G tending to 0 i.e. is coinitial.

DEFINITION 1.4.2 Let E be a vector space over a valued field K. A subset A of E is *absolutely convex* if it is a B_K-submodule of E, in other words, if $0 \in A$ and $x, y \in A$, $\lambda, \mu \in B_K$ implies $\lambda x + \mu y \in A$. A subset S of E is called *convex* if $x, y, z \in S$, $\lambda, \mu, \nu \in B_K$, $\lambda + \mu + \nu = 1$ implies $\lambda x + \mu y + \nu z \in S$.
It is easy to see that a nonempty $S \subset E$ is convex if and only if it is an additive coset of an absolutely convex set. The following Proposition describes the absolutely convex subsets of K.

PROPOSITION 1.4.3 ([19], 20.6, (5)) *Let K be a valued field with value group G. The sets $\{0\}$, K, $B(0,r^-) := \{\lambda \in K : |\lambda| < r\}$, $B(0,r) := \{\lambda \in K : |\lambda| \le r\}$ $(r \in G^\#)$ are absolutely convex. Each absolutely convex subset of K is of one of these forms.*

Proof. We only prove the second statement. Let $A \subset K$ be absolutely convex. If A is unbounded (i.e. if $\{|\mu| : \mu \in A\}$ is not bounded above), let $\lambda \in K$. Then there is a $\mu \in A$, $|\mu| > |\lambda|$. Then $\lambda = (\lambda \cdot \mu^{-1})\mu \in A$ and it follows that $A = K$. Now suppose that A is bounded above and contains at least one nonzero element. Then $r := \sup_{G^\#}\{|\lambda| : \lambda \in A\}$ exists. Clearly, if $\lambda \in A$, $|\lambda| < r$ then there is a $\mu \in A$, $|\mu| > |\lambda|$, so $\lambda = (\lambda\mu^{-1}) \cdot \mu \in A$. It follows that $B(0,r^-) \subset A \subset B(0,r)$. If the first inclusion is strict there is a $\mu \in A$, $|\mu| = r$. If $\lambda \in K$, $|\lambda| \le r$ then $\lambda = (\lambda\mu^{-1})\mu \in A$, so $A = B(0,r)$.

In the main part of this paper (Sections 3 and 4) we shall have to put a restriction upon K namely that each absolutely convex subset of K is countably generated as a B_K-module. In the following Proposition we describe the situation. For a linearly ordered set X the *interval topology* is defined to be the topology generated by the sets $\{x \in X : x > s\}$ and $\{x \in X : x < t\}$ $(s,t \in X)$.

PROPOSITION 1.4.4 *Let K be a valued field with value group G. The following are equivalent.*

(α) *Each absolutely convex subset of K is countably generated as a B_K-module.*

(β) *G has a cofinal sequence. For each $s \in G^\#$ there are $g_1, g_2, \ldots \in G$, $g_n < s$ for all n, such that $\sup_{G^\#}\{t \in G^\# : t < s\} = \sup_{G^\#}\{g_1, g_2, \ldots\}$.*

(γ) *G has a coinitial sequence. For each $s \in G^\#$ there exist $g_1, g_2, \ldots \in G, g_n > s$ for all n such that $\inf_{G^\#}\{t \in G^\# : t > s\} = \inf_{G^\#}\{g_1, g_2, \ldots\}$.*

(δ) *The interval topology on $G^\#$ satisfies the first axiom of countability. $G^\#$ has a cofinal sequence.*

Proof. (α) $\Rightarrow$ (β). Let K be generated as a B_K-module by $\alpha_1, \alpha_2, \ldots$ which we may suppose to be non-zero. We claim that $|\alpha_1|, |\alpha_2|, \ldots$ is cofinal in G. In fact, let $\lambda \in K$. Then there are $n \in \mathbb{N}$, $\xi_1, \ldots, \xi_n \in B_K$ such that $\lambda = \sum_{i=1}^n \xi_i\alpha_i$. Then $|\lambda| \le \max_{1\le i\le n} |\xi_i\alpha_i| \le \max_{1\le i \le n} |\alpha_i|$. Now let $s \in G^\#$. By Lemma 1.1.1 either $s_0 := \max\{t \in G^\# : t < s\}$ exists (then by Proposition 1.1.4 (iv) s_0, $s \in G$ and we can choose $g_n = s_0$ for each n), or $\sup\{t \in G^\# : t < s\} = s$. Let $\alpha_1, \alpha_2, \ldots \in K^\times$ generate $B(0,s^-)$. Then $|\alpha_n| < s$ for each n. To prove that $\sup_n |\alpha_n| = s$, let $t \in G^\#$, $t < s$. By 1.1.4 (iv) there is a $\lambda \in K$, $s > |\lambda| \ge t$. There are $n \in \mathbb{N}$ and $\xi_1, \ldots, \xi_n \in B_K$ such that $\lambda = \sum_{i=1}^n \xi_i\alpha_i$. Then $|\lambda| \le \max_{1\le i\le n} |\alpha_i|$. Hence, $\sup_n |\alpha_n| \ge t$ for each $t \in G^\#$, $t < s$, so $\sup_n |\alpha_n| = s$. The implications (β) $\iff$ (γ) can be proved by applying the antipode ω of Proposition 1.3.1. We now prove (β)&(γ) $\Rightarrow$ (δ). Let $s \in G^\#$. If $s_0 = \max\{t \in G^\# : t < s\}$ exists then $s_0, s \in G$ and $s_1 = \min\{t \in G^\# : t > s\}$ exists and $\{s\}$ is an open set, so trivially there exists a countable neighbourhood base at s. Now suppose $\sup\{t \in G^\# : t < s\} = s = \inf\{t \in G^\# : t > s\}$. Let $g_1, g_2, \ldots \in G$, $g_n < s$ for each n, $\sup\{g_1, g_2, \ldots\} = \sup\{t \in G^\#, t < s\}$. We may suppose $g_1 < g_2 < \cdots$. Let $h_1, h_2, \ldots \in G$, $h_n > s$ for each n, $h_1 > h_2 > \cdots$, $\inf\{h_1, h_2, \ldots\} = \inf\{t \in G^\# : t > s\}$. Then for each $n \in \mathbb{N}$, $U_n := \{x \in G^\# : g_n < x < h_n\}$ is an open neighbourhood of s in the interval topology; we prove the U_n to be a neighbourhood base. So let U be open in the interval topology, $s \in U$. As the sets $(a,b) := \{x \in G^{\#} : a < x < b\}$ $(a, b \in G^{\#})$ from a base for the interval topology of $G^{\#}$, we may suppose $U = (a,b)$. Then $a < s < b$. There is an n such that $a < g_n$ and $b > h_n$. We see that $U_n \subset (a,b)$ and we are done. Finally we prove $(\delta) \Rightarrow (\alpha)$. Let $s_1, s_2, \ldots$ be a cofinal sequence in $G^{\#}$. By cofinality of G in $G^{\#}$ we may suppose that $s_n \in G$ for each n. Choose $\lambda_n \in K$ such that $|\lambda_n| = s_n$ $(n \in \mathbb{N})$. It is easy to see that K is generated by $\{\lambda_1, \lambda_2, \ldots\}$ as a B_K-module. Obviously any set of the form $B(0,r)$ where $r \in G$ is generated by a single element, so to finish the proof we show that $B(0, s^-)$ is countably generated where $s \in G^{\#}$, $\sup\{t \in G : t < s\} = s$. Let $U_1 \supset U_2 \supset \cdots$ be a countable base of the interval topology at s, we may suppose that $U_n = \{t \in G^{\#} : a_n < t < b_n\}$ for some $a_n, b_n \in G^{\#}$. By assumption and 1.1.4 (iv) we may assume $a_n, b_n \in G$. Choose, for each n, a $\lambda_n \in K$ with $|\lambda_n| = a_n$. One proves easily that $B(0, s^-)$ is generated by $\{\lambda_1, \lambda_2, \ldots\}$.

REMARKS (i) If K is separable or, more generally, if G is countable we obviously have $(\alpha) - (\delta)$ of above. By 1.4.1 $(\alpha) - (\delta)$ imply ultrametrizability of K. Statement (δ) implies the first axiom of countability for G, and hence, since G is a group, metrizability of G (See [8], Problem O, p. 210). But we do not know if (δ) implies that $G^{\#}$ is metrizable.
(ii) It is not hard to see that $(\alpha) - (\delta)$ are equivalent to: each subset A of $G^{\#}$ that is bounded above has a countable subset S such that $\sup_{G^{\#}} A = \sup_{G^{\#}} S$. This property is known in Riesz space theory as 'super Dedekind completeness'.

1.5 G-modules

The G-modules we introduce below will serve as a natural range set for norms on K-vector spaces, see 2.1 and 2.2.

DEFINITION 1.5.1 Let G be a linearly ordered group. A linearly ordered set X is called a *G-module* if there exists a map $G \times X \to X$, written $(g, x) \mapsto gx$, called *multiplication*, such that for all $g, g_1, g_2, \in G$ and all $x, x_1, x_2 \in X$ we have

(i) $g_1(g_2x) = (g_1g_2)x$
(ii) $1x = x$
(iii) $g_1 \geq g_2 \Rightarrow g_1x \geq g_2x$
(iv) $x_1 \geq x_2 \Rightarrow gx_1 \geq gx_2$
(v) Gx is coinitial in X
(vi) X has no smallest element.

Thus, the requirements (i) - (iv) mean that G acts on X and that this action preserves the ordering $\leq$ in G and X. The pair (v)&(vi) is equivalent to "for each $\varepsilon \in X$ there is a $g \in G$ such that $gx < \varepsilon$". It follows that modules over the group $\{1\}$ do not exist. If X is a G-module we have for all $x_1, x_2 \in X$, $g \in G$

(iv)′ $x_1 > x_2 \Rightarrow gx_1 > gx_2$

(otherwise $gx_1 \leq gx_2$ hence $x_1 = g^{-1}gx_1 \leq g^{-1}gx_2 = x_2$ by (iv), a contradiction), but the formula $g_1 > g_2 \Rightarrow g_1 x > g_2 x$ does not hold in general. In fact, the semigroup $G^\#$ is, a fortiori, a G-module; if H is a proper convex subgroup $\neq \{1\}$ then $h \sup H = \sup H$ for all $h \in H$. Let X be a G-module. Then for each $x \in X$ the set Gx is cofinal in X. In fact, let $x, y \in X$. We just saw that there is a $g \in G$ with $gx < y$. Then $x < g^{-1}y$ by (iv)′. This proof also shows that X has no largest element.

For an element s of a G-module X, let $\mathrm{Stab}(s) := \{g \in G : gs = s\}$. It is a proper convex subgroup of G. If $\mathrm{Stab}(s) = \{1\}$ the element s is called *faithful.* Letting $\pi : G \to G/\mathrm{Stab}(s)$ be the canonical homomorphism, the G-module Gs becomes a $G/\mathrm{Stab}(s)$-module under the multiplication $\pi(g)s := gs$. It has only faithful elements.

Let X, Y be G-modules. A map $\phi : X \to Y$ is called a *G-module map* if ϕ is increasing and if $\phi(gs) = g\phi(s)$ for all $g \in G$, $s \in X$. Its extended map $X \cup \{0\} \to Y \cup \{0\}$ is called an *extended G-module map.*

PROPOSITION 1.5.2 *Let G be a linearly ordered group.*
(i) *(Extended) G-module maps are bicontinuous at 0.*
(ii) *Let X be a G-module. Then for a net $i \mapsto g_i$ in G we have*

$$\lim_i g_i s = 0 \iff \lim_i g_i = 0 \qquad (s \in X),$$

and for a net $i \mapsto s_i$ in X we have

$$\lim_i g\, s_i = 0 \iff \lim_i s_i = 0 \qquad (g \in G).$$

Proof. (i) Follows from Lemma 1.1.3 and 1.5.1 (v). Statement (ii) follows from (i) and the fact that $g \mapsto g\,s$ and $s \mapsto g\,s$ are G-module maps $G \to X$, $X \to X$ respectively.

PROPOSITION 1.5.3 *Let G be a linearly ordered group, let X be a G-module.*
(i) *Let $V \subset X$, $g \in G$. If $\sup V$ exists then $g \sup V = \sup gV$. If $\inf V$ exists then $g \inf V = \inf gV$. If V is not bounded above (below) then neither is gV.*
(ii) *Let $W \subset G$, $s \in X$. If $\sup_G W$ and $\sup_X Ws$ exist then $\sup Ws \le (\sup W)s$. If $\inf_G W$ and $\inf_X Ws$ exist then $\inf Ws \ge (\inf W)s$. If W is not bounded above (below) then neither is Ws, and conversely.*

Proof. (i) Let $s := \sup V$. Then $g\,s$ is an upper bound of gV. If $t \in X$, $t < gs$ then $g^{-1}t < s$, so there is a $v \in V$ with $g^{-1}t < v$ i.e. $t < gv$. We see that t is not an upper bound of gV. The proof of the second statement is similar. If s were an upper (lower) bound of gV then $g^{-1}s$ would be an upper (lower) bound of V, which finishes the proof of (i). (ii) The first two statements are obvious. Let W be not bounded above. Let $t \in X$. Since Gs is cofinal in X there is a $g \in G$ with $gs > t$. By unboundedness there is a $w \in W$ with $w > g$. Then $ws \ge gs > t$. Conversely, suppose Ws is not bounded above. Let $g \in G$. There is a $w \in W$ such that $ws > gs$. Then $w > g$, so W is not bounded above. The proof for the 'inf' case runs similarly.

REMARK (1) To express the fact that some subset V of a G-module X is not bounded below we sometimes write $\inf V = 0$ (this can be interpreted as the infimum taken in the linearly ordered set $X \cup \{0\}$).

(2) For an example in which the inequalities in (ii) above are strict, see 1.5.5 (c).

We now turn to the completion of G-modules and show that, unlike the (semi)group structure on G (see 1.3 Remark) the G-module structure on a set X (in particular, on G) can uniquely be extended to its completion.

THEOREM 1.5.4 *Let G be a linearly ordered group, let X be a G-module. Then the multiplication $G \times X \to X$ can uniquely be extended to a multiplication $G \times X^\# \to X^\#$ making $X^\#$ into a G-module. (In particular, $X^\#$ has no smallest or largest elements.)*

Proof. Let $g \in G$, $s \in X^\# \backslash X$. Then $A := \{gx : x \in X,\ x \leq s\}$ is a cut and $B := \{gy : y \in X :\ y \geq s\}$ is its complement in X. Clearly $\sup_{X^\#} A \notin X$, hence $\sup_{X^\#} A = \inf_{X^\#} B$ and we are forced to define

$$gs = \sup_{X^\#} A.$$

Straightforward verification shows that with respect to this extended multiplication $X^\#$ is a G-module.

EXAMPLES 1.5.5

(a) For every subgroup G of $X := (0, \infty) \subset \mathbb{R}$, X is in a natural way a (complete) G-module. Every element of X is faithful.

(b) If G has rank > 1 there are always G-modules having non-faithful elements. In fact, let H be a convex subgroup $\neq \{1\}$, $\neq G$. Then G/H is in a natural way a G-module and $h \cdot 1 = 1$ for each $h \in H$ so 1 (the unit element of G/H) is not faithful. To construct such G, let for each $n \in \mathbb{N}$, A_n be a subgroup of the multiplicative group $(0, \infty)$, $\neq \{1\}$ and take $G := \bigoplus_{n \in \mathbb{N}} A_n$, with the antilexicographic ordering. For each n, $A_1 \bigoplus \cdots \bigoplus A_n$ is a convex subgroup.

(c) We now construct a G-module X for which $\inf Ws > (\inf W)s$ for some $W \subset G$, $s \in X$ (see Proposition 1.5.3 (ii)). Let G be such that $1 = \inf\{g \in G : g > 1\}$, e.g. $G = (0, \infty)$. Let $G^- := \{g^- : g \in G\}$ be a copy of G, let $X := G \cup G^-$ be ordered by stating that

$$t < s^- < s$$

for all $s, t \in G$, $t < s$. (Thus, every $s \in G$ is given an immediate predecessor). X becomes a G-module by extending the multiplication by

$$gs^- := (gs)^- \qquad (g \in G, s \in G).$$

Now take $W := \{g \in G : g > 1\}$, $s := 1^-$. Then $\inf W = 1$, so $(\inf W) \cdot s = 1^-$. However $\inf Ws = \inf\{g^- : g > 1\} = 1$.

(d) In the sequel we will encounter the following situation. $G = \{|x| : x \in K, x \neq 0\}$, where K is some valued field, Γ is a linearly ordered group containing G as a cofinal (hence coinitial) ordered subgroup, $X := \Gamma^\#$. We will consider X sometimes as a G-module, sometimes as a Γ-module. Although often $G = \Gamma$, there are some cases where Γ contains G properly. A natural example in rank 1 case is given by $G := \{|x| : x \in \mathbb{Q}_p\}$, $\Gamma := \{|x| : x \in \mathbb{C}_p\}$. (Then $\Gamma^\# = (0, \infty)$.) In 1.6.3, 1.6.8, 4.4 we will meet the following example having infinite rank. For each n, let A_n be the free cyclic group generated by a_n (then $A_n \simeq \mathbb{Z}$ for each n) with the usual ordering, let $\sqrt{A_n}$ be the free group generated by, say, b_n, where $b_n^2 = a_n$. Let $G := \bigoplus_{n \in \mathbb{N}} A_n$, $\Gamma := \bigoplus_{n \in \mathbb{N}} \sqrt{A_n}$ with the antilexicographic ordering.

(e) A G-module is called *cyclic* if it has the form Gs for some element s. An arbitrary G-module is the disjoint union of its cyclic submodules. Conversely, if we are given a collection $\{Gs_i : i \in I\}$ of cyclic G-modules, one can form the (disjoint) union $X := \bigcup_{i \in I} G\, s_i$. We can extend the ordering on the subsets Gs_i to a linear ordering on X such that X becomes a G-module, for example, by putting a linear ordering on I and by declaring that $g\, s_i > g' s_j$ if either $g > g'$ or $g = g'$ and $i > j$.

(f) Let X be a G-module, let $H \subset G$ be a proper convex subgroup, let $\pi : G \to G/H$ be the natural map. For $s, t \in X$ define

$$s \sim t \quad \text{if} \quad s \in \operatorname{conv}_X(Ht)$$

where, for $Z \subset X$, $\operatorname{conv}_X(Z)$ the *X-convex hull of Z*, is the set

$$\{x \in X : \text{ there are } z_1, z_2 \in Z \text{ with } z_1 \le x \le z_2\}.$$

Then $\sim$ is an equivalence relation on X. Let $\rho : X \to X/{\sim}$ be the natural map. The requirement
'$v \le w$ if there exist $s, t \in X$ with $s \le t$, $\rho(s) = v$, $\rho(t) = w$'
defines a linear ordering on $X/{\sim}$ for which ρ is increasing and the formula

$$\pi(g)\rho(s) = \rho(gs) \qquad (g \in G, s \in X)$$

defines a multiplication $G/H \times X/{\sim} \to X/{\sim}$ making $X/{\sim}$ into a G/H-module. The proof consists of straightforward verification.

The following observation will be needed in 2.1.9.

THEOREM 1.5.6 *Let G be a linearly ordered group, let X be a G-module. Then there exists a G-module map $X \to G^\#$.*

Proof. Choose any $s_0 \in X$ and set

$$\phi(s) = \inf_{G^\#}\{g \in G : gs_0 \ge s\} \qquad (s \in X).$$

(The definition makes sense as Gs_0 is coinitial and cofinal in X so $\{g \in G : gs_0 \ge s\}$ is bounded below and non-empty). Obviously ϕ is increasing. By 1.5.3(i) we have for $g \in G$ that $g^{-1}\phi(gs) = g^{-1}\inf\{h \in G : hs_0 \ge gs\} = g^{-1}\inf\{h \in G : g^{-1}hs_0 \ge s\} = \inf\{g^{-1}h \in G : g^{-1}hs_0 \ge s\} = \phi(s)$.

REMARK The formula $\phi(s) = \sup_{G^\#}\{g \in G : gs_0 \le s\}$ would also have proved our theorem, likewise would $\phi(s) = \inf_{G^\#}\{g \in G : gs_0 > s\}$ and $\phi(s) = \sup_{G^\#}\{g \in G : gs_0 < s\}$.

1.6 The type condition

Let G be a linearly ordered group, let X be a G-module. The *algebraic type* of an element $s \in X$ is the set Gs or, equivalently, the element $\pi(s)$, where $\pi : X \to X/\sim$ is the canonical surjection and the equivalence relation $\sim$ is defined by $x \sim y$ if $x \in Gy$. Now choose $s_0 \in X$. (We may view s_0 as some sort of unit. If $G \subset X$ it is natural to put $s_0 := 1$.) The following constructions depend on the choice of s_0. For each $s \in X$ the set Gs is cofinal and coinitial so there are elements in Gs that are smaller than s_0 but also ones that are greater than s_0 and hence the definitions

$$\tau_\ell(s) = \sup_{X^\#}\{x \in Gs : x \leq s_0\}$$
$$\tau_u(s) = \inf_{X^\#}\{x \in Gs : x \geq s_0\}$$

make sense. It follows directly that $\tau_\ell(s) \leq \tau_u(s)$ and that $\tau_\ell(s)$ and $\tau_u(s)$ depend only on the algebraic type of s. It may happen that $\tau_\ell(s) < \tau_u(s)$. (In fact, let H be a proper convex subgroup of G, $H \neq \{1\}$, $X := G^\#$, $s_0 := 1$, $s := \sup_X H$, $t := \inf_X H$. Then $\tau_\ell(s) = t$, $\tau_u(s) = s$.) If $s_0 \in Gs$ then $\tau_\ell(s) = \tau_u(s) = s_0$.

DEFINITION 1.6.1 Let G, X, s_0 be as above. The *topological type* of an element $s \in X$ is the set $\tau(s) := \{h \in G : \tau_\ell(s) \leq hs_0 \leq \tau_u(s)\}$.

The following theorem shows that this definition ties in with the one given in [3], see Example 1.6.3. For the definition of conv_Γ see 1.5.5 (f).

THEOREM 1.6.2 *The topological type $\tau(s)$ of an element s of a G-module X is a proper convex subgroup of G. If $s_0 \in Gs$ then $\tau(s) = \{h \in G : hs_0 = s_0\}$, if $s_0 \notin Gs$ then $\tau(s)$ is the largest among the convex subgroups H of G for which* $\text{conv}_X(Hs_0) \cap Gs = \varnothing$.

Proof. To show the first statement we may suppose $s_0 \notin Gs$. The convexity is clear, as is properness, so it remains to prove that $\tau(s)$ is a group. Clearly $1 \in \tau(s)$. Now let $h_1, h_2 \in \tau(s)$. Let $g \in G$ be such that $gs \geq s_0$. From $h_2 \in \tau(s)$ it follows that $gs \geq h_2 s_0$ i.e. $h_2^{-1}gs \geq s_0$. This, combined with $h_1 \in \tau(s)$ yields $h_2^{-1}gs \geq h_1 s_0$ i.e. $gs \geq h_1 h_2 s_0$. This result holds for all $g \in G$ for which $gs \geq s_0$ i.e. $h_1 h_2 s_0 \leq \tau_u(s)$. Similarly one proves $h_1 h_2 s_0 \geq \tau_\ell(s)$ and it follows that $h_1 h_2 \in \tau(s)$. Now let $h \in \tau(s)$. To prove $h^{-1}s_0 \leq \tau_u(s)$, let again $g \in G$ be such that $gs \geq s_0$. If $h^{-1}s_0 > gs$ then $s_0 > hgs$ and, since $h \in \tau(s)$, $hs_0 \geq hgs$ or $s_0 \geq gs$ i.e. $s_0 > gs$ (since $s_0 \notin Gs$), contradiction. Hence, $h^{-1}s_0 \leq gs$ for all $g \in G$ with $gs \geq s_0$ i.e. $h^{-1}s_0 \leq \tau_u(s)$. Similarly one proves that $h^{-1}s_0 \geq \tau_\ell(s)$ and we are done. To prove the second statement we may assume $s_0 \notin Gs$. Let $H := \tau(s)$. Clearly $Hs_0 \cap Gs = \varnothing$, so let $t \in \text{conv}_X(Hs_0)\backslash Hs_0$. There exist $h_1, h_2 \in H$ with $h_1 s_0 < t < h_2 s_0$, so $\tau_\ell(s) < t < \tau_u(s)$ implying $t \notin Gs$ and we have proved $\text{conv}_X(Hs_0) \cap Gs = \varnothing$. Conversely, let H be a convex subgroup of G such that $\text{conv}_X(Hs_0) \cap Gs = \varnothing$; we must prove $H \subset \tau(s)$. Now $\text{conv}_X(Hs_0)$ contains s_0 and does not meet Gs. Hence by convexity it is contained in $\{t \in X : \tau_\ell(s) \leq t \leq \tau_u(s)\}$, implying $H \subset \tau(s)$.

EXAMPLE 1.6.3 Like in 1.5.5(d), let $G := \bigoplus_{n\in\mathbb{N}} A_n$, $\Gamma := \bigoplus_{n\in\mathbb{N}} \sqrt{A_n}$. Choose $X := \Gamma$, $s_0 := 1$. The definition of the topological type of an element $s \in \Gamma\backslash G$, given in [3], Def. 31 is the largest convex subgroup of Γ that does not meet Gs. According to 1.6.2, however, the topological type $\tau(s)$ is the largest convex subgroup H of G for which $\mathrm{conv}_\Gamma H$ does not meet Gs. The difference between these definitions is quite immaterial as there exists a 1–1 correspondence between the convex subgroups H of G and the convex subgroups S of Γ given by

$$\begin{array}{ccc} H & \longmapsto & \mathrm{conv}_\Gamma H \\ S\bigcap G & \longleftarrow & S. \end{array}$$

The verification is immediate.

Now we shall define the type condition for G-modules, thereby extending the definition given in [3], Def. 21, and prove a connection with the notion of type of 1.6.1. We will need all this in Section 4.

DEFINITION 1.6.4 Let G be a linearly ordered group, let X be a G-module and let $s_1, s_2, \ldots$ be a sequence in X.

(i) We say that $s_1, s_2, \ldots$ satisfies the *type condition* if, for any sequence $g_1, g_2, \ldots$ in G, boundedness above of $\{g_1 s_1, g_2 s_2, \ldots\}$ implies $\lim_n g_n s_n = 0$.

(ii) Let s_0, τ be as in 1.6.1. We say that $\lim_n \tau(s_n) = \infty$ if for each proper convex subgroup H of G we have $\tau(s_n) \supsetneq H$ for large n.

LEMMA 1.6.5 *Let G, X, s_0, τ be as above. Let $s_1, s_2, \ldots$ be a sequence in X satisfying the type condition. Then*

(i) *each subsequence of $s_1, s_2, \ldots$ satisfies the type condition;*

(ii) *if $g_1, g_2, \ldots \in G$ are such that $\{g_1 s_1, g_2 s_2, \ldots\}$ is bounded below then* $\lim_n g_n s_n = \infty$.

Proof. (i) Let $s_{n_1}, s_{n_2}, \ldots$ be a subsequence of $s_1, s_2, \ldots$. Let $g_{n_1}, g_{n_2}, \ldots \in G$ be such that $\{g_{n_1} s_{n_1}, g_{n_2} s_{n_2}, \ldots\}$ is bounded above, say by $t \in X$. For $j \in \mathbb{N}$, $j \notin \{n_1, n_2, \ldots\}$ we can choose, by coinitiality of $G s_j$, a $g_j \in G$ such that $g_j s_j < t$. Then $\{g_1 s_1, g_2 s_2, \ldots\}$ is bounded above and by assumption $\lim_n g_n s_n = 0$, so certainly $\lim_j g_{n_j} s_{n_j} = 0$. (ii) Let $\varepsilon \in X$ be such that $g_n s_n \geq \varepsilon$ for each n. Suppose not $\lim_n g_n s_n = \infty$. Then there are $n_1 < n_2 < \cdots$ in $\mathbb{N}$ such that $\{g_{n_1} s_{n_1}, g_{n_2} s_{n_2}, \ldots\}$ is bounded above. But then, by (i), $\lim_j g_{n_j} s_{n_j} = 0$, conflicting $g_n s_n \geq \varepsilon$ for each n.

THEOREM 1.6.6 *Let G be a linearly ordered group, let X be a G-module, let $s_0 \in X$ and let $\tau(s)$ be the corresponding topological type of X, defined in* 1.6.1. *Then, for a sequence $s_1, s_2, \ldots$ in X the following are equivalent.*

(α) *$s_1, s_2, \ldots$ satisfies the type condition.*

(β) $\lim_n \tau(s_n) = \infty$.

Proof. (α) $\Rightarrow$ (β). We first show that (α) implies that G does not have a maximal proper convex subgroup. Suppose it does, say $H \subset G$ is a maximal proper convex subgroup. Let $\varepsilon \in G\backslash H$, $\varepsilon < 1$. Then $1, \varepsilon, \varepsilon^2, \ldots$ is decreasing. If we had a $\delta \in G$,

$\delta \le \varepsilon^n$ for all $n \in \mathbb{N}$ then $H_1 := \{g \in G : \delta \le g^n \le \delta^{-1}$ for all $n \in \mathbb{N}\}$ is a convex subgroup containing H (since $\varepsilon < h < \varepsilon^{-1}$ for all $h \in H$) and ε, but not δ conflicting the maximality of H. Thus, $\lim_n \varepsilon^n = 0$ and $\{\varepsilon^n s_0 : n \in \mathbb{Z}\}$ is cofinal and coinitial. So, for each $m \in \mathbb{N}$ there is an $n_m \in \mathbb{Z}$ such that $\varepsilon^{n_m+1} s_0 \le s_m \le \varepsilon^{n_m} s_0$ i.e. $\varepsilon\, s_0 \le \varepsilon^{-n_m} s_m \le s_0$. By ($\alpha$) we would have $\lim_n \varepsilon^{-n_m} s_m = 0$, a contradiction. Now we come to the proof of (α) $\Rightarrow$ (β) proper. Suppose not $\lim_n \tau(s_n) = \infty$. Then there is a proper convex subgroup H of G and a subsequence $t_1, t_2, \ldots$ of $s_1, s_2, \ldots$ such that $\tau(t_n) \subset H$ for all n. By the first result of this proof there is a proper convex subgroup $H' \supset H$, $H' \neq H$. By 1.6.2 the intersection $\operatorname{conv}_X(H' s_0) \cap G\, t_n$ is non-empty for each n, so there exist $g_1, g_2, \ldots \in G$ with $g_n t_n \in \operatorname{conv}_X(H' s_0)$. Now H' is bounded above and below hence so are $\operatorname{conv}_X(H' s_0)$ and $\{g_1 t_1, g_2 t_2, \ldots\}$. By 1.6.5 (i), $\lim_n g_n t_n = 0$, a contradiction. (β) $\Rightarrow$ (α). Like in the previous part, we first observe that (β) implies that G does not have a maximal proper convex subgroup (this follows directly from Definition 1.6.4(ii)). Put $H_0 := \{g \in G : g\, s_0 = s_0\}$. By ($\beta$), $\tau(s_n)$ properly contains H_0 for large n, so without loss, to prove (α), we may assume that $\tau(s_n) \neq H_0$, for each n, implying $s_n \notin G\, s_0$ for each n (see 1.6.2). To prove (α), let $g_1, g_2, \ldots \in G$ be such that $\{g_n s_n : n \in \mathbb{N}\}$ is bounded above, say, by $t \in X$. If not $\lim_n g_n s_n = 0$ we could find a $u \in X$ and $n_1 < n_2 < \cdots$ in $\mathbb{N}$ such that, with $h_m := g_{n_m}$, $t_m := s_{n_m}$ we had $u \le h_n t_n \le t$ for all n. By cofinality and coinitiality of $G\, s_0$ there exists a $v \in G$ such that

$$v^{-1} s_0 \le h_n t_n \le v\, s_0 \tag{$*$}$$

for all n. Now let H be the smallest convex subgroup of G containing v. Then $H \neq G$ (If $H = G$, $V := \{h \in G : v^{-1} \le h^n \le v$ for each $n \in \mathbb{N}\}$ is a convex subgroup not containing v. If H_1 is a proper convex subgroup of H then it cannot contain v so $v^{-1} \le h \le v$ for all $h \in H_1$, so $H_1 \subset V$ and V is maximal, a contradiction.) By (β), $\tau(t_n) \supsetneq H$ for large n, i.e. $G\, t_n$ does not meet $\operatorname{conv}_X(H s_0)$ for large n. But ($*$) yields $h_n t_n \in \operatorname{conv}_X(H\, s_0)$ for all n, a contradiction.

REMARK We see that, although τ depends on the choice of s_0 (see the beginning of 1.6), property (β) does not.

THEOREM 1.6.7 *Let Γ be a linearly ordered abelian group containing G as a cofinal subgroup. Suppose in the G-module $\Gamma^\#$ the sequence $s_1, s_2, \ldots$ satisfies the type condition. Then so does $\omega(s_1), \omega(s_2) \ldots$, where $\omega : \Gamma^\# \to \Gamma^\#$ is the antipode (see Proposition* 1.3.1).

Proof. Let $g_1, g_2, \ldots \in G$ and $t \in \Gamma^\#$ be such that $g_n \omega(s_n) \le t$ for all n. Then $\omega(t) \le \omega(g_n \omega(s_n)) = g_n^{-1} \omega^2(s_n) = g_n^{-1} s_n$ for all n. By Lemma 1.6.5(ii), $\lim_n g_n^{-1} s_n = \infty$. By 1.3.1(v), $\lim_n g_n \omega(s_n) = \lim_n \omega(g_n^{-1} s_n) = 0$.

EXAMPLE 1.6.8 (Continuation of 1.3.2) **The topological type of an element $h \in \sqrt{G}$** Given $G = \bigoplus_{\iota \in \mathbb{N}} G_\iota$, as in 1.3.2, we describe $\sqrt{G}$ as the direct sum $\sqrt{G} = \bigoplus_{\iota \in \mathbb{N}} K_\iota$, where the cyclic group K_ι is generated by an element k_ι such that $k_\iota^2 = g_\iota$. We order $\sqrt{G}$ antilexicographically, the order of K_ι being the natural one. With the componentwise product $\sqrt{G}$ is a group, and identifying $g = (g_\iota^{n_\iota})_{\iota \in \mathbb{N}}$ with $(k_\iota^{2n_\iota})_{\iota \in \mathbb{N}}$ we can consider G as a subgroup of $\sqrt{G}$. Therefore $\sqrt{G}$ is a G-module. Taking

$s_0 := 1$ in the definition of topological type $\tau(k)$ of an element $k \in \sqrt{G}$, we have by Theorem 1.6.2 that if $1 \in Gk$, then $\tau(k) = H_0 = \{1\}$, but if $1 \notin Gk$ then $\tau(k)$ is the largest among the convex subgroups H_n of G for which $\operatorname{conv}_{\sqrt{G}}(H_n) \cap Gk = \varnothing$.

LEMMA 1.6.9 *Let $k \in \sqrt{G}$, $k = (k_i^{t_i})_{i \in \mathbb{N}}$.*
 i) *If $t_i \in 2\mathbb{Z}$ for all i, then $\tau(k) = H_0$.*
 ii) *If that is not the case, let $j = \max\{i \in \mathbb{N} : t_i \notin 2\mathbb{Z}\}$. Then $\tau(k) = H_{j-1}$.*

Proof. i) If for all i there exist $n_i \in \mathbb{Z}$ such that $t_i = 2n_i$, then $k \in \sqrt{G}$. Therefore $1 \in Gk$. ii) Since $H_n = \{(g_i^{n_i})_{i \in \mathbb{N}} : n_i = 0 \text{ if } i > n\}$, any element of $\operatorname{conv}_{\sqrt{G}}(H_n) = \{x \in \sqrt{G} : \text{ there are } z, w \in H_n \text{ with } z \le x \le w\}$ must be of the form $(k_i^{s_i})_{i \in \mathbb{N}}$ with $s_i = 0$ for $i > n$. Now let $g = (k_i^{2n_i})_{i \in \mathbb{N}} \in G$, then for $gk = (k_i^{2n_i + t_i})_{i \in \mathbb{N}}$ we have that $2n_j + t_j \notin 2\mathbb{Z}$, and gk does not belong to $\operatorname{conv}_{\sqrt{G}}(H_{j-1})$. Hence $\operatorname{conv}_{\sqrt{G}}(H_{j-1}) \cap Gk = \varnothing$. But there exists $g \in G$ such that $x = (k_1^{t_1}, \ldots, k_{j-1}^{t_j}, k_j^{t_j}, 1, \ldots) = gk$; since $1 \le x \le (1, \ldots, 1, k_j^2, 1, \ldots)$ we have $\operatorname{conv}_{\sqrt{G}}(H_n) \cap Gk \neq \varnothing$ for all $n \ge j$.

2 NORMED SPACES

In this Section we establish some basic theory of normed spaces; it is mainly the material we need for the main subject of this paper to be treated in Sections Three and Four. Many notations, statements and proofs will run similarly to the rank 1 case. However there are a few sharp deviations (2.1.5, 2.1.8, 2.4.10, 2.4.11, 2.4.18) which will of course get special attention.

Throughout K will be a valued field with a surjective valuation $|\ \ | : K \to G \cup \{0_G\}$, where G is a linearly ordered group and 0_G is a zero element adjoined to G having the properties $0_G < g$, $0_G \cdot g = 0_G \cdot 0_G = 0_G$ for all $g \in G$. More generally, to each G-module X we adjoin a zero element 0_X for which $0_X < x$, $0_G \cdot x = 0_G \cdot 0_X = 0_X$ for each $x \in X$. However, from now on we will omit the subscripts and write 0 for the zero element of any G-module.

2.1 Seminorms

DEFINITION 2.1.1 Let E be a K-vector space, let X be a G-module. An X-*seminorm* on E is a map $p : E \to X \cup \{0\}$ such that for all $x, y \in E$, $\lambda \in K$

(i) $p(0) = 0$

(ii) $p(\lambda x) = |\lambda| p(x)$

(iii) $p(x + y) \le \max(p(x), p(y))$. If, in addition, $p(x) = 0$ implies $x = 0$, then p is called an X-*norm*; in that case we often prefer the notation $\|x\|$ rather than $p(x)$. When there is no danger of confusion we often omit the prefix "X-" and just write "seminorm" ("norm").

REMARK In contrast to the requirements for valuations (see 1.4) we are not asking seminorms to be surjective. If p is an X-seminorm and Y is a G-module containing X, then p is, in a natural way, also a Y-seminorm. In particular this holds if $Y = X^{\#}$. It will turn out that at many instances it is useful to assume X to be complete; in general this will not restrict the problem we are dealing with.

Let X, Y be G-modules, let p be an X-seminorm, q a Y-seminorm on a K-vector space E. We say that *p is weaker than q* (or *q is stronger than p*) if, for each net $\iota \mapsto x_\iota$ in E, $\lim_\iota q(x_\iota) = 0$ implies $\lim_\iota p(x_\iota) = 0$, or, equivalently, if for each $\varepsilon \in X$ there exists a $\delta \in Y$ such that $x \in E$, $q(x) < \delta$ implies $p(x) < \varepsilon$. Otherwise stated, p is weaker than q if the topology induced by p in the usual way, is weaker than the topology induced by q. To express this notion in yet another way we introduce the concept of boundedness.

DEFINITION 2.1.2 Let p be an X-seminorm on a K-vector space E. A subset S of E is called **p-bounded** if $\{p(x) : x \in S\}$ is bounded above in $X \cup \{0\}$.

PROPOSITION 2.1.3 *Let p and q be seminorms on a K-vector space E. Then p is weaker than q if and only if each q-bounded set in E is p-bounded.*

Proof. Let p be an X-seminorm, let q be a Y-seminorm, where X and Y are G-modules. Suppose p is weaker than q and let S be a q-bounded set, say $q(x) \leq t \in Y$ for all $x \in S$. Choose $\varepsilon \in X$. There is a $\delta \in Y$ such that $q(x) < \delta$ implies $p(x) < \varepsilon$. There is a $g \in G$ with $gt < \delta$. Choose $\lambda \in K$ for which $|\lambda| = g$, and let $x \in S$. Then $q(\lambda x) = |\lambda| q(x) \leq |\lambda| t = gt < \delta$, so $p(\lambda x) < \varepsilon$. Then $p(x) < g^{-1}\varepsilon$, so S is p-bounded. Conversely, let each q-bounded set be p-bounded. Let $\varepsilon \in X$, choose $t \in Y$. There is an $s \in X$ such that $q(x) \leq t$ implies $p(x) \leq s$. Let $\lambda \in K^\times$ be such that $|\lambda| s < \varepsilon$. Then we see that for all $x \in E$, $q(x) \leq |\lambda| t$ implies $p(x) < \varepsilon$ i.e. p is weaker than q.

In the theory over fields with rank 1 valuation (i.e. G is a subgroup of $(0, \infty)$, see 1.3), a seminorm p is weaker than a seminorm q if and only if $p \leq Cq$ for some real constant (that can be taken in G). In our theory such a statement does not make sense if p is an X-seminorm, q is a Y-seminorm and X is not a subset of Y. If $X \subset Y$ we may p also consider as a Y-seminorm, so we may assume $X = Y$. Thus, we define the following.

DEFINITION 2.1.4 Let p, q be X-seminorms on a K-vector space E, where X is a G-module. We say that p is *q-Lipschitz* if there is a $g \in G$ such that $p(x) \leq g\, q(x)$ for all $x \in E$.

Clearly, if p is q-Lipschitz then p is weaker than q. The converse does not hold, see [1], [9], 3.7. But if p is weaker than q we do have an increasing function ϕ for which $p(x) \leq \phi\big(q(x)\big)$ for all $x \in E$, as is shown in the next Proposition.

PROPOSITION 2.1.5 *Let $p \neq 0$ be an X-seminorm, let q be a Y-seminorm on a K-vector space E, where X, Y are complete G-modules. Suppose p is weaker than q. Then there exist increasing functions $\xi : Y \cup \{0\} \to X \cup \{0\}$ such that $p \leq \xi \circ q$. Among them there is a smallest one, ϕ, given by the formula*

$$\phi(t) = \sup\{p(x) : \ x \in E, \ q(x) \leq t\} \qquad (t \in Y \cup \{0\}). \qquad (*)$$

Moreover, ϕ is an extended G-module map $Y \cup \{0\} \to X \cup \{0\}$ (and therefore bicontinuous at 0, *see* 1.5.2(i)).

Proof. By 2.1.3 the set $\{p(x) : q(x) \leq t\}$ is bounded above for each $t \in Y \cup \{0\}$, so $(*)$ defines a map $Y \cup \{0\} \to X \cup \{0\}$, which is obviously increasing. For each $x \in E$ we have $\phi(q(x)) = \sup\{p(z) : z \in E,\ q(z) \leq q(x)\} \geq p(x)$, so $p \leq \phi \circ q$. If $\xi : Y \cup \{0\} \to X \cup \{0\}$ is increasing and $p \leq \xi \circ q$ then we have for $t \in Y \cup \{0\}$ that $\phi(t) = \sup\{p(x) : q(x) \leq t\} \leq \sup\{\xi(q(x)) : q(x) \leq t\} = \xi(t)$. It remains to be shown that ϕ is an extended G-module map. Let $t \in Y \cup \{0\}$, $g \in G$. Choose $\lambda \in K$ such that $|\lambda| = g$. Then $g^{-1}\phi(gt) = |\lambda|^{-1}\phi(|\lambda|t) = |\lambda|^{-1} \sup\{p(x) : q(x) \leq |\lambda|t\}$. Now by 1.5.3 (i) this equals $\sup\{|\lambda|^{-1}p(x) : |\lambda|^{-1}q(x) \leq t\} = \sup\{p(y) : q(y) \leq t\} = \phi(t)$. Hence $\phi(gt) = g\phi(t)$ for all $g \in G$, $t \in Y \cup \{0\}$. To finish the proof we show $t = 0 \iff \phi(t) = 0$ for all $t \in Y \cup \{0\}$. By the above, if $t = 0$ then from $g\phi(0) = \phi(g \cdot 0) = \phi(0)$ and cofinality of Gs if $s \neq 0$ we conclude $\phi(t) = \phi(0) = 0$. Conversely, if $t \neq 0$ then on $\{x \in E : q(x) \leq t\}$, p must be not identically 0 (otherwise, p is zero on the whole of E against the assumption), so, by $(*)$, $\phi(t) \neq 0$.

In the same vein one can prove the following. We leave the proof to the reader.

PROPOSITION 2.1.6 *Let E, X, Y, p, q be as in* 2.1.5. *Then there exist increasing functions $\xi : X \cup \{0\} \to Y \cup \{0\}$ for which $\xi \circ p \leq q$. Among them there is a largest one, ψ, given by the formula*

$$\psi(s) = \inf\{q(x) : \ x \in E,\ p(x) \geq s\} \qquad (s \in X \cup \{0\}).$$

Moreover, ψ is an extended G-module map $X \cup \{0\} \to Y \cup \{0\}$ (and therefore bicontinuous at 0).

The following definition will not come as a suprise.

DEFINITION 2.1.7

(i) Two seminorms p and q on a K-vector space are called *equivalent* if p is weaker than q and q is weaker than p.

(ii) Let X be a G-module. Two X-seminorms p and q on a K-vector space are called *Lipschitz-equivalent* if p is q-Lipschitz and q is p-Lipschitz.

REMARK If p is an X-seminorm on a K-vector space E and $\phi : X \cup \{0\} \to Y \cup \{0\}$ is an extended G-module map then $\phi \circ p$ is an Y-seminorm on E that is equivalent to p. This follows from 1.5.2(i).

The previous Propositions 2.1.5 and 2.1.6 yield the following corollary. Observe that in (i) below $\phi_1 \circ q$ and $\phi_2 \circ q$ are equivalent to q.

Corollary 2.1.8 *Let X, Y be complete G-modules, let $p \neq 0$ be an X-seminorm, q a Y-seminorm on a K-vector space E. Then we have the following.*

(i) *p and q are equivalent if and only if there exist extended G-module maps $\phi_1, \phi_2 : Y \cup \{0\} \to X \cup \{0\}$ such that*

$$\phi_1(q(x)) \leq p(x) \leq \phi_2(q(x)) \qquad (x \in E).$$

(ii) *Let $X = Y$. Then p and q are Lipschitz-equivalent if and only if there exist $g_1, g_2 \in G$ such that*

$$g_1 q(x) \leq p(x) \leq g_2 q(x) \qquad (x \in E).$$

The following theorem shows that if one is interested in locally convex topologies rather than (geo)metrical properties it suffices to consider seminorms with values in $G^\#$.

THEOREM 2.1.9 *Each seminorm is equivalent to a $G^\#$-seminorm.*

Proof. Let p be an X-seminorm for some G-module X. Extend the map ϕ of 1.5.6 by $\phi(0) := 0$ so as to obtain an extended G-module map $\phi : X \cup \{0\} \to G^\# \cup \{0\}$. Then $\phi \circ p$ is a $G^\#$-seminorm equivalent to p.

2.2 Normed spaces

A *normed space*, more precisely, an *X-normed space*, is a pair $(E, \| \quad \|)$ where E is a K-vector space and where $\| \quad \|$ is an X-norm for some G-module X. The map $(x, y) \mapsto \|x - y\|$ is a scale (see 1.2) on E, the induced topology is a Hausdorff vector topology i.e. addition and scalar multiplication are continuous (use 1.5.2(ii)). Often we will write E (rather than $(E, \| \quad \|)$).

Of course one can define easy generalizations of well-known spaces from rank 1 theory in order to obtain examples of X-normed spaces (e.g. see 2.4.15). Typically infinite rank examples will appear in Sections 3 and 4. At the present stage it might be useful to consider the following example that is non-trivial, also in rank 1 case.

EXAMPLE 2.2.1 (Compare also [15]) Let X be a complete G-module. Let X^- be a copy of X and define the G-module $X \cup X^-$ in the spirit of 1.5.5(c) by requiring $y < x^- < x$ for all $x, y \in X$, $y < x$, and $gx^- = (gx)^-$. Let S be a topological space, let E be an X-normed space. The space $BC(S \to E)$, consisting of all continuous functions $S \to E$ that are bounded (above) is an X-normed space under

$$f \mapsto \|f\|_\infty = \sup_{X \cup \{0\}} \{\|f(x)\| : x \in S\}.$$

But it is also an $X \cup X^-$-normed space with respect to

$$f \mapsto \|f\|'_\infty = \sup_{X \cup X^- \cup \{0\}} \{\|f(x)\| : x \in S\}$$

i.e.

$$\|f\|'_\infty = \begin{cases} \|f\|_\infty & \text{if } \max\{\|f(x)\| : x \in S\} \text{ exists} \\ \|f\|^-_\infty & \text{otherwise.} \end{cases}$$

Let X, Y be G-modules, let E be an X-normed space, F a Y-normed space. We consider two types of maps $E \to F$.

1. The collection $\mathcal{L}(E, F)$ of all continuous linear maps $T : E \to F$ is a K-vector space. The seminorm $x \mapsto \|Tx\|$ is weaker than $\| \ \|$, so by 2.1.3 for each $s \in X$ the set $\{\|Tx\| : \|x\| \leq s\}$ is bounded above in Y and the formula

$$\|T\|_s := \sup\{\|Tx\| : \|x\| \leq s\}$$

defines a $Y^\#$-norm on $\mathcal{L}(E,F)$. It is easily seen that, for $t \in X$, the norms $\| \ \|_s$ and $\| \ \|_t$ are Lipschitz-equivalent. The induced topology on $\mathcal{L}(E,F)$ is the topology of uniform convergence on bounded sets. We call the norms $\| \ \|_s$ the *uniform norms.* If $E = F$ as normed spaces we write $\mathcal{L}(E)$ rather than $\mathcal{L}(E,E)$. The *dual space* E' of E is $\mathcal{L}(E,K)$ where K is assumed to be normed by the valuation.

2. If $X = Y$, the collection $\mathrm{Lip}(E,F)$ of all *linear Lipschitz maps* $T : E \to F$ (i.e. there is a $g \in G$ such that $\|Tx\| \leq g\|x\|$ for all $x \in E$; in most literature such T are called bounded maps) forms a K-linear subspace of $\mathcal{L}(E,F)$. The formula

$$\|T\| = \inf\{g \in G : \|Tx\| \leq g\|x\| \text{ for all } x \in E\}$$

defines a $G^\#$-norm on $\mathrm{Lip}(E,F)$, called the *Lipschitz norm.* Clearly, for each $s \in X$, $\| \ \|_s$ is weaker than $\| \ \|$. See [1] for an example of an element of $\mathcal{L}(E,F)\backslash\mathrm{Lip}(E,F)$. The terms 'linear homeomorphism', 'isometrical isomorphism' between normed spaces will need no explanation.

We now look into the forming of quotients in some detail because we will need this precise information later on. Let E be an X-normed space, where X is some G-module. Let $D \subset E$ be a closed subspace, let $\pi : E \to E/D$ be the canonical map. Like in the classical case one proves that the formula

$$\begin{aligned} \|\pi(a)\| &= \inf\{\|x\| : x \in E,\ \pi(x) = \pi(a)\} \\ &= \inf\{\|a - d\| : d \in D\} \qquad (a \in E) \end{aligned}$$

defines an $X^\#$-norm on E/D, the so-called *quotient norm.* We have $\|\pi(a)\| \leq \|a\|$ for each $a \in E$, so π is Lipschitz. The norm topology on E/D is the quotient topology induced by π. If F is a second normed space and $T \in \mathcal{L}(E,F)$ then the map T_1 in the factorization

$$\begin{array}{ccc} E & \xrightarrow{T} & F \\ \pi \searrow & & \nearrow T_1 \\ & E/\mathrm{Ker}\,T & \end{array}$$

(where $E/\mathrm{Ker}\ T$ is equipped with the quotient norm and π is the canonical map) is in $\mathcal{L}(E/\mathrm{Ker}\ T, F)$. For each $s, t \in X^\#$ with $s < t$ we have $\|T\|_s \leq \|T_1\|_s \leq \|T\|_t$. If T is Lipschitz then so is T_1 and $\|T\| = \|T_1\|$.

F is called *a quotient of* E if there is a $T \in \mathcal{L}(E,F)$ such that the map T_1 in the above diagram is an isometrical isomorphism; such a T is called *quotient map.* Obviously, the canonical map $\pi : E \to E/D$ of above is a quotient map. A surjective $T \in \mathcal{L}(E,F)$ is a quotient map if and only if for each $y \in F$ we have $\|y\| = \inf\{\|x\| : Tx = y\}$. A quotient map $T \in \mathcal{L}(E,F)$ is called *strict quotient map* (and F is called a *strict quotient of* E) if for all $y \in F$ we have $\|y\| = \min\{\|x\| : Tx = y\}$.

In the following lemma we characterize the (strict) quotient maps.

LEMMA 2.2.2 *Let X be a G-module, let E, F be X-normed spaces, let $\pi : E \to F$ be a linear map. Then π is a quotient map if and only if, for each $s \in X$, $\pi(B_E(0, s^-)) = B_F(0, s^-)$ while π is a strict quotient map if and only if, for each $s \in X$, $\pi(B_E(0, s)) = B_F(0, s)$.*

Proof. Suppose π is a quotient map. Let $s \in X$. Obviously $\pi(B_E(0, s^-)) \subset B_F(0, s^-)$. Conversely, if $y \in F$, $\|y\| < s$ there is by definition an $x \in E$ with $\|x\| < s$ and $\pi(x) = y$. Now suppose $\pi(B_E(0, s^-)) = B_F(0, s^-)$ for each $s \in X$. Then clearly π is surjective. Let $y \in F$, $\|y\| = t \in X$. By Lemma 1.1.1 there are two cases.

1. $s_1 := \min\{s \in X : s > t\}$ exists. Then $\|y\| < s_1$ so there is an $x \in B_E(0, s_1^-)$ with $\pi(x) = y$. Then $\|x\| \le t$. If $\|x\|$ were $< t$ then $\|\pi(x)\| < t$, a contradiction. Hence $\|x\| = t$, so $\|y\| = \min\{\|z\| : \pi(z) = y\}$.
2. $\inf\{s \in X : s > t\} = t$. For each $s \in X$, $s > t$, y is in $B_F(0, s^-)$ so there is an $x \in B_E(0, s^-)$ with $\pi(x) = y$. If $\|x\|$ were $< t$ then $\|\pi(x)\| < t$, a contradiction. Hence $\|x\| \ge t$, so

$$\|y\| = \inf\{s : s > \|y\|\} = \inf\{\|x\| : \pi(x) = y\}.$$

Now suppose π is a strict quotient map. Let $s \in X$. Obviously, $\pi(B_E(0, s)) \subset B_F(0, s)$. Conversely, let $y \in B_F(0, s)$. There exists an $x \in E$ with $\pi(x) = y$ and $\|x\| = \|y\|$. Hence $B_F(0, s) \subset \pi(B_E(0, s))$, and we have equality. Conversely, let $\pi(B_E(0, s)) = B_F(0, s)$ for each $s \in X$. Then π is surjective. If $y \in F$, $\|y\| = s \in X$ then there is an $x \in B_E(0, s)$ with $\pi(x) = y$. If $\|x\|$ were $< s$ then $\pi(x) \in B_F(0, \|x\|)$, so $\|\pi(x)\| \le \|x\| < s$. Hence, $\|x\| = s$ and the Lemma is proved.

The following corollaries obtain.

COROLLARY 2.2.3 *Let X be a G-module, let E, F be X-normed spaces, let $\pi : E \to F$ be a quotient map. If B is an 'open' ball in F with radius $s \in X$ then $\pi(B_E(a, s^-)) = B$ for each $a \in \pi^{-1}(B)$. If moreover, π is a strict quotient map and B is a 'closed' ball in F with radius $s \in X$ then $\pi(B_E(a, s)) = B$ for each $a \in \pi^{-1}(B)$.*

COROLLARY 2.2.4 *Let X be a G-module, let E, F be X-normed spaces, let $\pi : E \to F$ be a quotient map. Then, if $B_1 \supset B_2 \supset \cdots$ are 'open' balls in F there are 'open' balls $C_1 \supset C_2 \supset \cdots$ in E such that $\pi(C_n) = B_n$ for each n. If, in addition, π is a strict quotient map then, for any nest $\mathcal{C}$ of balls in F there is a nest $\mathcal{C}'$ in E such that $B \mapsto \pi(B)$ is a bijection $\mathcal{C}' \to \mathcal{C}$.*

Proof (of 2.2.4). The first assertion can be proved by induction and 2.2.3. To prove the second one we use Zorn's Lemma. We may assume that $\mathcal{C}$ is a maximal nest (by adding all balls in F that contain some element of $\mathcal{C}$). Let V be the collection of all non empty sets $\mathcal{D}$ of balls in E with the properties: 1. If $B \in \mathcal{D}$ and B' is a ball, $B' \supset B$ then $B' \in \mathcal{D}$. 2. $\pi(B) \in \mathcal{C}$ for all $B \in \mathcal{D}$. Order V by declaring that $\mathcal{D}_1 \le \mathcal{D}_2$ if $\mathcal{D}_1 \subset \mathcal{D}_2$. V is not-empty: choose any ball B in E with $\pi(B) \in \mathcal{C}$, the collection of all balls B' in E with $B' \supset B$ belongs to V. Clearly each chain in V has an upper bound, so by Zorn's Lemma, V has a maximal element $\mathcal{D}'$. It suffices to prove that $\mathcal{D}'$ is a maximal nest. If $\bigcap \mathcal{D}' = \varnothing$ this is true, so suppose $\bigcap \mathcal{D}' \neq \varnothing$. If $\mathcal{D}'$ has a smallest element B_0 then, since $\mathcal{C}$ has no smallest element, there is a $C \in \mathcal{C}$ such that $C \subset \pi(B_0)$ strictly and there is a ball $B' \subset B_0$ with $\pi(B') = C$ so $\{B : B \text{ ball in } E, B \supset B'\}$ is in V and strictly larger than $\mathcal{D}'$, a contradiction. If $\mathcal{D}'$ has no smallest element then $B_0 := \bigcap \mathcal{D}'$ is a ball of the form $B_E(a, s)$ for some $s \in X$, B_0 not in $\mathcal{D}'$. But then $\bigcap_{B \in \mathcal{D}'} \pi(B)$ is a 'closed' ball of radius s and contains $\pi(B_0)$, hence $\pi(B_0) = \bigcap_{B \in \mathcal{D}'} \pi(B)$ and $\mathcal{D}' \cup B_0$ is in V and strictly larger than $\mathcal{D}'$, a contradiction.

COROLLARY 2.2.5 *Strict quotients of spherically complete spaces are spherically complete. If K satisfies the countability conditions of* 1.4.4 *then all quotients of spherically complete spaces are spherically complete.*

Proof. The first assertion follows from the second assertion of 2.2.4. Now let F be a quotient of an X-normed spherically complete space E, let $\mathcal{C}$ be a nest of 'open' balls in F. To prove $\bigcap \mathcal{C} \neq \varnothing$ we may assume that $\mathcal{C}$ has no smallest element. Let $r := \inf\{\text{diam } B : B \in \mathcal{C}\}$. By assumption there are $r_1 > r_2 > \cdots$ in $X^{\#}$ such that $\inf_n r_n = r$. For each n, choose a $B_n \in \mathcal{C}$ with diameter $\leq r_n$. Then $\bigcap_n B_n = \bigcap \mathcal{C}$. Now the result follows after applying the first assertion of 2.2.4 and using the spherical completeness of E.

2.3 Linear operators with finite rank. Banach spaces

In 2.3 we extend results that were already observed in [11] for special normed spaces.

LEMMA 2.3.1 *Let E, F be onedimensional spaces, both equipped with an X-norm for some G-module X. Then every linear map $f : E \to F$ is Lipschitz.*

Proof. Such a map has the form $f : \lambda a \mapsto \lambda b$ $(\lambda \in K)$ for some non-zero $a \in E$ and some $b \in F$. Let $g \in G$ be such that $\|b\| \leq g\|a\|$. Then for each $\lambda \in K$ we have $\|f(\lambda a)\| = \|\lambda b\| \leq |\lambda| g \|a\| = g\|\lambda a\|$.

LEMMA 2.3.2 *Let E, F be normed spaces, let* $\dim F = 1$, *let $f : E \to F$ be linear. Then f is continuous if and only if* $\operatorname{Ker} f$ *is closed. If E, F are both X-normed for some (complete) G-module X we even have that f is Lipschitz if and only if* $\operatorname{Ker} f$ *is closed.*

Proof. By 2.1.9 we can choose on E and F equivalent $G^{\#}$-norms, so it suffices to prove the second statement. Suppose $\operatorname{Ker} f$ is closed. From the text following 2.2.1 it follows that in the canonical factorization

$$\begin{array}{ccc} E & \xrightarrow{f} & F \\ \pi \searrow & & \nearrow f_1 \\ & E/\operatorname{Ker} f & \end{array}$$

f_1 is continuous, hence Lipschitz by 2.3.1. Then so is $f_1 \circ \pi = f$.

DEFINITION 2.3.3 Let K be complete. A normed space over K is called *complete* (or a *Banach space*) if each Cauchy net converges.

We now prove the non-surprising theorem on finite-dimensional spaces. It extends 1.3 of [9].

THEOREM 2.3.4 *Let K be complete, let E be a finite-dimensional space over K. Then all norms are equivalent, E is a Banach space with respect to each norm. For a G-module X all X-norms on E are Lipschitz equivalent.*

Proof. We prove by induction on $\dim E$ that all X-norms are Lipschitz equivalent (then we are done since, for each n, K^n is complete under the norm $(\xi_1, \xi_2, \ldots, \xi_n) \mapsto \max_i |\xi_i|$, and by 2.1.9). If $\dim E = 1$ we have 2.3.1. Suppose the statement holds for spaces with dimension $\leq n-1$ and let E be an n-dimensional space, let $\| \quad \|$ be an X-norm on E. Let $e_1, \ldots, e_n$ be a base of E, we prove $\| \quad \|$ to be Lipschitz equivalent to $\| \quad \|_\infty : \xi_1 e_1 + \cdots + \xi_n e_n \mapsto \max_i |\xi_i| \, \|e_i\|$. Obviously $\| \quad \| \leq \| \quad \|_\infty$. To prove that $\| \quad \|_\infty$ is $\| \quad \|$-Lipschitz, let, for each $j \in \{1, \ldots, n\}$, $f_j : E \to K e_j$ be the map $\sum_{i=1}^n \xi_i e_i \mapsto \xi_j e_j$. Then $\dim \operatorname{Ker} f_j = n - 1$; from the induction hypothesis it follows that $\operatorname{Ker} f_j$ is complete, hence closed in $(E, \| \quad \|)$ and by Lemma 2.3.2 the f_j are Lipschitz. So there is a $g \in G$ such that $\|f_j(x)\| \leq g\|x\|$ $(x \in E, j \in \{1, \ldots, n\})$ and then for $x \in E$ we have $\|x\|_\infty = \|\sum_{i=1}^n f_i(x)\| \leq \max_i |f_i(x)| \leq g\|x\|$ and we are done.

COROLLARY 2.3.5 (Continuous linear operators of finite rank are Lipschitz). *Let K be complete, let X be a (complete) G-module. Then every continuous linear map of an X-normed space E into a finite-dimensional X-normed space F is Lipschitz.*

Proof. Let $T : E \to F$ be such a map; we may assume that T is surjective. In the canonical decomposition

$$\begin{array}{ccc} E & \xrightarrow{T} & F \\ & \pi \searrow \quad \nearrow T_1 & \\ & E/\operatorname{Ker} T & \end{array}$$

we have that T_1 is Lipschitz (by 2.3.4 the norm $x \mapsto \|T_1 x\|$ on $E/\operatorname{Ker} T$ is Lipschitz equivalent to the quotient norm), hence so is $T_1 \circ \pi = T$.

PROPOSITION 2.3.6 *Let E, F be normed spaces over K. If F is a Banach space then so is $\mathcal{L}(E, F)$. If, in addition, E, F are both X-normed for some (complete) G-module X then* $\operatorname{Lip}(E, F)$ *is a Banach space.*

Proof. Let $i \mapsto T_i$ $(i \in I)$ be a Cauchy net in $\mathcal{L}(E, F)$. Let $s \in X$. From $\lim_{i,j} \|T_i - T_j\|_s = 0$ and completeness of F it follows that $T := \lim_i T_i$ exists pointwise. From $\|Tx - T_i x\| \leq \max\{\|Tx - T_j x\|, \|T_j - T_i\|_s\}$ for all $x \in B_E(0, s)$ and $i, j \in I$ it follows easily that $T \in \mathcal{L}(E, F)$ and $\lim_i \|T - T_i\|_s = 0$. Now let $i \mapsto T_i$ be Cauchy in $\operatorname{Lip}(E, F)$. By the first part there is a $T \in \mathcal{L}(E, F)$ such that $\lim_i T_i = T$ in the topology of $\mathcal{L}(E, F)$. Now let $\varepsilon \in G$. There is an i_0 such that for

$i, j \geq i_0$

$$\|(T_i - T_j)x\| \leq \varepsilon \|x\| \qquad (x \in E)$$

which after taking $\lim_i$ becomes

$$\|(T - T_j)x\| \leq \varepsilon \|x\|$$

implying $\lim_j \|T - T_j\| = 0$.

COROLLARY 2.3.7 *Let K be complete, let E, F be normed spaces over K. Then, if F is finite-dimensional, $\mathcal{L}(E, F)$, in particular E', is a Banach space.*

2.4 The Hahn Banach Theorem. Orthogonality

In 2.4 we extend results in rank 1 theory [14] to X-normed spaces. This section contains no surprises, apart from the fact that the proof of 2.4.12 is somewhat more complicated than its rank 1 counterpart [14], Th. 5.4, and apart from Example 2.4.18 for non-metrizable K.

THEOREM 2.4.1 (Hahn Banach). *Let E be an X-normed space, let F be a Y-normed space over K where X, Y are G-modules. Suppose F is spherically complete with respect to the induced scale, let D be a subspace of E and let $T \in \mathcal{L}(D, F)$ be such that $\|Tx\| \leq \phi(\|x\|)$ $(x \in D)$ where $\phi : X \cup \{0\} \to Y \cup \{0\}$ is an extended G-module map. Then T can be extended to a $\widetilde{T} \in \mathcal{L}(E, F)$ for which $\|\widetilde{T}x\| \leq \phi(\|x\|)$ $(x \in E)$.*

Proof. (Basically classical) By a simple application of Zorn's Lemma we may assume $E = D + Ka$ where $a \in E \backslash D$. It suffices to find $\widetilde{T}a \in F$ such that

$$\|\lambda \widetilde{T}a - Td\| \leq \phi(\|\lambda a - d\|) \qquad (\lambda \in K^\times, d \in D).$$

Now for each $d \in D$, $\lambda \in K$ we have $\phi(\|\lambda a - \lambda d\|) = \phi(|\lambda|\,\|a - d\|) = |\lambda|\phi(\|a - d\|)$, so it is enough to find $\widetilde{T}a$ for which

$$\|\widetilde{T}a - Td\| \leq \phi(\|a - d\|) \qquad (d \in D),$$

in other words, we have to show that

$$\bigcap\nolimits_{d \in D} B_F(Td, \phi(\|a - d\|)) \neq \varnothing. \qquad (*)$$

Let $d_1, d_2 \in D$. Then, by increasingness of ϕ, $\|Td_1 - Td_2\| \leq \phi(\|d_1 - d_2\|) \leq \phi(\max \|d_1 - a\|, \|a - d_2\|) = \max(\phi(\|a - d_1\|), \phi(\|a - d_2\|))$, so $B_F(Td_1, \phi(\|a - d_1\|)) \cap B_F(Td_2, \phi(\|a - d_2\|)) \neq \varnothing$ showing that the balls in $(*)$ form a nest. By spherical completeness of F the intersection $(*)$ is not empty which is finishing the proof.

DEFINITION 2.4.2 Let $(E, \|\ \ \|)$ be a normed space over K. Two subspaces D_1 and D_2 of E are called *(norm) orthogonal* (notation $D_1 \perp D_2$) if for each $d_1 \in D_1$, $d_2 \in D_2$

$$\|d_1 + d_2\| = \max(\|d_1\|, \|d_2\|).$$

A subspace D is called *(norm)orthocomplemented in E* if there exists a subspace $S \perp D$ (called **an orthocomplement of** D) such that $D + S = E$. An operator $P \in \mathcal{L}(E)$ is called a *projection* if $P^2 = P$. If, in addition, $\|Px\| \le \|x\|$ $(x \in E)$, P is called a *(norm)orthogonal projection.*

LEMMA 2.4.3 *For an orthogonal projection P, Im P and Ker P are orthocomplements of each other. A subspace D of E is orthocomplemented in E if and only if there is an orthogonal projection $P \in \mathcal{L}(E)$ with $PE = D$.*

Proof. Left to the reader.

THEOREM 2.4.4 *A spherically complete subspace of a normed space is orthocomplemented.*

Proof. Let D be a spherically complete subspace of a normed space E. By 2.4.1 the identity $D \to D$ can be extended to a map $P \in \mathcal{L}(E, D)$ with $\|Px\| \le \|x\|$ $(x \in E)$. Then P, viewed as an element of $\mathcal{L}(E)$ is an orthogonal projection, $PE = D$. Now apply 2.4.3.

LEMMA 2.4.5 *If K is spherically complete then so is every onedimensional normed space over K.*

Proof. Let $E = Ke$ be a onedimensional normed space, let $\mathcal{C}$ be a nest of balls in Ke. For each $B \in \mathcal{C}$ the set

$$C_B := \{\lambda \in K : \lambda e \in B\}$$

is convex, $\neq \emptyset$. By spherical completeness of K, 1.4.3 and its preamble, $\bigcap C_B \neq \emptyset$ so $\bigcap \mathcal{C} \neq \emptyset$.

COROLLARY 2.4.6 *Let K be spherically complete. Then any onedimensional subspace of a normed space over K is orthocomplemented.*

Proof. Combine 2.4.4 and 2.4.5.

DEFINITION 2.4.7 A collection $\{e_i : i \in I\}$ of vectors of a normed space is called **(norm)orthogonal** if for each finite set $J \subset I$ and $\lambda_j \in K$

$$\|\sum\nolimits_{j \in J} \lambda_j e_j\| = \max_j \|\lambda_j e_j\|.$$

Clearly, $\{e_i : i \in I\}$ is orthogonal if and only if $[\, e_i \,] \perp [\, e_j : j \neq i \,]$ for each $i \in I$.

LEMMA 2.4.8 (Perturbation Lemma) *Let $\{e_i : i \in I\}$ be an orthogonal set in a normed space E, let $\{f_i : i \in I\} \subset E$ be such that $\|f_i - e_i\| < \|e_i\|$ for each i. Then $\{f_i : i \in I\}$ is orthogonal.*

Proof. (Compare [14], 5.B) Let $J \subset I$ be finite, let $\lambda_j \in K$ for each $j \in J$. To prove $\|\sum_{j\in J} \lambda_j f_j\| = \max_{j\in J} \|\lambda_j f_j\|$ we may assume $\lambda_j \neq 0$ for all $j \in J$. From $\|f_i - e_i\| < \|e_i\|$ it follows that $\|f_i\| = \|e_i\|$ for each $i \in I$. For each $j \in J$ we have $\|\lambda_j(f_j - e_j)\| < \|\lambda_j e_j\|$ so that $\|\sum_{j\in J} \lambda_j(f_j - e_j)\| < \max_{j\in J} \|\lambda_j e_j\| = \|\sum_{j\in J} \lambda_j e_j\|$, so that $\|\sum_{j\in J} \lambda_j f_j\| = \max(\|\sum_{j\in J} \lambda_j(f_j - e_j)\|, \|\sum_{j\in J} \lambda_j e_j\|) = \|\sum_{j\in J} \lambda_j e_j\| = \max_{j\in J} \|\lambda_j f_j\|$.

We will show in 2.4.14 that all maximal orthogonal sets in a normed space have the same cardinality. Because the situation is somewhat more complicated than in the rank 1 case (compare [14], 5.4) we shall develop some machinery.

Let H be a convex subgroup of G. Consider

$$D_H := \{\lambda \in K : \ |\lambda| \leq \sup H\}$$
$$D_H^- := \{\lambda \in K : \ |\lambda| < \inf H\}$$

(observe that, unless $H = \{1\}$, $\sup H$ and $\inf H$ belong to $G^{\#}\backslash G$). We have

$$D_H = \{\lambda \in K : \ |\lambda| \leq 1 \text{ or } |\lambda| \in H\}$$
$$D_H^- = \{\lambda \in K : \ |\lambda| < 1 \text{ and } |\lambda| \notin H\}$$
$$|D_H| = |D_H^-| \cup H, \ |D_H^-| \cap H = \varnothing.$$

D_H is a subring of K, D_H^- is a (unique) maximal ideal so we can define the field k_H by $k_H := D_H/D_H^-$.

DEFINITION 2.4.9 For each convex subgroup H we call the field k_H of above the **residue class field with respect to** H. The canonical map $D_H \to k_H$ is denoted $\lambda \mapsto \lambda^-$.

REMARK k_H is the residue class field (in the traditional sense) of the field K with respect to the valuation $\lambda \mapsto |\lambda| \bmod H$ (with value group G/H), but for our purpose we prefer the above point of view.

PROPOSITION 2.4.10 *Let E be an X-normed space where X is a G-module. For each $s \in X$ the balls $B_E(0,s)$ and $B_E(0,s^-)$ are modules over D_{H_s} when $H_s := \{g \in G : gs = s\}$. The quotient $\overline{E}_s := B_E(0,s)/B_E(0,s^-)$ is in a natural way a vector space over k_{H_s}.*

Proof. $B_E(0,s)$ and $B_E(0,s^-)$ are absolutely convex. If $x \in B_E(0,s)$, $|\lambda| \in H_s$ then $\|\lambda x\| = |\lambda| \, \|x\| \leq |\lambda| s = s$, so $B_E(0,s)$ is a D_{H_s}-module. If $x \in B_E(0,s^-)$, $|\lambda| \in H_s$ and $\|\lambda x\|$ were $\geq s$ then $\|x\| = |\lambda|^{-1}\|\lambda x\| \geq |\lambda|^{-1}s = s$ (as $|\lambda|^{-1} \in H_s$), a contradiction, so $B_E(0,s^-)$ is a D_{H_s}-module. If $\lambda \in D_{H_s}^-$ then $|\lambda| < 1$ and $|\lambda| \notin H_s$ so $|\lambda| s \leq s$ but not $|\lambda| s = s$ i.e. $|\lambda| s < s$. This implies $D_{H_s}^- B_E(0,s) \subset B_E(0,s^-)$ showing that $\overline{E}_s$ is a k_{H_s}-vector space under

$$\overline{\lambda} \cdot \overline{x} = \overline{\lambda x} \qquad (\lambda \in D_{H_s}, \quad x \in B_E(0,s))$$

where the canonical map $D_{H_s} \to D_{H_s}/D_{H_s}^- = k_{H_s}$ is denoted $x \mapsto \overline{x}$.

LEMMA 2.4.11 *Let E be an X-normed space for some G-module X, let $s \in X$ and let $\{e_i : i \in I\} \subset E$ be such that $\|e_i\| = s$ for all i. Then the following are equivalent.*

(α) *$\{e_i : i \in I\}$ is orthogonal.*

(β) *$\{\overline{e}_i : i \in I\}$ is linearly independent in $\overline{E}_s$.*

Proof. (α) $\Rightarrow$ (β). Let $J \subset I$ be finite, let $\lambda_j \in D_{H_s}$ for each $j \in J$ and suppose that $\sum_{j\in J} \overline{\lambda}_j \overline{e}_j = 0$. Then $\overline{\sum_{j\in J} \lambda_j e_j} = 0$ so $\|\sum_{j\in J} \lambda_j e_j\| < s$. By orthogonality, $\|\lambda_j e_j\| < s$ for each j. Then $|\lambda_j| < 1$ and $|\lambda_j| \notin H_s$, so $|\lambda_j| \in D_{H_s}^-$, i.e. $\overline{\lambda_j} = 0$. (β) $\Rightarrow$ (α). Let $J \subset I$ be finite, let $\lambda_j \in K$ for each $j \in J$; we show that $\|\sum_{j\in J} \lambda_j e_j\| = (\max|\lambda_j|) \cdot s$. To this end we may suppose that $\max|\lambda_j| = 1$. If $\|\sum_{j\in J} \lambda_j e_j\|$ were $< s$ then $\sum_{j\in J} \lambda_j e_j \in B_E(0, s^-)$ so $\overline{0} = \overline{\sum_{j\in J} \lambda_j e_j} = \sum_{j\in J} \overline{\lambda}_j \overline{e}_j$. By ($\beta$) $\overline{\lambda}_j = 0$ for each j i.e. $\lambda_j \in D_{H_s}^-$. But then $|\lambda_j| s < s$ for each j conflicting $\max|\lambda_j| = 1$.

COROLLARY 2.4.12 *For each $s \in X$ all maximal orthogonal sets in $\{x \in E : \|x\| \in G\, s\}$ have the same cardinality.*

Proof. Each such orthogonal set can, via suitable scalar multiplications, be transformed into an orthogonal set of which each vector has length s. Now use 2.4.11 and the fact that maximal linear independent sets in vector spaces have the same cardinality.

PROPOSITION 2.4.13 *Let E be an X-normed space, where X is a G-module, let $\{e_i : i \in I\}$ be a maximal orthogonal set of nonzero vectors in E. Then, for each $s \in X$, $\{e_i : \|e_i\| \in G\, s\}$ is a maximal orthogonal subset of $\{x \in E : \|x\| \in G\, s\}$.*

Proof. Suppose for some $s \in X$ we do not have maximality. Then there is an $f \in E$ with $\|f\| \in G\, s$ such that $\{f\} \cup \{e_i : \|e_i\| \in G\, s\}$ is orthogonal. We claim that $\{f\} \cup \{e_i : i \in I\}$ is orthogonal (which leads to a contradiction proving the Proposition). In fact, let $J \subset I$ be finite, $\lambda_j \in K$ for each $j \in J$, $\lambda_0 \in K$. Let $J_1 = \{j \in J : \|e_j\| \in G\, s\}$, $J_2 = \{j \in J : \|e_j\| \notin G\, s\}$. Then $\|\lambda_0 f + \sum_{j\in J_1} \lambda_j e_j\| = \max(\|\lambda_0 f\|, \max_{j\in J_1} \|\lambda_j e_j\|) \in G\, s$ while $\|\sum_{i\notin J_1} \lambda_j e_j\| = \max_{j\in J_2} \|\lambda_j e_j\| \notin G\, s$. We see that $\|\lambda_0 f + \sum_{j\in J} \lambda_j e_j\| = \max(\|\lambda_0 f\|, \max_{j\in J} \|\lambda_j e_j\|)$ and we are done.

COROLLARY 2.4.14 *In a normed space each two maximal orthogonal subsets of nonzero vectors have the same cardinality.*

We now introduce (norm)orthogonal bases. For the sequel we only need the concept of a countable orthogonal base.

DEFINITION 2.4.15 Let X be a G-module, let $s : \mathbb{N} \to X$. Then $c_0(\mathbb{N}, s)$ is the space of all sequences $(\lambda_1, \lambda_2, \ldots) \in K^{\mathbb{N}}$ for which $\lim_n |\lambda_n| s(n) = 0$ with coordinatewise operations and with X-norm $(\lambda_1, \lambda_2, \ldots) \mapsto \max_n |\lambda_n| s(n)$. If $X = G$, $s(n) = 1$ for all n we write c_0 rather than $c_0(\mathbb{N}, s)$. By c_{00} we denote the space of all $(\lambda_1, \lambda_2, \ldots) \in K^{\mathbb{N}}$ for which $\lambda_n = 0$ for large n.

If K is complete then the space $c_0(\mathbb{N}, s)$ is complete. The proof is standard.

DEFINITION 2.4.16 A sequence $e_1, e_2, \dots$ in a normed space E is called *Schauder base* of E if for each $x \in E$ there are unique $\lambda_1, \lambda_2, \dots \in K$ such that $x = \sum_{n=1}^{\infty} \lambda_n e_n$. An orthogonal Schauder base is simply called *orthogonal base*. Then, with x as above, $\|x\| = \max_n \|\lambda_n e_n\|$.

PROPOSITION 2.4.17 *Let E be an infinite-dimensional K-Banach space. For an orthogonal sequence $e_1, e_2, \dots$ the following are equivalent.*
(α) *$e_1, e_2, \dots$ is an orthogonal base.*
(β) *$e_n \neq 0$ for each n. The linear span of $e_1, e_2, \dots$ is dense in E.*

Proof. (α) $\Rightarrow$ (β). Obvious. To prove (β) $\Rightarrow$ (α) we define a linear map $T : c_0(\mathbb{N}, s) \to E$ as follows

$$T : (\lambda_1, \lambda_2, \dots) \mapsto \sum_{n=1}^{\infty} \lambda_n e_n,$$

where $s(n) := \|e_n\|$ for each n. (Since $\|\lambda_n e_n\| \to 0$ and E is complete $\sum_{n=1}^{\infty} \lambda_n e_n$ exists, so T is well-defined.) Clearly T is a linear isometry. By (β), Im T is dense, but also complete by completeness of $c_0(\mathbb{N}, s)$. Then, Im $T = E$ and the result follows.

EXAMPLE 2.4.18 (Weird spaces if K is nonmetrizable). Let K be complete and non-metrizable, let $E = c_0(\mathbb{N}, s)$ be as in 2.4.15.

1. *Every sequence in X is bounded below (and above).* In fact, let $s_1 > s_2 > \cdots$ be a strictly decreasing sequence in X. By coinitiality of G_{s_n} we can find $\lambda_1, \lambda_2, \dots \in K^\times$ such that $\lambda_n s_1 < s_n$ for each n. If $\lim_n s_n = 0$ then $\lim_n \lambda_n s_1 = 0$, so $\lim_n \lambda_n = 0$ (1.5.2), conflicting 1.4.1.
2. *For each $(\lambda_1, \lambda_2, \dots) \in E$, $\lambda_n = 0$ for large n.* This follows from 1. We see that $E = c_{00}$.
3. *E is complete but no Baire space.* Clearly $E = \bigcup_n D_n$ where

   $$D_n = \{(\lambda_1, \lambda_2, \dots, \lambda_n, 0, 0 \dots) : n \in \mathbb{N}, \lambda_i \in K \quad \text{for} \quad i \in \{1, \dots, n\}\}$$

   Each D_n is a finite-dimensional subspace hence complete (2.3.4) hence closed in E. However the interior of D_n is empty.
4. *All norms on c_{00} are equivalent!* Let $\| \ \|_1$ and $\| \ \|_2$ be X-norms on E. By 2.3.4 they are Lipschitz equivalent on D_n for each n so there are $g_1, g_2, \dots$ and $h_1, h_2, \dots \in G$ such that

   $$h_n\|x\|_1 \le \|x\|_2 \le g_n\|x\|_1 \qquad (x \in D_n).$$

   By nonmetrizability (1.4.1) there are $h, g \in G$ such that $h_n \ge h$ $g_n \le g$ for all n and we have

   $$h\|x\|_1 \le \|x\|_2 \le g\|x\|_1 \qquad (x \in E).$$

2.5 Metrizable normed spaces

Throughout 2.5 we will assume that K is complete and metrizable. Recall (1.4.1) that this implies that there is a sequence $\lambda_1, \lambda_2, \ldots$ in K such that $|\lambda_1| > |\lambda_2| > \cdots$, and $\lim_n \lambda_n = 0$. Then each G-module X has a coinitial (cofinal) sequence (let $s \in X$ and λ_n be as above. Then $|\lambda_1|s, |\lambda_2|s, \ldots$ is coinitial). If E is an X-normed space the balls $B_E(0, |\lambda_n|s)$ form a neighbourhood base at 0 for the norm topology. Hence the norm topology of any normed space over K is (ultra)metrizable; it is a Banach space if and only if each Cauchy sequence converges. Observe that a sequence $x_1, x_2, \ldots$ is Cauchy if and only if $\lim_n(x_{n+1} - x_n) = 0$, so that the question as to whether a normed space is Banach depends on the topology, not on the particular norm. Each Banach space is a Baire space (compare Example 2.4.18). The proofs in this section are basically classical.

PROPOSITION 2.5.1 *Let E be a Banach space, D a closed subspace. Then E/D is a Banach space.*

Proof. Let the norm have values in a complete G-module X, let $\varepsilon_1 > \varepsilon_2 > \cdots$ be a coinitial sequence in X. Let $z_1, z_2, \ldots$ be a Cauchy sequence in E/D. It has a subsequence $y_1, y_2, \ldots$ for which $\|y_{n+1} - y_n\| < \varepsilon_{n+1}$ for each n. There are $v_0, v_1, \ldots \in E$ such that, with $\pi : E \to E/D$ the natural map, $\pi(v_0) = y_1$, $\pi(v_n) = y_{n+1} - y_n$, $\|v_n\| < \varepsilon_n$ for each $n \in \mathbb{N}$. Then $m \mapsto \sum_{n=0}^{m} v_n$ is Cauchy, hence convergent. Set $x := \sum_{n=0}^{\infty} v_n$. Then $\pi(x) = \lim_m \pi(\sum_{n=0}^{m} v_n) = \lim_m y_m$. Thus, $y_1, y_2, \ldots$ converges and therefore so does $z_1, z_2, \ldots$.

We now prove the Open Mapping Theorem 2.5.4, generalizing the results of [11]. We use the easily proved fact that an additive subgroup (in particular, an absolutely convex subset) of E with a non-empty interior is open.

PROPOSITION 2.5.2 *Let E be a normed space, let F be a Banach space, let $T \in \mathcal{L}(E, F)$ be surjective. Then, for each ball B about 0 in E, $\overline{TB}$ is a zero neighbourhood in F.*

Proof. Suppose E is X-normed for some G-module X. Let $s_1 < s_2 < \cdots$ be a cofinal sequence in X. Set $B_n := \{x \in E : \|x\| \le s_n\}$ $(n \in \mathbb{N})$. Then $\bigcup_n \overline{TB_n} = F$, so by the Baire Category Theorem and absolute convexity, $\overline{TB_n}$ is open for some n. Then $\overline{TB_n}$ is open for all n and $\overline{TB}$ is open.

PROPOSITION 2.5.3 *Let E be a Banach space, let F be a normed space, let $T \in \mathcal{L}(E, F)$. If $B \subset E$ is a ball about 0 and $\overline{TB}$ is a zero neighbourhood in F then $\overline{TB} = TB$. In particular TB is clopen, T is surjective and open.*

Proof. Suppose $S := \{z \in F : \|z\| < s\} \subset \overline{TB}$. It suffices to prove $S \subset TB$. Let $z \in S$, let $\lambda_1, \lambda_2, \ldots \in K$ be such that $1 > |\lambda_1| > |\lambda_2| > \cdots$, $\lim_n \lambda_n = 0$. Set $\mu_1 := 1$, $\mu_n := \prod_{i=1}^{n-1} \lambda_i$ $(n \ge 2)$. Inductively we can select $b_1, b_2, \ldots \in B$ and $z_1, z_2, \ldots \in S$ such that for all $n \in \mathbb{N}$

$$z = \sum_{i=1}^{n} \mu_i T b_i + \mu_{n+1} z_n, \tag{$*$}$$

Now $\|\mu_n b_n\| \le |\lambda_{n-1}| \, \|b_n\| \to 0$ by boundedness of B so, by completeness of E, $a := \sum_{i=1}^{\infty} \mu_i b_i$ exists. We also have $\|\mu_{n+1} z_n\| \le |\mu_{n+1}| s \to 0$, so from $(*)$ we obtain $z = Ta$.

The following corollary is obtained from 2.5.2 and 2.5.3 by standard classical arguments.

COROLLARY 2.5.4 *Let E, F be Banach spaces.*

(i) (Open Mapping Theorem). *Let $T \in \mathcal{L}(E, F)$ be continuous and surjective. Then T is open.*

(ii) (Closed Graph Theorem). *Let $T : E \to F$ be linear. If the graph of T is closed in $E \times F$ then T is continuous.*

As an application we obtain the following.

THEOREM 2.5.5 *Let X be a complete G-module, let E, F be X-normed Banach spaces and suppose $\mathcal{L}(E, F) = \mathrm{Lip}(E, F)$* (e.g. if $\dim F < \infty$, see 2.3.5). *Then the uniform norms and the Lipschitz norm are equivalent.*

Proof. By 2.3.6 both $\mathcal{L}(E, F)$ and $\mathrm{Lip}(E, F)$ are Banach spaces. As the uniform norms are weaker than the Lipschitz norm, the identity $\mathrm{Lip}(E, F) \to \mathcal{L}(E, F)$ is continuous. Now apply 2.5.4 (i).

REMARK We failed to prove 2.5.5 directly (i.e. without using the Open Mapping Theorem).

THEOREM 2.5.6 (Uniform Boundedness Principle). *Let E be an X-normed Banach space, let F be a Y-normed space, where X, Y are G-modules. If $\{T_i : i \in I\} \subset \mathcal{L}(E, F)$ is pointwise bounded then it is bounded in the uniform topology of $\mathcal{L}(E, F)$.*

Proof. Let $t_1, t_2, \ldots$ be a cofinal sequence in Y. For each n let $A_n := \{x \in E : \|T_i x\| \le t_n \text{ for all } i \in I\}$. Then each A_n is closed and absolutely convex. By assumption $\bigcup A_n = E$, by the Baire Category Theorem A_n is open for some n, and hence it contains a ball $\{x \in E : \|x\| \le s\}$ for some $s \in X$. We see that $\|T_i\|_s \le t_n$ for each i.

COROLLARY 2.5.7 (Banach Steinhaus). *Let E, F be Banach spaces, let $T_1, T_2, \ldots$ be in $\mathcal{L}(E, F)$ such that, for each $x \in E$, $T_1 x, T_2 x, \ldots$ is Cauchy in F. Then $Tx := \lim_n T_n x$ exists for each $x \in E$ and $T \in \mathcal{L}(E, F)$. For each $s \in \|E\| \setminus \{0\}$ we have $\|T\|_s \le \sup_m \inf_{n \ge m} \|T_n\|_s$.*

Proof. The set $\{T_1, T_2, \ldots\}$ is pointwise bounded, hence, by 2.5.6, $n \mapsto \|T_n\|_s$ is bounded by, say, t. Then, for each $x \in B_E(0, s)$ we have either $Tx = 0$ or $\|Tx\| = \|T_n x\|$ for large n, so $\|T\|_s \le t$ i.e. $T \in \mathcal{L}(E, F)$. Obviously, for each $x \in B_E(0, s)$ there is an $m \in \mathbb{N}$ with $\|Tx\| = \inf_{n \ge m} \|T_n x\| \le \inf_{n \ge m} \|T_n\|_s$, hence $\|Tx\| \le \sup_m \inf_{n \ge m} \|T_n\|_s$ and we are done.

REMARK For a counterexample to the 'Lipschitz version' of 2.5.6 or 2.5.7 see [10].

3 SPACES OF COUNTABLE TYPE

From now on we assume that K is complete and satisfies the conditions $(\alpha)-(\delta)$ of Proposition 1.4.4 i.e. we assume that each absolutely convex subset of K is countably generated as a B_K-module. Then K is metrizable (Remark following 1.4.4).

3.1 Countably generated B_K-modules

As an algebraic introduction we prove that B_K-submodules of countably generated B_K-modules are themselves countably generated (Theorem 3.1.4).

Clearly, if B is a B_K-submodule of a countably generated B_K-module A, then A/B is countably generated. We also have the following.

LEMMA 3.1.1 *Let B be a submodule of a B_K-module A. If B and A/B are countably generated then so is A.*

Proof. Let $\pi : A \to A/B$ be the canonical homomorphism. Let $a_1, a_2, \ldots \in A$ be such that $\{\pi(a_1), \pi(a_2), \ldots\}$ generates A/B, let $b_1, b_2, \ldots \in B$ be such that $\{b_1, b_2, \ldots\}$ generates B. We prove that $\{a_1, a_2, \ldots\} \cup \{b_1, b_2, \ldots\}$ generates A. In fact, let $x \in A$. Then there exist an $m \in \mathbb{N}$ and $\lambda_1, \ldots, \lambda_m \in B_K$ such that $\pi(x) = \sum_{i=1}^{m} \lambda_i \pi(a_i) = \pi(\sum_{i=1}^{m} \lambda_i a_i)$. So $x - \sum_{i=1}^{m} \lambda_i a_i \in B$ and there exists an $n \in \mathbb{N}$ and $\mu_1, \mu_2, \ldots, \mu_n \in B_K$ such that $x - \sum_{i=1}^{m} \lambda_i a_i = \sum_{i=1}^{n} \mu_i b_i$, and we are done.

LEMMA 3.1.2 *Absolutely convex subsets of finite-dimensional vector spaces over K are countably generated.*

Proof. For onedimensional vector spaces this is just our assumption made at the beginning of this Section. Suppose the conclusion of the Lemma holds for absolutely convex subsets of vector spaces of dimension $\leq n-1$. Let A be an absolutely convex subset of an n-dimensional space E. To prove that A is countably generated we may suppose $A \neq \{0\}$, so let $a \in A$, $a \neq 0$ and $Ka := \{\lambda a : \lambda \in K\}$. We have the sequences

$$Ka \longrightarrow E \longrightarrow E/Ka$$

and

$$(Ka) \cap A \longrightarrow A \longrightarrow A/(Ka) \cap A$$

(the arrows indicating the natural maps) and the inclusions $(Ka) \cap A \to Ka$ and $A \to E$. There is a unique B_K-module homomorphism $\varphi : A/(Ka) \cap A \to E/Ka$

making the diagram

$$\begin{array}{ccccc} Ka & \longrightarrow & E & \longrightarrow & E/Ka \\ \uparrow & & \uparrow & & \uparrow \varphi \\ (Ka)\cap A & \longrightarrow & A & \longrightarrow & A/(Ka)\cap A \end{array}$$

commute. This φ is injective. Now $\dim E/Ka \leq n-1$, so by the induction hypothesis $A/(Ka)\cap A$ is countably generated and so is $(Ka)\cap A$. Now apply 3.1.1 to conclude that A is countably generated.

Let us denote the direct sum $\{(\lambda_1, \lambda_2, \ldots) \in B_K^{\mathbb{N}} : \lambda_n = 0 \text{ for large } n\}$ by $B_K^{(\mathbb{N})}$.

LEMMA 3.1.3 *Every B_K-submodule of $B_K^{(\mathbb{N})}$ is countably generated.*

Proof. For each $n \in \mathbb{N}$, let $D_n := \{(\lambda_1, \lambda_2, \ldots) \in K^{\mathbb{N}} : \lambda_m = 0 \text{ for } m > n\}$. If A is a submodule of $B_K^{(\mathbb{N})}$ then $A = \bigcup_n D_n \cap A$. By 3.1.2 each $D_n \cap A$ is countably generated, hence so is A.

THEOREM 3.1.4 *Any submodule of a countably generated B_K-module is countably generated.*

Proof. Let $\{e_1, e_2, \ldots\} \subset A$ generate A, let B be a submodule of A. The formula

$$\pi((\lambda_1, \lambda_2, \ldots)) = \sum_i \lambda_i e_i$$

defines a surjective homomorphism $\pi : B_K^{(\mathbb{N})} \to A$. By Lemma 3.1.3 $\pi^{-1}(B)$ is countably generated, hence so is $\pi(\pi^{-1}(B)) = B$.

3.2 Spaces of countable type, their subspaces and quotients

DEFINITION 3.2.1 (see [14] p. 66). A normed space over K is called *of countable type* if there is a countable set whose linear hull is dense.

If K is separable then 'of countable type' is identical to 'separable'. Spaces with a Schauder base (2.4.16) are of countable type, but we will see in 3.2.13 that the converse is not true (like in the complex case). Quotients (by closed subspaces) of spaces of countable type are of countable type. That subspaces of spaces of countable type are again of countable type is more difficult to prove. (If K is separable there is no problem: it is simply the fact that a subset of a separable space is separable. If the valuation is of rank 1 we have [14], 3.16 but that proof uses the existence of a Schauder base, which is no longer true in infinite rank case. We shall give a proof in 3.2.4 based upon ideas used in [12] to prove a similar theorem for locally convex B_K-modules. Observe that 3.2.4 only works for base fields satisfying 1.4.4. It is an intriguing open problem whether subspaces of spaces of countable type are of countable type in case the base field does not satisfy 1.4.4!). To this end we introduce the following. We will say that an absolutely convex subset A of a normed space E is a *B_K-module of countable type* if there is a countable set $S \subset A$

such that $\operatorname{co} S := \{\lambda_1 s_1, + \cdots + \lambda_n s_n : n \in \mathbb{N}, s_1, \ldots, s_n \in A, \lambda_1, \ldots, \lambda_n \in B_K\}$ is dense in A. For normed spaces this notion coincides with 3.2.1:

PROPOSITION 3.2.2 *A normed space is of countable type if and only if it is a B_K-module of countable type.*

Proof. We need to prove the 'only if' part. Suppose E is an X-normed space of countable type where X is a G-module, let $\{e_1, e_2, \ldots\} \subset E$ be such that its linear hull is dense in E. By our assumption made at the beginning of this Section there are $\lambda_1, \lambda_2, \ldots \in K$ such that $0 < |\lambda_1| < |\lambda_2| < \cdots$ is cofinal in G. Set $S := \{\lambda_i e_j : i, j \in \mathbb{N}\}$. Then S is countable. To show that co S is dense in E, let $x \in E$, $\varepsilon \in X$. There are $m \in \mathbb{N}$, $\mu_1, \ldots, \mu_m \in K$ such that $\|x - \sum_{i=1}^m \mu_i e_i\| < \varepsilon$. By cofinality there is an $n \in \mathbb{N}$ such that $|\lambda_n| > \max\{|\mu_i| : 1 \leq i \leq m\}$. Then $\lambda_n e_i \in S$, $\mu_i \lambda_n^{-1} \in B_K$ for each i and $\|x - \sum_{i=1}^m \mu_i \lambda_n^{-1}(\lambda_n e_i)\| < \varepsilon$.

PROPOSITION 3.2.3 *Let A be an absolutely convex subset of an X-normed space E. Then A is a B_K-module of countable type if and only if for each $\varepsilon \in X$ the module $A/A \cap B_E(0, \varepsilon)$ is countably generated.*

Proof. The 'only if' is obvious, so suppose $A/A \cap B_E(0, \varepsilon)$ is countably generated for each $\varepsilon \in X$. Let $\varepsilon_1 > \varepsilon_2 > \cdots$ be a coinitial sequence in X, let $S_1, S_2, \ldots$ be countable subsets of A such that, for each n, $\pi_n(S_n)$ generates $A/A \cap B_E(0, \varepsilon_n)$ (where $\pi_n : A \to A/A \cap B_E(0, \varepsilon_n)$ is the canonical map). Then $S := \bigcup_n S_n$ is countable. We show that co S is dense in A. In fact, let $a \in A$, let $\varepsilon \in X$, choose n such that $\varepsilon_n < \varepsilon$. There are $m \in \mathbb{N}$, $\lambda_1, \ldots, \lambda_m \in B_K$, $a_1, \ldots, a_m \in S_n$ such that

$$\pi_n(a) = \sum_{i=1}^m \lambda_i \pi_n(a_i).$$

Hence, $\pi_n(a - \sum_{i=1}^m \lambda_i a_i) = 0$ i.e. $\|a - \sum_{i=1}^m \lambda_i a_i\| \leq \varepsilon_n < \varepsilon$.

THEOREM 3.2.4 *Let E be a normed space of countable type. Then each subspace is of countable type. More generally, each absolutely convex subset of E is a B_K-module of countable type.*

Proof. Suppose E is X-normed for some G-module X. By 3.2.2 and 3.2.3, for each $\varepsilon \in X$ the B_K-module $E/B_E(0, \varepsilon)$ is countably generated. Now let $A \subset E$ be absolutely convex. There is a natural injective homomorphism $A/A \cap B_E(0, \varepsilon) \longrightarrow E/B_E(0, \varepsilon)$, so by 3.1.4 the B_K-module $A/A \cap B_E(0, \varepsilon)$ is countably generated for each $\varepsilon \in X$. Again by 3.2.3 we conclude that A is of countable type.

The following Proposition shows that being of countable type is a so-called 3-space property.

PROPOSITION 3.2.5 *Let E be a normed space, let D be a closed subspace. If D and E/D are of countable type then so is E.*

Proof. We may assume that $E, D, E/D$ are X-normed spaces for some complete G-module X. Let $S \subset D$, $T \subset E$ be countable sets such that the linear hulls of S

and $\pi(T)$ are dense in D, E/D respectively. (Here, $\pi : E \to E/D$ is the canonical map). We claim that the linear hull of $S \cup T$ is dense in E. In fact, let $a \in E$, let $\varepsilon \in X$. There is an element x in the linear hull of T such that $\|\pi(a) - \pi(x)\| < \varepsilon$. By 2.2.2 there is a $y \in E$, $\|y\| < \varepsilon$ with $\pi(a - x) = \pi(y)$, i.e. $a - x - y \in D$. There is an element z in the linear hull of S such that $\|a - x - y - z\| < \varepsilon$. Then $\|a - x - z\| < \varepsilon$ and we are done.

In [17] we described the strict quotients of c_0. In this paper we need the following characterization of all quotients of c_0. Clearly, if E is such a quotient it is a Banach space (2.5.1) of countable type and it is $G^{\#}$-normed. Surprisingly this turns out to be sufficient, as the following theorem shows.

THEOREM 3.2.6 *Let E be a $G^{\#}$-normed Banach space of countable type. Then E is a quotient of c_0.*

Proof. Let $B := \{x \in c_0 : \|x\| < 1\}$, $S := \{z \in E : \|z\| < 1\}$. By 3.2.4, S is of countable type as a B_K-module, so let $z_1, z_2, \ldots \in S\backslash\{0\}$ be such that $\mathrm{co}\{z_1, z_2, \ldots\}$ is dense in S. By 1.1.4 (iv) we can choose for each n a $\lambda_n \in K$ such that $\|z_n\| \le |\lambda_n| < 1$. Let $e_1, e_2, \ldots$ be the canonical base of c_0. The formula

$$\pi\Big(\sum_{n=1}^{\infty} \xi_n e_n\Big) = \sum_{n=1}^{\infty} \xi_n \lambda_n^{-1} z_n \qquad (\xi_n \in K, |\xi_n| \to 0)$$

defines a continuous linear map $\pi : c_0 \to E$. Obviously $\pi(B) \subset S$. For each n, $z_n = \pi(\lambda_n e_n) \in \pi(B)$, so $\mathrm{co}\{z_1, z_2, \ldots\} \subset \pi(B)$; it follows that $\pi(B)$ is dense in S. Proposition 2.5.3 tells us now that $\pi(B) = S$. Via scalar multiplication we arrive at $\pi\big(B_{c_0}(0, r^-)\big) = B_E(0, r^-)$ for all $r \in G$. If $r \in G^{\#}\backslash G$ observe that

$$\pi\big(B_{c_0}(0, r^-)\big) = \pi\Big(\bigcup_{\substack{g \in G \\ g < r}} B_{c_0}(0, g^-)\Big) = \bigcup_{\substack{g \in G \\ g < r}} B_E(0, g^-) = B_E(0, r^-).$$

Now apply 2.2.2 to conclude that π is a quotient map.

COROLLARY 3.2.7 *For each Banach space E of countable type there exists a linear continuous open surjection $c_0 \to E$.*

Proof. By 2.1.9 there is a $G^{\#}$-norm $\|\ \ \|'$ on E, equivalent to the initial norm $\|\ \ \|$. By 3.2.6 there is a quotient map $\pi : c_0 \to (E, \|\ \ \|')$. Then $\rho \circ \pi$, where $\rho : (E, \|\ \ \|') \to (E, \|\ \ \|)$ is the identity map, is the required surjection.

To prove the related result 3.2.11 we need some preparatory observations. As usual, $[X]$ is the linear span of $X \subset E$, $\overline{[X]}$ its closure. Further, $\|E\| := \{\|x\| : x \in E\}$.

PROPOSITION 3.2.8 *Let E be a normed space, $x_1, \ldots x_n \in E\backslash\{0\}$. If $\|x_i\| \notin G\|x_j\|$ whenever $i \neq j$ then $x_1, \ldots, x_n$ are orthogonal.*

Proof. Let $\lambda_1, \ldots, \lambda_n \in K$. If $i, j \in \{1, \ldots, n\}$, then either $\|\lambda_i x_i\| = \|\lambda_j x_j\| = 0$ or $\|\lambda_i x_i\| \neq \|\lambda_j x_j\|$. So, if not all λ_i are 0 there is a unique j for which $\max_i \|\lambda_i x_i\| = \|\lambda_j x_j\|$. It follows that $\|\sum_{i \neq j} \lambda_i x_i\| < \|\lambda_j x_j\|$ so $\|\sum_{i=1}^{n} \lambda_i x_i\| = \|\lambda_j x_j\| = \max_i \|\lambda_i x_i\|$.

PROPOSITION 3.2.9 *In a space of countable type each orthogonal subset of vectors is (at most) countable.*

Proof. Suppose E is a space of countable type, let $\{e_i : i \in I\}$ be an orthogonal set in E, where I is uncountable. Set $D := \overline{[\, e_i : i \in I\,]}$. Then D is of countable type (3.2.4), let $x_1, x_2, \ldots \in D$ be such that its linear hull is dense in D. Clearly for each n there is a countable set $I_n \subset I$ such that $x_n \in \overline{[\, e_i : i \in I_n\,]}$. It follows that $D = \overline{[\, e_i : i \in J\,]}$ where $J = \bigcup_n I_n$, a countable set, which is a contradiction.

COROLLARY 3.2.10 *Let E be an X-normed space of countable type, where X is some G-module. Then there is a countable set $S \subset X$ such that $\|E\| \backslash \{0\} = GS$.*

Proof. If the conclusion were false we could find an uncountable set $\{x_i : i \in I\} \subset E\backslash\{0\}$ for which $\|x_i\| \notin G\|x_j\|$ whenever $i \neq j$. By 3.2.8 the set $\{x_i : i \in I\}$ is orthogonal, so I is at most countable because of 3.2.9. This is a contradiction.

THEOREM 3.2.11 *Each Banach space E of countable type is a quotient of a Banach space with a (countable) orthogonal base.*

Proof. Let $S = \{r_1, r_2, \ldots\} \subset X$ be such that $\|E\|\backslash\{0\} = GS$ (3.2.10). Without loss, assume that $X = GS$. For each $n \in \mathbb{N}$ let $z_{n1}, z_{n2}, \ldots \in B_E(0, r_n^-)$ be nonzero vectors such that $\mathrm{co}\{z_{n1}, z_{n2}, \ldots\}$ is dense in $B_E(0, r_n^-)$. Let F be the space of all $(\xi_{nm}) \in K^{\mathbb{N}\times\mathbb{N}}$ for which $\lim_{m,n} |\xi_{nm}|\,\|z_{nm}\| = 0$, normed by $(\xi_{nm}) \mapsto \max_{n,m} |\xi_{nm}|\,\|z_{nm}\|$. Let (e_{nm}) be the natural orthogonal base of F. The formula

$$\pi((\xi_{nm})) = \pi\Big(\sum_{n,m=1}^{\infty} \xi_{nm} e_{nm}\Big) = \sum_{n,m=1}^{\infty} \xi_{nm} z_{nm}$$

defines a continuous linear map $\pi : F \to E$. Obviously, $\|\pi(x)\| \leq \|x\|$ for each $x \in F$ so $\pi\big(B_F(0, r_n^-)\big) \subset B_E(0, r_n^-)$ for each n. For each n, m we have $z_{nm} = \pi(e_{nm}) \in \pi\big(B_F(0, r_n^-)\big)$, so $\mathrm{co}\{z_{n1}, z_{n2}, \ldots\} \subset \pi\big(B_F(0, r_n^-)\big)$, it follows that $\pi\big(B_F(0, r_n^-)\big)$ is dense in $B_E(0, r_n^-)$. From 2.5.3 we obtain $\pi\big(B_F(0, r_n^-)\big) = B_E(0, r_n^-)$ for each n. Scalar multiplication shows that $\pi\big(B_F(0, r^-)\big) = B_E(0, r^-)$ for all $r \in GS = \|E\|\backslash\{0\}$.

To conclude this section we will present an example of a Banach space of countable type (in fact, a separable Banach space) without a Schauder base. To this end we first prove the following Proposition whose proof is basically classical.

PROPOSITION 3.2.12 *Let E be a Banach space with a Schauder base $e_1, e_2, \ldots$. Then E is linearly homeomorphic to a Banach space with an orthogonal base.*

Proof. Let $T : c_0(\mathbb{N}, s) \to E$ be the linear map given by

$$(\lambda_1, \lambda_2, \ldots) \mapsto \sum_{n=1}^{\infty} \lambda_n e_n$$

where $s(n) = \|e_n\|$ for each $n \in \mathbb{N}$. T is well-defined since $\|\lambda_n e_n\| \to 0$. For $x = (\lambda_1, \lambda_2, \ldots) \in c_0(\mathbb{N}, s)$ we have $\|Tx\| = \|\sum_n \lambda_n e_n\| \leq \max_n \|\lambda_n e_n\| = \|x\|$, so

T is continuous. Bijectivity follows from the fact that $e_1, e_2, \dots$ is a Schauder base. By the Open Mapping Theorem 2.5.4, T is a homeomorphism.

EXAMPLE 3.2.13 (A separable Banach space E without Schauder base). In classical analysis over $\mathbb{R}$ or $\mathbb{C}$ one has Enflo's famous example of a separable Banach space without a Schauder base. In contrast to this, in non-archimedean rank 1 theory any Banach space of countable type has a Schauder base [14]. The same conclusion holds in arbitrary rank case when the base field is spherically complete (3.4.2). Surprisingly we can construct a separable Banach space E without Schauder base over an infinite rank valued, non-spherically complete base field as follows. By [17] 2.2 there is such a (separable) field for which c_0 admits a closed subspace S and a $g \in S'$ that cannot be extended to an element of c_0'. Let $E := c_0/\mathrm{Ker}\ g$, let $\pi : c_0 \to c_0/\mathrm{Ker}\ g$ be the quotient map. Let $a \in S$ be such that $g(a) = 1$. If $\varphi \in (c_0/\mathrm{Ker}\ g)'$ and $\varphi(\pi(a)) = 1$ then $\varphi \circ \pi = g$ on $S = Ka + \mathrm{Ker}\ g$ conflicting the non-extendability of g. Thus, the elements of E' do not separate the points of E. If E had a Schauder base $e_1, e_2, \dots$ then the coordinate functions would be continuous by 3.2.12 so if $f(x) = 0$ for all $f \in E'$ and some $x = \sum_{n=1}^{\infty} \lambda_n e_n \in E$ then all $\lambda_n = 0$ i.e. $x = 0$, a contradiction. It follows that E has no Schauder base.

3.3 Finite-dimensional spaces with an orthogonal base

This section is a stepping stone for 3.4. The results do not differ from the ones in [14]. We start with a general lemma.

LEMMA 3.3.1 *Let E be a normed space for which every onedimensional subspace is orthocomplemented. Then so is every finite-dimensional subspace.*

Proof. (After [14] 4.35 (iii)). It suffices to prove the following. If $D_1 \subset D$ are subspaces, $\dim D/D_1 = 1$, D_1 has an orthogonal complement, then so has D. To prove this, let S_1 be an orthogonal complement of D_1. Then $D \cap S_1$ is an orthogonal complement of D_1 in D, so $\dim D \cap S_1 = 1$. By assumption $D \cap S_1$ has an orthogonal complement S_2. One verifies directly that $S_1 \cap S_2$ is an orthogonal complement of D. (Clearly $S_1 \cap S_2$ is orthogonal to D; from $E = S_2 + D \cap S_1$ it follows that $S_1 = S_1 \cap S_2 + D \cap S_1$, so $E = D_1 + S_1 = D_1 + S_1 \cap S_2 + D \cap S_1 = D + S_1 \cap S_2$.)

THEOREM 3.3.2 *For a finite-dimensional normed space E the following are equivalent.*

(α) *E has an orthogonal base.*
(β) *Every subspace has an orthogonal complement.*
(γ) *Every subspace has an orthogonal base.*
(δ) *Every orthogonal set of nonzero vectors can be extended to an orthogonal base of E.*

Proof. (α) $\Rightarrow$ (β). By Lemma 3.3.1 it suffices to prove that each onedimensional subspace has an orthogonal complement. To this end, let $e_1, \dots, e_n$ be an orthogonal base of E and let $a = \sum_{i=1}^{n} \lambda_i e_i$ ($\lambda_i \in K$) be a non-zero vector. There is an $m \in \{1, \dots, n\}$ for which $\|a\| = \|\lambda_m e_m\|$; we prove that $Ka \perp S := [\ e_i : i \neq m\ ]$;

it suffices to show that $\|a - s\| \geq \|a\|$ for all $s \in S$. So let $s = \sum_{i \neq m} \mu_i e_i \in S$. Then $\|a - s\| = \|\lambda_m e_m + \sum_{i \neq m} (\lambda_i - \mu_i) e_i\| \geq \|\lambda_m e_m\| = \|a\|$. To prove $(\beta) \Rightarrow (\delta)$, let $e_1, \ldots, e_m$ be a maximal orthogonal set of nonzero vectors in E, let $D = [\, e_1, \ldots, e_m \,]$. By (β), D has an orthogonal complement; by maximality this complement must be $\{0\}$. Hence $E = [\, e_1, \ldots, e_m \,]$ and we are done. Obviously $(\delta) \Rightarrow (\alpha)$, so at this stage we have proved the equivalence of $(\alpha), (\beta), (\delta)$. Now if (β) holds for E it holds for every subspace of E. But then also (α) holds for subspaces i.e. we have $(\alpha) \Rightarrow (\gamma)$. As trivially $(\gamma) \Rightarrow (\alpha)$ this completes the proof.

COROLLARY 3.3.3 *If K is spherically complete each finite-dimensional normed space has an orthogonal base and is spherically complete.*

Proof. Combining 2.4.6, 3.3.1 and 3.3.2 we conclude that a finite-dimensional normed space E has an orthogonal base, say $e_1, \ldots, e_n$. Spherical completeness can be proved inductively, using 2.4.5 and the easily proved fact that if D_1 and D_2 are orthogonal subspaces both spherically complete then $D_1 + D_2$ is spherically complete.

REMARK If K is not spherically complete there exist two-dimensional normed spaces without an orthogonal base (14, p. 69, 17, Lemma 1.4).

3.4 Spaces of countable type with an orthogonal base

THEOREM 3.4.1 *Let E be a Banach space of countable type. Then the following are equivalent.*

(α) *E has an orthogonal base.*
(β) *Each closed subspace has an orthogonal base.*
(γ) *Each finite-dimensional subspace has an orthogonal base.*
(δ) *Each onedimensional subspace has an orthogonal complement.*
(ε) *Each finite-dimensional subspace has an orthogonal complement.*
(ζ) *For each finite-dimensional subspace D and $a \in E$ the set $\{\|a - d\| : d \in D\}$ has a minimum.*

Proof. $(\alpha) \Rightarrow (\delta)$ (similar to $(\alpha) \Rightarrow (\beta)$ of 3.3.2). Let $e_1, e_2, \ldots$ be an orthogonal base of E, let $a = \sum_{i=1}^{\infty} \lambda_i e_i \quad (\lambda_i \in K, \quad \lambda_i e_i \to 0)$ be a nonzero vector. There is an $m \in \mathbb{N}$ such that $\|a\| = \|\lambda_m e_m\|$. Then $\overline{[e_i, i \neq m]}$ is an orthogonal complement of Ka. $(\delta) \Rightarrow (\varepsilon)$ is Lemma 3.3.1. Now we prove $(\varepsilon) \Rightarrow (\zeta)$. Let D have an orthogonal complement S, write $a = d_1 + s$ when $d_1 \in D$, $s \in S$. Then for each $d \in D$ we have $\|a - d\| = \|s + d_1 - d\| = \max(\|s\|, \|d_1 - d\|) \geq \|s\|$, so $\min\{\|a-d\| : d \in D\} = \|a-d_1\| = s$. To prove $(\zeta) \Rightarrow (\beta)$, let D be a closed subspace of E. By 3.2.4 D is of countable type, let $x_1, x_2, \ldots$ be linearly independent elements of D such that $D = \overline{[x_1, x_2, \ldots]}$. We construct inductively an orthogonal sequence $e_1, e_2, \ldots$ in D such that $[e_1, \ldots, e_n] = D_n := [x_1, \ldots, x_n]$ for each n. (Then we will be done by 2.4.17.) Set $e_1 := x_1$. Suppose we have constructed $e_1, \ldots, e_{m-1}$. By assumption there is a $d_1 \in D_{m-1}$ such that $\|x_m - d\| \geq \|x_m - d_1\|$ for all $d \in D_{m-1}$. Set $e_m := x_m - d_1$. Let $\lambda_1, \ldots, \lambda_{m-1} \in K$. Then $\|e_m - \sum_{i=1}^{m-1} \lambda_i e_i\| \geq \inf\{\|x_m - d\| : d \in D_{m-1}\} \geq \|x_m - d_1\| = \|e_m\|$, which proves orthogonality. As

$(\beta) \Rightarrow (\alpha)$ is trivial we now have established the equivalence of $(\alpha), (\beta), (\delta), (\varepsilon), (\zeta)$. Obviously $(\beta) \Rightarrow (\gamma)$. We complete the proof by showing $(\gamma) \Rightarrow (\zeta)$. From (γ) we obtain that $D + Ka$ has an orthogonal base so, by 3.3.2, D has an orthogonal complement, say, Kb in $D + Ka$. Then $a = \lambda b + d_1$ where $\lambda \in K$, $d_1 \in D$ and clearly $\{\|a - d\| : d \in D\} = \{\|\lambda b + d_1 - d\| : d \in D\} = \{\|\lambda b - d\| : d \in D\}$ has a minimum viz. $\|\lambda b\|$.

COROLLARY 3.4.2 *Each Banach space of countable type over a spherically complete field has an orthogonal base.*

Proof. 3.3.3 and 3.4.1 $(\gamma) \Rightarrow (\alpha)$.

REMARK Spaces of countable type with the property that every closed subspace has an orthogonal complement will be treated in Section 4.

We now will define and prove a canonical orthogonal decomposition of a Banach space with an orthogonal base.

DEFINITION 3.4.3 Let $E_1, E_2, \ldots$ be X-normed Banach spaces, where X is some G-module. The *orthogonal direct sum* $\bigoplus_n E_n$ of $E_1, E_2, \ldots$ is the subspace of $\prod_n E_n$ consisting of all $x = (x_1, x_2, \ldots)$ for which $\lim_n \|x_n\| = 0$, normed by $x \mapsto \max_n \|x_n\|$. (which makes $\bigoplus_n E_n$ into a Banach space.) In particular we say that a Banach space E is the *orthogonal direct sum of the subspaces* $E_1, E_2, \ldots$ if the map $\bigoplus_n E_n \to E$ given by $(x_1, x_2, \ldots) \mapsto \sum_{n=1}^{\infty} x_n$ is a bijective isometry.

Let E be an X-normed Banach space of countable type, where X is a G-module. Then $Y := \{\|x\| : x \in E, x \neq 0\}$ is a G-submodule of X. From 3.2.10 we know that there is a countable set S such that $Y = GS$. Let $\Sigma := Y/\sim$, where $s \sim t$ if and only if $s \in Gt$, be the collection of algebraic types of Y (see 1.6); Σ is countable.

DEFINITION 3.4.4 Let, as above, Σ be the collection of algebraic types of the G-module $\|E\| \backslash \{0\} = \{\|x\| : x \in E, x \neq 0\}$, where E is a Banach space of countable type. A *canonical* (*orthogonal*) *decomposition* of E is a decomposition into an orthogonal direct sum

$$E = \bigoplus_{\sigma \in \Sigma} E_\sigma$$

where each E_σ is a closed subspace and $\|E_\sigma\| \backslash \{0\} = \sigma$.

THEOREM 3.4.5 *Each Banach space E of countable type with an orthogonal base has a canonical decomposition. It is unique in the following sense. If $E = \bigoplus_{\sigma\in\Sigma} E_\sigma = \bigoplus_{\sigma\in\Sigma} F_\sigma$ are two canonical decompositions then, for each σ, E_σ and F_σ are isometrically isomorphic.*

Proof. Let $\{d_i : i \in V\}$ be an orthogonal base of E where either $V = \{1, 2, \ldots, n\}$ for some $n \in \mathbb{N}$ or $V = \mathbb{N}$. For each $\sigma \in \Sigma$ set $V_\sigma := \{i \in V : \|d_i\| \in \sigma\}$ and set $E_\sigma := [d_i : i \in V_\sigma]$. Since the V_σ $(\sigma \in \Sigma)$ form a partition of V we clearly have that $E = \bigoplus_{\sigma\in\Sigma} E_\sigma$ is a canonical decomposition which proves existence. To prove uniqueness, let $E = \bigoplus_{\sigma\in\Sigma} E_\sigma = \bigoplus_{\sigma\in\Sigma} F_\sigma$ be two canonical decompositions, let $\sigma \in \Sigma$, let $\{e_i : i \in W\}$ be an orthogonal base of E_σ, where W is $\{1, \ldots, m\}$ for some $m \in \mathbb{N}$ or $W = \mathbb{N}$. Decompose each e_i as follows

$$e_i = \sum_{\mu\in\Sigma} f_\mu^i$$

where $f_\mu^i \in F_\mu$ for each $\mu \in \Sigma$. Then $\|e_i\| = \max_{\mu\in\Sigma} \|f_\mu^i\|$. But $\|e_i\| \in \sigma$, $\|f_\mu^i\| \in \mu$, so $\|f_\mu^i\| \neq \|e_i\|$ i.e. $\|f_\mu^i\| < \|e_i\|$ if $\mu \neq \sigma$ and therefore we have $\|e_i\| = \|f_\sigma^i\|$ and even $\|e_i - f_\sigma^i\| < \|e_i\|$. By the Perturbation Lemma 2.4.8 the system $\{f_\sigma^i : i \in W\}$ is orthogonal in F_σ. By 2.4.14 each maximal orthogonal set in F_σ has cardinality $\geq \#W$. By symmetry we have equality. It follows that orthogonal bases of E_σ and F_σ have the same cardinality. As $\|E_\sigma\| = \|F_\sigma\| = \sigma \cup \{0\}$ one can construct a bijective isometry $E_\sigma \to F_\sigma$.

REMARK It is not difficult to see that each E_σ occuring in the canonical decomposition is either finite-dimensional or linearly homeomorphic to c_0.

3.5 Compactoids

In rank 1 theory the notion of compactoidity plays a fundamental role in Functional Analysis ([14], 133-146). We recall the definition.

DEFINITION 3.5.1 A subset A of a normed space E is called (a) *compactoid in E* if for each neighbourhood U of zero in E there exists a finite set $\{x_1, \ldots, x_n\} \subset E$ such that $A \subset \text{co}\{x_1, \ldots, x_n\} + U$. Here, $\text{co}\{x_1, \ldots, x_n\}$ is the *absolutely convex hull* of $\{x_1, \ldots, x_n\}$ i.e. $\{\lambda_1 x_1 + \cdots + \lambda_n x_n : \lambda_1, \ldots, \lambda_n \in B_K\}$.

PROPOSITION 3.5.2 *In* (i)-(viii), *E, F are normed spaces.*

(i) *Subsets of a compactoid are compactoid.*
(ii) *The absolutely convex hull of a compactoid is a compactoid.*
(iii) *The closure of a compactoid is a compactoid.*
(iv) *The sum of two compactoids is compactoid.*
(v) *If A is a compactoid in E, $T \in \mathcal{L}(E, F)$ then TA is compactoid in F.*
(vi) *A bounded finite-dimensional subset of E is compactoid in E.*
(vii) *If $Z \subset E$ is precompact then $\overline{\text{co}}\, Z$ is compactoid in E. In particular, if $x_1, x_2, \ldots \in E$, $\lim_{n\to\infty} x_n = 0$ then $\overline{\text{co}}\{x_1, x_2, \ldots\}$ is compactoid in E.*
(viii) *Compactoids are bounded.*

Proof. Only (vi) may need a proof. Let $A \subset K^n$ be bounded with respect to the norm $(\xi_1, \zeta_2, \ldots, \zeta_n) \mapsto \max_i |\xi_i|$. There is a $\lambda \in K$ such that $|\lambda| > \|a\|$ for each $a \in A$. Then $A \subset \mathrm{co}\{\lambda e_1, \lambda e_2, \ldots, \lambda e_n\}$ where $e_1, \ldots, e_n$ is the canonical base of K^n. It follows that A is a compactoid in K^n. Now let A be a bounded subset of an arbitrary normed finite-dimensional space F. By 2.3.4 there is a linear homeomorphism $T : F \to K^n$. By the above, TA is bounded in K^n, hence a compactoid. Then so is $A = T^{-1}TA$.

PROPOSITION 3.5.3 *Let A be a compactoid in a normed space E. Then there exists a space H of countable type with $A \subset H \subset E$ such that A is compactoid in H.*

Proof. Let E be X-normed for some G-module X. Choose $\delta_1 > \delta_2 > \cdots$ in X, $\inf \delta_n = 0$. There is a finite set $F_1 \subset E$ such that $A \subset \mathrm{co}\ F_1 + B_E(0, \delta_1)$. Then $A \subset \mathrm{co}\ F_1 + A_1$ where $A_1 = (A + \mathrm{co}\ F_1) \cap B_E(0, \delta_1)$, which is a compactoid by 3.5.2. There is a finite set $F_2 \subset E$ such that $A_1 \subset \mathrm{co}\ F_2 + B_E(0, \delta_2)$, etc.. Inductively we arrive at finite sets $F_1, F_2, \ldots \subset E$ such that for each n

$$A \subset \mathrm{co}\ (F_1 \cup \cdots \cup F_n) + B(0, \delta_n).$$

It follows that $A \subset \overline{[F]}$ where $F = \bigcup_n F_n$. We see that $H := \overline{[F]}$ is of countable type and that A is compactoid in H.

COROLLARY 3.5.4 *The (closed) linear hull of a compactoid is of countable type.*

Proof. 3.5.3 and 3.2.4.

REMARK In rank 1 theory we always have that if A is a compactoid in a normed space E then it is a compactoid in $\overline{[A]}$, or even in $[A]$. (See [4].) In our case we don't know this is in general, even if K is spherically complete. But we will prove 4.3.7 $(\beta) \Longleftrightarrow (\varepsilon) \Longleftrightarrow (\zeta)$.

THEOREM 3.5.5 *In a compactoid each orthogonal sequence tends to* 0.

Proof. Suppose not. Then we could find an orthogonal sequence $e_1, e_2, \ldots$ in some compactoid A and a $\delta \in \|E\| \backslash \{0\}$ such that $\|e_n\| \geq \delta$ for all n. Choose $\delta' \in \|E\| \backslash \{0\}$, $\delta' < \delta$. By compactoidity there is a finite-dimensional subspace D of E such that $A \subset D + B_E(0, \delta')$. For each n, write $e_n = d_n + \delta_n$ when $d_n \in D$, $\|\delta_n\| < \delta'$. Then $\|e_n - d_n\| < \|e_n\|$ for each n, so by the Perturbation Lemma 2.4.8, $d_1, d_2, \ldots$ are orthogonal, and non-zero hence linearly independent which conflicts the finite dimensionality of D.

THEOREM 3.5.6 *Let A be a compactoid in a Banach space E with an orthogonal base $e_1, e_2, \ldots$. Then there are absolutely convex subsets $C_1, C_2, \ldots$ in K such that $\operatorname{diam} C_n e_n \to 0$ and such that $A \subset \overline{C_1 e_1 + C_2 e_2 + \cdots}$.*

Proof. Let E be X-normed for some G-module X. For each $n \in \mathbb{N}$, let P_n be the canonical projection $E \to Ke_n$. We prove that $\sum_n P_n A$ is a compactoid and that diam $P_nA \to 0$ (which will finish the proof since we may assume that A is absolutely convex, so P_nA has the form C_ne_n for some absolutely convex set C_n in K, and since $A \subset \overline{\sum_n P_nA}$). Let $\varepsilon \in X$. There exists a finite set $F \subset E$ such that $A \subset \text{co } F + B_E(0,\varepsilon)$. There is an $m \in \mathbb{N}$ such that $F \subset [e_1,\ldots,e_m] + B_E(0,\varepsilon)$, so we may assume $F \subset [e_1,\ldots,e_m]$. For each n we have, since $\|P_n x\| \le \|x\|$ for all x, $P_nA \subset \text{co } P_nF + B_E(0,\varepsilon)$. Adding up and observing that $P_nF = \{0\}$ for $n > m$ we arrive at $\sum_n P_nA \subset \text{co}(\bigcup_{1\le n\le m} P_nF) + B_E(0,\varepsilon)$, proving that $\sum_n P_nA$ is a compactoid. We also see that $P_nA \subset B_E(0,\varepsilon)$ for $n > m$ implying diam $P_nA \to 0$.

REMARK In rank 1 theory one can even prove that there are $\lambda_1, \lambda_2, \ldots \in K$ with $\lim_n \lambda_n e_n = 0$ and $A \subset \overline{\text{co}}\{\lambda_1 e_1, \lambda_2 e_2, \ldots\}$, see [4]. However, we will show that this result no longer holds in our theory.

EXAMPLE 3.5.7 (A compactoid, not contained in the closed absolutely convex hull of a sequence tending to 0.) Let G be the union of a strictly increasing sequence of convex subgroups

$$\{1\} \subsetneqq H_1 \subsetneqq H_2 \subsetneqq \cdots$$

For each n, let $t_n := \inf_{G^\#} H_n$, $s_n := \sup_{G^\#} H_n$. Let (2.4.15) $E := c_0(\mathbb{N}, s)$ where $s(n) = t_n$ $(n \in \mathbb{N})$ i.e. the space of all sequences $(\lambda_1, \lambda_2, \ldots)$ in $K^{\mathbb{N}}$ for which $\lim_n |\lambda_n| t_n = 0$. For the canonical base $e_1, e_2, \ldots$ of E we have $\|e_n\| = t_n$ for each n. From 1.5.4 and its proof we have (in the G-module $G^\#$) for each $h \in H_n$ that $ht_n = \inf\{hg : g \in G : g > t_n\} = \inf H_n = t_n$, so, if $C_n := \{\lambda \in K : |\lambda| < s_n\}$ then diam $C_ne_n = t_n \to 0$. So $A := \sum_n C_ne_n$ is a compactoid. We now prove that, if $\lambda_n \in K$ $\lambda_n \notin C_n$ then $\|\lambda_n e_n\| \not\to 0$. In fact, by using 1.5.4 we have $\|\lambda_n e_n\| = |\lambda_n| t_n = \sup\{|\lambda_n| g : g \in G, g < t_n\} \ge |\lambda_n|\, |\lambda_n|^{-1} = 1$. We see that $\overline{A}$ is contained in $\overline{\text{co}}\{\lambda_n e_n : n \in \mathbb{N}\}$, for no λ_n for which $\lim_n \lambda_n e_n = 0$. We finish the proof by applying the next lemma.

LEMMA 3.5.8 *Let E be a Banach space with an orthogonal base $e_1, e_2, \ldots$. Then, for every sequence $x_1, x_2, \ldots$ in E tending to 0, there are $\lambda_1, \lambda_2, \ldots \in K$ such that $\overline{\text{co}}\{x_1, x_2, \ldots\} \subset \overline{\text{co}}\{\lambda_1 e_1, \lambda_2 e_2, \ldots\}$ and $\|\lambda_n e_n\| \to 0$.*

Proof. For each n, let

$$x_n = \sum_{i=1}^{\infty} \xi_i^n e_i$$

be the expansion of x_n. Then from $\|\xi_i^n e_i\| \le \|x_n\|$ for each i and n, and from $\lim_{n\to\infty} x_n = 0$ we obtain $\lim_{n\to\infty} \xi_i^n = 0$ (1.5.2), so there is a $\lambda_i \in K$ with $|\lambda_i| = \max_n |\xi_i^n|$. It is easily seen that $\lim_i |\lambda_i|\, \|e_i\| = 0$ and that $x_n = \sum_i \xi_i^n e_i \in \overline{\text{co}}\{\lambda_i e_i : i \in \mathbb{N}\}$ for each n.

The *weak topology* on a normed space E is the weakest topology for which all $f \in E'$ are continuous. In rank 1 theory it is shown that the weak and norm topology coincide on compactoids under the assumption that E is a so-called polar space (see [16], 5.12). The proof does not carry over to the arbitrary rank case. We have only the following corollary of 3.5.6.

COROLLARY 3.5.9 *In a Banach space with an orthogonal base the weak and norm topology coincide on compactoids.*

Proof. By 3.5.6 we may assume that $A = \overline{\sum_n C_n e_n}$ where $e_1, e_2, \ldots$ is an orthogonal base of E and $C_1, C_2, \ldots$ are absolutely convex subsets in K such that $s_n := \text{diam } C_n e_n := \sup_X \{\|\lambda e_n\| : \lambda \in C_n\} \to 0$, where E is X-normed for some complete G-module X. Then let $i \mapsto a_i \quad (i \in I)$ be a net in A converging weakly to 0. Let, for each i,

$$a_i = \sum_{n=1}^{\infty} \lambda_n^i e_n \qquad\qquad (\lambda_n^i \in C_n)$$

be the expansion of a_i. The coordinate maps are continuous so $\lim_i \lambda_n^i = 0$ for each n, hence $\lim_i \lambda_n^i e_n = 0$ for each n. Let $\varepsilon \in X$. There is an $m \in \mathbb{N}$ such that $s_n < \varepsilon$ for $n > m$. There is an $i_0 \in I$ such that $\|\lambda_n^i e_n\| < \varepsilon$ for $n \in \{1, \ldots, m\}$, $i \geq i_0$. It follows that $\|a_i\| < \varepsilon$ for $i \geq i_0$ proving that $\lim_i \|a_i\| = 0$.

4 HILBERT-LIKE SPACES

Recall that we assume throughout that K is complete and satisfies the conditions $(\alpha) - (\delta)$ of Proposition 1.4.4. In this Section we study the class of norm-Hilbert spaces over K (see Definition 4.1.1 below) and the -what will turn out to be- subclass of the form-Hilbert spaces introduced by H. Gross, H. Keller and U.M. Künzi in [5], [3], see 4.4.

4.1 Norm-Hilbert spaces

DEFINITION 4.1.1 A Banach space E of countable type will be called *norm-Hilbert space* if for each (norm-) closed subspace D of E there exists a linear surjective projection $P : E \to D$ for which $\|Px\| \leq \|x\| \quad (x \in E)$ (Compare the form-Hilbert spaces of 4.4.3).

We leave the proof of the following Proposition to the reader.

PROPOSITION 4.1.2 *Let E be a Banach space of countable type. Then the following are equivalent.*

(α) *E is a norm-Hilbert space.*

(β) *Each norm-closed subspace of E has a normorthogonal complement.*

(γ) *Every orthogonal system of nonzero vectors in E can be extended to an orthogonal base of E.*

(δ) *For every closed subspace D of E and every $a \in E$ the set $\{\|a - d\| : d \subset D\}$ has a minimum.*

Theorem 4.1.3 below characterizes the norm-Hilbert spaces. It is an extension of [14], 5.16 which treats the rank 1 case. Let us say, by abuse of language, that a sequence $x_1, x_2, \ldots$ in a normed space E is *decreasing* if $\|x_1\| \geq \|x_2\| \geq \cdots$, *strictly decreasing* if $\|x_1\| > \|x_2\| > \cdots$. In the same spirit we define (*strictly*) *increasing* sequences in E.

THEOREM 4.1.3 *For an infinite-dimensional Banach space of countable type E the following statements* (α) *and* (β) *are equivalent.*

(α) *E is a norm-Hilbert space.*

(β) (i) *E has an orthogonal base.*

(ii) *Every strictly decreasing orthogonal sequence in E tends to* 0.

(iii) *If G has a maximal proper convex subgroup H then $G/H \simeq \mathbb{Z}$.*

The proof runs in several steps. First two lemmas.

LEMMA 4.1.4 *If there exists an infinite-dimensional norm-Hilbert space over a field K whose valuation has rank* 1 *then the valuation of K is discrete.*

Proof. Let E be such a space, let E be X-normed for some G-module X. We may assume that $G \subset (0,\infty)$ (see [13]). Then $G^{\#} \subset (0,\infty)$. Now let $\phi : X \to G^{\#}$ be as in 1.5.6. Like in the proof of 2.1.9 it follows that its extension $\phi : X \cup \{0\} \to [0,\infty)$ is an extended G-module map and that $x \mapsto \phi(\|x\|)$ is equivalent to $\|\cdot\|$. If P is a linear projection onto a closed subspace and $\|Px\| \le \|x\|$ then $\phi(\|Px\|) \le \phi(\|x\|)$, so E is a norm-Hilbert space with respect to $\phi(\|\cdot\|)$. Now apply [14], 5.16 to conclude that K has discrete valuation.

LEMMA 4.1.5 *Let E be a Banach space.*

(i) *If D_1 and D_2 are spherically complete subspaces and D_1 and D_2 are normorthogonal then $D_1 + D_2$ is spherically complete.*

(ii) *If every strictly decreasing sequence in E tends to* 0 *then E is spherically complete.*

Proof. The proof of (i) is straightforward. To prove (ii) observe that to prove spherical completeness it suffices to show that any sequence of 'open' balls $B(a_1, r_1^-) \supset B(a_2, r_2^-) \supset \cdots$ has a nonempty intersection (see our assumption at the beginning of this Section). We may assume $B(a_n, r_n^-) \neq B(a_{n+1}, r_{n+1}^-)$ for all n. Choosing $b_n \in B(a_n, r_n^-) \backslash B(a_{n+1}, r_{n+1}^-)$ we obtain $\|b_1 - b_2\| > \|b_2 - b_3\| > \cdots$, so by assumption, $b_{n+1} - b_n \to 0$. By completeness $b := \lim_{n\to\infty} b_n$ exists and it follows easily that $b \in \bigcap_n B(a_n, r_n^-)$.

Proof of Theorem 4.1.3. (α) $\Rightarrow$ (β). Clearly we have (i) (every maximal orthogonal family in $E\backslash\{0\}$ is an orthogonal base). We proceed to prove (ii) (compare the proof of (α) $\Rightarrow$ (ι) of [14], 5.16). Let $e_1, e_2, \ldots$ be a strictly decreasing orthogonal sequence in E. Suppose $\|e_n\| > s$ for each $n \in \mathbb{N}$ and some nonzero norm value s. Let $D := [e_1, e_2, \ldots]$. The formula

$$\phi\Big(\sum_{\iota=1}^{\infty} \xi_\iota e_\iota\Big) = \sum_{\iota=1}^{\infty} \xi_\iota \qquad (\xi_\iota \in K, \|\xi_\iota e_\iota\| \to 0)$$

defines an element $\phi \in D'$. In fact, $\|\xi_\iota e_\iota\| \to 0$ is equivalent to $\xi_\iota \to 0$, so ϕ is a well-defined linear map $D \to K$. To prove continuity, let $n \mapsto x_n = \sum_{\iota=1}^{\infty} \xi_\iota^n e_\iota$ be a sequence in D tending to 0. Then since $\|x_n\| \ge (\max_\iota |\xi_\iota^n|)s$ we have $|\sum_\iota \xi_\iota^n| \le \max_\iota |\xi_\iota^n| \to 0$. Now D is a norm-Hilbert space so Ker ϕ has a normorthogonal complement. So there is an $a \in D$ such that Ka is normorthogonal to Ker ϕ,

$\phi(a) = 1$, $a = \sum_{n=1}^{\infty} \lambda_n e_n \quad (\lambda_n \in K, \lambda_n \to 0)$. From

$$1 = |\phi(a)| = |\sum_{n=1}^{\infty} \lambda_n| \leq \max |\lambda_n|$$

it follows that $|\lambda_i| \geq 1$ for some i and

$$\|a\| = \max \|\lambda_n e_n\| \geq \|\lambda_i e_i\| \geq \|e_i\| > \|e_{i+1}\|.$$

On the other hand, $\phi(a - e_{i+1}) = 0$ so a and $a - e_{i+1}$ are normorthogonal whence

$$\|e_{i+1}\| = \|a - (a - e_{i+1})\| \geq \|a\|,$$

a contradiction which proves (ii). To prove (iii), let X be the G-module $\|E\| \backslash \{0\}$, and let $X/\sim$ be the canonical G/H-module constructed in 1.5.5 (f), let $\pi : G \to G/H$, $\rho : X \to X/\sim$ be the natural maps. Let $N(x) := \rho(\|x\|) \quad (x \in E, x \neq 0)$, $N(0) := 0$, let $v(\lambda) = \pi(|\lambda|)$ $(\lambda \in K, \lambda \neq 0)$, $v(0) := 0$. Then v is a valuation on K of rank 1, equivalent to $|\ |$, and N is a norm on E. For any sequence $x_1, x_2, \ldots$ in E we have $\|x_n\| \to 0$ if and only if $N(x_n) \to 0$. If P is a linear projection onto a closed subspace D and $\|Px\| \leq \|x\| \quad (x \in E)$ then $N(Px) \leq N(x) \quad (x \in E)$. It follows that (E, N) is a norm-Hilbert space over (K, v). By 4.1.4 v is a discrete valuation i.e. $G/H \simeq \mathbb{Z}$.

Proof of Theorem 4.1.3. $(\beta) \Rightarrow (\alpha)$. Let $e_1, e_2, \ldots$ be a maximal orthogonal sequence of nonzero vectors in E, let $F := \overline{[e_1, e_2, \ldots]}$; we prove that $F = E$. To this end, let $F = \bigoplus_{\sigma \in \Sigma} F_\sigma$ be the canonical orthogonal decomposition of F in the sense of 3.4.4. Let $\sigma \in \Sigma$, let $s \in \sigma$ be a representative. We first prove (1) and (2) below.

(1) *If* $\mathrm{Stab}(s) = \{g \in G : gs = s\}$ *is not a maximal proper convex subgroup of* G *then* $\dim F_\sigma < \infty$.

(2) *If* $\mathrm{Stab}(s)$ *is a maximal convex subgroup then* F_σ *is spherically complete.*

Proof of (1). If $\dim F_\sigma = \infty$ it would have an orthogonal base $f_1, f_2, \ldots$ with $\|f_n\| = s$ for all n. We have rank $G/\mathrm{Stab}(s) > 1$, so there is a sequence $v_1 > v_2 > \cdots$ in $G/\mathrm{Stab}(s)$ with $v_n > 1$ for all n. Choose $\lambda_1, \lambda_2, \ldots \in K$ with $\pi(|\lambda_n|) = v_n$ for each n (where $\pi : G \to G/\mathrm{Stab}(s)$ is the canonical map). Then $|\lambda_1| > |\lambda_2| > \cdots$ so that $\|\lambda_1 f_1\| \geq \|\lambda_2 f_2\| \geq \ldots$. If, for some n, $\|\lambda_n f_n\| = \|\lambda_{n+1} f_{n+1}\|$ then $|\lambda_{n+1}^{-1} \lambda_n| s = s$, so $|\lambda_{n+1}^{-1} \lambda_n| \in \mathrm{Stab}(s)$ implying $v_{n+1} = v_n$, a contradiction. Hence $\|\lambda_1 f_1\| > \|\lambda_2 f_2\| > \cdots$ and therefore $\lim_{n\to\infty} \lambda_n f_n = 0$ i.e. $\lim_{n\to\infty} |\lambda_n| s = 0$, i.e. $|\lambda_n| \to 0$ or $v_n = \pi(|\lambda_n|) \to 0$, a contradiction.

Proof of (2). We prove (Lemma 4.1.5 (ii)) that every strictly decreasing sequence of norm values in F_σ tends to 0. Now, since $\|F_\sigma\| = Gs \cup \{0\}$ such a sequence has the form $|\lambda_1| s > |\lambda_2| s > \cdots$. Letting $\pi : G \to G/\mathrm{Stab}(s)$ be the canonical map we have $\pi(|\lambda_1|) > \pi(|\lambda_2|) > \cdots$. By (iii), $G/\mathrm{Stab}(s) \simeq \mathbb{Z}$, so that $\lim_{n\to\infty} \pi(|\lambda_n|) = 0$ implying $|\lambda_n| s \to 0$. With (1), (2) being proved, let $\Sigma = \{\sigma_1, \sigma_2, \ldots\}$, let $H_n := F_{\sigma_1} + F_{\sigma_2} + \cdots + F_{\sigma_n} \quad (n \in \mathbb{N})$. H_n is the orthogonal direct sum of a spherically complete space A (the sum of all F_{σ_i} which are spherically complete, 4.1.5 (i)) and a finite-dimensional space (the sum of the other F_{σ_i}). By 2.4.4, A has an orthogonal complement in E. We have $A \subset H_n$ and H_n/A is finite-dimensional so by the proof of 3.3.1 also H_n is orthocomplemented. Now let $x \in E$. To prove $x \in F$ we may assume that x is not in the union of the H_n. There exist $h_n \in H_n \quad (n \in \mathbb{N})$ such that $\|x - h_n\| = \operatorname{dist}(x, H_n)$. Then $x - h_n \perp H_n$ for each n. Now F_{σ_k} has an orthogonal base which is a maximal orthogonal set in $\{x \in E : \|x\| \in Gs_k\}$ for each k (2.4.13), hence $\|x-h_n\| \notin Gs_1 \cup Gs_2 \cup \cdots \cup Gs_n$. Thus the sequence $\|x-h_1\| \geq \|x-h_2\| \geq \cdots$ has a subsequence $i \mapsto \|x - h_{n_i}\|$ for which $\|x - h_{n_i}\| \notin G\|x - h_{n_j}\|$ whenever $i \neq j$. Then this subsequence is strictly decreasing. Orthogonality of $x - h_{n_1}, x - h_{n_2}, \ldots$ follows from 3.2.8. By (ii), $\lim_{i\to\infty} \|x - h_{n_i}\| = 0$ i.e. $x \in [h_1, h_2, \ldots] \subset F$.

4.2 Examples of norm-Hilbert spaces

We will present two groups of examples, namely in cases where G has or has not a maximal proper convex subgroup.

EXAMPLE 4.2.1 *An infinite-dimensional norm-Hilbert space over a field whose value group has a maximal proper convex subgroup.* Let G_1 be a linearly ordered abelian group satisfying the countability conditions of Proposition 1.4.4 (e.g., $G_1 = (0, \infty)$), let $G := G_1 \times \mathbb{Z}$, where $\mathbb{Z}$ is written multiplicatively with generator $a > 1$, G ordered antilexicographically and let K be a complete valued field with value group G. Choose $b_1, b_2, \ldots \in \mathbb{R}$ such that $1 < b_1 < b_2 < \cdots < a$ and put

$$X := \{(r, a^n b_m) : r \in G_1, \quad n \in \mathbb{Z}, \quad m \in \mathbb{N}\}.$$

With the ordering inherited from the antilexicographic ordering on $G_1 \times (0, \infty)$ and the structure map given by the formula

$$(s, a^k) \cdot (r, a^n b_m) = (sr, a^{n+k} b_m)$$

X becomes a G-module. Let $E := c_0$ but with the norm given by

$$\|x\| = \max_n |\xi_n|(1, b_n) \qquad (x = (\xi_1, \xi_2, \ldots))$$

(Indeed, since $(1,1) < (1, b_n) < (1, a)$ for all n we have that $\|\cdot\|$ is equivalent to the usual norm). We prove that $(E, \|\cdot\|)$ is a norm-Hilbert space by verifying (β) of Theorem 4.1.3. Clearly $(1, 0, 0, \ldots), (0, 1, 0, 0, \ldots)$ is an orthogonal base for E; $G_1 \times \{1\}$ is the maximal proper convex subgroup and $G/G_1 \times \{1\} \simeq \mathbb{Z}$, so it suffices to check (ii). To this end, let $f_1, f_2, \ldots$ be an orthogonal sequence such that, with $\|f_m\| = (r_m, a^{n_m} b_{s_m}) \quad (m \in \mathbb{N})$ we have

$$(r_1, a^{n_1} b_{s_1}) > (r_2, a^{n_2} b_{s_2}) > \cdots.$$

In the canonical orthogonal decomposition $E = \bigoplus_{\sigma\in\Sigma} E_\sigma$ of E each E_σ is onedimensional; therefore we must have that

$$a^{n_m} b_{s_m} \neq a^{n_{m+1}} b_{s_{m+1}}, \text{ hence } a^{n_m} b_{s_m} > a^{n_{m+1}} b_{s_{m+1}}$$

for each m. We see that $a^{n_m - n_{m+1}} > b_{s_{m+1}}/b_{s_m} > a^{-1}$, hence $n_m - n_{m+1} > -1$ or $n_m \geq n_{m+1}$. If $n_k = n_{k+1} = \ldots$ for some k we would have $b_{s_k} > b_{s_{k+1}} > \cdots$, which is impossible since the set $\{b_n : n \in \mathbb{N}\}$ is well-ordered. So $\lim_{k\to\infty} n_k = -\infty$ proving that $\lim_{m\to\infty} \|f_m\| = 0$.

REMARK In the above example G is of infinite rank, finite rank, rank 1 accordingly as G_1 is of infinite rank, of finite rank, $\{1\}$ respectively. We will see in 4.4.6 that there exist no infinite-dimensional form-orthogonal Hilbert spaces when G has a maximal convex proper subgroup.

EXAMPLE 4.2.2 *An infinite-dimensional norm-Hilbert space over a field whose value group does not have maximal proper convex subgroups.* Let G be the union of a strictly increasing sequence of convex subgroups $\{1\} = H_1 \subset H_2 \subset \ldots$. Set $s_n := \sup_{G^\#} H_n$, $t_n := \inf_{G^\#} H_n$ $(n \in \mathbb{N})$ and let E be the set of all $x = (\xi_1, \xi_2, \ldots) \in K^{\mathbb{N}}$ for which $\lim_{n\to\infty} |\xi_n| s_n = 0$ and where $\|x\| = \max_n |\xi_n| s_n$. Then clearly E is a $G^\#$-normed Banach space with orthogonal base $(1,0,0,\ldots), (0,1,0,\ldots), \ldots$. To prove E to be norm-Hilbertian, let $\xi_1, \xi_2, \ldots \in K$ be such that $n \mapsto |\xi_n| s_n$ is strictly decreasing; we show $\lim_{n\to\infty} |\xi_n| s_n = 0$. Let $\varepsilon \in G$; there is an m such that $t_m < \varepsilon$ and $|\xi_1| s_1 < s_m$. We claim that $|\xi_n| s_n \le t_m$ for all $n > m$. In fact, let $n > m$. If $|\xi_n| \in H_n$ then $|\xi_n| s_n = \sup_{h \in H_n} |\xi_n| h = s_n > s_m$, a contradiction. If $|\xi_n| > s_n$ then $|\xi_n| s_n \ge s_n > s_m$ which is again a contradiction. We see that $|\xi_n| < t_n$ for $n > m$. Then $|\xi_n| s_n = \sup_{h \in H_n} |\xi_n| h \le t_n \le t_m$ and we are done.

REMARK We are particularly interested in norm-Hilbert spaces over fields whose value group does not have maximal proper convex subgroups for two reasons. Firstly, because there do exist form-Hilbert spaces over such fields (see 4.4.9), secondly because such spaces have particular properties such as: each bounded set is a compactoid! (See 4.3.7). We devote the next section to the study of these so-called 'Keller spaces' named after the inventor of the first non-classical form-Hilbert space [5].

4.3 The Keller spaces

LEMMA 4.3.1 *The following statements on K are equivalent.*

(α) *The value group G does not have maximal proper convex subgroups.*

(β) *G is the union of a strictly increasing sequence of convex subgroups.*

(γ) *There is a G-module X and a sequence $s_1, s_2, \ldots$ in X satisfying the type condition.* (See 1.6.4)

Proof. $(\alpha) \Rightarrow (\beta)$. Let $g_1 < g_2 < \cdots$ be a cofinal sequence in G, let H_1 be the smallest convex subgroup containing g_1. If H_1 were equal to G then by [9], Prop. 3, page 14, G would have a maximal proper subgroup, so $H_1 \ne G$, and there exists an $n_2 > n_1 := 1$ such that $g_{n_2} \notin H_1$. Then the convex subgroup generated by g_{n_2} contains H_1 properly and is not equal to G for the same reason as above for H_1, etc.. We obtain a strictly increasing sequence $H_1 \subset H_2 \subset \ldots$ of convex subgroups, their union is cofinal so it must be equal to G. $(\beta) \Rightarrow (\gamma)$. We proved in 4.2.2 that the sequence $s_1, s_2, \ldots$ in $G^\#$ satisfies the type condition. (In fact, the conclusion can be drawn just from the assumption that $n \mapsto |\xi| s_n$ is bounded above, choose m such that $t_m < \varepsilon$ and $|\xi_n| s_n < s_m$ for each n.) $(\gamma) \Rightarrow (\alpha)$. This is the first part of the proof of 1.6.6, $(\alpha) \Rightarrow (\beta)$.

DEFINITION 4.3.2 Varying on 1.6.4 a sequence of non-zero vectors $x_1, x_2, \ldots$ in a normed space is said to satisfy *the type condition* if, for each sequence $\alpha_1, \alpha_2, \ldots$ in K, boundedness above of $\{\alpha_n x_n : n \in \mathbb{N}\}$ implies $\lim_{n\to\infty} \alpha_n x_n = 0$.

It is not hard to see that $x_1, x_2, \ldots$ satisfies the type condition in the sense of 4.3.2 if and only if $\|x_1\|, \|x_2\|, \ldots$ satisfies the type condition in the sense of 1.6.4.

DEFINITION 4.3.3 A Banach space E over K is called a *Keller space* if K satisfies $(\alpha), (\beta), (\gamma)$ of 4.3.1 and for each closed subspace D of E there is a linear surjective projection $P : E \to D$ for which $\|Px\| \leq \|x\| \quad (x \in E)$.

FROM NOW ON IN 4.3 WE ASSUME K TO SATISFY $(\alpha) - (\gamma)$ OF 4.3.1.

Before proving our Main Theorem we first prove that c_0 is not a Keller space (4.3.4) and that a Keller space is of countable type.

LEMMA 4.3.4 *c_0 has a closed subspace without closed complement.*

Proof. Let E be the Keller space constructed in 4.2.2. By Theorem 3.2.6 it is a quotient of c_0, so let $\pi : c_0 \to E$ be a quotient map. If Ker π had a closed complement D then it would be linearly homeomorphic to E by Banach's Open Mapping Theorem. It follows that D has a Schauder base $e_1, e_2, \ldots$ satisfying the type condition. But on the other hand, for each Schauder base $f_1, f_2, \ldots$ of D we can arrange that $\|f_n\| = 1$ for all n implying that no Schauder base of D can satisfy the type condition, a contradiction.

COROLLARY 4.3.5 *A Keller space is of countable type. In particular, it is a norm-Hilbert Space.*

Proof. Let E be an X-normed Keller space. X has a coinitial sequence $t_1 > t_2 > \cdots$ and a cofinal sequence $s_1 < s_2 < \cdots$. Then $B_n := \{x \in X : t_n < x < s_n\}$ is bounded above and below in X for each $n \in \mathbb{N}$. Now let $\{e_i : i \in I\}$ be a maximal orthogonal set of nonzero vectors and suppose I is uncountable; it suffices to derive a contradiction. For each $i \in I$ there is an $n(i) \in \mathbb{N}$ such that $\|e_i\| \in B_{n(i)}$. By uncountability there exists an $m \in \mathbb{N}$ for which $S := \{i \in I : n(i) = m\}$ is infinite, so assume $S \supset \mathbb{N}$. Then $t_m \leq \|e_n\| \leq s_m$ for all $n \in \mathbb{N}$, so

$$(\lambda_1, \lambda_2, \ldots) \mapsto \sum_{n=1}^{\infty} \lambda_n e_n$$

is a linear homeomorphism of c_0 onto a closed subspace of E. But a closed subspace of E is a Keller space and cannot be isomorphic to c_0 according to the previous lemma, a contradiction.

LEMMA 4.3.6 *Let E be an X-normed space for some G-module X, let $x_1, x_2, \ldots$ be a sequence in E satisfying the type condition. Let $M \in X$, let $\Lambda_M = \{(\lambda_1, \lambda_2, \ldots) \in$*

$K^{\mathbb{N}} : \|\lambda_n x_n\| \leq M$ *for all* $n \in \mathbb{N}\}$. *Then* $\lim_{n\to\infty} \lambda_n x_n = 0$ *uniformly on* $(\lambda_1, \lambda_2, \ldots) \in \Lambda_M$.

Proof. Suppose not. Then there would be an $\varepsilon \in X$ such that for all $n \in \mathbb{N}$ we could find a $(\lambda_1, \lambda_2, \ldots) \in \Lambda_M$ and $i > n$ such that $\|\lambda_i x_i\| > \varepsilon$, i.e. $\varepsilon < \|\lambda_i x_i\| \leq M$. This would imply that some subsequence of $x_1, x_2, \ldots$ does not satisfy the type condition which conflicts 1.6.5 (i).

We now formulate the Main Theorem characterizing Keller spaces in several ways.

THEOREM 4.3.7 *Let G be the union of a strictly increasing sequence of convex subgroups. Then, for an infinite-dimensional K-Banach space E with an orthogonal base* $e_1, e_2, \ldots$ *the following are equivalent.*

(α) *E is a Keller space.*
(β) *E is a norm-Hilbert space.*
(γ) $e_1, e_2, \ldots$ *satisfies the type condition.*
(δ) *Each orthogonal sequence in E satisfies the type condition.*
(ε) *Each bounded subset of E is a compactoid in E.*
(ζ) *Each bounded subset A of E is a compactoid in* $\overline{[A]}$.
(η) *Every closed subspace has a closed complement.*
(θ) *No subspace of E is linearly homeomorphic to* c_0.

Proof. Let us suppose that E is X-normed for some G-module X. Since E is of countable type, (α) $\Longleftrightarrow$ (β) is obvious. We prove (γ) $\Rightarrow$ (ε) $\Rightarrow$ (δ) $\Rightarrow$ (ζ) $\Rightarrow$ (β) $\Rightarrow$ (η) $\Rightarrow$ (θ) $\Rightarrow$ (γ). (γ) $\Rightarrow$ (ε). We prove that $B_E(0,r)$ is a compactoid in E for each $r \in X$. Let $\Lambda := \{(\lambda_1, \lambda_2, \ldots) \in K^{\mathbb{N}} : \lim_{n\to\infty} \lambda_n e_n = 0$ and $\|\sum_{n=1}^{\infty} \lambda_n e_n\| \leq r\}$. By orthogonality and the type condition we have $\Lambda = \{(\lambda_1, \lambda_2, \ldots) \in K^{\mathbb{N}} : \|\lambda_n e_n\| \leq r$ for each $n\}$. Now let $\varepsilon \in X$. By 4.3.6 there is an N such that $\|\lambda_n e_n\| < \varepsilon$ for all $n > N$ and all $(\lambda_1, \lambda_2, \ldots) \in \Lambda$. Now, let $x \in B_E(0,r)$ have expansion $x = \sum_{n=1}^{\infty} \lambda_n e_n$. Then $x \in \lambda_1 e_1 + \cdots + \lambda_N e_N + B_E(0,\varepsilon)$. Choose $\mu \in K$ such that $|\mu|\,\|e_i\| > r$ for all $i \in \{1, \ldots, N\}$. Then $|\mu|\,\|e_i\| > |\lambda_i|\,\|e_i\|$ hence $|\mu| > |\lambda_i|$ for $i \in \{1, \ldots, N\}$ and therefore $\lambda_1 e_1 + \cdots + \lambda_N e_N \in \mathrm{co}\{\mu e_1, \ldots, \mu e_N\}$. So $B_E(0,r) \subset \mathrm{co}\{\mu e_1, \ldots, \mu e_N\} + B_E(0,\varepsilon)$ proving compactoidity. To prove (ε) $\Rightarrow$ (δ), let $f_1, f_2, \ldots$ be an orthogonal sequence in E, let $\lambda_1, \lambda_2, \ldots \in K$ be such that $\{\lambda_n f_n : n \in \mathbb{N}\}$ is bounded (above). Then this set is a compactoid by assumption and orthogonality implies by 3.5.5 that $\lim_{n\to\infty} \lambda_n f_n = 0$. The implication ($\delta$) $\Rightarrow$ (γ) being trivial we have established the equivalence of (γ), (ε), (δ). We now prove (δ) $\Rightarrow$ (ζ). The space $\overline{[A]}$ satisfies (δ) (with E replaced by $\overline{[A]}$) and, by the equivalence of above, also (ε) (with E replaced by $\overline{[A]}$) which is (ζ). We proceed to prove (ζ) $\Rightarrow$ (β). By 4.1.3 it suffices to show that every strictly decreasing orthogonal sequence tends to 0. But this is clear from compactoidity and 3.5.5. The implication (β) $\Rightarrow$ (η) is obvious. We continue with (η) $\Rightarrow$ (θ). Let D be a closed subspace of E. Let F be a closed subspace of D. By (η) F has a closed complement C in E. Then $C \cap D$ is a closed complement of F in D, proving that D satisfies (η), and from 4.3.4 it follows that D cannot be linearly homeomorphic to c_0. Finally we prove (θ) $\Rightarrow$ (γ). Let $\lambda_1, \lambda_2, \ldots \in K$ be such that $\{\|\lambda_n e_n\| : n \in \mathbb{N}\}$ is bounded above, say by $M \in X$. If not $\lambda_n e_n \to 0$ we would have a $\delta \in X$, a subsequence $n_1 < n_2 < \cdots$ of 1,2,3,... such that

$$\delta \leq \|\lambda_{n_i} e_{n_i}\| \leq M \qquad (i \in \mathbb{N}).$$

But then the formula

$$T((\xi_1, \xi_2, \ldots)) = \sum_{i=1}^{\infty} \xi_i \lambda_{n_i} e_{n_i}$$

defines a linear homeomorphism of c_0 onto $Tc_0 \subset E$ conflicting (θ).

COROLLARY 4.3.8 *Closed subspaces and quotients of Keller spaces are Keller spaces. If two Banach spaces with an orthogonal base are linearly homeomorphic and one is a Keller space then so is the other.*

The canonical decomposition (see 3.4.4) is suited to characterizing spaces (containing subspaces that are) isomorphic to c_0 or a Keller space. Recall that the topological type of an element of a G-module X depends only on the algebraic type, i.e. (see the introduction of 1.6 and 1.6.1) $\tau(gs) = \tau(s)$ for all $s \in X$, $g \in G$. For the next theorem it is more convenient to define the type function as a map defined on the collection Σ of all algebraic types (with values in the collection of all proper convex subgroup of G) via the formula

$$Gs \longmapsto \tau(s).$$

We shall denote this type function by $\overline{\tau}$. We will say that $\lim_\sigma \overline{\tau}(\sigma) = \infty$ if for each proper convex subgroup H of G we have $\overline{\tau}(\sigma) \subset H$ for only finitely many $\sigma \in \Sigma$. (This ties in with Definition 1.6.4 (ii)). We will say that $\overline{\tau}$ is bounded if there is a proper convex subgroup H of G such that $\overline{\tau}(\sigma) \subset H$ for all $\sigma \in \Sigma$.

We will say that a Banach space E *contains* a Banach space F if there exists a linear homeomorphism of F onto a subspace of E.

THEOREM 4.3.9 *Let E be an infinite-dimensional Banach space of countable type with an orthogonal base. Then*

(i) *E is a Keller space if and only if it does not contain c_0,*

(ii) *E is linearly homeomorphic to c_0 if and only if it does not contain an infinite-dimensional Keller space.*

Proof. (i) Follows from 4.3.7 (α) $\Longleftrightarrow$ (θ); (ii) is a consequence of the next theorem.

THEOREM 4.3.10 *Let E be an infinite-dimensional Banach space of countable type with an orthogonal base, let Σ be the set of algebraic types of $Y := \{\|x\| : x \in E, x \neq 0\}$, let*

$$E = \bigoplus_{\sigma \in \Sigma} E_\sigma$$

be the canonical decomposition of E. Then we have the following.

(i) *E contains c_0 if and only if E is bounded or dim $E_\sigma = \infty$ for some $\sigma \in \Sigma$.*

(ii) *E contains an infinite-dimensional Keller space if and only if $\overline{\tau}$ is unbounded.*

(iii) *E is linearly homeomorphic to c_0 if and only if $\overline{\tau}$ is bounded.*

(iv) *E is a Keller space if and only if each E_σ is finite-dimensional and* $\lim_\sigma \overline{\tau}(\sigma) = \infty$.

Proof. We prove (ii), (iii) and (iv) ((i) follows from (iv) and 4.3.9 (i)). If E contains an infinite-dimensional Keller space it contains an orthogonal sequence $e_1, e_2, \ldots$ satisfying the type condition; so by Theorem 1.6.6 we have $\lim_n \tau(\|e_n\|) = \infty$ i.e. $\lim_n \overline{\tau}(\sigma_n) = \infty$ for some sequence $\sigma_1, \sigma_2, \ldots \in \Sigma$. We see that $\overline{\tau}$ is unbounded. If, conversely, $\overline{\tau}$ is unbounded we can find mutually distinct $\sigma_1, \sigma_2, \ldots \in \Sigma$ with $\lim_n \overline{\tau}(\sigma_n) = \infty$. Choose, for each n, a vector $x_n \in E$ for which $\|x_n\| \in \sigma_n$. Then $x_1, x_2, \ldots$ is orthogonal (3.2.8) and satisfies the type condition by 1.6.6. So $[x_1, x_2, \ldots]$ is a Keller space. This proves (ii). To prove (iii), let E be linearly homeomorphic to c_0. If it contained an infinite-dimensional Keller space D then by 4.3.9 (i) D does not contain c_0, in particular, D is not linearly homeomorphic to c_0, a contradiction. By (ii), $\overline{\tau}$ is not unbounded, i.e. bounded. Conversely, if $\overline{\tau}$ is bounded, say $\overline{\tau}(\sigma) \subset H$ for all $\sigma \in \Sigma$ and some proper convex subgroup H, then take $g_1, g_2 \in G$ for which $g_1 < h < g_2$ for all $h \in H$. Then for each $s \in Y$, Gs intersects $\text{conv}(H s_0)$ (1.6.2), so, if $e_1, e_2, \ldots$ is an orthogonal base of E there are $\lambda_1, \lambda_2, \ldots \in K$ such that $g_1 s_0 \leq |\lambda_n| \, \|e_n\| \leq g_2 s_0$ for all n. Then $(\xi_1, \xi_2, \ldots) \mapsto \sum_{n=1}^{\infty} \xi_n \lambda_n e_n$ is a linear homeomorphism of c_0 onto some subspace of E. Finally we prove (iv). If E is a Keller space then so is its subspace E_σ, so by Remark 3.4.6 and 4.3.9 (i), $\dim E_\sigma < \infty$. If not $\lim_\sigma \overline{\tau}(\sigma) = \infty$ there were mutually distinct $\sigma_1, \sigma_2, \ldots \in \Sigma$ such that $n \mapsto \overline{\tau}(\sigma_n)$ is bounded. Choose $e_n \in E$ with $\|e_n\| \in \sigma_n$. Then $e_1, e_2, \ldots$ is orthogonal, so it satisfies the type condition by 4.3.7 $(\alpha) \iff (\delta)$. But then $\lim_n \tau(\|e_n\|) = \lim_n \overline{\tau}(\sigma_n) = \infty$ by 1.6.6, a contradiction. Hence, $\lim_\sigma \overline{\tau}(\sigma) = \infty$. Conversely suppose that each E_σ is finite-dimensional and that $\lim_\sigma \overline{\tau}(\sigma) = \infty$. By choosing an orthogonal base of E_σ for every $\sigma \in \Sigma$ and by taking the union we obtain an orthogonal base $e_1, e_2, \ldots$ of E. By finite-dimensionality, $\{n : \|e_n\| \in \sigma\}$ is finite for each $\sigma \in \Sigma$, so $\{\|e_n\| : n \in \mathbb{N}\}$ meets infinitely many $\sigma \in \Sigma$, so since $\lim_\sigma \overline{\tau}(\sigma) = \infty$ we have $\lim_n \tau(\|e_n\|) = \infty$. Applying 1.6.6 we obtain that $e_1, e_2, \ldots$ satisfies the type condition i.e. that E is a Keller space.

We like to end this section with a discussion on reflexivity of Keller spaces. First some remarks on duality for general normed spaces E. If E is X-normed but $G \not\subset X$ then the Lipschitz norm (see 2.2) $\|f\| = \inf\{g \in G : |f(x)| \leq g\|x\| \text{ for all } x \in E\}$ is meaningless for $f \in E'$. The topology on E' of uniform convergence on bounded sets is perfectly defined but again, there is no canonical norm that describes this topology: for each $\delta \in X$ one may take $\|f\|_\delta = \sup\{|f(x)| : x \in B_E(0, \delta)\}$ (see 2.2). Clearly, E' is always a normable space (if $G \subset X$ the Lipschitz norm is equivalent to $\| \ \|_\delta$ for each δ, according to 2.5.5) but there is no natural device to define a norm on E', that is valid for each E. In any case, the bidual E'' is also well-defined as a normable space and we can define the following concept.

DEFINITION 4.3.11 A normed space is called (*topologically*) **reflexive** if the natural map $j_E : E \to E''$ (given by $j_E(x)(f) = f(x) \quad (f \in E',\ x \in E)$) is a linear homeomorphism.

LEMMA 4.3.12 *Let E be an X-normed Keller space where X is some G-module. Then there exists an equivalent $G^{\#}$-norm on E for which it is again a Keller space.*

Proof. Define, like in 1.5.6 and 2.1.9, a map $\phi : X \cup \{0\} \to G^{\#} \cup \{0\}$ by $\phi(0) := 0$ and $\phi(s) := \inf_{G^{\#}}\{g \in G : gs_0 \geq s\}$ $(s \in X)$ where $s_0 \in X$ is fixed. Then $N : x \mapsto \phi(\|x\|)$ $(x \in E)$ is a $G^{\#}$-norm equivalent to $\|\ \ \|$ by 2.1.9. Let $e_1, e_2, \ldots$ be an orthogonal base in $(E, \|\ \ \|)$. To show that it is an orthogonal base in (E, N) it suffices to prove orthogonality. Let $\lambda_1, \ldots, \lambda_n \in K$. Then $N(\sum_{i=1}^n \lambda_i e_i) = \phi(\|\sum_{i=1}^n \lambda_i e_i\|) = \phi(\max_i \|\lambda_i e_i\|) = \max_i N(\lambda_i e_i) = \max_i |\lambda_i| N(e_i)$, where we have used increasingness of ϕ. Thus, (E, N) has an orthogonal base and is linearly homeomorphic to a Keller space. Then it is itself a Keller space by 4.3.8.

Thanks to the above Lemma, to prove that Keller spaces are reflexive it suffices to show that $G^{\#}$-normed Keller spaces are. To be able to describe the reflexivity of the first Keller space in history (see [5]) in a more geometric way we shall prove slightly more. For topological reflexivity only the reader may take $\Gamma = G$ in the next Proposition and Theorem.

PROPOSITION 4.3.13 *Let E be a $\Gamma^{\#}$-normed Keller space where Γ is a linearly ordered group containing G as a cofinal subgroup. For $f \in E'$ set*

$$\|f\| := \inf\{g \in \Gamma : |f(x)| \leq g\|x\| \text{ for all } x \in E\}.$$

Let $e_1, e_2, \ldots$ be an orthogonal base of E and let $f_1, f_2, \ldots \in E'$ be the coordinate functions given by

$$f_n(\sum_{m=1}^{\infty} \lambda_m e_m) = \lambda_n.$$

Then $\|\ \ \|$ induces the topology of uniform convergence on bounded sets and $E' = (E', \|\ \ \|)$ is a Keller space with orthogonal base $f_1, f_2, \ldots$. We have $\|f_n\| = \omega(\|e_n\|)$ for each n (where ω is the antipode $\Gamma^{\#} \to \Gamma^{\#}$, defined in 1.3.1).

Proof. It is easily seen that $\|\ \ \|$ is equivalent to the 'ordinary' Lipschitz norm $f \mapsto \inf\{g \in G : |f(x)| \leq g\|x\| \text{ for all } x \in E\}$ (by using the fact that, if $g_1, g_2, \ldots \in \Gamma$, $\inf_n g_n = 0$, there exist, by coinitiality, $h_1, h_2, \ldots \in G$ for which $h_n < g_n$ for each n and so $\inf_n h_n = 0$). Then 2.5.5 shows that $\|\ \ \|$ induces the usual topology on E'. We now prove that $\|f_n\| = \omega(\|e_n\|)$ for all $n \in \mathbb{N}$. By definition we have

$$\|f_n\| = \inf\{g \in \Gamma : |f_n(x)| \leq g\|x\| \text{ for all } x \in E\}.$$

Now the expression '$|f_n(x)| \leq g\|x\|$ for all $x \in E$' is equivalent to '$|f_n(e_m)| \leq g\|e_m\|$ for all $m \in \mathbb{N}$' which is in turn equivalent to '$|f_n(e_n)| \leq g\|e_n\|$' i.e. to '$1 \leq g\|e_n\|$'. We see that $\|f_n\| = \inf\{g \in \Gamma : 1 \leq g\|e_n\|\} = \omega(\|e_n\|)$ by 1.3.1 (i). From 1.6.7 it follows that $f_1, f_2, \ldots$ satisfies the type condition. So, it remains to be shown that $f_1, f_2, \ldots$ is an orthogonal base for E'. To prove orthogonality, let $\lambda_1, \ldots, \lambda_n \in K$; we show that $\|\sum_{i=1}^n \lambda_i f_i\| \geq \max_i \|\lambda_i f_i\|$. Writing $f = \sum_{i=1}^n \lambda_i f_i$ we have $f(e_i) = \lambda_i$ for each $i \in \{1, \ldots, n\}$. If $g \in \Gamma$, $g \geq \|f\|$ then by definition $|f(x)| \leq g\|x\|$ for all $x \in E$, so in particular for $x = e_1, e_2, \ldots, e_n$ yielding $|\lambda_i| \leq g\|e_i\|$ for $i \in \{1, \ldots, n\}$. If $\lambda_i \neq 0$ we have $1 \leq |\lambda_i|^{-1} g\|e_i\|$ implying $|\lambda_i|^{-1} g \geq \omega(\|e_i\|) = \|f_i\|$ i.e. $|\lambda_i|\|f_i\| \leq g$. The latter formula is also trivially valid for $\lambda_i = 0$ and we find $\max_i |\lambda_i|\,\|f_i\| \leq g$. This result holds for each $g \in \Gamma$, $g \geq \|f\|$, so $\max_i |\lambda_i|\,\|f_i\| \leq \|f\|$ and orthogonality is proved. Now let $f \in E'$; we prove that $f = \sum_{n=1}^{\infty} \lambda_n f_n$, where $\lambda_n := f(e_n)$ for each n. In fact, we have for each n that $|\lambda_n| \leq g\|e_n\|$ for all $g \in \Gamma$,

$g \geq \|f\|$. So, $|\lambda_n|^{-1}g \geq \omega(\|e_n\|) = \|f_n\|$ i.e. $|\lambda_n|\,\|f_n\| \leq g$ for all $g \in \Gamma$, $g \geq \|f\|$, for all $n \in \mathbb{N}$. Thus, $\{\|\lambda_n f_n\| : n \in \mathbb{N}\}$ is bounded (above) and by the type condition $\|\lambda_n f_n\| \to 0$. Therefore $\widetilde{f} := \sum_{n=1}^{\infty} \lambda_n f_n$ makes sense (E' is complete by 2.3.7). But $\widetilde{f}(e_n) = f(e_n)$ for each n so $f = \widetilde{f}$ and we are done.

THEOREM 4.3.14 *A Keller space is reflexive. In particular, let $E, \Gamma, \|\ \ \|$ be as in 4.3.13 and define on E'' the norm*

$$\theta \mapsto \inf\{g \in \Gamma : |\theta(f)| \leq g\|f\| \text{ for all } f \in E'\}.$$

Then the natural map $j_E : E \to E''$ is a surjective isometry.

Proof. Thanks to 4.3.12 it suffices to prove the second statement. Let $e_1, e_2, \ldots$ be an orthogonal base of E. From 4.3.13 we obtain that $(E', \|\ \ \|)$ is a Keller space with orthogonal base $f_1, f_2, \ldots$ where the f_n are the coordinate functions, and where $\|f_n\| = \omega(\|e_n\|)$ for each n. By the same token E'' is a Keller space with the coordinate functions $\delta_n : \Sigma\lambda_m f_m \mapsto \lambda_n$ as an orthogonal base where $\|\delta_n\| = \omega(\|f_n\|) = \omega^2(\|e_n\|) = \|e_n\|$ (1.3.1) for each n. We complete the proof by showing that j_E maps e_n into δ_n for each n. But that is clear as for each m we have $j_E(e_n)(f_m) = f_m(e_n) = \delta_{mn} = \delta_n(f_m)$.

4.4 Form-Hilbert spaces

Throughout 4.4, let $\lambda \mapsto \lambda^*$ be an isometrical involution in K (that is allowed to be the identity). Also, assume $|2| = 1$; this technicality is needed for 4.4.2. A *Hermitean form* on a K-vector space E is a map $(\ ,\ ) : E \times E \to K$ satisfying

$$(x+y, z) == (x, z) + (y, z)$$
$$(\lambda x, y) = \lambda(x, y)$$
$$(x, y) = (y, x)^*$$

for all $x, y \in E$, $\lambda \in K$.

Let Γ be the divisible hull of G. It is known ([13]) that there is precisely one way to extend the ordering of G so as to let Γ become an linearly ordered group. Let $\sqrt{G} = \{s \in \Gamma : s^2 \in G\}$. Then $\sqrt{G}$ is a linearly ordered group; we consider it as a G-module.

DEFINITION 4.4.1 ([3], Def. 15). A normed space E is called a *definite space* if there is a Hermitean form $(\ ,\ )$ such that $\|x\| = \sqrt{|(x, x)|}$ for all $x \in E$.

A definite space has its norm values in $\sqrt{G} \cup \{0\}$.

PROPOSITION 4.4.2 *Let E be a definite Banach space of countable type. Then we have the following.*

(i) *If $x, y \in E$, $(x, y) = 0$ then $\|x + y\| = \max(\|x\|, \|y\|)$ (Form-orthogonality implies norm-orthogonality).*

(ii) $|(x,y)| \le \|x\|\,\|y\| \quad (x,y \in E)$ *(Cauchy-Schwarz).*
(iii) *E has a form-orthogonal base* $e_1, e_2, \ldots$. *For* $x \in E$ *we have* $x = \sum_{i=1}^{\infty} (x, e_i)\,(e_i, e_i)^{-1} e_i$, $\|x\| = \max_i |(x, e_i)|\,\|e_i\|^{-1}$.

Proof. For (i), (ii), see [3], Lemma 14. To prove (iii), let $x_1, x_2, \ldots$ be a linearly independent sequence whose linear hull is dense in E. The well-known Gram-Schmidt process

$$e_1 := x_1$$

$$e_2 := x_2 - \frac{(x_2, e_1)}{(e_1, e_1)} e_1$$

$$e_3 := x_3 - \frac{(x_3, e_2)}{(e_2, e_2)} e_2 - \frac{(x_3, e_1)}{(e_1, e_1)} e_1$$

$$\vdots$$

leads to a form orthogonal sequence $e_1, e_2, \ldots$ for which $[e_1, e_2, \ldots, e_n] = [x_1, x_2, \ldots, x_n]$ for each n. Thus, $e_1, e_2, \ldots$ is norm orthogonal with dense linear hull and therefore norm orthogonal base by 2.4.17. For $x \in E$, let $x = \sum_i \xi_i e_i$ be its expansion. Then for each j we have $(x, e_j) = (\xi_j e_j, e_j)$ and (iii) follows.

DEFINITION 4.4.3. A definite Banach space of countable type is called a *form-Hilbert space* if for every closed subspace $D \subset E$ we have $D + D^{\perp} = E$ where $D^{\perp} := \{x \in E : (x, d) = 0 \text{ for all } d \in D\}$.

Thus, norm-(form-)Hilbert spaces are characterized by the fact that every closed subspace has a norm-(form-)orthogonal complement. Form Hilbert spaces (also called GKK-spaces in [11]) have been extensively studied in [3] and [11].

An immediate consequence of the Definition is

PROPOSITION 4.4.4 *Every form-Hilbert space is a norm-Hilbert space.*

Proof. 4.4.2 (i).

However we can say more.

THEOREM 4.4.5. *Let E be an infinite-dimensional definite space of countable type. The following are equivalent.*
(α) *E is a form-Hilbert space.*
(β) *E is a norm-Hilbert space and* $\sqrt{G}$ *has a sequence with the type condition.*
(γ) *A subspace D is closed if and only if* $D^{\perp\perp} = D$.
(δ) *E is a Keller space.*

Proof. The equivalence of $(\alpha), (\gamma)$ and (δ) follows from [3], Th. 28 and Theorem 4.3.7 $(\alpha) \iff (\gamma)$. If $\sqrt{G}$ admits a sequence with the type condition we have by 4.3.1 that G has no maximal convex proper subgroups. So we have $(\beta) \Rightarrow (\delta)$. Conversely if (δ) holds then $\|E\| \subset \sqrt{G} \cup \{0\}$ by definiteness and each orthogonal base of E has the type condition (4.3.7 $(\alpha) \Rightarrow (\gamma)$). So we have (β).

COROLLARY 4.4.6 *If G admits a maximal proper convex subgroup (in particular, if the valuation of K has finite rank) then there do not exist infinite-dimensional form-Hilbert spaces over K.*

Proof. See the previous proof.

COROLLARY 4.4.7 *If G has no maximal proper convex subgroup then a definite Banach space of countable type is norm-Hilbert space if and only if it is a form-Hilbert space.*

Proof. 4.4.4 and 4.4.5 $(\delta) \Rightarrow (\alpha)$.

COROLLARY 4.4.8 *Let G have no proper maximal convex subgroup. Let E be a norm-Hilbert space with canonical decomposition $\bigoplus_{\sigma\in\Sigma} E_\sigma$. Then E is a form-Hilbert space if and only if each E_σ is one. In particular, if each E_σ is onedimensional then E is a form Hilbert space if and only if $\|x\| \in \sqrt{G}$ for each nonzero $x \in E$.*

Proof. The "only if" parts are clear. To prove the first "if" part, suppose that each E_σ is a form-Hilbert space with Hermitean form $(\ ,\ )_\sigma$. It suffices to define a Hermitean form $(\ ,\ )$ on E for which $\|x\|^2 = |(x,x)|$ $(x \in E)$ (4.4.7). To this end, let $x \in E$, $x = \sum_{\sigma\in\Sigma} x_\sigma$ where $x_\sigma \in E_\sigma$ for each $\sigma \in \Sigma$: similarly, let $y \in E$, $y = \sum_{\sigma\in\Sigma} y_\sigma$. We have $x_\sigma \to 0$, $y_\sigma \to 0$ so $|(x_\sigma, y_\sigma)| \le \|x_\sigma\|\,\|y_\sigma\| \to 0$ and the definition

$$(x,y) := \sum\nolimits_\sigma (x_\sigma, y_\sigma)_\sigma$$

makes sense. If $x_\sigma \neq 0$, $x_\tau \neq 0$ for some $\sigma, \tau \in \Sigma$, $\sigma \neq \tau$ we have $\|x_\sigma\| \notin G\|x_\tau\|$, so $\|x_\sigma\| \neq \|x_\tau\|$ and therefore $\|x_\sigma\|^2 \neq \|x_\tau\|^2$. Thus, if $\sigma \neq \tau$ and $(x_\sigma, y_\sigma)_\sigma$ and $(x_\tau, y_\tau)_\tau$ are not both 0 then $|(x_\sigma, x_\sigma)_\sigma| \neq |(x_\tau, x_\tau)_\tau|$ so that for $x \in E$ we have

$$|(x,x)| = |\sum\nolimits_\sigma (x_\sigma, x_\sigma)_\sigma| = \max_\sigma |(x_\sigma, x_\sigma)_\sigma| = \max_\sigma \|x_\sigma\|^2 = \|x\|^2.$$

Now suppose that each E_σ is onedimensional, say $E_\sigma = K a_\sigma$. By assumption there is a $c_\sigma \in K$ such that $|c_\sigma| = \|a_\sigma\|^2$. Then the formula $(\lambda a_\sigma, \mu a_\sigma) = \lambda\mu c_\sigma$ defines a form on $K\, a_\sigma$ making it into a form-Hilbert space. Now apply the first part of the proof to conclude that E is a form-Hilbert space.

COROLLARY 4.4.9 *The following conditions on K are equivalent.*

(α) *There exists an infinite-dimensional form-Hilbert space over K.*

(β) *G has no maximal proper convex subgroups. The G-module $\sqrt{G}$ admits a sequence having the type condition.*

Proof. $(\alpha) \Rightarrow (\beta)$. Corollary 4.4.6 furnishes the first part of (β). Then by 4.3.7 every orthogonal base has the type condition which yields a sequence in $\sqrt{G}$ with the type condition. Conversely, let $s_1, s_2, \ldots \in \sqrt{G}$ satisfy the type condition. By taking a suitable subsequence we may assume $s_n \notin G s_m$ whenever $n \neq m$. Let $E :=$

$\{(\xi_1, \xi_2, \ldots) \in K^{\mathbb{N}} : \lim_n |\xi_n| s_n = 0\}$ and for $x = (\xi_1, \xi_2, \ldots)$, $y = (\eta_1, \eta_2, \ldots) \in E$ set

$$(x, y) = \sum_{n=1}^{\infty} \xi_n \eta_n a_n$$

where $a_n \in K$ are such that $|a_n| = s_n^2$. We see that $|(x,x)| = |\Sigma \xi_n^2 a_n| = \max_n(|\xi_n| s_n)^2$. So E is a Keller space and definite hence form-Hilbert.

REFERENCES

[1] A Fässler-Ullmann. On non-classical Hilbert spaces. Exposition Math 3:275–277, 1983.

[2] L Fuchs. Partially Ordered Algebraic Systems. Oxford: Pergamon, 1963.

[3] H Gross, UM Künzi. On a class of orthomodular quadratic spaces. Enseign Math 31:187–212, 1985.

[4] AK Katsaras. On compact operators between non-archimedean spaces. Ann Soc Sci Bruxelles Sér I 96:129–137, 1982.

[5] H Keller. Ein nicht-klassischer Hilbertscher Raum. Math Z 172:41–49, 1980.

[6] H Keller, H Ochsenius. An algebra of self-adjoint operators on a non-archimedean orthomodular space. In: WH Schikhof, C Perez-Garcia, J Kakol, ed. p-Adic Functional Analysis. New York: Marcel Dekker, 1997, pp 253-264.

[7] H Keller, H. Ochsenius. Residual spaces and operators on orthomodular spaces. In: WH Schikhof, C Perez-Garcia, J Kakol, ed. p-Adic Functional Analysis. New York: Marcel Dekker, 1997, pp 265–274.

[8] J Kelley. General Topology. Toronto: Van Nostand, 1955.

[9] H Ochsenius. Los espacios ortomodulares y el teorema de Kakutani-Mackey. PhD dissertation, Universidad Católica, Chile, 1990.

[10] H Ochsenius. Non-archimedean analysis when the value group has a non-archimedean order. In: N De Grande-De kimpe, S Navarro, WH Schikhof, ed. p-Adic Functional Analysis. Editorial Universidad de Santiago, Chile, 1994, pp 87–98.

[11] H Ochsenius. A characterization of Orthomodular Spaces. Boll Un Math Ital 10-A:575–585, 1996.

[12] S Oortwijn, WH Schikhof. Locally convex modules over the unit disk. In: WH Schikhof, C Perez-Garcia, J Kakol, ed. p-Adic Functional Analysis. New York: Marcel Dekker, 1997, pp 305–326.

[13] P Ribenboim. Théorie des valuations. Les Presses de l'Université de Montreal, 1965.

[14] ACM van Rooij. Non-archimedean Functional Analysis. New York: Marcel Dekker, 1978.

[15] WH Schikhof. Distinguishing non-archimedean seminorms. Report 8409, Department of Mathematics, University of Nijmegen, The Netherlands: 1-10, 1984.

[16] WH Schikhof. Locally convex spaces over non-spherically complete fields. Bull Soc Math Belg Sér B 38:187–224, 1986.

[17] WH Schikhof. A scalar field for which C-zero has no Hahn-Banach property. Ann Math Blaise Pascal 2:267–273, 1995.

[18] MP Solèr. Characterization of Hilbert spaces by orthomodular spaces. Comm Algebra 23:219–243, 1995.

[19] S Warner. Topological Fields. North-Holland, 1989.

The p-adic Banach-Dieudonné theorem and semi-compact inductive limits

C PEREZ-GARCIA* Department of Mathematics, Facultad de Ciencias, Universidad de Cantabria, 39071 Santander, Spain.

WH SCHIKHOF Department of Mathematics, University of Nijmegen, Toernooiveld 6525 ED Nijmegen, The Netherlands.

1 INTRODUCTION

In [1] a systematic treatment of the theory of locally convex inductive limits over complete non-archimedean rank 1 valued fields K has been initiated. The following question was not considered in [1] and has been the starting point of the present paper. Let $E_1 \subset E_2 \subset \ldots$ be a semicompact inductive sequence of Banach spaces (i.e. the unit ball of E_n is a compactoid in E_{n+1} for each n), let τ be the strongest locally convex topology on $E := \bigcup_n E_n$ making all inclusions $E_n \to E$ continuous. Does it follow that a (not necessarily convex) subset of E is τ-open if and only if its intersection with E_n is open for each n? In other words, is τ the strongest topology on E for which the inclusions $E_n \to E$ are continuous? For the case of a real or complex scalar field an affirmative answer was obtained by J.S. e Silva in [11], giving a direct proof to get the answer, and by H. Komatsu in [5], as a consequence of the Banach-Dieudonné Theorem. It is not hard to see that their proofs can be carried over to the case of a locally compact base field K. For all remaining K the answer is negative as is shown by the following.

EXAMPLE *Let K be not locally compact, and let, for each $n \in \mathbb{N}$, $E_n := \{(\lambda_1, \lambda_2, \ldots) \in K^{\mathbb{N}} : \lambda_m = 0$ for $m > n\}$ have the usual topology. Let τ be the locally convex inductive limit topology on $E := \bigcup_n E_n$. Then there is a set $X \subset E$ for which $X \bigcap E_n$ is closed in E_n for all n, but such that X is not τ-closed.*

* Research partially supported by the Spanish Direccion General de Investigacion Cientifica y Tecnica (DGICYT PB95-0582)

Proof. By finite-dimensionality the inductive sequence $(E_n)_n$ is semicompact and τ is the strongest locally convex topology on E. Since K is not locally compact there is a $\pi \in K$ and there are $\lambda_1, \lambda_2, \ldots \in K$ such that $0 < |\pi| < |\lambda_n| \leq 1$, $|\lambda_n - \lambda_m| > |\pi|$ $(n, m \in \mathbb{N}, n \neq m)$. Let V be the set of all finite sequences in $\mathbb{N}$ of the form $(n_1, n_2, \ldots, n_{n_1})$ where $n_1 < n_2 < \cdots < n_{n_1}$. By countability there exists a bijection $\sigma : V \to \{\lambda_1, \lambda_2, \ldots\}$. For each $v = (n_1, n_2, \ldots, n_{n_1}) \in V$ set $x_v := (\sigma(v)\pi^{n_1}, \pi^{n_2}, \ldots, \pi^{n_{n_1}}, 0, 0, \ldots) \in E$, let $X := \{x_v : v \in V\}$. We first prove that $X \bigcap E_s$ is closed in E_s for each $s \in \mathbb{N}$. Let $\| \;\; \|$ be the maximum norm on E_s, let $v, w \in V$, $v \neq w$ be such that $x_v, x_w \in E_s$. Then $v = (n_1, n_2, \ldots, n_{n_1})$, $w = (m_1, m_2, \ldots, m_{m_1})$ where $n_1 \leq s$ and $m_1 \leq s$. If $n_1 = m_1$ then $\|x_v - x_w\| \geq |\sigma(v) - \sigma(w)|\, |\pi|^{n_1} \geq |\pi|\, |\pi|^s = |\pi|^{s+1}$. If $n_1 > m_1$ then $|\sigma(v)|\, |\pi|^{n_1} \leq |\pi|^{n_1}$ and $|\sigma(w)\pi^{m_1}| > |\pi|\, |\pi|^{m_1} \geq |\pi|^{n_1}$ implying $|\sigma(v)\pi^{n_1}| < |\sigma(w)\pi^{m_1}|$ and $\|x_v - x_w\| \geq |\sigma(v)\pi^{n_1} - \sigma(w)\pi^{m_1}| = |\sigma(w)\pi^{m_1}| \geq |\pi|^{s+1}$. We conclude that $X \bigcap E_s$ is discrete, hence closed. Clearly $0 \notin X$; we complete the proof by showing that $0 \in \overline{X}^{\tau}$ i.e. that for every non-archimedean seminorm p on E there exists an $x \in X$ for which $p(x) \leq 1$. To this end, let $e_1, e_2, \ldots$ be the natural base of E, choose $n_1 < n_2 < \cdots$ in $\mathbb{N}$ for which $|\pi|^{n_k} p(e_k) \leq 1$ for all $k \in \mathbb{N}$ and let $v = (n_1, n_2, \ldots, n_{n_1}) \in V$. Then $x = \sigma(v)\pi^{n_1} e_1 + \pi^{n_2} e_2 + \cdots + \pi^{n_{n_1}} e_{n_1} \in X$ and $p(x) \leq \max(|\sigma(v)|\, |\pi|^{n_1} p(e_1), |\pi|^{n_2} p(e_2), \ldots, |\pi|^{n_{n_1}} p(e_{n_1})) \leq 1$.

The above example can be generalized (see Proposition 4.6) and gives rise to related questions. For example if X is a convex subset of E and $X \bigcap E_n$ is closed in E_n for each n, does it follow that X is closed? Being inspired by the classical case (see [5] and [11]), our answers to those questions will follow two different kind of proofs: as corollaries of certain non-archimedean translations of the classical Banach-Dieudonné Theorem (section 4), and direct proofs in which basic properties of (c-)compact and compactoid sets are involved (section 5).

2 PRELIMINARIES

Throughout K is a non-archimedean non-trivially valued field that is complete under the metric induced by its valuation $| \;\; | : K \to [0, \infty)$. Let E be a K-vector space. If $X_1, X_2, \ldots$ are subsets of E we write $\sum_n X_n$ for the set of all sums $x_1 + x_2 + \cdots + x_m$, where $m \in \mathbb{N}$, $x_i \in X_i$ for $i \in \{1, \ldots m\}$. A subset A of E is called *absolutely convex* if A is a module over the valuation ring, *convex* if A is either empty or an additive coset of an absolutely convex set. For an absolutely convex subset A of E we define $A^e := A$ if the valuation of K is discrete, $A^e = \bigcap\{\lambda A : \lambda \in K, |\lambda| > 1\}$ otherwise. A is called *edged* if $A = A^e$. The linear hull of a set $X \subset E$ is written $[X]$, its absolutely convex hull $\operatorname{co} X$. The *Minkowski function* associated to an absorbing absolutely convex set $A \subset E$ is the seminorm defined on E by the formula $x \mapsto \inf\{|\lambda| : \lambda \in K, x \in \lambda A\}$ $(x \in E)$.

Let E be a locally convex space over K. The closure of a set $X \subset E$ is written $\overline{X}$, instead of $\overline{\operatorname{co} X}$ we write $\overline{\operatorname{co}}\, X$. On the dual space E' we consider the following three locally convex topologies. The topology of *pointwise convergence* $\sigma(E', E)$, the topology of *precompact convergence* $\tau_{pc}(E', E)$ and the topology of *bounded convergence* $\beta(E', E)$ (also called strong topology, we sometimes write $E'_b = (E', \beta(E', E))$ and call it the *strong dual* of E). The canonical map $E \to (E'_b)'$ is denoted by j_E.

For subsets $X \subset E$, $Y \subset E'$ we define $X^0 := \{f \in E' : |f(x)| \leq 1 \text{ for all } x \in X\}$ and $Y_0 := \{x \in E : |g(x)| \leq 1 \text{ for all } g \in Y\}$. $X \subset E$ is a *polar set* if $(X^0)_0 = X$. The space E is called a *polar space* if its topology has a neighbourhood base at 0 consisting of polar sets. E is called *strongly polar* if each closed edged set is polar.

A *compactoid* in a locally convex space E over K is a set $X \subset E$ such that for each zero neighbourhood U in E there is a finite set $F \subset E$ such that $X \subset U + \text{co } F$. If E is metrizable then there exist $e_1, e_2, \ldots \in E$ with $\lim_{n\to\infty} e_n = 0$ such that $X \subset \overline{\text{co}}\{e_1, e_2, \ldots\}$ ([6], 8.2). Therefore, the topology on E' of uniform convergence on compactoids equals $\tau_{pc}(E', E)$.

Let U be a zero neighbourhood in E. Then U^0 is an edged $\sigma(E', E)$-complete and compactoid set. Thus, U^0 is c-compact if K is spherically complete, compact if K is locally compact ([7]). A subset X of E' is equicontinuous if and only if it is contained in U^0 for some zero neighbourhood U in E. On such a set X the topology $\sigma(E', E)$ coincides with the topology of uniform convergence on compactoids ([6], 10.6). Also, like in the real or complex case one proves that, if E is Fréchet, X is equicontinuous if and only if X is $\sigma(E', E)$-bounded.

For a normed space E over K and $r > 0$ we denote the ball $\{x \in E : \|x\| \leqslant r\}$ by $B_E(r)$. Also, if A and B are non-empty subsets of E, by $\text{dist}_E(A, B)$ we will mean the distance between A and B defined in the usual way.

A bounded and absolutely convex set B in a locally convex space E over K is called a *Banach disk* if $[B]$ is complete with respect to the topology induced by the associated Minkowski function. A linear map $T : E \to F$ (where F is a second locally convex space over K) is called *semicompact* if there exists a compactoid Banach disk D in F such that $T^{-1}(D)$ is a zero neighbourhood in E. It is easy to see that every semicompact operator is continuous. An *inductive sequence* is an increasing sequence $E_1 \subset E_2 \subset \cdots$ of subspaces of a K-vector space E such that $E = \bigcup_n E_n$ and where, for each n, E_n is provided with a locally convex topology τ_n such that each inclusion $E_n \to E_{n+1}$ is continuous. The inductive sequence is called *proper* if all the inclusions $E_n \to E_{n+1}$ are strict, *semicompact* if all inclusions $E_n \to E_{n+1}$ are semicompact. The *inductive limit topology* on E is the strongest locally convex topology on E for which all inclusions $E_n \to E$ are continuous, and (E, τ) is called the *locally convex inductive limit* of the sequence $(E_n)_n$ which is denoted by $(E, \tau) = \varinjlim E_n$.

For terms that are still unexplained, see [1], [6].

3 THE BANACH-DIEUDONNE THEOREM

The classical Banach-Dieudonné Theorem ([2], Theorem 3.10.1) in its original form reads as follows.

Let E be a metrizable locally convex space over $\mathbb{R}$ or $\mathbb{C}$. Then the strongest topology on E' coinciding with $\sigma(E', E)$ on equicontinuous sets is the (locally convex) topology of uniform convergence on the precompact subsets of E.

First, we verify that the conclusion of this theorem holds when the scalar field is locally compact. This fact will be a consequence of the next result, whose proof follows the same ideas as the corresponding one of [2], 3.10.1.

THEOREM 3.1 *Let E be a metrizable locally convex space over K. Let $A \subset E'$ be such that $A \bigcap V^0$ is $\sigma(E', E)$-compact for each zero neighbourhood V in E. Then A is $\tau_{pc}(E', E)$-closed.*

Proof. Let $E = U_1 \supset U_2 \supset \ldots$ be a zero neighbourhood base in E. It is enough to see that if $0 \notin A$ then $W := E \setminus A$ is a neighbourhood of 0 in $(E', \tau_{pc}(E', E))$. For that we prove the existence of finite sets $F_n \subset U_n$ $(n \in \mathbb{N})$ such that $(F_1 \bigcup \cdots \bigcup F_n)^0 \bigcap U_{n+1}^0 \subset W$ for each $n \in \mathbb{N}$. (This will complete the proof since $\bigcup_n F_n$ is precompact and $(\bigcup_n F_n)^0 \subset W$). Suppose for some $n \in \mathbb{N}$, $n > 1$, the sets F_k $(k < n)$ have been constructed with the required properties and suppose there is no finite set $F_n \subset U_n$ for which $(F_1 \bigcup \cdots \bigcup F_n)^0 \bigcap U_{n+1}^0 \subset W$. Then, for every finite subset F of U_n the set $T_F := (F_1 \bigcup \cdots \bigcup F_{n-1} \bigcup F)^0 \bigcap U_{n+1}^0 \bigcap A$ is nonempty, and $\sigma(E', E)$-compact by compactness of $A \bigcap U_{n+1}^0$ and closedness of $(F_1 \bigcup \cdots \bigcup F_{n-1} \bigcup F)^0$. By compactness and finite intersection property there is an $a \in T_F$ for all finite subsets F of U_n implying $a \in (F_1 \bigcup \cdots \bigcup F_{n-1})^0 \bigcap U_n^0 \bigcap A$ conflicting $(F_1 \bigcup \cdots \bigcup F_{n-1})^0 \bigcap U_n^0 \subset W$.

Now, since for locally compact fields V^0 is $\sigma(E', E)$-compact for every zero neighbourhood V in E (see the Preliminaries), we can apply 3.1 to conclude the desired Banach-Dieudonné Theorem.

THEOREM 3.2 (Banach-Dieudonné Theorem for locally compact fields) *Let K be locally compact, let E be a metrizable locally convex space over K. Then the strongest topology on E' that coincides with $\sigma(E', E)$ on equicontinuous subsets of E' is the topology $\tau_{pc}(E', E)$ of uniform convergence on precompact subsets of E.*

We will see in Corollaries 4.7, 4.8 that the conclusion of Theorem 3.2 fails if K is not locally compact. But that does not mean that all is lost; we shall consider a few weaker versions of the Banach-Dieudonné Theorem. First of all we have the following result that was proved in [4], 4.5.

THEOREM 3.3 (Locally convex version of the Banach-Dieudonné Theorem) *Let E be a polar metrizable locally convex space over K. Then the strongest locally convex topology on E' that coincides with $\sigma(E', E)$ on equicontinuous subsets of E' is the topology $\tau_{pc}(E', E)$. Otherwise stated, an absolutely convex set $U \subset E'$ is $\tau_{pc}(E', E)$-open if and only if $U \bigcap M$ is $\sigma(E', E)$-open in M for each equicontinuous set $M \subset E'$.*

Moreover, we can give the following extension of Theorem 3.3.

THEOREM 3.4 (Linear version of the Banach-Dieudonné Theorem) *Let E be a polar metrizable locally convex space over K. Then the strongest linear topology on E' that coincides with $\sigma(E', E)$ on equicontinuous subsets of E' is the topology $\tau_{pc}(E', E)$.*

Proof. Let $U_1 \supset U_2 \supset \cdots$ be a zero neighbourhood base in E where each U_n is absolutely convex. For each $n \in \mathbb{N}$ set $A_n := U_n^0$.

Let V_0 be a zero neighbourhood for a linear topology τ on E' which agrees with

$\sigma(E', E)$ on each A_n. Choose a sequence $(V_n)_n$ of balanced τ-neighbourhoods of zero in E' and a sequence $(W_n)_n$ of absolutely convex $\sigma(E', E)$-neighbourhoods of zero in E' such that $V_n + V_n \subset V_{n-1}$ and $W_n \bigcap A_n \subset V_n \bigcap A_n$ for each $n \in \mathbb{N}$. If we set $D := \bigcup_{n \in \mathbb{N}}(W_1 \bigcap A_1 + \cdots + W_n \bigcap A_n)$, then $D \subset V_0$ and, by [3], 5.2.2, V_0 is a zero neighbourhood for the finest locally convex topology on E' agreeing with $\sigma(E', E)$ on each A_n. Now the conclusion follows from Theorem 3.3.

To find another versions, on closedness of convex sets, we observe that if K is spherically complete and we take $A \subset E'$ convex such that for each zero neighbourhood V in E $A \bigcap V^0$ is $\sigma(E', E)$-c-compact (or equivalently $\sigma(E', E)$-closed, see the Preliminaries) then the sets T_F considered in the proof of Theorem 3.1 are convex and $\sigma(E', E)$-c-compact. So, this modification of the proof of 3.1 leads to the following.

THEOREM 3.5 (Banach-Dieudonné Theorem for spherically complete fields) *Let E be a metrizable locally convex space over a spherically complete field K. If $A \subset E'$ is convex and $A \bigcap M$ is $\sigma(E', E)$-closed in M for each equicontinuous set $M \subset E'$ then A is $\tau_{pc}(E', E)$-closed.*

Also observe that if K is spherically complete and E is a space over K, from the property $(E', \tau_{pc})' = E = (E', \sigma(E', E))'$ proved in [4], 4.7, we deduce that a convex subset A of E' is $\sigma(E', E)$-closed if and only if A is $\tau_{pc}(E', E)$-closed. This fact, together with Theorem 3.5, allows to obtain the following extension to Fréchet spaces of the Krein-Šmulian Theorem given in [9], 5.1 for Banach spaces over spherically complete fields, whose statement is about general convex sets rather edged ones (see Proposition 3.7 for this last case).

COROLLARY 3.5 (Krein-Šmulian Theorem for spherically complete fields) *Let E be a Fréchet space over a spherically complete field K. If $A \subset E'$ is convex and $A \bigcap M$ is $\sigma(E', E)$-closed in M for each equicontinuous set $M \subset E'$ then A is $\sigma(E', E)$-closed.*

To find a Banach-Dieudonné statement for arbitrary (i.e. possibly non-spherically complete) K we first generalize the Krein-Šmulian Theorem about edged sets, proved in [10], 2.2 for Banach spaces, to Fréchet spaces.

PROPOSITION 3.7 (Krein-Šmulian Theorem for non-spherically complete fields) *Let E be a strongly polar Fréchet space. Let $A \subset E'$ be edged and suppose that $A \bigcap M$ is $\sigma(E', E)$-closed in M for each equicontinuous set $M \subset E'$. Then A is $\sigma(E', E)$-closed.*

Proof. By Corollary 3.6 we may assume the valuation to be dense. Choose $\lambda \in K$, $|\lambda| > 1$. There are $\lambda_1, \lambda_2, \ldots \in K$ such that $|\lambda_n| > 1$ for each n and $\prod_n |\lambda_n| \leq |\lambda|$. Furthermore, let $U_1 \supset U_2 \supset \cdots$ be a zero neighbourhood base in E, consisting of absolutely convex sets U_n. By strong polarness of E and $(E', \sigma(E', E))$ we have $Z = (Z^0)_0$ for each closed edged set $Z \subset E$ and $Y = (Y_0)^0$ for each $\sigma(E', E)$-closed edged subset Y of E'. For each $n \in \mathbb{N}$ set $X_n := (A \bigcap U_n^0)_0$ and $X := \bigcap_n X_n$. We prove that $A = X^0$. If $f \in A$ then $f \in A \bigcap U_n^0$ for some n so $|f| \leq 1$ on

$(A\bigcap U_n^0)_0 = X_n \supset X$. Thus, $A \subset X^0$. To prove the converse, let $f \in X^0$. We have $f \in U_n^0$ for certain n. Choose $x \in X_n$; we prove $|f(x)| \leq 1$ (then observing that A and $A\bigcap U_n^0$ are edged, we have $f \in X_n^0 = A\bigcap U_n^0 \subset A$ and we are done). Now $X_n = (A\bigcap U_n^0)_0 = (A\bigcap U_{n+1}^0\bigcap U_n^0)_0 = (X_{n+1}^0\bigcap U_n^0)_0 = (X_{n+1}+U_n)_0^0 \subset \lambda_1(X_{n+1}+U_n)$. The last inclusion follows from the fact that $(X_{n+1}+U_n)^e$ is closed and edged hence polar by strong polarness of E, therefore

$$(X_{n+1}+U_n)_0^0 = (X_{n+1}+U_n)^e \subset \lambda_1(X_{n+1}+U_n).$$

So, $X_n \subset \lambda_1 X_{n+1} + \lambda U_n$. Similarly, $X_{n+1} \subset \lambda_2(X_{n+2}+U_{n+1})$ so $\lambda_1 X_{n+1} \subset \lambda_1\lambda_2 X_{n+2} + \lambda U_{n+1}$, etc.. So we can find an $x_{n+1} \in \lambda_1 X_{n+1}$ with $x - x_{n+1} \in \lambda U_n$, an $x_{n+2} \in \lambda_1\lambda_2 X_{n+2}$ with $x_{n+1} - x_{n+2} \in \lambda U_{n+1}$, etc.. By completeness $a := \lim_{k\to\infty} x_{n+k}$ exists, $a \in \lambda(X_{n+1}\bigcap X_{n+2}\bigcap\cdots) = \lambda X$ and $a - x \in \lambda U_n$. We have $|f(x)| \leq \max(|f(x-a)|, |f(a)|) \leq |\lambda|$. So, $|f(x)| \leq |\lambda|$ for all $\lambda \in K$, $|\lambda| > 1$ implying $|f(x)| \leq 1$ since the valuation of K is dense.

COROLLARY 3.8 *A subspace of the dual of a strongly polar Fréchet space E is $\sigma(E',E)$-closed if and only if its intersection with any equicontinuous set M in E' is $\sigma(E',E)$-closed in M.*

A subspace of the dual of a strongly polar Banach space E is $\sigma(E',E)$-closed if and only if its intersection with the unit ball of E' is $\sigma(E',E)$-closed ([10, 2.4).

To derive the announced form of the Banach-Dieudonné Theorem we need one more lemma.

LEMMA 3.9 *Let E be a metrizable space with a dense subspace D. Then, for each compactoid X in E there exists a compactoid Y in D such that $\overline{Y} \supset X$.*

Proof. By [6], 8.2 there exist $e_1, e_2, \ldots \in E$ with $\lim_{m\to\infty} e_m = 0$ such that $X \subset \overline{\text{co}}\{e_1, e_2, \ldots\}$. Let $U_1 \supset U_2 \supset \ldots$ be a zero neighbourhood base in E consisting of absolutely convex sets U_n. For each $n, m \in \mathbb{N}$, choose $d_{nm} \in D$ such that $d_{nm} - e_m \in U_{n+m}$, and set $Y := \text{co}\{d_{nm} : n, m \in \mathbb{N}\}$. Then $Y \subset D$ and since $\lim_{n\to\infty} d_{nm} = e_m$ we have $\overline{Y} \supset X$.

To prove that Y is compactoid fix $k \in \mathbb{N}$. Since $\lim_{m\to\infty} e_m = 0$ there exists $m_k \in \mathbb{N}$ with $m_k \geqslant k$ such that $d_{nm} \in U_k$ for all $n \in \mathbb{N}$ and all $m \in \mathbb{N}$ with $m \geqslant m_k$. On the other hand, for each $m < m_k$ there exists $n_{mk} \in \mathbb{N}$ such that $n_{mk} + m \geqslant k$ and hence $d_{nm} \in e_m + U_k$ for all $n \geqslant n_{mk}$. Then $F := \{d_{nm} : m < m_k, n < n_{mk}\}\bigcup\{e_m : m < m_k\}$ is a finite set in E for which $Y \subset U_k + \text{co}\, F$.

THEOREM 3.10 (Banach-Dieudonné Theorem for non-spherically complete fields) *Let E be a strongly polar metrizable locally convex space over K. If $A \subset E'$ is edged and $A\bigcap M$ is $\sigma(E',E)$-closed in M for each equicontinuous set $M \subset E'$ then A is $\tau_{pc}(E',E)$-closed.*

Proof. Let $E^\wedge$ be the completion of E. Then the restriction map $R : (E^\wedge)' \longrightarrow E'$ is a bijection and induces a $1-1$ correspondence between the equicontinuous subsets of $(E^\wedge)'$ and of E'. For an equicontinuous set $\hat{M}$ of $(E^\wedge)'$ the map $R|\hat{M}$ is a homeomorphism of $\hat{M}$ onto $R(\hat{M})$ with respect to the topologies $\sigma((E^\wedge)', E^\wedge)|\hat{M}$

and $\sigma(E', E)|R(\hat{M})$, respectively. So, $R^{-1}(A)$ is $\sigma((E^\wedge)', E^\wedge)$-closed in $(E^\wedge)'$ by Proposition 3.7 (observe that $E^\wedge$ is strongly polar by [6], 4.6), hence certainly $\tau_{pc}((E^\wedge)', E^\wedge)$-closed. But by Lemma [3.9], R is a homeomorphism $(E^\wedge)' \longrightarrow E'$ with respect to the τ_{pc}-topologies. Thus, A is $\tau_{pc}(E', E)$-closed.

REMARKS 1. We have made a proof leading from Proposition 3.7 to Theorem 3.10. Conversely, assume Theorem 3.10 is true and let E be a strongly polar Fréchet space. It follows from Theorem 3.3 and [4], 3.2 that (E', τ_{pc}) is of countable type (and hence strongly polar [6], 4.4). Also, by [4], 4.7 we have $(E', \tau_{pc})' = E = (E', \sigma(E', E))'$. Hence if $A \subset E'$ is edged, A is $\sigma(E', E)$-closed if and only if A is $\tau_{pc}(E', E)$-closed, and therefore Proposition 3.7 holds.

Analogously, we have proofs leading from Theorem 3.5 to Corollary 3.6 and conversely.

2. The following example shows that there is no Krein-Šmulian Theorem for non-complete metrizable spaces.

EXAMPLE Let E be the normed space of all sequences in K that are eventually zero endowed with the canonical maximum norm. E is a space of countable type (and hence strongly polar) whose dual is canonically isomorphic to ℓ^∞. Choose $a \in c_0 \setminus E$ and define $f \in (\ell^\infty)'$ by $f(x) = (x, a)$ $(x \in \ell^\infty)$, where $(\, , \,)$ is the natural bilinear form. Then the edged set $A := \mathrm{Ker} f$ is not $\sigma(E', E)$-closed but it is $\sigma(\ell^\infty, c_0)$-closed and so its intersection with every equicontinuous (= bounded) set $M \subset E'$ is $\sigma(E', E)$-closed in M (see the proof of Theorem 3.10).

4 CONNECTION WITH SEMICOMPACT INDUCTIVE LIMITS

Let us assign to each K-vector space E a family $\mathcal{P}_E$ of subsets of E in such a way that if $T : E \to F$ is a linear bijection between two K-vector spaces E and F then $A \mapsto TA$ is a bijection $\mathcal{P}_E \to \mathcal{P}_F$. In fact, to be less mysterious, we will consider for $\mathcal{P}_E$ choices like: all subsets of E, all convex subsets of E, all complements of convex subsets of E, all edged subsets of E.

The next Theorem 4.1 connects the various versions of the Banach-Dieudonné Theorem (α) with the question on semicompact inductive limits (β), posed in the Introduction. The proof uses several results of [1].

THEOREM 4.1 *The following statements on K are equivalent.*

(α) *Let E be a nuclear Fréchet space over K. Let $A \in \mathcal{P}_{E'}$ be such that $A \bigcap M$ is $\sigma(E', E)$-closed in M for each equicontinuous set $M \subset E'$. Then A is $\tau_{pc}(E', E)$-closed.*

(β) *Let $E_1 \to E_2 \to \ldots$ be a semicompact inductive sequence of locally convex spaces E_n over K, let $(E, \tau) := \varinjlim E_n$. If $A \in \mathcal{P}_E$ is such that $A \bigcap E_n$ is closed in E_n for each n, then A is τ-closed.*

Proof. (α) $\Rightarrow$ (β) To prove that A is closed we may assume, by [1], 3.1.4 that all E_n are Banach spaces. By [1], 3.1.7 E is reflexive and $F := E'_b$ is nuclear

Fréchet. Then the strong topology on F' is the one of uniform convergence on compactoids, which, in turn, is the topology of uniform convergence on precompact sets i.e. $\tau_{pc}(F', F)$ (see the Preliminaries). Thus, the natural bijection $j_E : E \to (E'_b)'$ is a linear homeomorphism of (E, τ) onto $(F', \tau_{pc}(F', F))$. So, τ-closedness of A is equivalent to $\tau_{pc}(F', F)$-closedness of $j_E(A)$, which, in turn, by (α) and the fact that $j_E(A) \in \mathcal{P}_{F'}$, is equivalent to $\sigma(F', F)$-closedness in M of $j_E(A) \bigcap M$ for each equicontinuous ($= \sigma(F', F)$-bounded, see the Preliminaries) set $M \subset F'$. In other words (identifying F' and E) we have to show that $A \bigcap B$ is $\sigma(E, E')$-closed in B for every τ-closed and bounded set $B \subset E$. By [1], 3.1.7.(iii) B is a complete metrizable compactoid, $B \subset E_n$ for some n and B is a compactoid with respect to the topology τ_n of E_n. Then $\overline{B}^{\tau_n}$ is a complete compactoid in the Banach space E_n, hence by [1], 3.1.7.(i) and [7], 10.(ii), $\tau|B = \tau_n|B$, so B is τ_n-complete. We have $A \bigcap B = (A \bigcap E_n) \bigcap B$ is τ_n-complete, hence τ-closed. By [6], 5.13 (ii), $A \bigcap B$ is $\sigma(E, E')$-closed and the proof is complete.

$(\beta) \Rightarrow (\alpha)$ Let $U_1 \supset U_2 \supset \cdots$ be a zero neighbourhood base in E where each U_n is absolutely convex. Let $E'_{U_n^0} := [U_n^0]$ be normed by the Minkowski function induced by U_n^0. Then $E'_{U_n^0}$ is a K-Banach space for each n and $E' = \bigcup_n E'_{U_n^0}$. Let $(E', \tau_e) := \varinjlim E'_{U_n^0}$ (called Berezanskii dual in [1]). By [1], 3.1.13 the inductive sequence $(E'_{U_n^0})_n$ is semicompact. Now E is a Montel space in the sense of [6], 10.1, so by [1], 2.5.9 we have $\tau_e = \beta(E', E)$. On the other hand, by nuclearity and metrizability, each bounded set in E is contained in the absolutely convex hull of a precompact set, so $\beta(E', E) = \tau_{pc}(E', E)$.

Now let $A \in \mathcal{P}_{E'}$ be such that $A \bigcap M$ is $\sigma(E', E)$-closed in M for each equicontinuous set $M \subset E'$. To prove that A is $\tau_{pc}(E', E)$-closed (i.e. τ_e-closed by what we just have proved) it suffices by (β) to prove that $A \bigcap E'_{U_n^0}$ is closed in the Banach space $E'_{U_n^0}$, for each n. So let $f_1, f_2, \ldots \in A \bigcap E'_{U_n^0}$ be such that $f_m \to f$ in the norm of $E'_{U_n^0}$ for some $f \in E'_{U_n^0}$. Then this sequence is bounded so there is a $\lambda \in K$ such that $f_m \in \lambda U_n^0$, $f \in \lambda U_n^0$ for all m. Now λU_n^0 is equicontinuous and $f_m \to f$ with respect to $\tau_e = \beta(E', E)$, so certainly $f_m \to f$ with respect to $\sigma(E', E)$. The set $A \bigcap \lambda U_n^0$ is $\sigma(E', E)$-closed in λU_n^0, hence in E' (as λU_n^0 itself is $\sigma(E', E)$-closed in E'). It follows that $f \in A \bigcap E'_{U_n^0}$ and we are done.

The following corollaries obtain.

COROLLARY 4.2 (See also Proposition [4.6]) *Let K be locally compact, let $(E_n)_n$ be a semicompact inductive sequence of locally convex spaces, let $(E, \tau) = \varinjlim E_n$. Then τ is the strongest (not necessarily locally convex) topology on E making all inclusions $E_n \to E$ continuous.*

Proof. Combine Theorem 3.2 with Theorem 4.1 $(\alpha) \Rightarrow (\beta)$ for the case where $\mathcal{P}_E$ resp. $\mathcal{P}_{E'}$ is the collection of all subsets of E resp. E'.

COROLLARY 4.3 *Let K be spherically complete, let $(E_n)_n$ be a semicompact inductive sequence of locally convex spaces, let $(E, \tau) = \varinjlim E_n$. Let $A \subset E$ be convex such that $A \bigcap E_n$ is closed in E_n for each n. Then A is τ-closed.*

Proof. Combine Theorem 3.5 with Theorem 4.1 $(\alpha) \Rightarrow (\beta)$ for the case where $\mathcal{P}_E$ resp. $\mathcal{P}_{E'}$ is the collection of all convex subsets of E resp. E'.

COROLLARY 4.4 *Let $(E_n)_n$ be a semicompact inductive sequence of locally convex spaces, let $(E,\tau) = \varinjlim E_n$. Let $A \subset E$ be edged such that $A \bigcap E_n$ is closed in E_n for each n. Then A is τ- closed.*

Proof. Combine Theorem 3.10 with Theorem 4.1 $(\alpha) \Rightarrow (\beta)$ for the case where $\mathcal{P}_E$ resp. $\mathcal{P}_{E'}$ is the collection of all edged subsets of E resp. E'.

Also, applying Theorem 3.1 and with similar argument as in the proof of $(\alpha) \Rightarrow (\beta)$ of Theorem 4.1 we derive the following version of Corollary 4.2 on closedness of (not-necessarily convex) subsets of semicompact inductive limits.

COROLLARY 4.5 *Let $(E_n)_n$ be a semicompact inductive sequence of locally convex spaces, let $(E,\tau) = \varinjlim E_n$. Let $A \subset E$ be such that $A \bigcap E_n$ is compact in E_n for each n. Then A is τ-closed.*

REMARK Clearly (β) of Theorem 4.1 holds by definition of the locally convex inductive limit if for $\mathcal{P}_E$ is chosen the set of all complements of absolutely convex sets. So, 4.1 $(\beta) \Rightarrow (\alpha)$ furnishes an alternative proof of Theorem 3.3 (but only for those E that are nuclear!).

To find another application of the implication $(\beta) \Rightarrow (\alpha)$ of Theorem 4.1 we first extend the Example of the Introduction.

PROPOSITION 4.6 *Let K be not locally compact, let $E = \varinjlim E_n$ be the locally convex inductive limit of a proper inductive sequence of Banach spaces $(E_n)_n$. Then there exists a non-closed subset X of E for which $X \bigcap E_n$ is closed in E_n for each n.*

Proof. If E is not Hausdorff, take $X = \{0\}$. If E is Hausdorff, by [1], 2.1.4 the space E is not metrizable, so by [1], 2.1.5, E contains ϕ (i.e. the space $K^{(\mathbb{N})}$ equipped with the strongest locally convex topology). Since ϕ is complete, the spaces $E_n \bigcap \phi$ are finite-dimensional and hence $\phi = \varinjlim E_n \bigcap \phi$. On the other hand, in the Introduction we have constructed a set $X \subset \phi$, not closed in ϕ, such that $X \bigcap E_n$ is closed in $E_n \bigcap \phi$ for each n. Further, the spaces $E_n \bigcap \phi$ are closed in E_n by finite-dimensionality. It follows that X meets the requirements.

COROLLARY 4.7 (Failure of the Banach-Dieudonné Theorem for non-locally compact fields) *Let K be not locally compact, let E be an infinite-dimensional nuclear Fréchet space. Then there is a set $X \subset E'$ for which $X \bigcap M$ is $\sigma(E',E)$-closed in M for each equicontinuous set $M \subset E'$ but such that X is not $\tau_{pc}(E',E)$-closed.*

Proof. From the proof of Theorem 4.1 $(\alpha) \Rightarrow (\beta)$ it follows that if (α) holds for some nuclear Fréchet space E when $\mathcal{P}_{E'}$ is the collection of all subsets of E', then $(\beta)'$ holds for $\mathcal{P}_{E'}$ where $(\beta)'$ is (β), but with $E'_{U_n^0}$ in place of E_n and (E',τ_e) in place of (E,τ), where $U_1 \supset U_2 \supset \ldots$ is a zero neighbourhood base on E consisting of absolutely convex sets. By infinite-dimensionality E is not a Banach space so the sets U_n are not (weakly) bounded implying that the U_n^0 are not absorbing and so we can assume that the inductive sequence $(E'_{U_n^0})_n$ is proper. By Proposition 4.6, statement $(\beta)'$ does not hold, so neither does (α).

Even more, putting together Theorem 3.4 and Corollary 4.7 we deduce the following.

COROLLARY 4.8 *Let K be not locally compact, let E be an infinite-dimensional nuclear Fréchet space. Then the finest topology on E' that coincides with $\sigma(E',E)$ on equicontinuous subsets of E' is not a linear topology.*

5 DIRECT PROOFS FOR SEMICOMPACT INDUCTIVE LIMITS

The study of the p-adic Banach-Dieudonné Theorem carried out in the above sections constitutes an interesting subject by itself turning out to be also a useful tool to obtain interesting information on closedness of sets in semicompact inductive limits, and conversely. In this section we are going to show that by using basic properties of (c-)compact and compactoid sets it is also possible to give alternative direct proofs of the results containing that information (4.2, 4.3, 4.4 and 4.5).

Firstly, for locally compact fields the proof of 4.2 does not differ much from the classical one in [11], Theorem 1.

Direct proof of Corollary 4.2.

Let $A \subset E$ be a non-empty subset of E such that $A \bigcap E_n$ is closed in E_n for each n. To prove that A is τ-closed we can assume that all E_n are Banach spaces ([1], 3.1.4) and that for each n the closed unit ball of E_n is closed (and hence compact) in E_{n+1} (see the proof of [1], 3.1.16).

Let $x \in E \setminus A$ and choose $p \in \mathbb{N}$ such that $x \in E_p$ and $A \bigcap E_p \neq \emptyset$. Without loss of generality we can take $p = 1$. We prove the existence of $\varepsilon_n \in |K| \setminus \{0\}$ $(n \in \mathbb{N})$ such that for all $m \in \mathbb{N}$

$$(x + B_{E_1}(\varepsilon_1) + \ldots + B_{E_m}(\varepsilon_m)) \bigcap A = \emptyset \qquad (*)$$

(This will complete the proof since $x + \sum_n B_{E_n}(\varepsilon_n)$ is a neighbourhood of x in E which does not intersect A).

The existence of ε_1 follows directly from the fact that $A \bigcap E_1$ is τ_1-closed and $x \notin A \bigcap E_1$. Suppose for some $n \in \mathbb{N}$ the numbers ε_k $(k \leqslant n)$ have been constructed satisfying (*) for $m = n$. Since $x + B_{E_1}(\varepsilon_1) + \ldots + B_{E_n}(\varepsilon_n)$ is a τ_{n+1}-compact set not intersecting the τ_{n+1}-closed set $A \bigcap E_{n+1}$, we have that

$$d_{n+1} := \operatorname{dist}_{E_{n+1}}(x + B_{E_1}(\varepsilon_1) + \ldots + B_{E_n}(\varepsilon_n), A \bigcap E_{n+1}) > 0.$$

Then by taking as ε_{n+1} any element of $|K| \setminus \{0\}$ with $\varepsilon_{n+1} \leqslant d_{n+1}/2$ we derive property $(*)$ for $m = n + 1$.

For a spherically complete field K the same proof as above works for Corollary 4.3 on τ-closedness of convex sets $A \subset E$. Indeed, we have just to substitute "compact" by "c-compact" and recall the property stated in [12] , 6.24, assuring that if B and C are closed convex subsets of a Banach space E over K such that $B \bigcap C = \emptyset$ and one of which is compactoid, then $\text{dist}_E(B, C) > 0$. Although this last property is not true in general for Banach spaces over arbitrary K ([12], 6.28), it can be partially saved, as the following lemma shows, which will be crucial for the direct proof of Corollary 4.4.

LEMMA 5.1 *Let A and B be closed absolutely convex subsets of a Banach space E over K, one of which is compactoid. Let $\lambda \in K$, $|\lambda| > 1$ if the valuation is dense, $\lambda = 1$ if the valuation is discrete. If $(a + \lambda A) \bigcap \lambda B = \emptyset$ for some $a \in E$, then* $\text{dist}_E(a + A, B) > 0$.

Proof. By [12], 6.24 we can assume the valuation of K to be dense. Since $(a + \lambda A) \bigcap \lambda B = \emptyset$ we clearly have that $a \notin (A + B)^e$. Also, it follows from [8], 1.4.(ii) that $(A + B)^e$ is closed, from which we deduce that $a \notin \overline{A + B}$ and we are done.

Direct proof of Corollary 4.4.

By Corollary 4.3 we can assume that the valuation of K is dense.

Let $A \subset E$ be an edged subset of E such that $A \bigcap E_n$ is closed in E_n for each n. To prove that A is τ-closed we can again assume that all E_n are Banach spaces and that for each n the closed unit ball of E_n is closed (and hence compactoid and complete) in E_{n+1}.

Let $x \in E \setminus A$ and suppose $x \in E_1$ and $A \bigcap E_1 \neq \emptyset$. As in the direct proof of Corollary [4.2] it is enough to construct $\varepsilon_n \in |K| \setminus \{0\}$ $(n \in \mathbb{N})$ satisfying $(*)$. To this end, let $\mu \in K$ with $|\mu| > 1$ for which $x \notin \mu A$ and let $\lambda_1, \lambda_2, \ldots \in K$ such that $|\lambda_n| > 1$ for all n and $|\mu| \prod_n |\lambda_n^{-1}| \geqslant 1$. We prove the existence of real numbers $\varepsilon_n \in |K| \setminus \{0\}$ $(n \in \mathbb{N})$ such that for all $m \geqslant 2$

$$\begin{aligned}[x + \lambda_m^{-1}\lambda_{m-1}^{-1} \cdots \lambda_1^{-1}\mu B_{E_1}(\varepsilon_1) + \lambda_m^{-1}\lambda_{m-1}^{-1} \cdots \lambda_2^{-1}\mu B_{E_2}(\varepsilon_2) + \cdots + \\ \lambda_m^{-1}\lambda_{m-1}^{-1}\mu B_{E_{m-1}}(\varepsilon_{m-1}) + \lambda_m^{-1}\mu B_{E_m}(\varepsilon_m)] \bigcap \lambda_m^{-1}\lambda_{m-1}^{-1} \cdots \lambda_1^{-1}\mu A = \emptyset\end{aligned} \qquad (**)$$

(which clearly implies $(*)$).

Firstly observe that since $x \notin \lambda_1^{-1}\mu(A \bigcap E_1)$ and this last set is τ_1-closed, there exists $\varepsilon_1 \in |K| \setminus \{0\}$ such that

$$(x + \lambda_1^{-1}\mu B_{E_1}(\varepsilon_1)) \bigcap (\lambda_1^{-1}\mu A \bigcap E_2) = \emptyset.$$

Hence, by Lemma 5.1,

$$d_2 := \text{dist}_{E_2}(x + \lambda_2^{-1}\lambda_1^{-1}\mu B_{E_1}(\varepsilon_1), \lambda_2^{-1}\lambda_1^{-1}\mu A \bigcap E_2) > 0.$$

Then by taking as ε_2 any element of $|K| \setminus \{0\}$ with $\varepsilon_2 \leqslant (d_2 |\lambda_2|)/(2|\mu|)$ we derive property (**) for $m = 2$.

Suppose for some $n \in \mathbb{N}$, $n \geqslant 2$, the numbers ε_k $(k \leqslant n)$ have been constructed satisfying (**). Choose $\alpha_{n+1}, \beta_{n+1} \in K$ with $|\alpha_{n+1}|, |\beta_{n+1}| > 1$ such that $\lambda_{n+1} = \alpha_{n+1}\beta_{n+1}$. Since (**) holds for $m = n$ we have

$$(x + \beta_{n+1}A_{n+1}) \bigcap \beta_{n+1}(\lambda_{n+1}^{-1}\lambda_n^{-1} \cdots \lambda_1^{-1}\mu A \bigcap E_{n+1}) = \emptyset,$$

where

$$\begin{aligned} A_{n+1} :=&(\lambda_{n+1}^{-1}\lambda_n^{-1} \cdots \lambda_1^{-1}\mu B_{E_1}(\varepsilon_1) + \lambda_{n+1}^{-1}\lambda_n^{-1} \cdots \lambda_2^{-1}\mu B_{E_2}(\varepsilon_2) + \cdots + \\ &\lambda_{n+1}^{-1}\lambda_n^{-1}\lambda_{n-1}^{-1}\mu B_{E_{n-1}}(\varepsilon_{n-1}) + \lambda_{n+1}^{-1}\lambda_n^{-1}\mu B_{E_n}(\varepsilon_n))^e \end{aligned}$$

is an absolutely convex compactoid subset of E_{n+1} which, by [7],10.(ii) and [8],1.4.(ii), is τ_{n+1}-complete. On the other hand, since $B_{n+1} := \lambda_{n+1}^{-1}\lambda_n^{-1} \cdots \lambda_1^{-1}\mu A \bigcap E_{n+1}$ is an absolutely convex τ_{n+1}-closed set, we can apply Lemma 5.1 to obtain that

$$d_{n+1} := \mathrm{dist}_{E_{n+1}}(x + A_{n+1}, B_{n+1}) > 0.$$

Then by taking as ε_{n+1} any element of $|K| \setminus \{0\}$ with $\varepsilon_{n+1} \leqslant (d_{n+1} |\lambda_{n+1}|)/(2|\mu|)$ we derive property (**) for $m = n + 1$.

Similarly, the direct proof of Corollary 4.5 can be carried out by following the techniques used for Corollary 4.4.

REFERENCES

[1] N De Grande-De Kimpe, J Kakol, C Perez-Garcia and WH Schikhof. p-Adic locally convex inductive limits. In: WH Schikhof, C Perez-Garcia, J Kąkol. ed. p-Adic Functional Analysis. New York: Marcel Dekker, 1997, pp 159–222.

[2] J Horváth. Topological Vector Spaces and Distributions. Vol I. Reading, Massachusetts: Addison-Wesley, 1966.

[3] AK Katsaras. Spaces of non-archimedean valued functions. Bolletino UMI, (6) 5-B:603–621, 1986.

[4] AK Katsaras, A Beloyiannis. On the topology of compactoid convergence in non-archimedean spaces. Ann Math Blaise Pascal 3:135–153, 1996.

[5] H Komatsu. Proyective and injective limits of weakly compact sequences of locally convex spaces. J Math Soc Japan 19:366–383, 1967.

[6] WH Schikhof. Locally convex spaces over non-spherically complete valued fields I-II. Bull Soc Math Belg B 38:187–224, 1986.

[7] WH Schikhof. Compact-like sets in non-archimedean functional analysis. Proceedings of the Conference on p-adic analysis, Hengelhoef, Belgium, 1986, pp 137–147.

[8] WH Schikhof. The continuous linear image of a p-adic compactoid. Proc Kon Ned Akad Wet A92:119–123, 1989.

[9] WH Schikhof. The complementation property of ℓ^∞ in p-adic Banach spaces. In: F Baldassarri, S Bosch, B Dwork, ed. p-Adic Analysis. Berlin: Springer-Verlag, 1990, pp 342–350.

[10] WH Schikhof. The p-adic Krein-Šmulian Theorem. In: JM Bayod, N De Grande-De Kimpe, J Martinez-Maurica, ed. p-Adic Functional Analysis. New York: Marcel Dekker, 1992, pp 177–189.

[11] JS e Silva. Su certe classi di spazi localmente convessi importanti per le applicazioni. Rend Mat e Appl 14:388–410, 1955.

[12] ACM van Rooij. Notes on p-adic Banach spaces. Report 7725, Department of Mathematics, Catholic University, Nijmegen, The Netherlands:1–52, 1977.

Mahler's and other bases for p-adic continuous functions

STANY DE SMEDT Vrije Universiteit Brussel, Faculteit Toegepaste Wetenschappen, Pleinlaan 2, B-1050 Brussel, Belgium.

Abstract. In this paper we prove Mahler's expansion with remainder for p-adic continuous functions in several variables. These results are similar to those obtained by Van Hamme ([5]) for the case of one variable. In a second part we give, with the help of infinite matrix theory, a sufficient condition to construct orthonormal bases for the Banach space of p-adic continuous functions by taking a linear combination of a given orthonormal base. This extends an earlier theorem ([3]) on this subject.

1 INTRODUCTION

Let p be a prime number. $\mathbb{Z}_p$ and $\mathbb{Q}_p$ denote, respectively, the ring of p-adic integers and the field of p-adic numbers. The valuation on $\mathbb{Q}_p$ will be denoted $|\ |$. As usual, we write $C(\mathbb{Z}_p \to K), ||\ ||_s$ for the Banach space of continuous functions from $\mathbb{Z}_p$ to K, where K is an algebraic extension of $\mathbb{Q}_p$ and $||\ ||_s$ is the sup-norm defined as $||f||_s = \sup_{x \in \mathbb{Z}_p} |f(x)|$. We have the following well-known bases for $C(\mathbb{Z}_p \to K)$. On one hand, we have the Mahler base $\binom{x}{n}$ $(n \in \mathbb{N})$, consisting of polynomials of degree n. On the other hand we have the van der Put base $\{e_n \mid n \in \mathbb{N}\}$ consisting of locally constant functions e_n defined as follows : $e_0(x) = 1$ and for $n > 0$, e_n is the characteristic function of the ball $\{\alpha \in \mathbb{Z}_p \mid |\alpha - n| < 1/n\}$. For every $f \in C(\mathbb{Z}_p \to K)$ we have the following uniformly convergent series:

$$f(x) = \sum_{n=0}^{\infty} a_n \binom{x}{n} \quad \text{where} \quad a_n = \sum_{j=0}^{n} (-1)^{n-j} \binom{n}{j} f(j)$$

$$f(x) = \sum_{n=0}^{\infty} b_n e_n(x) \quad \text{where} \quad b_0 = f(0) \quad \text{and} \quad b_n = f(n) - f(n_-).$$

Here n_- is defined as follows. For every $n \in \mathbb{N}_0$, we have a Hensel expansion $n = n_0 + n_1 p + ... + n_s p^s$ with $n_s \neq 0$. Then $n_- = n_0 + n_1 p + ... + n_{s-1} p^{s-1}$.

For the Banach space $C(\mathbb{Z}_p \times \mathbb{Z}_p \to \mathbb{Q}_p), || \ ||_s$ of all continuous functions from $\mathbb{Z}_p \times \mathbb{Z}_p$ to $\mathbb{Q}_p$ there exists the well-known orthonormal base $\binom{x}{i}\binom{y}{j}$ $(i, j \in \mathbb{N})$. Recall that for a continuous function $f : \mathbb{Z}_p \times \mathbb{Z}_p \to \mathbb{Q}_p$ the sup-norm $||f||_s$ is defined as $\sup_{x,y \in \mathbb{Z}_p} |f(x,y)|$.

Every continuous function $f : \mathbb{Z}_p \times \mathbb{Z}_p \to \mathbb{Q}_p$ can be written as

$$f(x,y) = \sum_{n=0}^{\infty} \sum_{m=0}^{\infty} a_{n,m} \binom{x}{n}\binom{y}{m}$$

with

$$a_{n,m} = \sum_{i=0}^{n} \sum_{j=0}^{m} (-1)^{n+m-i-j} \binom{n}{i}\binom{m}{j} f(i,j) = \Delta_1^n \Delta_2^m f(0,0)$$

where Δ_1 is the difference operator defined by $\Delta_1 f(x,y) = f(x+1,y) - f(x,y)$ and Δ_2 is the difference operator defined by $\Delta_2 f(x,y) = f(x,y+1) - f(x,y)$. This is called the Mahler expansion of the continuous function f.
(see [2] and [4] for more information)

2 MAHLER'S EXPANSION WITH REMAINDER

To prove the Mahler expansion with remainder, we will need a convolution on $C(\mathbb{Z}_p \times \mathbb{Z}_p \to \mathbb{Q}_p)$. One way to define this is as follows.

DEFINITION The convolution of two double sequences $f(m,n)$ and $g(m,n)$ is the double sequence given by

$$(f * g)(m,n) = \sum_{k=0}^{m} \sum_{l=0}^{n} f(k,l) g(m-k, n-l).$$

In order to extend this definition to continuous functions defined on $\mathbb{Z}_p^2$, we need the following lemma.

LEMMA *If the double sequences $f(m,n)$ and $g(m,n)$ are the restrictions to $\mathbb{N}^2$ of two continuous functions from $\mathbb{Z}_p^2$ to $\mathbb{Q}_p$ then* $\lim_{m+n \to \infty} \Delta_1^m \Delta_2^n (f * g)(0,0) = 0$.

COROLLARY *The double sequence $(m,n) \to (f * g)(m,n)$ is the restriction of a continuous function $f * g$ from $\mathbb{Z}_p^2$ to $\mathbb{Q}_p$.*

The proof of the lemma is based on a special case of the following theorem which generalizes Mahler's expansion for p-adic continuous functions in two variables.

THEOREM *If $f : \mathbb{Z}_p \times \mathbb{Z}_p \to \mathbb{Q}_p$ is a continuous function and $g : \mathbb{N} \times \mathbb{N} \to \mathbb{Q}_p$ is a bounded double sequence then*

$$\sum_{i,j=0}^{\infty} \binom{x}{i}\binom{y}{j} g(i,j)\Delta_1^i\Delta_2^j f(0,0) = \sum_{i,j=0}^{\infty} \binom{x}{i}\binom{y}{j} \Delta_1^i\Delta_2^j g(0,0)\Delta_1^i\Delta_2^j f(x-i,y-j).$$

This is a generalization of Mahler's expansion since if $g(i,j) = 1$ for all $i,j \in \mathbb{N}$ then $\Delta_1^i\Delta_2^j g(0,0) = 0$ for $i \neq 0$ or $j \neq 0$ and the formula reduces to Mahler's expansion.

Proof. Since $\lim_{i+j\to\infty} \Delta_1^i\Delta_2^j f(0,0) = 0$ uniformly, the series on the left hand side and on the right hand side are uniformly convergent, which means that their sums are continuous functions of (x,y). Hence it is sufficient to prove the theorem for $(x,y) = (m,n) \in \mathbb{N}^2$.
Starting from the identity

$$(1+v_1u_1-v_1)^m(1+v_2u_2-v_2)^n = ((v_1-1)(u_1-1)+u_1)^m((v_2-1)(u_2-1)+u_2)^n$$

we obtain

$$\sum_{k=0}^{m}\binom{m}{k} v_1^k(u_1-1)^k \sum_{l=0}^{n}\binom{n}{l} v_2^l(u_2-1)^l$$
$$= \sum_{k=0}^{m}\binom{m}{k}(v_1-1)^k(u_1-1)^k u_1^{m-k}\sum_{l=0}^{n}\binom{n}{l}(v_2-1)^l(u_2-1)^l u_2^{n-l}. \qquad (1)$$

Define a linear map $L_u : \mathbb{Q}_p[u_1,u_2] \to \mathbb{Q}_p : u_1^k u_2^l \to f(k,l)$ then L_u maps $(u_1-1)^k(u_2-1)^l$ on $\Delta_1^k\Delta_2^l f(0,0)$.
Similar, define a linear map $L_v : \mathbb{Q}_p[v_1,v_2] \to \mathbb{Q}_p : v_1^k v_2^l \to g(k,l)$ that maps $(v_1-1)^k(v_2-1)^l$ on $\Delta_1^k\Delta_2^l g(0,0)$.
If we first apply L_u and then L_v to (1) we get

$$\sum_{k=0}^{m}\sum_{l=0}^{n}\binom{m}{k}\binom{n}{l} g(k,l)\Delta_1^k\Delta_2^l f(0,0)$$
$$= \sum_{k=0}^{m}\sum_{l=0}^{n}\binom{m}{k}\binom{n}{l}\Delta_1^k\Delta_2^l g(0,0)\Delta_1^k\Delta_2^l f(m-k,n-l).$$

SPECIAL CASE *For $|S|,|T| \leqslant 1$ take $f(x,y) = (1+S)^x(1+T)^y$.*
This gives

$$\sum_{i,j=0}^{\infty}\binom{x}{i}\binom{y}{j} g(i,j)S^iT^j$$
$$= (1+S)^x(1+T)^y \sum_{i,j=0}^{\infty}\binom{x}{i}\binom{y}{j}\Delta_1^i\Delta_2^j g(0,0)\left(\frac{S}{1+S}\right)^i\left(\frac{T}{1+T}\right)^j$$

Putting $x = -1, y = -1$ and replacing S by $-S$ and T by $-T$, we get the expansion

$$\sum_{i,j=0}^{\infty} g(i,j)S^iT^j = \sum_{i,j=0}^{\infty} \Delta_1^i\Delta_2^j g(0,0)\frac{S^i}{(1-S)^{i+1}}\frac{T^j}{(1-T)^{j+1}} \tag{2}$$

Proof of the lemma.

$$\sum_{i,j=0}^{\infty}(f*g)(i,j)S^iT^j$$

$$= \sum_{i,j=0}^{\infty} f(i,j)S^iT^j \sum_{i,j=0}^{\infty} g(i,j)S^iT^j$$

$$= \frac{1}{(1-S)^2}\frac{1}{(1-T)^2}\sum_{i,j=0}^{\infty}\Delta_1^i\Delta_2^j f(0,0)\left(\frac{S}{1-S}\right)^i\left(\frac{T}{1-T}\right)^j$$

$$\sum_{i,j=0}^{\infty}\Delta_1^i\Delta_2^j g(0,0)\left(\frac{S}{1-S}\right)^i\left(\frac{T}{1-T}\right)^j$$

$$= \frac{1}{(1-S)^2}\frac{1}{(1-T)^2}\sum_{i,j=0}^{\infty} a_{i,j}\left(\frac{S}{1-S}\right)^i\left(\frac{T}{1-T}\right)^j$$

$$= \frac{1}{1-S}\frac{1}{1-T}\left(1+\frac{S}{1-S}\right)\left(1+\frac{T}{1-T}\right)$$

$$\sum_{i,j=0}^{\infty} a_{i,j}\left(\frac{S}{1-S}\right)^i\left(\frac{T}{1-T}\right)^j$$

$$= \frac{1}{1-S}\frac{1}{1-T}\left(\sum_{i,j=0}^{\infty} a_{i,j}\left(\frac{S}{1-S}\right)^i\left(\frac{T}{1-T}\right)^j + \sum_{i,j=0}^{\infty} a_{i,j}\left(\frac{S}{1-S}\right)^{i+1}\left(\frac{T}{1-T}\right)^j\right.$$

$$\left. + \sum_{i,j=0}^{\infty} a_{i,j}\left(\frac{S}{1-S}\right)^i\left(\frac{T}{1-T}\right)^{j+1} + \sum_{i,j=0}^{\infty} a_{i,j}\left(\frac{S}{1-S}\right)^{i+1}\left(\frac{T}{1-T}\right)^{j+1}\right)$$

$$= \frac{1}{1-S}\frac{1}{1-T}\left(a_{0,0} + \sum_{i=1}^{\infty}(a_{i,j}+a_{i-1,j})\left(\frac{S}{1-S}\right)^i + \sum_{j=1}^{\infty}(a_{i,j}+a_{i,j-1})\left(\frac{T}{1-T}\right)^j\right.$$

$$\left. + \sum_{i,j=1}^{\infty}(a_{i,j}+a_{i-1,j}+a_{i,j-1}+a_{i-1,j-1})\left(\frac{S}{1-S}\right)^i\left(\frac{T}{1-T}\right)^j\right)$$

with $a_{i,j} = \sum_{k=0}^{i}\sum_{l=0}^{j}\Delta_1^k\Delta_2^l f(0,0)\Delta_1^{i-k}\Delta_2^{j-l}g(0,0)$.

Hence $\Delta_1^i\Delta_2^j(f*g)(0,0) = a_{i,j} + a_{i-1,j} + a_{i,j-1} + a_{i-1,j-1}$ for $i,j \geqslant 1$.

Since $\lim_{i+j\to\infty}\Delta_1^i\Delta_2^j f(0,0) = \lim_{i+j\to\infty}\Delta_1^i\Delta_2^j g(0,0) = 0$, we see that $\lim_{i+j\to\infty} a_{i,j} = 0$ which proves the lemma.

The definition above does not seem to be very useful, therefore we define a convolution on each variable separately.

DEFINITION Given two double sequences $f(m,n)$ and $g(m,n)$ we define the following double sequences.

The first convolution $(f *_1 g)$ is the double sequence

$$(f *_1 g)(m,n) = \sum_{i=0}^{m} f(i,n) g(m-i,n).$$

The second convolution $(f *_2 g)$ is the double sequence

$$(f *_2 g)(m,n) = \sum_{j=0}^{n} f(m,j) g(m,n-j).$$

It can be easily shown (similar as in [5] and the proof above) that the double sequences $(m,n) \to (f *_1 g)(m,n)$ and $(m,n) \to (f *_2 g)(m,n)$ are respectively the restriction of continuous functions $(f *_1 g)$ and $(f *_2 g)$ from $\mathbb{Z}_p^2$ to $\mathbb{Q}_p$.

In order to prove our main theorem, we now introduce a slightly different convolution obtained from the previous one using a shift.

DEFINITION Let f and g be two continuous functions from $\mathbb{Z}_p^2$ to $\mathbb{Q}_p$ and let $x, y \in \mathbb{Z}_p$

$$(f \circledast_1 g)(x,y) = (f *_1 g)(x-1,y)$$

$$(f \circledast_2 g)(x,y) = (f *_2 g)(x,y-1)$$

THEOREM (Mahler's expansion with remainder) *If $f : \mathbb{Z}_p \times \mathbb{Z}_p \to \mathbb{Q}_p$ is continuous, then*

$$f(x,y) = \sum_{i=0}^{n} \sum_{j=0}^{m} \Delta_1^i \Delta_2^j f(0,0) \binom{x}{i} \binom{y}{j} + R$$

with

$$R = \binom{x}{n} \circledast_1 \Delta_1^{n+1} f(x,y) + \binom{y}{m} \circledast_2 \Delta_2^{m+1} f(x,y)$$
$$- \binom{x}{n} \circledast_1 \binom{y}{m} \circledast_2 \Delta_1^{n+1} \Delta_2^{m+1} f(x,y).$$

Proof. From the case of one variable, we know (see [5])

$$f(x) = f(0) + \Delta f(0) \binom{x}{1} + \ldots + \Delta^n f(0) \binom{x}{n} + \binom{x}{n} \circledast \Delta^{n+1} f(x)$$

We will now apply this to our function $f : \mathbb{Z}_p \times \mathbb{Z}_p \to \mathbb{Q}_p : (x,y) \to f(x,y)$ in which we consider y as a constant. This gives

$$f(x,y) = f(0,y) + \Delta_1 f(0,y) \binom{x}{1} + \ldots + \Delta_1^n f(0,y) \binom{x}{n} + \binom{x}{n} \circledast_1 \Delta_1^{n+1} f(x,y)$$
$$= \sum_{i=0}^{n} \Delta_1^i f(0,y) \binom{x}{i} + \binom{x}{n} \circledast_1 \Delta_1^{n+1} f(x,y)$$

And also (taking $x = 0$ as a constant)

$$f(0,y) = f(0,0) + \Delta_2 f(0,0)\binom{y}{1} + \ldots + \Delta_2^m f(0,0)\binom{y}{m} + \binom{y}{m} \circledast_2 \Delta_2^{m+1} f(0,y)$$
$$= \sum_{j=0}^{m} \Delta_2^j f(0,0)\binom{y}{j} + \binom{y}{m} \circledast_2 \Delta_2^{m+1} f(0,y).$$

So,

$$\Delta_1^i f(0,y) = \sum_{j=0}^{m} \Delta_1^i \Delta_2^j f(0,0)\binom{y}{j} + \binom{y}{m} \circledast_2 \Delta_1^i \Delta_2^{m+1} f(0,y)$$

for every i. Substituting this last one in the previous gives

$$f(x,y) = \sum_{i=0}^{n}\sum_{j=0}^{m} \Delta_1^i \Delta_2^j f(0,0)\binom{x}{i}\binom{y}{j} + R$$

with

$$R = \binom{y}{m} \circledast_2 \sum_{i=0}^{n} \Delta_1^i \Delta_2^{m+1} f(0,y)\binom{x}{i} + \binom{x}{n} \circledast_1 \Delta_1^{n+1} f(x,y).$$

Applying Mahler's expansion with remainder for the case of one variable on the continuous function $\Delta_2^{m+1} f(x,y)$ with y constant gives

$$\Delta_2^{m+1} f(x,y) = \sum_{i=0}^{n} \Delta_1^i \Delta_2^{m+1} f(0,y)\binom{x}{i} + \binom{x}{n} \circledast_1 \Delta_1^{n+1}\Delta_2^{m+1} f(x,y)$$

Elimination of $\sum_{i=0}^{n} \Delta_1^i \Delta_2^{m+1} f(0,y)\binom{x}{i}$ between the last two equations gives

$$R = \binom{x}{n} \circledast_1 \Delta_1^{n+1} f(x,y) + \binom{y}{m} \circledast_2 \Delta_2^{m+1} f(x,y)$$
$$- \binom{x}{n} \circledast_1 \binom{y}{m} \circledast_2 \Delta_1^{n+1}\Delta_2^{m+1} f(x,y)$$

which ends our proof.

REMARK This formula is not similar to the usual form of Taylor's formula for two variables in classical analysis, the latter proceeding by homogeneous polynomials in x and y. It is still an open problem whether or not we can derive a second form of our theorem to obtain

$$f(x,y) = f(0,0) + \sum_{i+j=1} \Delta_1^i \Delta_2^j f(0,0)\binom{x}{i}\binom{y}{j}$$
$$+ \ldots + \sum_{i+j=n} \Delta_1^i \Delta_2^j f(0,0)\binom{x}{i}\binom{y}{j} + R_n$$

with an explicit expression for the remainder R_n.

Mahler's expansion with remainder can be easily derived for three or more variables by the same method. In the case of three variables the theorem becomes.

THEOREM *If $f : \mathbb{Z}_p^3 \to \mathbb{Q}_p$ is continuous, then*

$$f(x,y) = \sum_{i=0}^{n}\sum_{j=0}^{m}\sum_{k=0}^{l} \Delta_1^i \Delta_2^j \Delta_3^k f(0,0,0) \binom{x}{i}\binom{y}{j}\binom{z}{k} + R$$

with

$$\begin{aligned} R = & \binom{x}{n} \circledast_1 \Delta_1^{n+1} f(x,y,z) + \binom{y}{m} \circledast_2 \Delta_2^{m+1} f(x,y,z) + \binom{z}{l} \circledast_3 \Delta_3^{l+1} f(x,y,z) \\ & - \binom{x}{n} \circledast_1 \binom{y}{m} \circledast_2 \Delta_1^{n+1}\Delta_2^{m+1} f(x,y,z) \\ & - \binom{x}{n} \circledast_1 \binom{z}{l} \circledast_3 \Delta_1^{n+1}\Delta_3^{l+1} f(x,y,z) \\ & - \binom{y}{m} \circledast_2 \binom{z}{l} \circledast_3 \Delta_2^{m+1}\Delta_3^{l+1} f(x,y,z) \\ & + \binom{x}{n} \circledast_1 \binom{y}{m} \circledast_2 \binom{z}{l} \circledast_3 \Delta_1^{n+1}\Delta_2^{m+1}\Delta_3^{l+1} f(x,y,z). \end{aligned}$$

3 INFINITE MATRICES

A p-adic infinite matrix is a twofold table $A = (a_{i,j})$ $(i,j = 1,2,\ldots,\infty)$ of p-adic numbers, with addition and multiplication defined by $A + B = (a_{i,j} + b_{i,j})$, $\lambda.A = (\lambda a_{i,j})$ where λ is any p-adic number and $A.B = \left(\sum_{k=1}^{\infty} a_{i,k}b_{k,j}\right)$ if $\sum_{k=1}^{\infty} a_{i,k}b_{k,j}$ converges for every i and j.

As in the case of finite matrices we also have here notions of column vector (which we might describe as a matrix of order $\infty \times 1$), row vector (which we might describe as a matrix of order $1 \times \infty$) and the transpose of a matrix A obtained from A by interchanging rows and columns and denoted by A', so that for the matrix A' we have $a'_{i,j} = a_{j,i}$.

The zero matrix 0 is the table $a_{i,j} = 0$ for every i and j, and the unit matrix I is the table $a_{i,j} = \delta_{i,j}$ so that $\delta_{i,j} = 0$ if $i \neq j$ and $\delta_{i,i} = 1$. We then have

$0.A = 0 = A.0$ and $I.A = A = A.I$

The matrix ${}^{-1}A$ such that ${}^{-1}A.A = I$ is called a left-hand reciprocal of A, the matrix A^{-1} such that $A.A^{-1} = I$ is called a right-hand reciprocal of A. If A and B are both different from 0 and if $A.B = 0$, then B is called a right-hand zero-divisor of A, denoted by A^0, and A is a left-hand zero-divisor of B, denoted by ${}^0B$.

In the classical case (see [1]) the multiplication of infinite matrices is not in general associative, but in the non-archimedean case it is since

$$\sum_{l=1}^{\infty}\sum_{k=1}^{\infty} a_{i,l}b_{l,k}c_{k,j} = \sum_{k=1}^{\infty}\sum_{l=1}^{\infty} a_{i,l}b_{l,k}c_{k,j}$$

if these series converges. (see [4]). Thus $A.B.C = A.(B.C) = (A.B).C$

If $a_{i,j} = 0$ for $j > i$, A is called a lower semi-matrix; if $a_{i,j} = 0$ for $j < i$, A is called an upper semi-matrix.

If $(a_{i,j}), (b_{i,j})$ are both lower semi-matrices then

$$(a_{i,j}).(b_{i,j}) = (c_{i,j}) \text{ where } c_{i,j} = \begin{cases} \sum_{k=j}^{i} a_{i,k} b_{k,j} & \text{if } i \geqslant j \\ 0 & \text{otherwise} \end{cases}$$

Hence the product of two lower semi-matrices is a lower semi-matrix.

Suppose we are given a system of an infinity of linear equations in an infinity of unknowns $x_1, x_2, \dots$, say

$$\sum_{k=1}^{\infty} a_{i,k} x_k = b_i.$$

This can then be written as $A.X = B$ where $A = (a_{i,j}), X = (x_i), B = (b_i)$.

It is our aim to solve such infinite systems, therefore we will treat the reciprocals of infinite matrices in the next section. In the finite theory determinants play a fundamental part to do so, but their value is lost in the theory of infinite matrices.

4 RECIPROCALS OF INFINITE MATRICES

THEOREM *The lower semi-matrix A has no right-hand reciprocal if $a_{i,i} = 0$ for one or more values of i. But if $a_{i,i} \neq 0$ holds for every i then A has a unique right-hand reciprocal, which is a lower semi-matrix, and whose leading diagonal elements are $\frac{1}{a_{i,i}}$.*

Proof. Since A is a lower semi-matrix, $A.X = I$ becomes

$$\sum_{k=1}^{i} a_{i,k} x_{k,j} = \delta_{i,j}.$$

When $i = 1$, we obtain $a_{1,1} x_{1,1} = 1, a_{1,1} x_{1,j} = 0$ if $j > 1$.

If $a_{1,1} = 0$, the first equation has no solution in $x_{1,1}$, so that in this case X does not exist. If $a_{1,1} \neq 0$ then $x_{1,1} = \frac{1}{a_{1,1}}$ and $x_{1,j} = 0$ for $j > 1$.

When $i = 2$ we have

$$a_{2,1} x_{1,1} + a_{2,2} x_{2,1} = 0, a_{2,2} x_{2,2} = 1, a_{2,2} x_{2,j} = 0 \quad \text{for } j > 2.$$

If $a_{2,2} = 0$, the second equation has no solution in $x_{2,2}$, so that X does not exist. If $a_{2,2} \neq 0$, this equation gives $x_{2,2} = \frac{1}{a_{2,2}}$. Since $x_{1,1} = \frac{1}{a_{1,1}}$, the first equation determines $x_{2,1} = \frac{-a_{2,1}}{a_{1,1} a_{2,2}}$ and the last equations show that $x_{2,j} = 0$ for $j > 2$.

Proceeding on this way, we have a direct row-by-row determination of X, which is unique.

The proposition that a finite matrix A has an inverse matrix if and only if $\det(A) \neq 0$ is not usable in the infinite matrix theory, but it can be replaced by the two following propositions.

PROPOSITION *If A has a zero row, A has no right-hand reciprocal; if A has a zero column, A has no left-hand reciprocal.*

Proof. Let $a_{n,k} = 0$ for a fixed n and for every $k \geqslant 1$. Then $\sum_{k=1}^{\infty} a_{n,k}x_{k,j} = \delta_{n,j}$ reduces to the contradiction $0 = 1$ when $j = n$, so that A^{-1} does not exist. Similarly for the case of a zero column.

PROPOSITION *If the elements of a row (column) in A are the same multiple of the corresponding elements of another row (column), A has no right-hand (left-hand) reciprocal.*

Proof. If $a_{i_1,k} = c.a_{i_2,k}$ for every k, we have $\delta_{i_1,j} = c.\delta_{i_2,j}$ which leads to the contradiction $1 = 0$ when $j = i_1$. Similarly for columns.

The relation between left-hand and right-hand reciprocals is illustrated by the following theorem.

THEOREM *If A^{-1} is unique, then A^{-1} is also a left-hand reciprocal of A. If ${}^{-1}A$ is unique, then ${}^{-1}A$ is also a right-hand reciprocal of A.*
If A has both a right-hand reciprocal A^{-1} and a left-hand reciprocal ${}^{-1}A$, then $A^{-1} = {}^{-1}A$ and A has no other two-sided reciprocal.

Proof. Note that if A^{-1} is a right-hand reciprocal of A, then so is $A^{-1}+A^0$ and every solution of $A.X = I$ is of this form, since $A.X = I$ is equivalent to $A.(X - A^{-1}) = 0$. Thus $X - A^{-1} = A^0$.
If C is a solution of $A.X = B$, then all solutions are of the form $C + A^0$, which can be proven as above. In particular, the solutions of $A.X = A$ are all of the form $I + A^0$. If A^{-1} is unique, there is no A^0, since $A^{-1} + A^0$ would be another right-hand reciprocal, and hence $X = I$ is the only solution of $A.X = A$.
Now $A.(A^{-1}.A) = (A.A^{-1}).A = A$ so that $A^{-1}.A = I$ and hence A^{-1} is also a left-hand reciprocal of A.
Similarly for the second part. There remains to prove the last part.
From ${}^{-1}A.A = I$ and $A.A^{-1} = I$, we have

$${}^{-1}A = {}^{-1}A.(A.A^{-1}) = ({}^{-1}A.A).A^{-1} = A^{-1}.$$

Fixing A^{-1} (say), every ${}^{-1}A$ is equal to this fixed A^{-1}; i.e. there is one ${}^{-1}A$ only and similarly for A^{-1}.

REMARK Let A be a lower semi-matrix for which $a_{i,i} \neq 0$ holds for every i, so that A has a unique right-hand reciprocal X. Then X is also a left-hand reciprocal of A and is the only two-sided reciprocal of A.

Before proceeding further and proving two general theorems on reciprocals, we have to define the norm of a matrix and row-zero-convergent matrices.

DEFINITION For an infinite matrix $A = (a_{i,j})$ we define the norm as $||A|| = \max_{i,j} |a_{i,j}|$

It can be easily seen that

$$||A|| = 0 \Leftrightarrow A = 0$$
$$||cA|| = |c|.||A||$$
$$||A + B|| \leqslant \max(||A||, ||B||)$$
$$||A.B|| \leqslant ||A||.||B||$$

DEFINITION A matrix A is said to be row-zero-convergent if every row converges to zero.

A matrix A is said to be column-zero-convergent if every column converges to zero.

THEOREM *If A is a row-zero-convergent matrix such that $||A|| < 1$, then $B = I + \sum_{i=1}^{\infty}(-1)^i A^i$ exists (is convergent) and is the unique, and two-sided, reciprocal of $I + A$. Moreover B is also row-zero-convergent.*

Proof. $(A^2)_{i,j} = \sum_{k=1}^{\infty} a_{i,k}a_{k,j}$ exist for every i and j since $||A|| < 1$ and hence, $\lim_{k\to\infty} a_{i,k}a_{k,j} = 0$ so the sum converges. Similarly also $(A^n)_{i,j} = \sum_{k=1}^{\infty} a_{i,k}(A^{n-1})_{k,j}$ exist for every i and j since

$\lim_{k\to\infty} a_{i,k}(A^{n-1})_{k,j} = 0$.

Since $||A|| < 1$ we get that $\lim_{i\to\infty}(-1)^i A^i = 0$ so B exists.

That B is the reciprocal of $I + A$ can be easily seen as

$$\begin{aligned}(I + A).\left(I + \sum_{i=1}^{\infty}(-1)^i A^i\right) &= I + A + \sum_{i=1}^{\infty}(-1)^i A^i + \sum_{i=1}^{\infty}(-1)^i A^{i+1} \\ &= I + \sum_{i=1}^{\infty}(-1)^i A^i + \sum_{i=1}^{\infty}(-1)^{i-1} A^i \\ &= I\end{aligned}$$

Similarly

$$\left(I + \sum_{i=1}^{\infty}(-1)^i A^i\right)(I + A) = I.$$

So B is a two-sided reciprocal of $I + A$, and thus unique.

That B is row-zero-convergent can be proven by induction as follows :

Suppose A^n is row-zero-convergent.

Since $(A^{n+1})_{i,j} = \sum_{k=1}^{\infty} a_{i,k}(A^n)_{k,j}$, also $\lim_{j\to\infty}(A^{n+1})_{i,j} = 0$ for every i. This proves that A^i is row-zero-convergent for all i, but then also $B = I + \sum_{i=1}^{\infty}(-1)^i A^i$ is row-zero-convergent.

LEMMA *If A^{-1} and $\left(I + BA^{-1}\right)^{-1}$ exists, then $A^{-1}.\left(I + BA^{-1}\right)^{-1}$ is a right-hand reciprocal of $A + B$ if this product exists.*

Proof. Trivial.

THEOREM *If A and B are such that a right-hand reciprocal A^{-1} of A exists and BA^{-1} is a row-zero-convergent matrix such that $\|BA^{-1}\| < 1$, then $A^{-1}.\left(I + \sum_{i=1}^{\infty}(-1)^i(BA^{-1})^i\right)$ is a right-hand reciprocal of $A + B$. If moreover A^{-1} is the only right-hand reciprocal of A, then $A^{-1}.\left(I + \sum_{i=1}^{\infty}(-1)^i(BA^{-1})^i\right)$ is the unique and two-sided reciprocal of $A + B$.*

Proof. $I + \sum_{i=1}^{\infty}(-1)^i(BA^{-1})^i$ is the right-hand reciprocal of $I + BA^{-1}$, as is shown in the previous theorem. So we get

$$(A+B).A^{-1}.\left(I + \sum_{i=1}^{\infty}(-1)^i(BA^{-1})^i\right) = (I+BA^{-1}).\left(I + \sum_{i=1}^{\infty}(-1)^i(BA^{-1})^i\right) = I$$

which proves our theorem in the first case.
If A^{-1} is unique, then so is $A^{-1}.\left(I + \sum_{i=1}^{\infty}(-1)^i(BA^{-1})^i\right)$, and

$$A^{-1}.\left(I + \sum_{i=1}^{\infty}(-1)^i(BA^{-1})^i\right).(A+B) = A^{-1}.\left(I + BA^{-1}\right)^{-1}.(I + BA^{-1}).A = I.$$

COROLLARY *The row-zero-convergent matrix $A = (a_{i,j})$ such that*

$$\begin{cases} |a_{i,j}| \leqslant 1 & \text{for } i > j \\ |a_{i,j}| = 1 & \text{for } i = j \\ |a_{i,j}| < 1 & \text{for } i < j \end{cases}$$

has a unique two-sided reciprocal.

Proof. Let $A = B + C$ with

$$b_{i,j} = \begin{cases} a_{i,j} & \text{for } i \geqslant j \\ 0 & \text{otherwise} \end{cases}$$

and

$$c_{i,j} = \begin{cases} a_{i,j} & \text{for } i < j \\ 0 & \text{otherwise} \end{cases}.$$

B is a lower semi-matrix, so our first theorem tells that B has a unique right-hand reciprocal B^{-1}. B^{-1} is also a lower semi-matrix, say $B^{-1} = (d_{i,j})$ with $d_{i,j} = 0$ for $j > i$; and $\|B^{-1}\| = 1$ as can be easily seen from the construction of B^{-1} in the proof of the first theorem of section 4.

$CB^{-1} = \left(\sum_{k=\max\{j,i+1\}}^{\infty} c_{i,k} d_{k,j} \right)$ exists and is row-zero-convergent since $\lim_{k\to\infty} c_{i,k} = 0$

Further $||CB^{-1}|| \leqslant ||C||.||B^{-1}|| < 1$.

So the previous theorem assures us that $B^{-1}.\left(I + \sum_{i=1}^{\infty} (-1)^i (CB^{-1})^i \right)$ is the unique and two-sided reciprocal of $B + C = A$.

5 ORTHONORMAL BASES FOR P-ADIC CONTINUOUS FUNCTIONS

As we mentioned in the introduction there exist two well-known bases for $C(\mathbb{Z}_p \to K)$: on one hand, we have the Mahler base $\binom{x}{n}$ $(n \in \mathbb{N})$, consisting of polynomials of degree n and on the other hand we have the van der Put base $\{e_n \mid n \in \mathbb{N}\}$ consisting of locally constant functions e_n defined as follows : $e_0(x) = 1$ and for $n > 0$, e_n is the characteristic function of the ball $\{\alpha \in \mathbb{Z}_p \mid |\alpha - n| < 1/n\}$. One can construct other orthonormal bases of $C(\mathbb{Z}_p \to K)$ by generalizing the procedure used to define the Mahler base as did Y. Amice. In general, we have (see [3]) the following characterization of the polynomial sequences $e_n \in K[x], n \geqslant 0$ such that $deg(e_n) = n$ and which are orthonormal bases of the space $C(\mathbb{Z}_p \to K)$.

DEFINITION For a polynomial $f(x) = \sum_{i=0}^{n} a_i x^i \in K[x]$, the Gauss-norm is defined as $||f||_G = \max_{i \leqslant n} |a_i|$.

THEOREM *Let $(e_n)_{n \geqslant 0}$ be a sequence of polynomials in $K[x]$ of degree n. They form an orthonormal base of $C(\mathbb{Z}_p \to K)$ if and only if $||e_n||_s = 1$ and $||e_n||_G = |coeff\ x^n|$.*

We proved earlier (see [3]) that we can construct other orthonormal bases by taking a certain linear combination of a given base, as is stated in the following theorem.

THEOREM *Let $e_n (n \in \mathbb{N})$ be an orthonormal base of $C(\mathbb{Z}_p \to K)$ and put $p_n = \sum_{j=0}^{n} a_{n,j} e_j$, where $a_{n,j} \in K$ and $a_{n,n} \neq 0$. The $p_n (n \in \mathbb{N})$ form an orthonormal base for $C(\mathbb{Z}_p \to K)$, if and only if $|a_{n,j}| \leqslant 1$ for all $j \leqslant n$ and $|a_{n,n}| = 1$.*

With the help of the theory of infinite matrices developed in the previous sections, we can now generalize this to.

THEOREM *Let $e_n (n \in \mathbb{N})$ be an orthonormal base of $C(\mathbb{Z}_p \to K)$, and put $p_n = \sum_{j=0}^{\infty} a_{n,j} e_j$ where $a_{n,j} \in K$ and $a_{n,n} \neq 0$. The $p_n (n \in \mathbb{N})$ form an orthonormal base for $C(\mathbb{Z}_p \to K)$ if $|a_{n,j}| \leqslant 1$ for all $j < n$, $|a_{n,n}| = 1$, $|a_{n,j}| < 1$ for all $j > n$ and $\lim_{j\to\infty} a_{n,j} = 0$ for every n.*

Proof. $\|p_n\| = \left\| \sum_{j=0}^{\infty} a_{n,j} e_j \right\| = \max_{j \geqslant 0} |a_{n,j}| = 1$ since the (e_n) form an orthonormal base.

Let $f = \sum_{i=0}^{n} \alpha_i p_i$. For the orthogonality we have to prove that $\|f\| \geqslant |\alpha_n|$. (see [4])
We may suppose that $|\alpha_i| = |\alpha_j|$ for every i and j since if $|\alpha_i| \neq |\alpha_j|$ for $i \neq j$ then $\left\| \sum_{i=k}^{n} \alpha_i p_i \right\| = \max |\alpha_i|$ as follows immediately from the definition of a non-archimedan norm (see [4]). Now

$$\left\| \sum_{i=0}^{n} \alpha_i p_i \right\| = \left\| \sum_{i=0}^{n} \alpha_i \sum_{j=0}^{\infty} a_{i,j} e_j \right\| = \left\| \sum_{j=0}^{\infty} \left(\sum_{i=0}^{n} \alpha_i a_{i,j} \right) e_j \right\|$$
$$= \max_j \left| \sum_{i=0}^{n} \alpha_i a_{i,j} \right| \geqslant \left| \sum_{i=0}^{n} \alpha_i a_{i,n} \right| = |\alpha_n|$$

since $|a_{i,n}| < 1$ for $i < n$. So (p_n) are orthonormal.
It remains to prove that the $p_n (n \in \mathbb{N})$ form a base for $C(\mathbb{Z}_p \to K)$. To do this, it suffices to write the e_n as a linear combination of the p_n.
$p_n = \sum_{j=0}^{\infty} a_{n,j} e_j$ can be written as $P = A.E$, where $P = (p_n)$, $A = (a_{n,j})$ and $E = (e_j)$. From the conditions in the theorem, we conclude that A is a row-zero-convergent matrix satisfying the conditions of the corollary of section 4. So A has a unique two-sided reciprocal. This means that $E = A^{-1}.P$ and thus that we can write the e_n as a convergent linear combination of the p_n which ends our proof.

REMARKS
1) The condition $|a_{n,j}| < 1$ for all $j > n$ cannot be replaced by $|a_{n,j}| \leqslant 1$ for all $j > n$ (as in the case $j < n$), since if we take

$$p_i(x) = \binom{x+1}{i+1} = \binom{x}{i} + \binom{x}{i+1}$$

then we would get

$$\binom{x}{i} = p_i(x) - p_{i+1}(x) + p_{i+2}(x) - \cdots + (-1)^n p_{i+n}(x) + \cdots$$

which is a divergent series. So the (p_i) do not form a base for $C(\mathbb{Z}_p \to K)$
2) We do not have here a sufficient and necessary condition as in the first theorem of this section which can be seen as follows. Let

$$p_0(x) = \binom{x}{0} + \binom{x}{1} + \binom{x}{2}$$

$$p_1(x) = \binom{x}{1} + \binom{x}{3}$$
$$p_2(x) = \binom{x}{2} - \binom{x}{3}$$
$$p_n(x) = \binom{x}{n} \quad \text{for } n \geqslant 3.$$

It can be easily seen that the (p_n) $(n \geqslant 0)$ are orthonormal. They also form an orthonormal base for $C(\mathbb{Z}_p \to K)$, since

$$\binom{x}{0} = p_0(x) - p_1(x) - p_2(x)$$
$$\binom{x}{1} = p_1(x) - p_3(x)$$
$$\binom{x}{2} = p_2(x) + p_3(x)$$
$$\binom{x}{n} = p_n(x) \quad \text{for } n \geqslant 3$$

so every continuous function from $\mathbb{Z}_p$ to K can be written as a convergent combination of the $p_\imath$.
However the $p_\imath$ do not satisfy the conditions of our theorem.

The two theorems above are also valid for other Banach spaces, for example for the Banach space of n-times continously differentiable functions or the Banach space of continuous functions in several variables (see also [3]).

REFERENCES

[1] RG Cooke. Infinite matrices and sequence spaces, New York: Dover publications, 1950.

[2] S De Smedt. p-adic continuously differentiable functions of several variables, Coll Math 45(2): 137–152, 1994.

[3] S De Smedt. Orthonormal bases for p-adic continuous and continuously differentiable functions, Ann Math Blaise Pascal 2(1): 275–282, 1995.

[4] WH Schikhof. Ultrametric Calculus: an introduction to p-adic analysis, Cambridge University Press, 1984.

[5] L Van Hamme. Three generalizations of Mahler's expansion for continuous functions on $\mathbb{Z}_p$. In: Lecture notes in mathematics 1454: p-adic analysis, Springer-Verlag, 1990, pp 356–361.

[6] L Van Hamme. The p-adic Z-transform, Ann. Math. Blaise Pascal 2(1): 131–146, 1995.

Orthonormal bases for non-archimedean Banach spaces of continuous functions

ANN VERDOODT Vrije Universiteit Brussel, Faculty of Applied Sciences, Pleinlaan 2, B-1050 Brussels, Belgium.

Abstract. Our aim is to find orthonormal bases for the Banach space $C(M \to K)$ of continuous functions from M to K, where K is a local field and M is a regular compact in K with diameter 1. Very well distributed sequences in M are used to find the orthonormal bases.

1 INTRODUCTION

The main aim of this paper is to find orthonormal bases for non-archimedean Banach spaces of continuous functions. We start by recalling some definitions and some previous results. All these results can be found in [1], chapter 1, sections 1 and 2, and chapter 2, sections 5 and 6. (we remark that the notations used here are sometimes different from the notations used in [1]). For additional information we refer the reader to [1]. Throughout this paper, N denotes the set of natural numbers, and N_0 is the set of natural numbers without zero.

A *countable projective system of finite sets* $(M_i, \varphi_{i,j})_{i \leqslant j \in N}$ consists of
1) a sequence (M_n) of finite sets
2) mappings $\varphi_{k,n}$ of M_n in M_k defined for $k \leqslant n$ and such that for $k \leqslant n \leqslant m$ we have $\varphi_{k,m} = \varphi_{k,n} \circ \varphi_{n,m}$ and $\varphi_{n,n}$ is the identity mapping on M_n.

Let $M = \lim\limits_{\leftarrow} - M_k$ denote the projective limit of this system.

M, as the projective limit of the sequence of sets M_n equipped with the discrete topology, is compact and ultrametric.

A countable projective system of finite sets $(M_i, \varphi_{i,j})_{i \leqslant j \in N}$ is called a *regular projective system* if it satisfies
1) $\# M_0 = 1$,
2) for all $i \leqslant j \in N : \varphi_{i,j}$ surjective,
3) there exist $q_1, q_2, \ldots \in N \setminus \{0, 1\}$ such that for all $n \in N_0$, for all $\omega \in M_{n-1}$: $\#\{\varphi_{n-1,n}^{-1}(\omega)\} = q_n$ (where # means "cardinality").

In this case we put $N_0 = \#M_0 = 1$, $N_n = \#M_n = q_1 q_2 \dots q_n$.
Recall that $M = \{(x_\imath) \in \prod_{\imath \in \mathbb{N}} M_\imath : x_\imath = \varphi_{\imath\jmath}(x_\jmath) \text{ for all } i \leqslant j \in \mathbb{N}\}$. On $M \times M$ the function $v(x,y)$ defined as follows $v(x,y) = \sup\{i \in \mathbb{N} | x_k = y_k \text{ for all } k \leqslant i\}$ if $x \neq y$; $v(x,x) = +\infty$; satisfies the ultrametric inequality $v(x,z) \geqslant \inf\{v(x,y), v(y,z)\}$ and this for all $x, y, z \in M$. The function $d(x,y) = \alpha^{v(x,y)}$ ($\alpha \in (0,1)$, α fixed) is an ultrametric on M, which induces the projective limit topology. From 3) it immediately follows that

(∗) *Each closed ball in M with radius α^n is the finite disjoint union of q_{n+1} closed balls with radius α^{n+1}.*

Now let M be a set with an integer valued function v defined on $M \times M$ which satisfies for all $x, y, z \in M$
$v(x,y) = v(y,x)$,
$v(x,y) = +\infty \Leftrightarrow x = y$,
$v(x,z) \geqslant \inf\{v(x,y), v(y,z)\}$.
We say that M is valued by v. The function $d(x,y) = \alpha^{v(x,y)}$ ($\alpha \in (0,1)$, α fixed) is an ultrametric on M. If M is compact, then each closed ball with radius α^n is a finite disjoint union of balls with radius α^{n+1}. We then say that M is a *valued compact.*
Let M be a valued compact such that the closed balls satisfy (*). Let $B_b(r)$ denote the 'closed' ball with center b and radius r. We introduce
- the equivalence relation π_k on M, defined for $k \geqslant 0$ by

$$x \pi_k y \iff B_x(\alpha^k) = B_y(\alpha^k),$$

- the quotient M_k of M by π_k, and the canonical projection pr_k of M on M_k,
- the mapping $\varphi_{k,n}$ of M_n on M_k defined for $k \leqslant n$ by

$$\varphi_{k,n}(pr_n(y)) = pr_k(y) \quad for \quad y \in M.$$

The system $(M_\imath, \varphi_{\imath,\jmath})_{\imath \leqslant \jmath \in \mathbb{N}}$ is then a regular projective system, and its projective limit is isomorphic to M.
If the closed balls of a valued compact M satisfy condition (*), then we call M a *regular valued compact.*

EXAMPLE 1 Let $A_1, \dots, A_n, \dots$ be non-empty finite sets and let $q_n \geqslant 2$ be the the cardinality of A_n. The products $M_n = A_1 \times \dots \times A_n$ (M_0 consisting of one element), equipped with the canonical projections from M_n on M_k defined by $(x_1, \dots, x_n) \to (x_1, \dots, x_k)$, $k \leqslant n$, $x_\imath \in A_\imath$, form a regular projective system.
Conversely, every regular projective system is isomorphic to such a system.

EXAMPLE 2 Let p be a prime number. $\mathbb{Z}_p = \varprojlim \mathbb{Z}/p^n\mathbb{Z}$ is a regular valued compact, with $q_n = p$, $N_n = p^n$.

EXAMPLE 3 Let p be an odd prime number. The unit circle $\{x \in \mathbb{Z}_p | \, |x| = 1\}$, where $|.|$ denotes the p-adic valuation, is also a regular valued compact, with $q_1 = p - 1$, $q_n = p$ if $n \geqslant 2$. To see this put in example 1 $A_1 = \{1, 2, \dots, p-1\}$, $A_n = \{0, 1, \dots, p-1\}$ if $n \geqslant 2$.

Let A be a finite set, $\#A = N$, and $M = \underleftarrow{\lim} - M_k$ a regular valued compact. We say that a sequence u: $\mathrm{N} \to A$ is *well distributed* if for all $n \in \mathrm{N}_0$, for all $a \in A$: $\#\{i < nN | u_i = a\} = n$. A sequence u: $\mathrm{N} \to M$ is called *very well distributed* if for all $n \in \mathrm{N}$: $\varphi_n \circ u$: $\mathrm{N} \to M_n$ is well distributed, where $\varphi_n : M \to M_n$ is the canonical projection. A very well distributed sequence u in M is always injective, and lays dense in M.

EXAMPLE 4 Let p be a prime number. $\mathrm{Z}_p = \underleftarrow{\lim} - \mathrm{Z}/p^n\mathrm{Z}$ is a regular valued compact, with $q_n = p$, $N_n = p^n$. The sequence (u_i), $u_i = i$ for all i, is a very well distributed sequence in Z_p.
For more details concerning regular valued compacts and very well distributed sequences, we refer the reader to [1].
Let K be a local field (i.e. a locally compact, non-trivially, non-archimedian valued field) with valuation $|.|$ and logarithmic valuation v and let k be the finite residue class field of K. We have $|x| = \alpha^{v(x)}$, where $\alpha = |\pi|$, with $\pi \in K$, $0 < |\pi| < 1$, and $|\pi|$ the generator of the value group of K. K is an ultrametric space by putting $v(x, y) = v(x - y)$, $d(x, y) = |x - y|$. The closed unit ball of K is a regular valued compact. If a compact part M of the closed unit ball satisfies (*), then we call M a *regular compact* of K. We will always assume that the diameter of M equals 1.
$C(M \to K)$ denotes the non-archimedean Banach space of continuous functions from M to K, equipped with the supremum norm $||.||_\infty$. A sequence $e_0, e_1, e_2, \ldots$ of elements of $C(M \to K)$ is called an *orthonormal basis* for $C(M \to K)$ if every element f of $C(M \to K)$ has a unique representation $f = \sum_{i=0}^{+\infty} x_i e_i$ where $x_i \in K$ and $|x_i| \to 0$ if $i \to \infty$, and if $||f||_\infty = \max_{0 \leqslant i}\{|x_i|\}$.
The aim of this paper is to find orthonormal bases for $C(M \to K)$. In section 2 we prove some preliminary lemmas, and finally in section 3 we prove the main theorem of this paper.

2 PRELIMINARY LEMMAS

Let K be a local field with valuation $|.|$ and finite residue class field k and let M be a regular compact in K. Let $\alpha = |\pi|$, with $\pi \in K, 0 < |\pi| < 1$, and $|\pi|$ the generator of the value group of K. Throughout sections 2 and 3, (u_n) denotes a very well distributed sequence in M. First we introduce some functions on M which are going to play an important role in this paper.
We define sequences (q_n) and (ψ_n) of functions on M as follows:

$$q_0(x) = \psi_0(x) = 1,$$

$$q_n(x) = \frac{(x - u_0) \ldots (x - u_{n-1})}{(u_n - u_0) \ldots (u_n - u_{n-1})} \quad n \geqslant 1,$$

$\psi_n(x) = 1$ *if* $x \in B_{u_n}(r_n)$, *where* $r_n = \alpha^{i+1}$ *if* $N_i \leqslant n < N_{i+1}$, *otherwise* $\psi_n(x) = 0 \;\; n \geqslant 1$.
The functions q_n and ψ_n are clearly continuous. The functions q_n were introduced by Amice (see [1], sections 2.4 and 6). We immediately have $q_n(u_n) = \psi_n(u_n) = 1$, $q_n(u_j) = 0$ if $j < n$. Further, $\psi_n(u_j) = 0$ if $j < n$. To see this, suppose that

$N_\imath \leqslant n < N_{\imath+1}$. Then ψ_n is the characteristic function of the ball $B_{u_n}(r_n)$ where $r_n = \alpha^{\imath+1}$. There are $N_{\imath+1}$ disjoint balls with radius $\alpha^{\imath+1}$, namely the balls with centers $u_0, u_1, \ldots, u_\jmath, \ldots, u_n, \ldots, u_{N_\imath+1-1}$. So $|u_\jmath - u_n| > \alpha^{\imath+1}$ and we conclude that $\psi_n(u_\jmath) = 0$ if $j < n$. It is clear that $||\psi_n||_\infty = 1$ for all n. From [1], theorem 1, p. 143 it follows that $||q_n||_\infty = 1$ for all n. The sequence (q_n) forms an orthonormal basis for $C(M \to K)$ ([1], section 6.2).
The following lemma can be found in [1] (p. 135, lemma 4).

LEMMA 1 *If* $x, y \in M$, $|x - y| \leqslant \alpha^t$ *then* $|q_n(x) - q_n(y)| \leqslant \alpha$ *if* $0 \leqslant n < N_t$.
For the functions ψ_n we can prove something analogous.

LEMMA 2 *If* $x, y \in M$, $|x - y| \leqslant \alpha^t$ *then* $\psi_n(x) = \psi_n(y)$ *if* $0 \leqslant n < N_t$.

Proof. Let $x, y \in M$ such that $|x - y| \leqslant \alpha^t$ and let $0 \leqslant n < N_t$. Then $|x - y| \leqslant \alpha^t \leqslant r_n$. So the elements x and y either belong both to $B_{u_n}(r_n)$ and then $\psi_n(x) = \psi_n(y) = 1$, or none of them is in $B_{u_n}(r_n)$ and then $\psi_n(x) = \psi_n(y) = 0$.
We introduce the following:
For each $k \in \mathbb{N}$, let I_k be a subset of the set $\{0, 1, \ldots, k\}$ (I_k can also be empty or can be equal to $\{0, 1, \ldots, k\}$). Let p be a (continuous) function on M of the following form:

$$p = \sum_{\imath \in I_n} a_\imath q_\imath + \sum_{\imath \in \{0,1,\ldots,n\} \setminus I_n} a_\imath \psi_\imath,$$

where $a_\imath \in K$ for all i.
For functions of this type we can prove the following lemmas.

LEMMA 3 *Let* p *be a function of the form*

$$p = \sum_{\imath \in I_n} a_\imath q_\imath + \sum_{\imath \in \{0,1,\ldots,n\} \setminus I_n} a_\imath \psi_\imath, \quad a_\imath \in K \quad for \quad all \quad i.$$

Then 1) *and* 2) *are equivalent:*
1) $|p(u_n)| = 1$ *and* $|p(u_k)| < 1$ *if* $0 \leqslant k < n$.
2) $|a_n| = 1$ *and* $|a_k| < 1$ *if* $0 \leqslant k < n$.
Furthermore, if p *satisfies 1) or 2), then* $||p||_\infty = 1$.

Proof. For $n = 0$ we have $p(u_0) = a_0$, so $|p(u_0)| = 1 \Leftrightarrow |a_0| = 1$. If $n > 0$, then we show 1) $\Rightarrow$ 2) by induction. If $|p(u_0)| < 1$ then $|a_0| < 1$. Now suppose that $|a_k| < 1$ if $0 \leqslant k < n - 1$. Then $|\sum_{\imath \in I_n \bigcap \{0,1,\ldots,k+1\}} a_\imath q_\imath(u_{k+1}) + \sum_{\imath \in \{0,1,\ldots,k+1\} \setminus I_n} a_\imath \psi_\imath(u_{k+1})| = |p(u_{k+1})| < 1$ and by the induction hypothesis it follows that $|a_{k+1}| < 1$ since $\psi_{k+1}(u_{k+1}) = q_{k+1}(u_{k+1}) = 1$, and we can conclude that $|a_k| < 1$ for all k, $0 \leqslant k < n$. Since $|\sum_{\imath \in I_n} a_\imath q_\imath(u_n) + \sum_{\imath \in \{0,1,\ldots,n\} \setminus I_n} a_\imath \psi_\imath(u_n)| = |p(u_n)| = 1$ we have $|a_n| = 1$, since $\psi_n(u_n) = q_n(u_n) = 1$. 2) $\Rightarrow$ 1) is obvious since $||\psi_n||_\infty = ||q_n||_\infty = 1$ for all n. Also, if p satisfies 1) or 2), then, since $||\psi_n||_\infty = ||q_n||_\infty = 1$ for all n, we immediately have that $||p||_\infty = 1$.

LEMMA 4 *Suppose that* p *is a function of the form*

$$p = \sum_{i \in I_n} a_i q_i + \sum_{i \in \{0,1,\ ,n\} \setminus I_n} a_i \psi_i, \quad a_i \in K \quad for \quad all \quad i$$

such that $|a_n| = 1$ *and* $|a_k| < 1$ *if* $0 \leqslant k < n$.
If $x, y \in M$, $|x - y| \leqslant \alpha^t$ *then if* $0 \leqslant n < N_t$, $j \in N$ *we have*

$$|p(x)^j - p(y)^j| \leqslant \alpha.$$

Proof. We have $|p(x) - p(y)| \leqslant \max_{i \in I_n} \{|a_i| \cdot |q_i(x) - q_i(y)|\} \leqslant \alpha$ by lemmas 1 and 2. If $j > 1$ then

$$|p(x)^j - p(y)^j| = |p(x) - p(y)|.|\sum_{s=0}^{j-1} p(x)^s p(y)^{j-1-s}| \leqslant \alpha$$

So the lemma holds for all $j \in$ N (the case j $= 0$ is trivial).
In lemma 5, we have that for each $k \in$ N, J_k is a subset of the set $\{0, 1, \ldots, k\}$.

LEMMA 5 *Let* p *and* r *be functions of the form*

$$p = \sum_{s \in I_n} a_s q_s + \sum_{s \in \{0,1,\ .\ ,n\} \setminus I_n} a_s \psi_s, \quad a_s \in K \quad for \quad all \quad s$$

$$r = \sum_{s \in J_n} b_s q_s + \sum_{s \in \{0,1,\ ..,n\} \setminus J_n} b_s \psi_s, \quad b_s \in K \quad for \quad all \quad s$$

such that $|a_n| = |b_n| = 1$ *and* $|a_k| < 1, |b_k| < 1$ *if* $0 \leqslant k < n$.
If $x, y \in M$, $|x - y| \leqslant \alpha^t$ *then if* $i, j \in N, 0 \leqslant n < N_t$ *we have*

$$|r(x)^i p(x)^j - r(y)^i p(y)^j| \leqslant \alpha.$$

Proof.

$$\begin{aligned} &|r(x)^i p(x)^j - r(y)^i p(y)^j| \\ &\leqslant \max\{|r(x)^i p(x)^j - r(x)^i p(y)^j|, |r(x)^i p(y)^j - r(y)^i p(y)^j|\} \\ &\leqslant \max\{|r(x)^i||p(x)^j - p(y)^j|, |p(y)^j||r(x)^i - r(y)^j|\} \leqslant \alpha \end{aligned}$$

by lemmas 3 and 4.
In lemmas 6 and 7, we introduce a function g which is a linear combination of powers of functions of the previous type.

LEMMA 6 *Let* (p_n) *be a sequence of continuous functions, where for each* n, p_n *is of the form*

$$p_n = \sum_{i \in I_n} a_{n,i} q_i + \sum_{i \in \{0,1,\ldots,n\} \setminus I_n} a_{n,i} \psi_i, \quad a_{n,i} \in K,$$

with $|a_{n,n}| = 1$ *and* $|a_{n,i}| < 1$ *if* $0 \leqslant i < n$, *and let* g *be a function of the form*

$$g = \sum_{i=0}^{n} c_i (p_i)^{m_i}, \quad c_i \in K,$$

with $|c_n| = 1,\ |c_i| < 1$ *if* $0 \leqslant i < n,\ m_i \in N_0$.
Then $|g(u_n)| = 1,\ |g(u_i)| < 1$ *if* $0 \leqslant i < n$. *Furthermore,* $||g||_\infty = 1$.

Proof. From lemma 3 it follows that $|p_j(u_j)| = 1$, $||p_j||_\infty = 1$ and $|p_j(u_i)| < 1$ for all j and for all i, $0 \leqslant i < j$. So we have $|g(u_n)| = \max\{|\sum_{i=0}^{n-1} c_i p_i(u_n)^{m_i}|, |c_n p_n(u_n)^{m_n}|\}$ $= 1$ since $\max_{0\leqslant i<n}\{|c_i p_i(u_n)^{m_i}|\} < |c_n p_n(u_n)^{m_n}| = 1$. Further, if $j < n$, $|g(u_j)| =$ $|\sum_{i=0}^{n} c_i p_i(u_j)^{m_i}| \leqslant \max_{0\leqslant i\leqslant n}\{|c_i p_i(u_j)^{m_i}|\} < 1$. Also, it follows immediately that $||g||_\infty = 1$, since $||p_j||_\infty = 1$ for all j.

LEMMA 7 *Let the sequence* (p_n) *and the function* g *be as in lemma 6, and let* r *be as in lemma 5. If* $x, y \in M$, $|x - y| \leqslant \alpha^t$ *then if* $j, k \in N$, $0 \leqslant n < N_t$ *we have*

$$|g(x)^j r(x)^k - g(y)^j r(y)^k| \leqslant \alpha.$$

Proof. $|g(x) - g(y)| \leqslant \max_{0\leqslant i\leqslant n}\{|c_i| \cdot |p_i(x)^{m_i} - p_i(y)^{m_i}| \leqslant \alpha$ by lemma 4. As usual, if $j > 1$ then

$$|g(x)^j - g(y)^j| = |g(x) - g(y)| \cdot |\sum_{i=0}^{j-1} g(x)^i g(y)^{j-1-i}| \leqslant \alpha$$

since $||g||_\infty = 1$. So $|g(x)^j - g(y)^j| \leqslant \alpha$ for all $j \in$ N (the case $j = 0$ is trivial). Finally, $|g(x)^j r(x)^k - g(y)^j r(y)^k| \leqslant \alpha$ can be shown in an analogous way as in the proof of lemma 5.

3 ORTHONORMAL BASES FOR $C(M \to K)$

Using the lemmas in section 2, we can make orthonormal bases for $C(M \to K)$ with the aid of the following theorem. Let as before for each n, I_n and J_n be subsets of the set $\{0, 1, \ldots, n\}$.
For f in $C(M \to K)$ with $||f||_\infty \leqslant 1$ let $\overline{f}$ be the canonical projection of f on $C(M \to k)$. Then we have the following ([5], lemme 1)
A sequence (e_i) of elements of $C(M \to K)$ forms an orthonormal basis for $C(M \to K)$ if and only if
1) $||e_i||_\infty \leqslant 1$ for all i,
2) $(\overline{e}_i)$ forms an algebraic basis for the k-vector space $C(M \to k)$.
Now we can prove the main result of this paper .

THEOREM *Let* (p_n), (r_n) *and* (g_n) *be sequences of functions of the following form: for each* n, p_n, r_n *and* g_n *are of the form*

$$p_n = \sum_{i\in I_n} a_{n,i} q_i + \sum_{i\in\{0,1,\ldots,n\}\setminus I_n} a_{n,i} \psi_i,$$

$$r_n = \sum_{i\in J_n} b_{n,i} q_i + \sum_{i\in\{0,1,\ldots,n\}\setminus J_n} b_{n,i} \psi_i$$

and

$$g_n = \sum_{i=0}^{n} c_{n,i}(p_i)^{m_{n,i}},$$

with $|a_{n,n}| = |b_{n,n}| = |c_{n,n}| = 1$ *and* $|a_{n,i}| < 1$, $|b_{n,i}| < 1$, $|c_{n,i}| < 1$ *if* $0 \leqslant i < n$, $a_{n,i}, b_{n,i}, c_{n,i} \in K$, $m_{n,i} \in N_0$.
If (k_n) *is a sequence in* N *and if* (j_n) *is a sequence in* N_0, *then*
1) *the sequence* $((r_n)^{k_n}(p_n)^{j_n})$ *and*
2) *the sequence* $((r_n)^{k_n}(g_n)^{j_n})$
form orthonormal bases for $C(M \to K)$.

Proof. We remark that $|p_n(u_n)| = |r_n(u_n)| = |g_n(u_n)| = 1$, $|p_n(u_i)| < 1$, $|r_n(u_i)| < 1$ and $|g_n(u_i)| < 1$, for all n in N and for all i, $0 \leqslant i < n$, and $||(r_n)^{k_n}(p_n)^{j_n}||_\infty = 1$, $||(r_n)^{k_n}(g_n)^{j_n}||_\infty = 1$ (lemmas 3 and 6). By the remark above, it suffices to prove that $(\overline{(r_n)^{k_n}(p_n)^{j_n}})$ and $(\overline{(r_n)^{k_n}(g_n)^{j_n}})$ form algebraic bases for $C(M \to k)$. Let C_t be the subspace of $C(M \to k)$ of the functions constant on closed balls with radius α^t. Since $C(M \to k) = \bigcup_{t \geqslant 0} C_t$ it suffices to prove that $(\overline{(r_n)^{k_n}(p_n)^{j_n}}|n < N_t)$ and $(\overline{(r_n)^{k_n}(g_n)^{j_n}}|n < N_t)$ form algebraic bases for C_t. M is the union of N_t disjoint balls with centers u_n, $0 \leqslant n < N_t$, radius α^t. Let χ_i be the characteristic function of the ball with center u_i. Using lemma 5, we have

$$\begin{aligned}\overline{(r_n(x))^{k_n}(p_n(x))^{j_n}} &= \sum_{i=0}^{N_t-1} \chi_i(x)\overline{(r_n(u_i))^{k_n}(p_n(u_i))^{j_n}} \\ &= \sum_{i=n}^{N_t-1} \chi_i(x)\overline{(r_n(u_i))^{k_n}(p_n(u_i))^{j_n}}\end{aligned}$$

since $|(r_n(u_i))^{k_n}(p_n(u_i))^{j_n}| < 1$ if $i < n$ and hence the transition matrix from $(\chi_n|n < N_t)$ to $(\overline{(r_n)^{k_n}(p_n)^{j_n}}|n < N_t)$ is triangular since $|(r_n(u_n))^{k_n}(p_n(u_n))^{j_n}| = 1$, so $(\overline{(r_n)^{k_n}(p_n)^{j_n}}|\ n < N_t)$ forms a basis for C_t. This proves 1).
Using lemma 7, we have

$$\begin{aligned}\overline{(r_n(x))^{k_n}(g_n(x))^{j_n}} &= \sum_{i=0}^{N_t-1} \chi_i(x)\overline{(r_n(u_i))^{k_n}(g_n(u_i))^{j_n}} \\ &= \sum_{i=n}^{N_t-1} \chi_i(x)\overline{(r_n(u_i))^{k_n}(g_n(u_i))^{j_n}}\end{aligned}$$

since $|(r_n(u_i))^{k_n}(g_n(u_i))^{j_n}| < 1$ if $i < n$ and hence the transition matrix from $(\chi_n|n < N_t)$ to $(\overline{(r_n)^{k_n}(g_n)^{j_n}}|n < N_t)$ is triangular since $|(r_n(u_n))^{k_n}(g_n(u_n))^{j_n}| = 1$, so $(\overline{(r_n)^{k_n}(g_n)^{j_n}}|\ n < N_t)$ forms a basis for C_t. This proves 2).

REMARK 1 If we look at the proof of the theorem and at the proofs of the lemmas in section 2, it is not difficult to extend the theorem and to find more orthonormal bases for $C(M \to K)$.

REMARK 2 Let K be a non-archimedean valued field which contains $\mathbb{Q}_p$, and suppose that K is complete for the valuation $|.|$, which extends the p-adic valuation.

V_q is the closure of the set $\{aq^n | n = 0, 1, 2, \ldots\}$ where a and q are two units of Z_p, q not a root of unity (for a description of the set V_q we refer to [7]). A result analogous to 1) of the theorem for the space $C(V_q \to K)$ can be found in [6].
Let us finally consider the following examples.

EXAMPLE 5 Let for all n, p_n be a polynomial of degree n defined as follows $p_n = \sum_{i=0}^{n} a_{n,i} q_i$ with $|a_{n,n}| = 1$ and with $|a_{n,i}| < 1$ if $0 \leqslant i < n$ $(a_{n,i} \in K)$. Then the sequence $((p_n)^{j_n})$ $(j_n \in N_0)$ forms an orthonormal basis for $C(M \to K)$. To see this, put in the theorem $I_n = \{0, 1, \ldots n\}$ and $k_n = 0$ for all n. In particular, the sequence $((q_n)^{j_n})$ forms an orthonormal basis. If we put $j_n = 1$ for all n then the sequence (q_n) also forms an orthonormal basis (see also [1], p. 143, theorem 1).

EXAMPLE 6 Let for all n, p_n be a function defined as follows $p_n = \sum_{i=0}^{n} a_{n,i} \psi_i$ with $|a_{n,n}| = 1$ and with $|a_{n,i}| < 1$ if $0 \leqslant i < n$ $(a_{n,i} \in K)$. Then the sequence $((p_n)^{j_n})$ $(j_n \in N_0)$ forms an orthonormal basis for $C(M \to K)$. To see this, put in the theorem $I_n = \phi$ and $k_n = 0$ for all n. In particular, the sequence (ψ_n) forms an orthonormal basis.

EXAMPLE 7 Put $K = Q_p$, the field of the p-adic numbers and put $M = Z_p$, the ring of the p-adic integers, and let $|.|$ be the p-adic valuation on Q_p. $M = Z_p$ is a regular valued compact, and the sequence (u_n), $u_n = n$ for all n, is a very well distributed sequence in Z_p (see Example 4). From example 5 above it follows that the sequence of polynomials $(\binom{x}{i})$ defined by $\binom{x}{0} = 1$, $\binom{x}{k} = \frac{x(x-1)\ (x-k+1)}{k!}$ if $k \geqslant 1$, forms an orthonormal basis for $C(Z_p \to Q_p)$ which is known as Mahler's basis ([4]). It also follows from the theorem that for every $s \in N_0$, the sequence $(\binom{x}{i}^s)$ forms an orthonormal basis for $C(Z_p \to Q_p)$ ([2]). Furthermore, from example 6 it follows that the sequence (ψ_n) defined on M as follows:
$\psi_0(x) = 1$ for all x in M,
$\psi_n(x) = 1$ if and only if $x \in B_{u_n}(r_n)$ $(n \geqslant 1)$, where $r_n = \alpha^{i+1}$ if $p^i \leqslant n < p^{i+1}$ (otherwise $\psi_n(x) = 0$),
forms an orthonormal basis for $C(Z_p \to Q_p)$, which is known as van der Put's basis ([3], example 7.2).

EXAMPLE 8 If we put for al n, $k_n = 0$, then the sequence $((g_n)^{j_n})$ $(j_n \in N_0)$ forms an orthonormal basis for $C(M \to K)$. In particular, (g_n) forms an orthonormal basis.

ACKNOWLEDGEMENT I want to thank Professor Caenepeel for giving me some ideas to prepare this paper.

REFERENCES

[1] Y Amice. Interpolation p-adique. Bull Soc Math France 92: 117–180, 1964.

[2] S Caenepeel. About p-adic Interpolation of Continuous and Differentiable Functions. Groupe d' étude d'analyse ultramétrique (Y. Amice, G. Christol, P. Robba), 9e année, 1981/82, no. 25, 8 p.

[3] L Gruson, M van der Put. Banach Spaces. Table Ronde d' Analyse non archimédienne (1972 Paris), Bulletin de la Société Mathématique de France, Mémoire 39-40, 1974, p. 55 - 100.

[4] K Mahler. An Interpolation Series for Continuous Functions of a p-adic Variable. Journal für reine und angewandte Mathematik 199:23–34, 1958.

[5] JP Serre. Endomorphismes Complètement Continus des Espaces Banach p-adiques. Presses Universitaires de France, Paris, 1962, Institut Hautes Etudes Scientifiques, Publications Mathématiques, 12, p. 69-85.

[6] A Verdoodt. Normal Bases for the Space of Continuous Functions defined on a Subset of Z_p. Publicacions Matemàtiques, vol 38, no 2, 1994, p. 371-380.

[7] A Verdoodt. Normal Bases for Non-Archimedean Spaces of Continuous Functions. Publicacions Matemàtiques, Vol. 37, 1993, p. 403-427.